CLERKS AND CRAFTSMEN IN
CHINA AND THE WEST

CLERKS AND CRAFTSMEN IN CHINA AND THE WEST

LECTURES AND ADDRESSES
ON THE HISTORY OF SCIENCE AND
TECHNOLOGY

BY

JOSEPH NEEDHAM, F.R.S.

Master of Gonville & Caius College, Cambridge
Foreign Member of Academia Sinica

BASED LARGELY ON
COLLABORATIVE WORK WITH

WANG LING

Professor of Chinese Literature, Australian
National University, Canberra, A.C.T.

LU GWEI-DJEN

Fellow of Lucy Cavendish College
Cambridge

AND

HO PING-YÜ

Professor of Chinese and Dean of the
Faculty of Arts, University of
Malaya, Kuala Lumpur

CAMBRIDGE · AT THE UNIVERSITY PRESS
1970

Published by the Syndics of the Cambridge University Press
Bentley House, 200 Euston Road, London N.W.1
American Branch: 32 East 57th Street, New York, N.Y.10022

© Cambridge University Press 1970

Library of Congress Catalogue Card Number: 69–10218

Standard Book Number: 521 07235 2

Printed in Great Britain
at the University Printing House, Cambridge
(Brooke Crutchley, University Printer)

三人行必有我師

'Where there are three men walking together, one or other of them will certainly be able to teach me something.'

Lun Yü VII, xxi
(Conversations and Discourses
[of Confucius], 5th century B.C.)

CONTENTS

CONTENTS

ILLUSTRATIONS

(The plates are grouped together in a single series following p. 440)

ILLUSTRATIONS

PREFACE

IT is now twenty-six years since the project of *Science and Civilisation in China*, a thorough-going account of the history of science, scientific thought, and technology in Chinese culture, was planned; twenty years since the writing of it first got under way, and fourteen years since the first printing of the first volume of the projected seven (in about a dozen parts) saw the light of day. Such a 'travail de longue haleine' (as my Unesco colleagues used to call it when I told them of it in the years just after the Second World War) does not, it seems, get itself done without generating a sort of 'by-product' literature, partly consisting of monographs on unearthed material too elaborate and detailed to find a place as such in the main work, partly of occasional lectures and other publications. Of 'by-product' monographs there have been two, *The Development of Iron and Steel Technology in China* prepared with the assistance of Dr Wang Ling (Wang Ching-Ning),[1] and *Heavenly Clockwork*, produced with the collaboration of Dr Wang Ling and Professor Derek de Solla Price.[2] Of 'by-product' lectures and papers there have been many, and it is a select collection of these that the present book reprints.

What is presented here is not addressed to one audience only; on the contrary I hope that all sorts of people working in many different subjects, and at many different levels, will find information and profit in it. Needless to say, no knowledge of the Chinese language is required, and each reader will be able to form his own impression of what Chinese culture contributed to world science and technology by starting from the subject with which he or she happens to be most familiar. The papers in this book fall indeed into two broad divisions: (i) those that are particularly easy reading, not much documented (nos. 1, 2–8, 12–14), (ii) those that are more solid stuff, annotated and referenced for specialist interlocutors (nos. 9–11, 15–18 and 19). If allegorical subtitles were appropriate, the former group could be called 'General Compass-Bearings', and the latter 'Logs of Special Voyages', voyages of course of research in waters uncharted by previous investigators—for traditionally sinologists have been literary and philological men, unconcerned with the history of science and technology. The first and

[1] Second Dickinson lecture, Newcomen Society, London, 1956, pub. 1958, repr. Heffer, Cambridge, 1964.
[2] Cambridge, 1960, Antiquarian Horological Society Monographs, no. 1; prelim. pub. *Antiq. Horol.* 1956, 1, 153; *Nature*, 1956, **177**, 600.

the last papers escape somewhat from this classification. The opening one might be headed 'Welcoming Friends from Distant Quarters' (*Lun Yü*, I, i, 2) for it was a narration in the London Planetarium in the presence of a delegation of Chinese scientists from Academia Sinica visiting the Royal Society. The last might be captioned 'Hua I Chhang Chiang; a Scroll-Painting of the River of Time'—for it is a preliminary attempt to contrast the general course of scientific advance in the two civilisations of China and Europe, the slow and steady rise of the former being overpassed by the latter after the Renaissance and the Scientific Revolution of the seventeenth century. This paper started as an address at the opening of the permanent exhibition of Chinese medicine at the Wellcome Historical Medical Museum and Library, and ended as a Sectional Presidential Address at the British Association; the ideas which it contains will certainly be presented in a more extended way in vol. 7 of *Science and Civilisation in China*.

The nature of the various papers herein also varies, as may be seen in the following list. They represent: (i) lectures requested on particular occasions (nos. 1, 2, 3, 5), (ii) discourses on the methods of scholarship used (nos. 2, 6, 12, 13, 19), (iii) abstracts or résumés of by-product monographs (nos. 8, 11), (iv) abbreviations of chapters in *Science and Civilisation in China* (nos. 2, 4, 12, 13, 14), (v) material that came to light too late for incorporation in *Science and Civilisation in China* (nos. 7, 10), (vi) drafts for chapters in *Science and Civilisation in China* (nos. 16, 17, 19), and (vii) by-product material probably too detailed for incorporation in *Science and Civilisation in China* (nos. 10, 16, 18). Although it has not been possible, therefore, to avoid completely all overlaps with certain paragraphs of the main series of volumes, considerable effort has been made to avoid unnecessary duplications between the writings here reprinted. Certain paragraphs have been transferred and certain omissions made. The inevitable consequence of this has been that not all the lectures in this book are absolutely identical with their texts as delivered and subsequently printed in various places, but such discrepancies will, it is hoped, justify themselves by the avoidance of repetition as far as possible in the present volume. Certain subjects (such as the technique of printing) do come up, it is true, in several successive places in this book, but the remarks on them differ, and these can all be found by means of the general index. Sometimes additional footnotes have proved necessary, either for cross-reference, or for updating or modifying statements in the text, but all those introduced for this book are easily distinguished from original ones by the square brackets in which they are enclosed.

The illustrations presented a more difficult problem, for one or two important ones were out of date and some were repetitious. To remove the latter was easy, but to substitute up-to-date ones for the former would have destroyed the historicity of the various discourses themselves, interfering, as it were, with the history of the history of science; they have therefore been left in place, with the addition of footnotes to indicate where the latest and most correct diagrams or illustrations can be consulted. Then the removal of unnecessary pictures has given the opportunity for the insertion of a larger number of illustrations, including some important ones not known to us (or in some cases to anyone) at the time of printing of the various volumes of *Science and Civilisation in China*. Among these are unpublished photographs which justify some of the statements in the 'stop press news' Addendum on p. 758 of vol. 4, pt. 2; e.g. one which takes back the oldest known spinning-wheel and driving-belt from about +1270 to +1035; and others again which prove the existence of the standard method of interconversion of rotary and longitudinal motion in China in about +970 already rather than only in +1313. So for the illustrations also there is something here which is not found in *Science and Civilisation in China*.

The reader must not expect to find in this book a systematic exposé of subject after subject such as is contained in the main series of volumes, for chance has brought it about that only some of them are represented. Once again we can list the papers, this time according to the topics which they take up: (i) general (nos. 5, 6, 19), (ii) East–West contacts, confrontations and transmissions (nos. 2, 3, 4), (iii) astronomy and meteorology (nos. 1, 9), (iv) metallurgy (no. 7), (v) engineering, mechanical and horological (nos. 8, 10, 11), (vi) the sea and ships (nos. 4, 12, 13), (vii) alchemy, chemistry and pharmacy (nos. 2, 15), and (viii) medicine (nos. 14, 16, 17, 18). Regrettably therefore there is almost nothing on mathematics, physics, civil engineering, botany, zoology or agriculture; but reference may of course be made by those particularly interested in these subjects to the relevant chapters of *Science and Civilisation in China*, the volumes of which are divided as follows:

    i General Orientations (geography, history, East–West contacts and transmissions).

    ii History of Scientific Thought (in Chinese philosophy).

    iii Astronomy, Meteorology and the Earth Sciences (geography, cartography, geology, mineralogy, palaeontology, seismology, etc.).

    iv (in three parts) Physics, Mechanical and Civil Engineering, Nautical Technology. [All so far published.]

What unites the papers in the present collection is that they all have to do with scientific and technical matters. Two separate collections of 'by-product' papers of a more sociological and philosophical nature are being prepared for publication elsewhere under the titles *Within the Four Seas—The Dialogue of East and West*, and *The Grand Titration—Science and Society in East and West*.

This may be the place to mention some minor conventions adopted in the present book. As in *Science and Civilisation in China* itself (abbreviated uniformly as *SCC*) I use + and − signs instead of A.D. and B.C., and the additional *h* instead of the aspirate apostrophe of the Wade-Giles system in the romanisation of Chinese names and technical terms. Thus we say that Chang Chhiu-Chien wrote a mathematical book about +470, not that Chang Ch'iu-Chien did it about A.D. 470. With the single exception of paper no. 6, the linguistic subject of which demands them, no Chinese characters are included in the present book, for the few readers who will want them will find them abundantly either in *Science and Civilisation in China* itself or in the original places of publication of the several articles. In order to avoid interruption of the texts, the reference bibliographies have been all grouped together at the end of the book, yet still kept separately since the conflation of them would have been an unnecessary labour. Nor have we attempted to unify their conventions. The footnotes, however, where they occur, have been kept on the pages to which they belong, and not relegated, according to an opprobrious modern habit, to a special section at the back of the book.

At the beginning I called this collection of papers a 'select' one. This is because it excludes at least as many which were felt to deal with subjects of too specialised technical interest for the general public—such as an elucidation of the methods the Han mathematicians used for calculating square roots and solving equations,[1] an account of the 2,000-mile meridian survey conducted in the early eighth cen-

---

[1] 'Horner's Method in Chinese Mathematics; its Origin in the Root-Extraction Procedures of the Han Dynasty' (with Wang Ling), *T'oung Pao*, 1955, **43**, 345.

tury by the astronomers of the Thang,[1] a description of a screen with star-maps and astronomical diagrams made for the royal palace at Seoul in Korea,[2] a verification of medieval Chinese records of magnetic declination by remanent magnetism data,[3] the proof of 'dishing' by ancient Chinese wheelwrights,[4] the detailed analysis of harness types depicted in the Tunhuang cave-temple frescoes,[5] the story of the invention of the pound-lock on canals in the Sung,[6] careful studies of the apparatus used by the medieval Chinese alchemists[7] and of some of their methods[8] and theories,[9] a survey of ancient and medieval Chinese ideas on biological evolution,[10] the demonstration of the empirical knowledge of deficiency diseases in Sung and Yuan times,[11] the astonishing discovery that Chinese iatro-chemists were making steroid sex-hormone preparations in the late middle ages,[12] and an attempt to identify the earliest mentions and descriptions of specific diseases in ancient China.[13] Other papers again have been pre-empted by simultaneous publication elsewhere, such as an account of the optical virtuosi of seventeenth-century Suchow, contributed to the Victor Purcell Memorial Volume.[14] And lastly others have been omitted because paralleling too closely particular chapters of *Science and Civilisation in China*.

[1] 'An Eighth-Century Meridian Line; I-Hsing's Chain of Gnomons and the Pre-History of the Metric System' (with Arthur Beer, Ho Ping-Yü, Lu Gwei-Djen, E. G. Pulleyblank & G. I. Thompson), *Vistas in Astronomy*, 1964, **4**, 3.

[2] 'A Korean Astronomical Screen of the Mid-Eighteenth Century from the Royal Palace of the Yi Dynasty (Chosŏn Kingdom, +1392 to 1910)' (with Lu Gwei-Djen), *Physis*, 1966, **8**, 137.

[3] 'Magnetic Declination in Mediaeval China' (with Peter J. Smith), *Nature*, 1967, **214**, 1213.

[4] 'The Wheelwright's Art in Ancient China, I, The Invention of "Dishing"; II, Scenes in the Workshop' (with Lu Gwei-Djen & Raphael A. Salaman), *Physis*, 1959, **1**, 103, 196.

[5] 'Efficient Equine Harness; the Chinese Inventions' (with Lu Gwei-Djen), *Physis*, 1960, **2**, 121. 'A Further Note on Efficient Equine Harness; the Chinese Inventions' (with Lu Gwei-Djen), *Physis*, 1965, **7**, 70.

[6] 'China and the Invention of the Pound-Lock', *Trans. Newcomen Soc.* 1963, **36**, 85.

[7] 'The Laboratory Equipment of the Early Mediaeval Chinese Alchemists' (with Ho Ping-Yü), *Ambix*, 1959, **7**, 57.

[8] 'An Early Mediaeval Chinese Alchemical Text on Aqueous Solutions' (with Tshao Thien-Chhin & Ho Ping-Yü), *Ambix*, 1959, **7**, 122.

[9] 'Theories of Categories in Early Mediaeval Chinese Alchemy' (with Ho Ping-Yü), *Journ. Warburg & Courtauld Institutes*, 1959, **22**, 173.

[10] 'Ancient and Mediaeval Chinese Thought on Evolution' (with Donald Leslie), *Bull. Nat. Inst. Sci. India*, 1952, **7**, 1.

[11] 'A Contribution to the History of Chinese Dietetics' (with Lu Gwei-Djen), *Isis*, 1951, **42**, 13 (submitted 1939 and 1942 but lost by enemy action, finally submitted, 1948).

[12] 'Mediaeval Preparations of Urinary Steroid Hormones' (with Lu Gwei-Djen), *Med. History*, 1964, **8**, 101; *Nature*, 1963, **200**, 1047.

[13] 'Records of Diseases in Ancient China' (with Lu Gwei-Djen), art. in *Diseases in Antiquity*, ed. D. R. Brothwell.

[14] 'The Optick Artists of Chiangsu' (with Lu Gwei-Djen), art. in *Historical Aspects of Microscopy*, Roy. Mic. Soc. London, 1967, 113.

Lastly the place of honour, at the episcopal rear of the procession, as it were, belongs to my collaborators and advisers without whom nothing whatever could have been accomplished. Already the title-page has borne witness to the three closest, Dr Wang Ling, the first to labour with me nine fat years in Cambridge,[1] Dr Lu Gwei-Djen, my oldest friend and interpreter-general of the Chinese world-view,[2] and Professor Ho Ping-Yü, gentlest of scholars learned in the science of both China and Frankistan.[3] These have been my *thung chuang* friends, studying 'under the same window' day after day, year after year, but we have been in our turn indebted to many others too. Apart from hundreds of Chinese 'clerks and craftsmen' in their own country I would like to mention the help received from Dr Tshao Thien-Chhin, a former Fellow of Caius, in the history of alchemy and chemistry, Professor Wu Shih-Chhang of Oxford and Peking, in Chinese philosophy, and Professor Lo Jung-Pang in civil engineering and nautical technology. If I were myself Chinese I would perhaps have mentioned first the late Gustav Haloun, the only sinologist I could claim as my teacher, though for the most part I confess that I have been (as my Unesco colleagues would have said) 'auto-didact', hence all the faults that this must mean; and with his name I would certainly couple that of the late Charles Singer, whose friendship and inspiration during forty years gave me what understanding I have of the techniques and aims of the historian of science. A special place is occupied by Dr Kenneth Robinson, Deputy Director of Education in Sarawak, who drafted the chapter on physical acoustics in *Science and Civilisation in China*, but in the field of horological engineering two others collaborated equally closely, Professor Derek de S. Price already mentioned, and Mr J. H. Combridge of the G.P.O. Engineering Division, while on such things as vehicle wheels and equine harness Dr Lu and I had great pleasure in working with Mr R. A. Salaman, the authority on all craftsmen's tools. It would be hard to mention all those who have helped us in particular fields of detail, but we are deeply grateful to a great number such as Professor Kurt Mahler in mathematics, Dr Arthur Beer, Dr D. W. Dewhirst, Professor Yabuuchi Kiyoshi and Dr Nakayama Shigeru in astronomy, Mr E. G. Sterland in mechanical engin-

---

[1] Without giving a full list of the organisations which have helped to make our work financially possible, I should like to name here the British Council, the Spalding Trust, the Universities China Committee, the Leverhulme Foundation and the Holt family fund.

[2] Dr Lu's actual role is primarily that of specialist in the history of medicine and the biological sciences, in which capacity she is and has been most generously supported by the Wellcome Trust.

[3] Professor Ho's support was from the University of Malaya. In this connection I should like to record generous gifts from Dato Lee Kong-Chian of Singapore, and a long-continuing subvention from the Bollingen Foundation.

eering, Professor A. W. Skempton in civil engineering and Professor Bryan Thwaites in the physics of sailing craft. Every chapter of *Science and Civilisation in China* is of course submitted to the criticism of experts in the relevant field, but it is not always that they respond with such helpful or encouraging emendations as did the late Étienne Balazs in Taoist philosophy or Lt.-Cdr. David Waters in nautical technology. Moreover every volume is vetted on the orientalist side by kind advisers, Professor D. M. Dunlop for Arabic, Dr Shackleton Bailey for Sanskrit, Dr Charles Sheldon for Japanese and Professor Gari Ledyard for Korean. Lastly every sentence is critically read by my wife Dr Dorothy Needham FRS, to whom an incalculable debt of understanding and support is owing. It is true that all this applies primarily to *Science and Civilisation in China*, but the writings here collected have enjoyed just the same benefits, so the opportunity of a general and grateful acknowledgement could not be passed by.

Since some paragraphs of what is here reprinted were first written as long as twenty-two years ago there are certainly places which seem to me amateurish, some other passages doubtless 'off the cuff', others again which I should now phrase differently. None of my collaborators bears any responsibility for such things, nor for mistakes that I may have made. The purpose of this volume is to stimulate interest in the fascinating achievements of science and technology in a non-European civilisation, that of the Central Floriate Empire, one of the three greatest of the Old World. If anyone's curiosity and sympathy is kindled by dipping into the articles, some simple, some more complex, here presented, I shall be wholly and entirely satisfied.

CAMBRIDGE, 1968   J. N.

ASTRONOMY IN CLASSICAL CHINA[1]

[London, 1961]

THE object of this discourse is to render homage to some of our far distant predecessors in the study of the heavens, the astronomers of classical (ancient and medieval) Chinese culture.[2] Too often people have taken an unduly narrow view of the history of astronomy, confining it to the discoveries of the Greeks and the more ancient nations from whom they learnt, but we must be more oecumenical and celebrate the achievements of their contemporaries beyond the Himalayas in the East of Asia. Though Chinese civilisation is not so old as those of Babylonia and ancient Egypt, it is much older than that of Greece and Rome, and already in the middle of the − 2nd millennium the Chinese were keeping records of astronomical matters on fragments of bone still extant today.

Knowledge of the heavens grew up in China step by step with its development in the Middle East and in Hellenic culture. That there were learned contacts between East and West in those most ancient times we have no doubt, but it is certain that they were never sufficient to deprive Chinese astronomy of a highly original character. Though the phenomena were the same, there was more than one way of considering them, and by the + 17th century, when all the rivers of science flowed into the sea of modern science, Chinese astronomy as a system was quite different from that of the West in its emphases and understandings. I shall hope to explain some of these contrasts. Their practical consequences were very great, for the characteristic Chinese way of looking at things led directly to certain

---

[1] Reprinted from the *Quarterly Journal of the Royal Astronomical Society*, 1962, **3**, 87. A narration delivered in the London Planetarium in collaboration with Mr Leonard Clarke, Senior Narrator of that institution, and sponsored by the Royal Society and the Royal Astronomical Society, on the occasion of the visit of a Scientific Delegation to the Royal Society from Academia Sinica, Peking, headed by Dr Chu Kho-Chen, Vice-President of the Chinese National Academy, 19 October 1961.

[2] Without intending any elaborate documentation for this paper, I cannot refrain from recalling the excellent survey of the late Herbert Chatley, 'Ancient Chinese Astronomy', contributed to the *Occasional Notes of the Royal Astronomical Society*, 1939 (no. 5), 65; at the same time placing on record the debt which my collaborators and I owe to the constant kindness and encouragement of a former Chief Engineer of the Huangpo Conservancy. For the rest, readers are referred to vol. 3 of *SCC*, which contains full explanations, references and Chinese characters relating to the present text.

outstanding advances in instrumentation and scientific apparatus which we shall also try to describe in what follows. The fact is that for many centuries Chinese astronomy was far in advance of European in many important respects, while in others the Greek tradition had the advantage.

We might dedicate our evening to the memory of two among the greatest of Chinese astronomers, the Confucian scholar Kuo Shou-Ching in the +13th century, and the Tantric Buddhist monk I-Hsing in the +8th; and perhaps we might associate with them the name of Antoine Gaubil, a religious of another faith, who spent a lifetime in China in the eighteenth century revealing the history of Chinese astronomy to the West through the dark curtain which separates the ideographic from the alphabetic languages. This evening's occasion is a very special one, since we meet in honour of the scientific delegation to the Royal Society from the Chinese National Academy in Peking (Academia Sinica), headed by one of its Vice-Presidents, Dr Chu Kho-Chen, himself a most distinguished meteorologist and historian of the astronomical sciences.

We want now to try to show something of what China contributed to the world growth of astronomy. Let us first deal with the remarkable fruits of the fact that the Chinese were the most persistent and accurate observers of celestial phenomena in any culture before the Renaissance. Why is it that radio-astronomers today, for instance, scan the Chinese records of two thousand years ago with intense interest? This feature is a relatively simple one; it involved no abstract theory, it needed no engineering skill, it simply happened that China's age-old way of government was not merely feudal as we understand the word, but feudal-bureaucratic, so that the Bureau of Astronomy was an integral part of the civil service, and the Bureau of Historiography could be relied upon by and large (in spite of all tragedies of civil strife and invasion) to hand the records down. Individual genethliacal astrology, fortune-telling for ordinary people, was not a typical Chinese idea, but they did believe in a kind of State astrology, that (for example) 'comets do foretell the death of princes'; so celestial events were very carefully recorded, and later historians were accustomed to moralise systematically on them. Hence today we all benefit. Thus the first Chinese eclipse records date from as far back as −1361, and this is by far the most ancient verifiable eclipse in the history of any people. Much work has been done on the precision and reliability of the Chinese records, and if good collaboration between sinologists and astronomers may be assumed, they possess a quite unique value, covering as they do long periods for which no other records have come down to us

A particularly interesting series is that of the lists of novae and super-novae. The most recent catalogue, prepared under the auspices of Academia Sinica by Dr Hsi Tsê-Tsung, gives details of ninety between − 1400 and + 1690. As we now know from astrophysics, when a star like our own Sun reaches what is called 'Point M on the Main-sequence line' of stellar evolution, its properties begin to undergo a rapid change, leading it in relatively short time to a critical point O where it may or may not explode like a colossal atomic bomb. If it does this, then a 'new star', a nova, appears in the heavens, intensely bright, and if it is of the highest intensity we know it as a super-nova. Here is a fragment of oracle-bone from approximately − 1300; the ancient Chinese writing on it says: 'On the 7th day of the month, a *chi-ssu* day, a great new star appeared in company with Antares.' As for the supernovae, recorded history has only three of them; one was that new star of Tycho Brahe observed in + 1572 which did so much to ruin the Ptolemaic hypothesis, a second was that seen by his pupil Kepler in + 1604, while the third, that of + 1054, was recorded only by the astronomers of China and Japan. This was the origin of the Crab Nebula in Taurus. All these observations are of interest to the radio-astronomers because they help to elucidate the origin and nature of the radio-sources which are now being plotted in the heavens as distinctly as the visible stars.

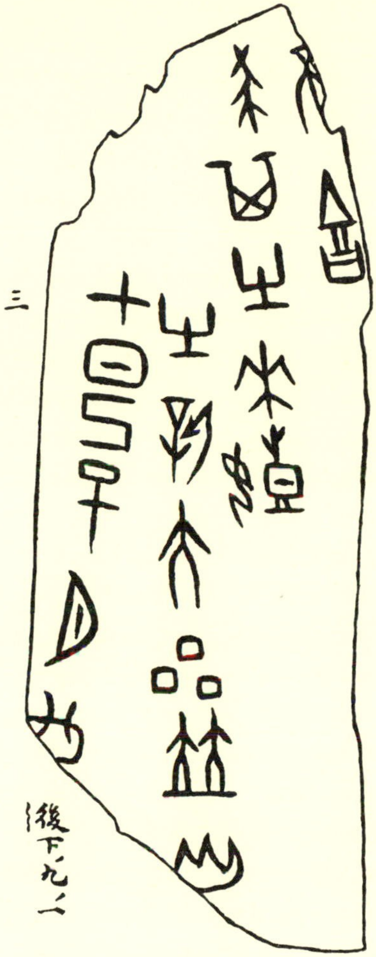

Chinese records of comets, the latest catalogue of which (by one of my own collaborators, Dr Ho Ping-Yü) includes 581 entries between − 1600 and + 1600, are also of great value. That the tails of comets always point away from the Sun was stated as early as + 635. The first Chinese observation of Halley's comet was

Fig. 1. The oldest record of a nova or supernova in any civilisation. The inscription on this Shang oracle-bone, dating from about − 1300, reads (in the two central columns of characters): 'On the 7th day of the month, a *chi-ssu* day, a great new star appeared in company with Antares.'

3

recorded in −467, and its many reappearances in these records helped modern astronomers to calculate its approximate orbit. We have also an abundance of Chinese records of meteors and meteorites, but perhaps the most extraordinary fact is that systematic records of sun-spots were kept from as early as −28. The Renaissance astronomers who disputed among themselves the priority of the discovery about +1615 might have been somewhat abashed if they had known of this. Very likely European tardiness here was due to a preconception of the Aristotelian–Ptolemaic system that the Sun was necessarily perfect and could not have spots.

Now we must come to greater contrasts. How else did Chinese astronomy differ from the contributions of the Greeks and their successors in Islam? There were many things, of course, which ran parallel in the different systems. Both China and Greece early developed star catalogues; both were interested in calendrical studies; both followed attentively the periods of revolution of the planets and their retrogradations. The classical stellar catalogue in China, the *Hsing Ching*, seems to go back to the astronomers Shih Shen and Kan Tê in the −4th century; if so, it will be anterior to that of Hipparchus (−134). Even if the data on star-positions which it now contains belong rather to the epoch of Ptolemy (+2nd century), it is one-third larger than that in his *Almagest*. In any case the definition of the star-positions in degrees presupposes a clear conception of the great circles of the celestial sphere, and an apparatus capable of measuring along them with some accuracy. In the −4th century this may have been only a single armillary ring held steady in different directions of space, but by +100, a little before Ptolemy, the armillary sphere was already fully developed in China in the hands of his younger contemporary Chang Hêng. But the most ancient astronomical instrument of China was the simple vertical pole. With this gnomon, usually 8 ft in height, one could measure the length of the Sun's shadows by day to determine the solstices and equinoxes, and note the times of transits of the stars by night to observe the revolution of the sidereal year. It is highly probable that this was already in use by the −12th century. From that time at least began the computation of the calendar, so vitally important for a primarily agrarian culture, and its progressive improvement continued all down through the ages.

Where Chinese astronomy profoundly differed from the Greek, however, was in the fact that it extended the Babylonian algebraic tradition, computing and predicting the positions of Sun, Moon and planets without any desire for a concrete geometrical model such as was constituted by the *machina coelestis* of Eudoxus,

Aristotle and Ptolemy, 'cycle on epicycle, orb on orb'. Euclidean geometry did not develop in China; it travelled thither only in the ͵+ 13th century. The absence of Ptolemaic planetary astronomy had, however, strange to say, its good side, for (apart from the fact that the Ptolemaic system was objectively wrong) the Chinese of the Middle Ages, unlike the Europeans, were not shut up in that prison of the solid crystalline celestial spheres which it took the herculean strength of Bruno, Gilbert and Galileo to break. For the Chinese, the stars were un-explained lights floating great distances apart in infinite empty space, and millions of years had passed since the last general conjunction—a very much wider outlook than the idea that creation had occurred in '4004 B.C. at six o'clock in the evening'. There is evidence that these Chinese conceptions had much influence in Europe in the + 17th century. Moreover, because the Chinese did not have Ptolemaic geometrical astronomy one must not assume that they had no part in the back-ground of the Galilean revolution where universal modern science was born. For they and they alone in the Middle Ages made deep investigations into the pheno-mena of magnetism, and this knowledge, transmitted through Peter of Maricourt (Kuo Shou-Ching's contemporary), stimulated Gilbert and Kepler to liken gravi-tational and magnetic attractions, thence helping to achieve the great synthesis of Isaac Newton.

Another fundamental contrast between Chinese and Greek astronomy arose from the same difference in interest. Greek attention was always concentrated on the ecliptic, where the activities of the great luminaries went on, and the Greeks solved the problem of finding out where the Sun is among the stars when no stars are to be seen, by noting, like the ancient Egyptians, the heliacal risings and settings of asterisms (i.e. what rises just before dawn, and what sets just after dusk). Chinese attention, however, was always concentrated on the equator and on the circumpolar region, that disc of perpetual visibility the stars in which never rise and never set. We can see it here as it was seen from Yang-chhêng in Honan, for more than two thousand years the central observatory of China. As de Saussure said, technically and epigrammatically, Greek astronomy was 'ecliptic, angular, true and annual', while Chinese astronomy was 'equatorial, horary, mean and diurnal'. By keying certain circumpolar stars to fixed mark-points upon the equator, and by observing systematically the transits of the former, the Chinese were never in doubt as to the position of the constellations along the latter hour by hour, and thus they could fix the position of the invisible Sun and Moon. Prolongation of the directions indicated by the Great Bear, for example, fixed a number of equatorial points. One of the most ancient Chinese astronomical instruments was a discoidal piece of jade indented at the rim so as to fit the circumpolar stars and held at right angles to the polar axis—this was the 'circumpolar constellation template'. Its probable use was to determine the position of the true celestial pole and the solsticial colure, and also to serve as a simple orienting instrument for the study of the positions of more equatorial constellations. It would have been the ancestor of the + 17th-century instrument known as the 'nocturnal'.

The Chinese sky was divided into four great palaces (*kung*) like the four main segments into which we cut an apple. Each one of these had its ancient symbolical animal—the Caerulean Dragon for the east and spring, the Vermilion Bird for the south and summer, the White Tiger for the west and autumn, and the Black Tortoise (the 'Sombre Warrior') for the north and winter. The northern circumpolar region counted, however, as a separate central Yellow palace, directly surrounding the imperial pole-star, in accordance with the symbolic correlations in groups of five, analogous with the Five Elements, which run through all ancient Chinese natural philosophy. But there was a much more important division than this.

From high antiquity the Chinese equator (the 'Red Road' as opposed to the 'Yellow Road' of the ecliptic) was divided into 28 segments called 'lunar mansions' (*hsiu*), seven for each of the palaces. Each of these mansions was defined by a particular constellation, and took its beginning from a particular determinative star therein, so that the distances covered by each along the equator differed considerably. The hour-circles dividing each *hsiu* from its neighbours and radiating from the poles thus netted the heavens like the segments of an orange. Some mansion constellations lay on the equator, while others were a good deal south or north of it.

Medieval Chinese star-maps show how they depicted it. The splendid Suchow planisphere, inscribed on stone in + 1193, is a polar projection. The central circumpolar palace is seen in the middle, and then the hour-circles shown as straight lines radiating from the pole and meeting the equator at the periphery, with very different widths between them. Or we may take one of the maps published by Su Sung in his book of + 1094 which we shall mention again presently. It shows one half of the sky on Mercator's projection (though, let us remember, this was 500 years before Mercator), with the equator running horizontally along the middle and the unequal segments shown as vertical rectangles (Fig. 2). Hemispheres are also depicted in scroll-paintings of Ming and early Chhing date.

As for the time of origin of the system of the *hsiu*, some indications of it are already present on the oracle-bones of the − 14th century and it was certainly well established by the − 4th century. The intriguing scatter of the *hsiu* asterisms has invited attempts to date them having regard to precessional change, attempts in

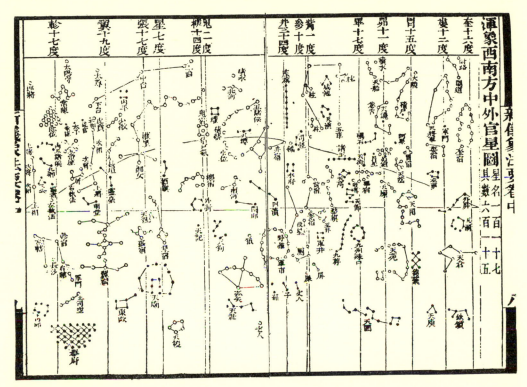

Fig. 2. Star-chart from Su Sung's *Hsin I Hsiang Fa Yao* of + 1094 showing fourteen of the twenty-eight *hsiu*, on 'Mercator's' projection, with many of the Chinese constellations contained in them. The equator is marked by the central horizontal line; the ecliptic arches upward above it. The legend on the right-hand side reads: 'Map of the asterisms North and South of the Equator in the S.W. part of the heavens, as shown on our celestial globe; 615 stars, in 117 constellations.' The *hsiu*, reading from right to left, are Khuei, Lou, Wei, Mao, Pi, Tsui, Shen (Orion), Ching, Kuei, Liu, Hsing, Chang, I and Chen. The unequal equatorial extensions are well seen.

which Dr Chu Kho-Chen has himself prominently participated, and the general result indicates the middle of the − 3rd millennium. Since this is rather early for Chinese culture, it may well be that further research will find the system to have been originally Babylonian, spreading thence in several directions, to India as well as China.

We then find this astronomical cosmology depicted on medieval Chinese art objects such as bronze mirrors. On the back of one of these dating from the Thang period (about the + 8th century), one can see in the centre the symbolical animals of the four palaces, and all round the edge in the outermost circle but one the array of the twenty-eight lunar mansion asterisms diagrammatically drawn. The outermost circle is inscribed with a poem which runs as follows:

> This metal mirror has the virtue of the Evening Star
> And the essence of the White Tiger of the West,
> The mutual endowments of Yin and Yang are present in it,
> And the mysterious spirituality of mountains and rivers.
> With due observance of the recurrent Heavenly motions
> And due regard to the tranquillity of Earth,
> The Eight Trigrams are exhibited upon it,
> And the Five Elements disposed in order on it.
> Let none of the hundred spiritual beings hide their face from it,
> Let none of the myriad things withhold their reflection from it.
> Whoever possesses this mirror and treasures it
> Will meet with good fortune and achieve exalted rank.

You will already have noticed, from what has been said about the lunar mansion constellations, that the Chinese sky was filled with patterns entirely different from those which were recognised by the Greeks and have become generally used in modern astronomy. Of all the thousands of star-patterns, not more than a dozen at the most were seen in the same way, and named accordingly, by Chinese and Westerners. Among these are the Great Bear and Orion; Auriga, Corona Australis, and the Southern Cross. As just one example of the difference we may take Cygnus, the Swan. The Chinese saw no swan; instead of that they pictured a rectangle of nine stars flung across the Milky Way at the point where it branches, and they called it Thien-chin, the 'Heavenly Ford', because they thought of the Milky Way as the River of Heaven. Just to the east of this was a group of seven stars known as Chhê-fu, the 'Garage', where the Sun's chariot was parked for the winter. Finally north of the Ford was a little tassel of four stars bearing the name of Chi-chung, the Sun's charioteer, probably because it looked like a bunch of reins held in the hand. This 'other world' of celestial topography is one of the most convincing arguments for the independent and original development of Chinese astronomy.[1]

But if our modern skies are Greek in nomenclature, they are entirely Chinese

---

[1] It is set forth in great detail by G. Schlegel in his *Uranographie Chinoise* (Brill, Leiden, 1875), one of the few timeless works of scholarship.

in an equally important respect, namely the way in which we reckon the precise position of the stars. There have been three classical ways of doing this. The Greeks, fully in accordance with their main preoccupation, made everything depend on the ecliptic and the pole of the ecliptic, defining a star-position in terms of celestial latitude and celestial longitude. The Muslims later on preferred the terrestrial horizon as a parameter, speaking in terms of azimuth and altitude. But the Chinese used from the beginning the equator and the hour-circle, exactly as we do today when we speak of right ascension and declination. It is true that they measured in terms of the number of degrees within each *hsiu*, and took the north polar distance instead of the distance from the equator, but their system was nonetheless identical with the modern one. When, one may ask, did the changeover take place in the West? It was at the Renaissance and in the great observational work of Tycho Brahe during the latter half of the + 16th century. It seems that he was stimulated to choose the modern system partly by certain Arabic influences, and the Arabs knew well what the Chinese system was.[1]

From the Chinese polar-equatorial emphasis two great instrumental consequences flowed, the invention of the equatorial mounting of sighting-tubes and telescopes, and the invention of the clock-drive and the mechanical clock. With these we shall have to conclude our exposition. To the armillary sphere we have already alluded; here we need only add the word of explanation that it was essentially a system of great circles which modelled those of the celestial sphere, and cradled a sighting-tube so that it could be pointed as accurately as possible to any given point on the apparent dome of the sky. It survives to this day in reduced and dissected form as the mounting of all telescopes, relatively insignificant in comparison with their bulky tubes, and the development of it led to the mounting of the projector in the middle of this hall. All through Chinese history it was improved and modified—the best pictures of it perhaps are Su Sung's own illustrations dating from the end of the + 11th century. But the cardinal invention which marked the transition from medieval to modern instruments was the mounting of the sighting-tube in the polar axis, freed from the prison of the great circles, and this came about not at the Renaissance in the West but under the aegis of Kuo Shou-Ching, the Astronomer-Royal of the Yuan dynasty in + 1276.

How this happened is a fascinating story. About + 1170 far away in Muslim Spain, the first dissection of the armillary sphere into its component circles or discs was carried out by an astronomer named Jābir ibn Aflaḥ, chiefly to make a

[1] [Cf. *SCC*, vol. 3, p. 267.]

computing machine which would facilitate passage from one set of spherical co-ordinates to another. This became known in Europe later on as the 'torquetum' or Turkish instrument; one is illustrated in the book of Peter Apianus, +1540. It has a remote descendant today in the form of the astro-compass used for aerial navigation. Knowledge of the torquetum probably reached China in +1267 on the occasion of the friendly scientific delegation from the Ilkhan of Persia to his brother of Cathay, led by a Marāghah astronomer, Jamāl al-Dīn. But as the ecliptic components were relatively unimportant for his purpose, Kuo Shou-Ching left them out in the great 'Simplified Instrument' (*chien i*) which he built for the re-equipment of the Peking Observatory a decade later, and thus became the originator of that equatorial mounting which long afterwards proved the most valuable suspension for the great telescopes of the modern post-Newtonian age. Here is a photograph (Fig. 3, pl.) of the Simplified Instrument taken a couple of years ago at the Purple Mountain Observatory near Nanking, where it is preserved today (in the form of an exact +15th-century recasting by Huangfu Chung-Ho). Compare with this a picture such as that of the 73-in. Reflector at Victoria, B.C., and the identity of the equatorial mountings can at once be seen.

But our Chinese predecessors of the Middle Ages were not only the first to take the observational armillary sphere to pieces and so to usher in the modern world of astronomical instruments, they also led the way in mechanising demonstrational and observational apparatus so that it would turn slowly round keeping pace with the apparent nightly revolution of the heavens. In other words, they made an armillary sphere rotate automatically. It will at once be clear that this must have involved nothing less than the invention of the mechanical clock itself, an invention hitherto generally believed to have occurred in Western Europe about the beginning of the +14th century. Modern research has shown, however, that the first escapement device—'the soul of the mechanical clock', as it has been called—was due to the Buddhist monk I-Hsing and his collaborators, working in the College of All Sages in the Thang capital Chhang-an (Sian) about +723.

This Chinese contribution was a missing link (in more senses than one).[1] It bridges the gap between time-measurement by the mere flow of a liquid (as in the clepsydras of ancient times), and time-measurement by rotation of wheels periodically arrested by purely mechanical means (as in the clocks of the Renaissance).

[1] [Cf. *SCC*, vol. 4, pt. 2, pp. 435 ff.]

The power source of the Chinese medieval clocks[1] was not, as in later Europe, a falling weight suspended on a cord, but an actual vertical water-wheel like that of a mill. Each of the scoops on its periphery took a precisely defined time to fill with water from a constant-level tank delivered at constant rate, and each scoop was during this time prevented from falling by an arrangement of weighbridges and link-work levers which periodically released the driving-wheel to move forward by one scoop. The device can best be understood, perhaps, by pictures. Here (Fig. 4) is an imaginative reconstruction by Mr John Christiansen of the great astronomical clock-tower erected by Su Sung and his collaborators at the Sung capital, Khaifêng, in + 1090. Below is the driving-wheel, with its escapement mechanism, above on the right are the water-tanks, the first floor houses a celestial globe automatically rotated, and on the platform there is the clock-driven observational armillary sphere. We now believe that this had two movements effected by gearing, an annual one and a diurnal one. The drive was transmitted by means of endless chains and even some oblique gearing, not only to the automatically turning astronomical instruments but also to a set of tiers of wheels carrying puppets or 'jacks' which announced the passing of the hours and quarters, supplying the place of a dial. Lastly we see the escapement itself in rather simplified form, with its two weighbridges or counter-weights, its chain and its arresting levers (Fig. 73 on p. 218). Beautiful working models of this medieval Chinese astronomical clockwork (Figs. 5, pl. and 64, pl.), which keep time within ±20 seconds per hour, have been made by Mr John Combridge of the G.P.O., and he is continuing experimental investigation.

How could it be, one may well ask, that the Chinese were able to push so far ahead with the mechanisation of instruments? It was, we think, precisely because of their emphasis on polar-equatorial co-ordinates. If one thinks in terms of celestial latitude and celestial longitude, the grid based on the ecliptic, one is thinking in terms of a purely artificial network along the lines of which nothing ever actually moves. Well did John Donne say (in + 1611):

> For of Meridians and Parallels
> Man hath weav'd out a net, and this net throwne
> Upon the Heavens, and now they are his owne.

But the declination parallels, on the other hand, marching with the equator, are precisely the natural lines along which all the company of the stars does seem to

---

[1] See J. Needham, Wang Ling & D. J. de S. Price, *Heavenly Clockwork* (Cambridge, 1960); and J. Needham, *Proc. Roy. Soc.* A, 1959, **250**, 14 (a Wilkins lecture, reproduced on pp. 203 ff. below).

Fig. 4. Reconstruction by John Christiansen of the great astronomical clock erected by Su Sung, Han Kung-Lien and their collaborators at the Imperial Palace, Khaifêng (Honan province) +1088 to +1092. The water-wheel rotated a celestial globe and an armillary sphere, and also an elaborate series of jack figures in a pagoda-like time-annunciator without a dial. As the sphere was used for observational purposes, this was the first clock-drive, such as is used for modern telescopes. The escapement employed had been invented by I-Hsing (a Buddhist monk) and Liang Ling-Tsan in +723, six centuries before the first appearance of mechanical clocks in the West. The water-tanks were replenished by manually-operated norias (Needham, Wang & Price).

move. This would most readily have inspired men with the idea of demonstrating such movement upon earthly armillaries or spheres automatically rotated —if that could be done. Of course it was not so easy. Rough approximations were constructed in China in every century from the + 2nd onwards, but not until early in the + 8th came definitive success with the work of I-Hsing.

The development of time-keeping machinery has been so important in so many ways for all modern science, and for astronomical science in particular, that we can hardly overrate the importance of this invention. The accurate measurement of time is one of the fundamental requirements of all science, like the measurement of spatial distances, or of temperature and pressure. In a way, mankind has always lived inside a clock. In a German Renaissance broadsheet there is a picture of a + 16th-century European peering through the curtain of surmise and inference at the revolving wheels of the Ptolemaic sphere. No one believes any longer in these, but if we imagine ourselves in a space-ship far out in our Galaxy watching the revolutions of the planets round our Sun, would it not remind us of the wheels of a time-keeping machine, perishable perhaps like everything that the Tao has wrought, but accurate over vast aeons of time? Within our lesser history, the discovery of the Chinese contribution to horological engineering six pre-European centuries long has now made it possible to tell a continuous story of time-keeping from the most ancient Babylonian and Egyptian leaking water-pots to the wrist-watches we are all wearing here, and to the caesium and ammonia clocks which check cosmic time against atomic and molecular vibration frequency time. But 'the Quincunx of Heaven runs low, and 'tis time to close the five ports of knowledge'—Time, 'which antiquates antiquities and hath an art to make a dust of all things' brings this little lecture also in its turn now to an end.

May I have your permission to add a few words addressed to our distinguished guests in their own language, the language of Chang Hêng, I-Hsing and Kuo Shou-Ching:[1]

Dear Colleagues and Friends,

We are delighted that this occasion has presented itself to do honour in your presence to the ancestors of astronomical science of your great country. We have for them the highest admiration, and feel sure that as time goes on the services of Chinese culture to man's knowledge of the universe will be more and more appreciated everywhere. We wish also all success to the astronomical work of our contemporary colleagues of Academia Sinica, and hope that you will all fully enjoy your visit among us. Long live international scientific co-operation for the peace and benefit of mankind!

[1] What follows was spoken in Chinese.

2

THE UNITY OF SCIENCE;
ASIA'S INDISPENSABLE CONTRIBUTION[1]

[Beirut, 1948]

MODERN SCIENCE and technology, as all the world knows, grew up in Western Europe, in the work of such men as Galileo, Vesalius, Harvey and Newton; as part and parcel of that vast social change which we know as the Renaissance, the Reformation and the rise of capitalism. Franklin and Priestley are symbols of its extension to the North American continent and its subsequent vigorous growth there. When, therefore, the history of science and scientific thought began to be written it was perhaps natural that attention was concentrated on the achievements of the ancient Mediterranean peoples who stood at the threshold of European history, especially the Greeks, and even their debt (which they themselves frequently acknowledged) to the earlier civilisations of the Mediterranean basin and the Fertile Crescent (the Egyptians, Babylonians, Hittites, Phoenicians, etc.) was overlooked. William Whewell's pioneer book, *History of the Inductive Sciences*, of 1837, may stand as an example of this. Still more in shadow were the scientific and technical achievements of East Asia. To name another Cambridge author, J. B. Bury's *Idea of Progress* (1920) has much to say about the arguments of those who at the time of the Renaissance defended the 'Moderns' against the 'Ancients', often successfully, on the grounds of the inventions of gunpowder, printing, and the compass, which had not been known to European antiquity—but nowhere is there even a footnote referring to the Far Eastern origin of these inventions. In later times, however, much attention has been given to Arabic and Middle Eastern science, as in the remarkable book of Mieli, *La Science Arabe*.

The truth here is, of course, that the science of Asia has a dividing line running north and south through Bactria and the opening of the Persian Gulf. The science and scientific thought of Arabic civilisation forms in one sense a unity with European science; not only because at the furthest extension of Islam, the Mediter-

[1] Unesco Month Lecture, Beirut, 1948.

14

ranean was a Muslim lake, and Spanish no less than Persian Muslims contri buted to the progress of science; but also because, as everyone knows, Arabic was the channel through which the Greek writings of the Mediterranean ancients reached the medieval Europeans. All the important, and most of the less important, Greek scientific texts were translated into Arabic between the + 7th and the + 11th centuries, and then were translated back into Latin. Direct translations hardly begin until the + 12th century with James the Venetian (fl. + 1128), Robert Grossteste of Suffolk (b. + 1175) and William Moerbeke the Fleming (b. + 1215). In this remarkable phenomenon of transmission, other Middle Eastern languages, such as Syriac and Hebrew, played a minor but still important part.

Into this system, however, East Asian science was not incorporated, and it is for this reason that only in our own time are we beginning to have any appreciation of the fundamental contributions, indispensable I have rightly called them in the title, which the scientists of China and India made throughout the centuries to the scientific patrimony of mankind.

Here we come upon a very interesting fact. It is not that there was no contact between Arabic civilisation and East Asian science; quite the contrary. But for some reason or other, when the translations were being made from Arabic into Latin, it was always the famous authors of Mediterranean antiquity who were chosen, and not the books of Islamic scholars concerning the science of India and China. I would like at this point to give you some examples of the way in which this knowledge was made available to Arabic readers but did not penetrate through to the Franks and Latins.

As early as the middle of the + 9th century 'Alī al-Ṭabarī, son of a Persian Christian astronomer, who lived at Baghdad, wrote his great medical work *Firdaus al-Ḥikma* (The Paradise of Wisdom), and it is striking that he quoted from Indian physicians such as Caraka, Suśruta, and Vāgbhaṭa II, no less than from Hippocrates, Galen and Dioscorides. But after a thousand years al-Ṭabarī's work has not yet been translated into a Western language. Similarly the great al-Khwārizmī, whose work *Ḥisāb al-Jabr wa'l-Muqābalah*—on algebra—was written about + 820, introduced a knowledge of the Indian numeral system. Fifty years earlier, al-Fazārī had certainly been acquainted with parts at least of the Indian astronomical work *Sūrya Siddhānta*.

Representative of this trend stands above all the great al-Bīrūnī, who having followed Maḥmūd of Gaznah in his conquest of India, returned and wrote very early in the + 11th century, about + 1012, his admirable work *Ta'rīkh al-Hind*.

15

This is not only a history and geography of India in the ordinary sense, but a profound examination of all the sciences of the Indians. But it was not translated into any European language until 1888! In the +13th century Muslim geographers gave many similar accounts of China and Chinese science, for example the cosmographers Muḥammad ibn Ibrāhīm al Dimashqī (+1256 to +1326) and Ahmad ibn 'Abd al-Wahhāb al-Nuwairī (+1279 to +1332). Further information came through the *Taqwīm al-Buldān* of Abū'l-Fidā al-Aiyūbī written in +1321, and by the work of the Persian geographer Ḥamdallah al-Mustaufī al-Qazwīnī (+1281 to +1340). Some travelled to China to see for themselves, as we know from the accounts of the lovable Ibn Baṭṭūṭah (+1304 to +1377), called by Sarton the greatest traveller of Islam, and the greatest, not excepting Marco Polo, of all medieval times. Ibn Baṭṭūṭah describes the construction of Chinese ships, the making of porcelain, the institution of old-age pensions, paper money, trade inspection, and so on.

Indeed, the personal contacts between scientists from the ends of Asia in those days have not been sufficiently appreciated. After the Mongol Hūlāgu Khan sacked Baghdad in +1258 and put an end to the Abbasid caliphate, he entrusted the illustrious Naṣīr al-Dīn al-Tūsī with the formation of an astronomical observatory at Marāghah in Azerbaidjan south of Tabriz. The observatory was equipped with the best instruments constructed up to that time, and the library is said to have contained over 400,000 volumes. Hūlāgu sent astronomers from China to collaborate in the work, and we know the name of one of them (Fu Mêng-Chi, unfortunately we do not have the characters). At Marāghah they met with men from as far west as Spain; al-Maghribī al-Andalusī, for instance, who published astronomical tables and many other books from Marāghah, including his *Risālat al-Khiṭā wa'l-Īghūr*, a monograph on the astronomy and calendar of the Chinese and the Uighurs. None of the astronomical instruments of Marāghah, used throughout the last half of the +13th century, have survived, though we know a good deal about them since al-'Urḍī al-Dimashqī, the Syrian, gave an elaborate description of them. Fortunately we do still possess some of the contemporary astronomical instruments made during the Yuan dynasty in China under the supervision of Kuo Shou-Ching in +1279, for the observatory which still exists at the south-east corner of the Tartar city-wall at Peking. I myself have had the good luck to visit this holy place of science, though it now contains only the Jesuit instruments of the +17th century, the Mongol ones having been removed to the Purple Mountain at Nanking. It would be extremely interesting to compare the various types of armillary spheres, mural quadrants and apparatus for determining sines and azimuths, at Marāghah and Peking, but time does not permit.

In + 1362 'Aṭā ibn Aḥmad al-Samarqandī wrote an astronomical treatise with lunar tables for a Mongol prince of the Yuan dynasty, Chen-Hsi-Wu-Ching. The MS is in Paris, and, as illustrated by Sarton, shows a title-page bearing both Chinese and Arabic writing. The contents, however, have not yet been investigated.

Passing from the field of astronomy to that of the medical and biological sciences, we come upon the remarkable work of Rashīd al-Dīn al-Hamadanī (+ 1247 to + 1318), Persian physician and patron of learning, prime minister under the greatest of the Mongol rulers of Persia, Ghāzān Maḥmūd Khān. His 'Universal History' (*Jami' al-Tawārīkh*) contains much information on China, especially on the use of paper money. About + 1313 he caused to be prepared an encyclopaedia of Chinese medicine, *Tanksuqnāmah-i Īlkhān dar funūn-i 'ulūm-i Khiṭāi* (Treasures of the Ilkhan on the Sciences of Cathay). Sphygmology (pulse lore), anatomy, embryology, gynaecology, etc., are dealt with, and most remarkably—the Chinese ideographic language is considered as superior to alphabetic ones for science because of its independence of phonetics and consequent internationalism. Under the name Wank-shu-khu we can recognise, as Sarton says, Wang Shu-Ho, the famous Chin dynasty physician (+ 265 to + 317) who wrote the principal classic on pulse lore, the *Mo Ching*. Rashīd al-Dīn was also interested in alchemy—a particularly Chinese characteristic.

The examples which have been given show that there was no lack of contacts between Arabic and East Asian science, but it remains true that East Asian science did not filter through to the Franks and Latins, i.e. to precisely that part of the world where, by a series of historical accidents perhaps (though their geographical and social determinism remain to be worked out), modern science and technology were later to develop. This barrier or filter, however, was operative only in the case of the abstract or pure sciences; it was definitely not effective for technology. Technical inventions show a slow but massive infiltration from east to west throughout the Christian era. Before describing some of these astonishing invasions, however, it is necessary to speak of certain other barriers and borrowings.

First, it seems that although the Chinese gave so abundantly to the rest of the world in inventions, their own sciences were relatively little affected by those of the rest of the world. Chinese medicine, for instance, is characterised by many special features, the theories of the two principles (Yin and Yang), the five elements (*wu hsing*), stasis (*yü*), pneuma (*chhi*), etc.; great elaboration of pulse lore (*mo hsüeh*), some of which may have reached the West through Ibn Sīnā; acupuncture (*pien chen*), moxa (*chiu*), the use of mineral drugs such as mercury and antimony long before the West, etc., etc. It is really hard to find in it any western influences. Hence the great interest of the following account, which we find in the *Fihrist al-ʿulūm* (Index of the Sciences) of Abūʾl-Faraj ibn Abū Yaʿqūb al-Nadīm, written in +988. The al-Rāzī referred to is the great Rhazes, physician and alchemist (*c.* +850 to +925), Muḥammad ibn Zakarīyā al-Rāzī:

al-Rāzī said, 'A Chinese scholar came to my house, and remained in the town [probably Baghdad] about a year. In five months he learnt to speak and write Arabic, attaining indeed eloquence in speech and calligraphy in writing. When he decided to return to his country, he said to me a month or so beforehand, "I am about to leave. I would be very glad if someone would dictate to me the 16 books of Galen before I go." I told him that he had not sufficient time to copy more than a small part of it, but he said, "I beg you to give me all your time until I go and to dictate to me as rapidly as possible. You will see that I shall write faster than you can dictate." So together with one of my students, we read Galen to him as fast as we could, but he wrote still faster. We did not believe that he was getting it until we made a collation and found it exact throughout. I asked him how this could be, and he said, "We have in our country a way of writing which we call shorthand, and this is what you see. When we wish to write very fast, we use this style, and then afterwards transcribe it into the ordinary characters at will." And he added that an intelligent man who learns quickly cannot learn this style in under twenty years.'

From this fascinating glimpse of Arab–Chinese contact, it is clear that the Chinese scholar, whose name, unfortunately, has not been preserved, was using 'grass-writing' (*tshao shu*). While the story was told by al-Nadīm *à propos* of the writing methods of the Chinese, it strongly indicates, if it does not absolutely prove, that there was at least one translation of Galen into Chinese in the +10th century. No perceptible influence of Greek or Hellenistic tradition upon Chinese medicine can however be found. This has been noted also by Sarton in remarking upon the presence of Muslim and Nestorian physicians at the Chinese court.

There is a field, it is true, in which it may turn out that the Chinese continued a living scientific tradition after it had been completely lost in the West. I refer to quantitative geography and cartography.[1] It is generally known that Greek and Hellenistic cartographers (symbolised by Eratosthenes and Ptolemy) set up grids

[1] [Cf. *SCC*, vol. 3, pp. 525 ff.]

of latitude and longitude, the former being comparatively accurate, but the latter greatly distorted because they overestimated the angle subtended by the Eurasian land-mass on the globe's surface. We know also that while the Arabs conserved Greek geography and added to it, the whole science was completely lost in the West, descending indeed to a deplorable level with the wheel-maps and T-maps of early medieval religious cosmography. Now Ptolemy of Alexandria finished his observations in +151, and the next place in which we find a quantitative cartography is in China with Phei Hsiu (+224 to +271) who was Minister of Public Works to the first emperor of the Chin dynasty. Phei Hsiu started the system of grids in which the side of each square represented a stated number of *li* (approximately half kilometres), which was continued by such eminent geographers as Chia Tan of the Thang (+730 to +805) and culminated in the magnificent stone-engraved maps of +1137 which are still in the Pei Lin museum at Sian. The Chinese had of course their own distortion, since their orthogonal mesh-net projection did not allow for the curvature of the earth, but the fact remains that between the time of Ptolemy and the Renaissance, the cartography of China was at an immeasurably higher level than that of Europe.

As might possibly be expected, in spite of the geographical difficulties in communications, Indian–Chinese mutual influences were much stronger. The deep penetration of an Indian religion into Chinese life and thought, though Buddhism was quite transformed in the process, brought with it not only much attention to Sanskrit–Chinese philology, but also considerable scientific exchange. The official history of the Sui dynasty, completed in +636 by Wei Chêng, contains, in the usual bibliographical catalogue, the titles of a large number of books, now lost, beginning with the words 'po-lo-mên' or Brahmin. Thus we have *Polomên Thien-Wên Ching* (Brahmin Astronomy), *Polomên Suan Fa* (Brahmin Mathematics), *Polomên Yin-Yang Suan Li* (Brahmin Calendrical Methods), *Polomên Yao Fang* (Brahmin Drugs and Prescriptions), etc., etc. It may be added, remembering the Galen translation mentioned above, that the same bibliography also lists a *Hsi-Yü Ming I so chi Yao Fang* (Drugs and Prescriptions collected by the most famous Physicians of the Western Countries). But in spite of these facts, evidence of a lasting influence either Indian or Western on Chinese science and medicine is hard to adduce. It would need a careful comparison of the traditional pharmacopoeias of China and India to trace out the pharmacological borrowings involved. It is usually supposed that drugs such as chaulmoogra oil for leprosy, which have been for many centuries in the Chinese pharmocopoeia, were of

Indian origin. Owing to the work of Bretschneider and many other scholars, the materials for making such a comparison are now largely available in Western languages. But all such studies are rendered excessively difficult by the almost complete absence of a settled chronology for India, and the great uncertainty which exists in the dating of even the most important scientific texts of that culture-area.[1]

Conversely evidences of Chinese influence on Hindu mathematics have been several times detected.[2] A proof used by Chao Chün-Chhing in his + 2nd-century commentary on the *Chou Pei* (the oldest mathematical classic) appears again in the work of Bhāskara (+ 1150). The rule for the area of the segment of a circle given in the *Chiu Chang Suan Shu* (Arithmetic in Nine Sections, + 1st century) appears again in the + 9th-century work of Mahāvīra. Problems of the *Sun Tzu Suan Ching* (+ 1st century) are found in Brahmagupta (+ 7th century).

I shall end these few words on the Chinese–Indian contacts by giving you what I believe must be one of the earliest passages on mineral acids. Hitherto it has been generally believed that the mineral acids were first known in Europe in the + 13th century, and Partington, our greatest authority, finds the first mention of them in the *Pro Conservanda Sanitate* of the French Franciscan Vital du Four about + 1295. However, listen now to Tuan Chhêng-Shih, writing in his *Yu-Yang Tsa Tsu* of + 863 about events which took place between + 647 and + 649:

Wang Hsüan-Tshê captured an Indian prince named A-Lo-Na-Shun. He had with him a scholar versed in curious arts named Na-Lo-Mi-So-Pho, who said he could make people live for two hundred years. The Emperor (Thai Tshung) was very astonished, and invited him to live in the Chin Yen Mên Palace, to make the drugs for prolonging life...The Indian said that in India there is a substance called Pan-chha-cho Water, which is produced from minerals in the mountains, has seven varieties of different colours, is sometimes hot, sometimes cold, can dissolve herbs, wood, metals, and iron; indeed if it is put into a person's hand, it will melt and destroy it.

This gives colour to the hints about mineral acids in P. C. Ray's *History of Hindu Chemistry*. The *Rasārṇava Tantra* (dated by Renou & Filliozat as of the + 12th century) speaks of the 'killing' of iron and other metals by a *viḍa* (solvent?) which is prepared from green vitriol (*kāsīsa*), pyrites, etc. From the *Rasaratna-samuchchaya*, which according to Renou & Filliozat may go back to about + 1300, the process of 'killing' certainly seems to be the formation of salts from metals. In any case, the present passage, which does not seem to have been noted pre-

---

[1] [Cf. J. Needham, *Nature*, 1951, **168**, 64, 1048 and S. L. Hora, *Nature*, 1951, **168**, 1047, *Proc. Nat. Inst. Sci. India*, 1952, **18**, 323.]
[2] [Cf. *SCC*, vol. 3, pp. 146 ff.]

viously, clearly suggests a knowledge of mineral acids in India in the + 7th century.[1]

Before passing to technological matters, this may be the place to point out that the more one compares East Asian with Occidental science, the more one is impressed with certain striking differences of emphasis. Thus in general it is true to say that while the Greek mathematical genius lay largely in the direction of geometry, that of the Chinese lay in the direction of algebra. We have to come to a relatively late period—that of Diophantus towards the end of the + 3rd century —before we find Greek algebra, and even so it remained somewhat isolated. Conversely, although the early Chinese mathematical books contain a certain amount of geometrical reasoning (e.g. the 'Pythagoras' problem), China hardly had geometry until the translation of the first six books of Euclid by Fr Matteo Ricci and Hsü Kuang-Chhi, the *Chi Ho Yuan Pên* of + 1607; though it is strongly suspected that a lost book of + 1273, the *Ssu-Pi Suan Fa Tuan Shu* by 'Wu-Hu-Lieh-Ti', was a translation of fifteen chapters of Euclid arranged by Naṣīr al-Dīn, the founder of the Marāghah observatory. On the other hand, the Sung and Yuan algebraists, such as Chhin Chiu-Shao (fl. + 1247), Li Yeh (fl. + 1259), Yang Hui (fl. + 1280), and above all Chu Shih-Chieh (fl. + 1303) constituted, as is now generally recognised, the leading school of mathematics any-where at that time. This algebra, which seems to have owed nothing to influences from outside China, was called *thien yuan shu*, the 'method of the radiating powers and coefficients'.[2]

Another contrast of the same kind may lie between particle theory and wave theory, though here the dividing line will come between China and Graeco-India rather than between the Mediterranean and Asia. It is needless to dwell upon the important school of Greek and Roman atomic speculations (Democritus, Epi-curus, Lucretius), which probably had some connection with, if it was not in fact originally based upon, similar Indian speculations previous to the *Vaiśesika* (+ 1st century) and the *Sāṃkhya-kārikā* (+ 4th century) on atoms (*paramāṇu*). By com-parison the remarkable thing is that in Chinese thought we can hardly find any traces at all of atomism. A very few fragments here and there show that even if it was introduced more than once, it never found a fertile soil in Chinese thought. It is possible that this may have been because of the deep preference of Chinese ideas for what amounts to a wave conception. From about the − 4th century onwards, Chinese theory of Nature was dominated by the Yang–Yin dualism;

---

[1] [See further Tshao Thien-Chhin, Ho Ping-Yü & J. Needham, *Ambix*, 1959, **7**, 140 ff.]
[2] [See more fully *SCC*, vol. 3, pp. 129 ff.]

the two forces or influences (light and dark, male and female, up and down, convex and concave, sun and moon, prince and minister) which controlled all phenomena by their regular and predictable course, alternately waxing and waning, the one being inversely proportional to the other. The earliest pre-Cartesian European graphs, which show on co-ordinates this waxing and waning of celestial bodies, would have been quite applicable to the Yang–Yin conception, and we need not yet despair of finding some similar pictorial representation in old Chinese literature. The physics of the medieval Chinese (as distinguished from practical mechanics) was indeed one of the weakest branches of their science, but nevertheless I believe there is a sense in which we may regard modern particle-theory as Graeco-Indian in origin, and modern wave-theory as Chinese.[1]

The time has now come to consider a few of those technical inventions which regularly crept through the filters or barriers of which we have already spoken, and came from East Asia to Europe in a continuous stream from the early part of the Christian era onwards. I must reserve for another occasion the history of iron-casting,[2] which was regularly practised among the Chinese at least as early as the − 1st century, but not in Europe (extraordinary historical paradox!) until the + 14th century. Nor can we follow the history of the humble wheelbarrow. I hesitated to pause over the techniques of paper-making and printing, the passage of which from China to Europe, through many intermediate stages, has been so brilliantly pin-pointed by Carter, but since a certain Arabic passage demands quotation, let us read it. First we should remember that the earliest Chinese printing which has come down to us is a Buddhist *sūtra* found in the Tunhuang cave-temples and dated + 868.[3] Now here is Liu Phien speaking (in his *Chia Hsün Hsü*):

In the summer of the third year of the Chung-Ho reign-period (+883), when the Imperial court had been for three years in Szechuan, I was a departmental head in the Grand Secretariat (Chung-Shu Shih-Jen). Whenever I was off duty, every tenth day, I used to look over the books (exposed for sale on stalls) at the south-east corner of the Inner City (Chung Chhêng)

[1] [Cf. *SCC*, vol. 4, pt. 1, pp. 3 ff., or J. Needham & K. Robinson, *Sciences*, 1960, **1** (no. 4), 65.]

[2] [Cf. p. 107 below.]

[3] [It is however tolerably certain that printing must have begun in China towards the end of the +7th century. For many years back the oldest extant printed texts from any civilisation have been the charms ordered by the Japanese Empress Shōtoku between +764 and +770 and distributed in a million model pagodas. Very recently, however, an even earlier text has been found in Korea, a Buddhist *sūtra* printed not later than +751, the date of the *stupa* in which it was laid up. Indeed it may be as old as the period +684 to +704, the years when its translator, the Tokharian monk Mi-Tho-Hsien, was living and working in the Chinese capital, under the reign of the Empress Wu Tsê-Thien, whose own tabu character forms are found in the text. See L. Carrington Goodrich, *Technology and Culture*, 1967, **8**, 376, and Gari Ledyard, *Columbia Library Columns*, 1967, **16**, 3. Buddhism, Taoism and Confucianism all in their several ways encouraged the spread of printing.]

(in Chhêngtu). These books consisted mostly of texts on the interpretation of dreams, the principles of geomancy, and astrological prognostications by the Nine Palaces or the Five Wefts (the planets), with other miscellaneous tractates of natural magick (Yin Yang); but there were also dictionaries, encyclopaedias and school books. For the most part they had been engraved on (wood) blocks and printed on paper, but sometimes the ink had smudged and one could not quite make out everything.[1]

We shall all sympathise with this *flâneur* of bookstalls who lived at the very dawn of printed books, especially those of us who have been fortunate enough, as I have, to visit the bookshops in that same Szechuanese city. A fragment of one of these very printed sheets has been found among the Tunhuang library collection at the British Museum, a calendar dated +882; and another such sheet belongs by internal evidence to +877.

Here is another glimpse of that time. Wang Chung-Yen says in his *Hui Chhen Hou Lu* (+1194):

When Wu Chao-I was poor, he often used to borrow the *Wên Hsüan* anthology from friends, but sometimes they were embarrassed and reluctant to lend it. So he said to himself, 'If ever I come into a position of power, I shall have such books cut in blocks, so that all scholars may have an opportunity of reading them.' And eventually he did become minister of the State of Shu (Szechuan) under the reigning family of Wang [in fact it was Mêng, +935], so that he was able to fulfil his resolve and print them. This was the beginning of printed books... After the Emperor Ming Tsung of the Later Thang had subdued Shu, he commissioned the professor Li Ê to write out the text of the Five Classics, and following the example (of Wu Chao-I) blocks were prepared in the Imperial University for printing them. This was the beginning of printing in the University.[1]

And in fact the printing of the Confucian classics was achieved in +953, after which many other books became available. As Carter has shown, printing spread to the Uighurs in Central Asia early in the +13th century and then to Egypt. Gutenberg's accepted date is +1436.

The Arabic mention to which I wished to refer is that of the Persian scholar Dāwūd al-Banākitī in +1317. He appreciated particularly the service of printing to the standardisation of correct texts. In his *Raudat ūlī'l-Albāb* (Garden of the Intelligent) he wrote:

The Chinese are wont to make copies from books in such wise that no change or alteration can find its way into the text. And when they thus desire, they order a skilful calligrapher to copy a page of the book on a tablet in a fair hand, and then all the men of learning carefully correct it, inscribing their names on the back of the tablet. Then skilled and expert engravers are

[1] [Retranslated here.]

ordered to cut out the letters. And when they have thus taken a copy of all the pages of the book, numbering all the blocks consecutively, they place them in sealed bags, like dies in a mint, and entrust them to reliable persons appointed for the purpose, keeping them securely in special offices on which they set a particular seal. When anyone wants a copy of the book he goes before this committee and pays the dues and charges fixed by the government, after which they bring out the tablets, impose them on leaves of paper like the dies are imposed on gold for coins, and so deliver the sheets to him. Thus it is impossible that there should be any omission or addition in any of their books, on which, therefore, they place complete reliance; and thus is the transmission of their histories effected.[1]

One must remember that this was written a century before Gutenberg's time, yet something like six centuries at least after printing started in China.

Al-Banākitī was over-optimistic about the critical status of Chinese texts, for many changes did occur with different editions, and serious corruption had often set in before the first printing; nevertheless it is certainly true that the sinologist is much less dependent upon unreliable manuscripts of fitful and sporadic location than students of other civilisations. There is a widely held idea that sinologists pore much over manuscripts, but on the contrary the situation is that since the making of paper began about + 100 and printing about + 700, practically everything in Chinese is either printed or lost. Only very rarely does a great cache of manuscripts come to light, as at the Tunhuang cave-temples, and even then some of the texts only duplicate printed versions we already have. All the same the losses of Chinese literature have been incalculably great, partly because the flimsy replication medium of paper was invented so long before the mass replication technique of printing, but partly also because the great inter-dynastic upheavals and foreign conquests destroyed even whole editions of printed books. There are tens of thousands of books of which we know only the titles. Without printing there would have been a million. So much for printing.

Paper, printing, gunpowder and the magnetic compass have been so fully recognised as Chinese inventions and discoveries that it will be better to turn to other examples of the westward advancing wave of technology. Let us take the art of deep drilling.[2] The province of Szechuan, some twelve hundred miles from the sea, possesses enormous natural deposits of brine and natural gas, the most important of which are situated between Chungking and Chiating, in the district of Tzu-liu-ching. Had this not been the case, the kingdom of Shu, so often independent during Chinese history, and forming as it did the citadel of defence

[1] [Al-Banākitī here copied closely the *Jami'al-Tawārīkh* of Rashīd al-Dīn, written seven years earlier.]
[2] [Cf. p. 33 below.]

against the Japanese invaders during World War II, could never have maintained itself. Lacking salt, that inescapable essential of human diet, its population would have had to capitulate, or else to migrate *en masse*, as did the Communist enclave in Chiangsi between the two world wars. Now we know, from references in the *Chhien Han Shu* and the *Hua Yang Kuo Chih*, that salt wells began to be exploited at least as early as the Former Han dynasty (− 1st century), and at Chhêngtu there are Han bricks in relief which show the derricks and the pans in which the brine was evaporated. We have a particularly detailed description of the wells from the pen of the poet Su Tung-Pho, who was born in Szechuan in + 1036. The borings are remarkable in that they descend as far as 3,000 ft in some cases, and are drilled by ancient methods.

Another technological complex which it is fascinating to study from the comparative point of view, is that of the mill water-wheel and paddle-wheel propulsion. Let us work backwards. Feldhaus, in his *Technik d. Vorzeit*, reproduces an illustration from the great Chinese Encyclopaedia of + 1726, *Thu Shu Chi Chhêng*, of a paddle-boat with two paddle-wheels on each beam, but adds that of course such an idea must have been copied by the Chinese from the Jesuit missionaries during the + 17th century. However, in truth there was no 'of course' about it. We know that the first Western illustrations of manpower-operated paddle-boats occur in Renaissance engineering works such as those of Guido da Vigevano (+ 1335), and Roberto Valturio (engineer to Sigismondo Malatesta) in his *De Re Militari* of + 1472. We know further that Blasco de Garay actually constructed one in Emperor Charles V's harbour at Barcelona in + 1543. Later eighteenth-century experiments paved the way for the advent of really applicable power with Fitch, Symington and Fulton. But nearly a thousand years before Valturio's sketch, there had been paddle-propelled boats in China.[1]

Leaving out of account references in the *Nan Shih* and the *Nan Chhi Shu* to a 'thousand-*li* boat' invented by the mathematician Tsu Chhung-Chih (between + 494 and + 497), we find the following in the official history of the Thang dynasty (*Thang Shu*) with reference to the years + 782 to + 785:

Prince Tshao (Li Kao) gave instructions that war-ships should be made carrying two wheels worked by a treadmill, which stirred the water and caused the boat to move as rapidly as the wind.

Again in Yo Fei's biography in the *Sung Shih*, about + 1130, we find:

[1] [Cf. *SCC*, vol. 4, pt. 2, pp. 413 ff., and Lo Jung-Pang, *Chhing-Hua Journal*, 1960, **2**, 189.]

Yang Yao [the leader of a peasant rebellion] launched ships in the lake that were moved by wheels which stirred the water. A ram was attached to their bows, so that vessels carrying government officers were destroyed when they met these ships.

Finally, in the *Mêng Liang Lu* of + 1275 (a kind of guide-book to the wonders of Hangchow) it is said:

There are also the 'wheel boats' (such as those) belonging to the great house of Chia Chhiu-Ho. On the deck above the cabin there are no men poling or rowing, for these craft move by means of wheels worked by a treadmill, and the speed is like flying.

Evidently it was rather rash to suppose that the Jesuits introduced the idea of paddle-boats to the Chinese.

There is another reference in European history to a builder or proposer of a boat fitted with paddle-wheels earlier than that of Valturio, namely Konrad Kyeser, who in + 1405 may have used the system for hauling boats upstream, the wheels working a winch to wind up a tow-rope fixed to a point ashore higher up the river. It is possible that at some time there may have been a parallel to this in China, for I myself saw and photographed at Ping-lo in Kuangsi boats hauling themselves upstream by a similar method, the winch however being worked directly by hand. It is at this point that we see the connection between paddle-boats and mill water-wheels, since the same contrivance may obviously be used either to supply motive power from the force of rushing water, or to give forward motion by the application of motive power to a wheel acting in still water. Here an important transitional case is that of the corn-mills mounted on boats and operated by water-wheels (paddle-wheels) turning in the current of the River Tiber, by the Byzantine general Belisarius during the siege of Rome in + 536, as fully described by Procopius (*De Bello Gothico*). The same procedure was used in China, but I have not yet succeeded in establishing from what date; the description occurs in a late Ming author Wang Shih-Chên, who says (in his *Shu Tao I Chhêng Chi*):

In Liang-chiang there were many ship-mills (*wei chhuan*) anchored in the rushing current. The grinding, pounding, and sifting were all carried on by the force of the water, and the machinery made a noise 'kha yao, kha yao, kha yao' incessantly.

Such boats are still found today at Fou-ling, some sixty miles downstream from Chungking.

Water-wheels of all kinds may thus be divided into two classes, for which we might coin the words 'ad-aqueous' where energy is transmitted to water, and 'ex-aqueous' where energy is transmitted from moving water. The simplest possible example of the former class is that of the *kua chhê*, a hand-operated paddle-wheel pushing water along a flume, and this seems to be very ancient in China, used as a method of water-raising for irrigation when the lift required is small. At present, however, we cannot adduce any description earlier than the *Nung Shu* of +1313. It is quite impossible to say whether this simple device led to the mounting of an ad-aqueous wheel on a boat for forward motion, though the adaptation may well have been made in Tsu Chhung-Chih's time (+5th century). Alternatively, the mounting of ex-aqueous wheels on moored boats for milling may have led to the substitution of ad-aqueous wheels also on boats.

The earliest history of stationary ex-aqueous wheels (mill water-wheels) in the West is still, in spite of the brilliant work of Bloch and Curwen, very obscure. They seem to appear simultaneously in China and upon the eastern edge of Europe. The water-mills of King Mithridates of Pontus at the south-east of the Black Sea, and their capture by Pompey in −65, are mentioned as a notable curiosity by Strabo. By −30 water-wheels working trip-hammer batteries were common in China. Then in +31 we have a most circumstantial account in the official history of the Han dynasty of water-wheels used not only to pound cereals, but also to operate bellows for metal-working, thus converting rotary to longitudinal motion.[1] It is impossible to believe that such a complicated piece of machinery had not had a considerable history already behind it.

Available evidence makes it quite likely that these first-century water-wheels were of the horizontal type with vertical shaft (*wo lun* in Chinese). The vertical water-wheel with horizontal shaft (necessary before ad-aqueous wheels could be used) was first described by Vitruvius about the same time in the −1st century (−27), but did not displace the horizontal wheel; for certain European regions, especially Gaelic and Norse, continued to use the latter until the present time. The horizontal wheel has also remained the commonest type in China, although there we know that the vertical wheel was no importation of the Jesuits since it is illustrated in the *Nung Shu* of +1313 and was even used at that time to drive elaborate textile machinery.

Perhaps the most ingenious application of the ex-aqueous wheel was that of fixing buckets sideways to its periphery, so arranged as to discharge at the top of the wheel into a flume. This provided a perfectly automatic water-raising device for irrigation, capable of very high lifts; its usual name in the West is noria from

[1] [See further, p. 176 below.]

the Arabic *al-nā'ūra*; in China it is called *thung chhê*. Unfortunately, although Vitruvius describes a related machine, the chain of pots or *sāqīya*, we cannot trace back the noria before the + 12th century, either in the Middle East, in Europe, or in China; we know that the Egyptian engineer Qayṣar ibn Abī al-Qāsim al-Ḥanafī (+ 1168 to + 1251) built norias on the Orontes at Hama in Syria, and that the Iraqi physicist Ismā'īl ibn al-Razzāz al-Jazarī wrote about them in his book on hydraulics and mechanics of + 1206; while in China Wang Chên in the *Nung Shu* of + 1313 describes them as if they had long been in use.

It will be seen that throughout this part of the history of applied science it is rather difficult as yet to name definite priorities for Eastern or Western Asia; all we can say is that in spite of Vitruvius and the Alexandrian theorists Heron, Philon and Ctesibius, Asia was greatly in advance of Europe from the earliest times down to about the + 15th century. I believe that further intensive research will clarify greatly the story of the original inventions and of their transmission. We always have to bear in mind that inventions may have been made in Central Asia or the Iranian culture-area, about the scientific and technological history of which comparatively little is known, and spread in both directions. Such seems to have been the case, for instance, with the biological invention of falconry. And there is much evidence that certain ancient scientific ideas radiated from Meso-potamia.

It is now time to look back over the ground we have covered.

We speak of the unity of science, and the phrase is just. Even the most isolated discoveries, such as the invention of the mathematical zero by the Mayas, and that of the wheel (though applied only to toys) by the Aztecs, take their place as contributions to the scientific patrimony of mankind. But when we come to genetic relationships, there can be no doubt at all that the work of Asian peoples was at least as important in the history of science and technology as that of the Europeans, until the time of the Renaissance. The science of Arabic culture, we may say, was focal; it gathered in East Asian science, pure and applied, just as it built upon the work of Mediterranean antiquity. But, as we have seen, while on the one hand East Asian applied science penetrated to Europe in a continuous flow for the first fourteen centuries of the Christian era, East Asian pure science was filtered out; it came into Arabic culture but no further west. Obviously this is a historical phenomenon of much interest and importance.

In spite of the schoolboy ideas which we of the West acquired about the great gulf between Crusaders and Saracens, one feels, on better acquaintance, very reluctant to regard Arabic civilisation as 'oriental' at all. The culture of Islam, whatever its desert origins, was really much more closely allied to the European culture of mercantile city-states than to the Chinese culture of agrarian bureaucratism. This is the direction in which we have to look if we want to reveal the social and economic causes for the different courses of science and technology taken in Occident and Orient. But that is another story, and, as the Szechuanese tea-garden story-tellers say, 'If you want to know how this ended, you should come back here at the same time tomorrow evening.'

3

CENTRAL ASIA AND THE HISTORY
OF SCIENCE AND TECHNOLOGY[1]

[London, 1949]

I AM to speak of the role of Central Asia in the history of science and technology. This does not mean that Central Asia (by which I have particularly in mind Sinkiang, Tibet, and the countries bordering thereon) was ever itself in history a place where vital scientific discoveries were made. It has never been a centre of scientific or technological advance in the narrow sense of the words, but it does take a very high place as an area of transmission, which is really what I would like to speak about on this occasion.

The map [Fig. 6] will remind you of the routes across Central Asia. We are all familiar with the Tarim basin in Sinkiang and with the Old Silk Road running up from Lanchow city on the Yellow River, crossing the Gobi desert and going on to Kashgar and through Bactria into Persia. To the north of it lies the steppe country. There is a famous record of the ambassador Chang Chhien finding Szechuanese bamboo products and cloth in Bactria in the − 2nd century; technical methods may have travelled as well, slower than export merchandise, much faster than abstract science. One of the 'mysteries' which travelled over these routes from east to west, probably partly through Persian lands,[2] was iron-working in its more developed forms.

It is an extraordinary historical paradox that while modern Western European civilisation, which has so much influenced world civilisation today, is so dependent upon the working of iron and steel, the Chinese were 1,300 years ahead of the West in regard to cast iron.[3] Cast iron was a rarity in Europe until the + 14th century. Already, however, in the − 1st century in China (−28 to be exact) we have a record of a cupola furnace exploding, so the Chinese were habitually engaged at

---

[1] A lecture to the Royal Central Asian Society, London, Feb. 1949; reprinted from *Journ. Roy. Centr. As. Soc.* 1950, p. 135.

[2] [Cf. the recent discussion on iron smelting by J. R. Spencer, C. S. Smith, T. A. Wertime, J. Needham & L. C. Eichner in *Technology and Culture*, 1963, **4**, 201, 1964, **5**, 386, 391, 398 and 404.]

[3] [See further below, p. 107.]

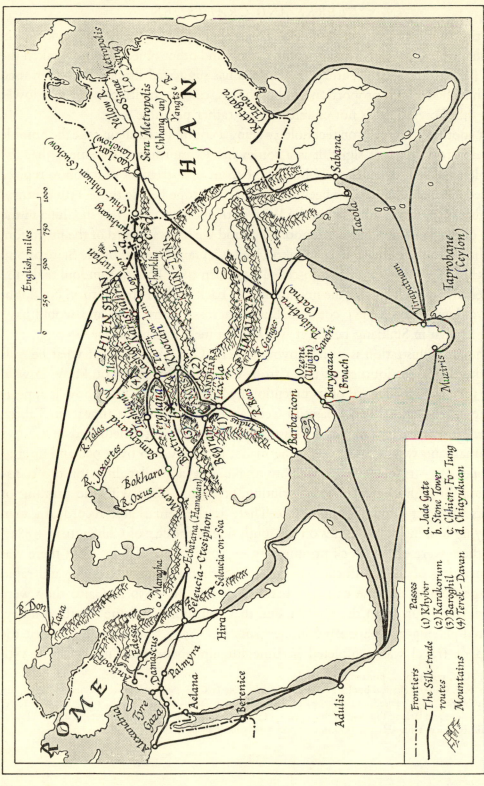

Fig. 6. Routes of trade between China and the West from the +1st and +2nd centuries onwards. From *SCC*, vol. I, fig. 32, based on Hudson.

that time in the foundry technique of iron metallurgy. Archaeological evidence in the shape of actual cast-iron agricultural tools and other objects has now taken the date back to the − 4th century. T. T. Read has suggested that the ores then used in China had a high proportion of phosphorus, causing melting some 200° lower than other ores. That is not established but rather plausible.

At another extreme, iron technology being a complicated affair, take the humble wheelbarrow. At first sight, it is a somewhat extraordinary statement that the simple expedient of putting a wheel at the front of a stretcher or a hod to replace the second man should have been unknown in Europe before the + 13th century, but that is so. It was not used in the West until about + 1250. But in China in the time of the San Kuo (Three Kingdoms) period we have a record of the invention, attributed to the military hero Chuko Liang, Captain-General of the State of Shu (Szechuan), about + 232, of a so-called 'wooden ox'. This was undoubtedly the wheelbarrow. But in fact the invention goes back to the − 1st century. No doubt such a device would have come to Europe across Central Asia and would have been known in Sinkiang before it was known in Western Europe.[1]

The next illustration is well known, namely, Carter's picture of what he calls the thousand-year journey of paper from China to Europe.[2] You know how in + 105 we have the record of its invention by Tshai Lun in China; how it appears in the + 3rd century at the oases in Central Asia; how later it appears in Tabriz, in Persia, where paper money was issued in + 1294. An intermediate date is + 751 at Samarqand where paper was being made; then in Cairo by + 900 and up into Europe by + 1150. It is possible to make out a similar line for the transmission through Central Asia of printing, because it starts with the printing of books in West China about + 870. There is the remarkable record of a man enjoying looking at bookstalls on the south side of the city of Chhêngtu in + 883,[3] and there were earlier block printings by + 770 in Japan; also in + 953 there was the printing of the Confucian classics. Altogether a more interesting and rapid passage than in the case of paper because already in + 950 in Cairo they were printing excerpts from the Qu'rān, and that continues until the + 14th century. Printed playing-cards appeared in Germany in + 1377, and we may say that the chain of travel was completed in time for alphabetic printing in the + 15th

---

[1] [Cf. SCC, vol. 4, pt. 2, pp. 258 ff.]

[2] [See T. F. Carter's classical book, *The Invention of Printing in China, and its Spread Westwards* (Columbia, New York, 1925), revised and enlarged by L. C. Goodrich (Ronald, New York, 1955) but with fewer illustrations and lacking the map of the journey of paper. See also the valuable monograph of Tsien Tsuen-Hsuin (Chhien Tshun-Hsün), *Written on Bamboo and Silk* (Chicago, 1962).]

[3] [P. 22 above.]

century, about + 1440. We know also of the independent invention of metal type in Korea in + 1240 and of its intensive use there after + 1390. The original invention of movable type, made of earthenware (later of tin), was due to Pi Shêng, about + 1045.

Those facts are all familiar to the members of this Society, but to pass to what may not be so familiar I will speak, first, of deep drilling. There are three technological advances on which I should like to dwell and which I think came over Central Asia. These three technical advances are, first, the art of deep drilling; secondly, the invention of the efficient horse-harness; and, thirdly, the invention of iron-chain suspension bridges.

As regards the art of deep drilling, the province of Szechuan, which is some 1,200 miles from the sea, possesses great deposits of brine probably associated with old oil deposits. A very important district which some of us visited during the war was that of Tzu-liu-ching in Szechuan. I often show modern photographs depicting the derricks for these wells, which are frequently 3,000 ft deep; also a picture taken from the *Thien Kung Khai Wu*, the great technological encyclopaedia of + 1637, showing the derricks, the bamboo bucket with valves ascending and descending, and the winding gear—a horizontal drum worked by water buffaloes or oxen. Another picture from the same book shows the fire wells, and you see the natural gas escaping from one of them, also gas being conveyed to the different vessels which are evaporating the brine.

A modern picture shows the salt being evaporated in pans. If Szechuan had not had its salt industry it could not have been an independent kingdom; if the salt had not been available it would have been very much more difficult for China to have held out during the last war when the Japanese had control of the whole coast. We know from literary references that these salt wells began to be exploited about the – 2nd century. In Chhêngtu I was given a rubbing of a Han dynasty moulded brick which shows a derrick and dates from the – 1st century [Fig. 7, pl.].

The method of making the boreholes is remarkable in that it includes the jumping on and off a beam by a team of men, giving thus the necessary impacts while the boring tool is rotated. The process of drilling a single well may take as long as ten years. The same technique was used in the first petroleum wells bored in California, a process known as 'kicking her down'. It seems to me possible that some hints of the Chinese technique may have been derived by the Californians from knowledge on the part of the Chinese workers who were brought in to build the railways in the beginning of the nineteenth century in California. That, however, remains as yet a conjecture. What one does begin to wonder is whether the boring of artesian wells derives from the same source. Artesian wells are so called because the first in Europe was bored for a monastery in Lillers, in Artois, in +1126. They are self-flowing wells because they gush out owing to the existence of a porous layer between two impermeable strata, the pressure-head in which is higher than the depth of the well. Artesian wells need not be very deep, though in fact they often are, e.g. in the Paris basin, where early nineteenth-century ones reach over 1,500 ft. Famous self-flowing wells of Mediterranean antiquity (Aemilius Paulus, Olympiodorus, etc.) never exceeded 100.

A pamphlet was printed in Nantes in 1829 dealing with artesian wells, and I was attracted to it and bought it because the author refers specifically to the deep drilling carried out in China and suggests that the knowledge of artesian well drilling may have been brought to Europe from China. One may doubt whether the actual technique is likely to have come from there, but it is possible that when the first artesian wells were tried in the +12th century in Europe somebody was emboldened to make the effort because he had learned that it had been done successfully elsewhere. Hence it is noteworthy that al-Bīrūnī at Damascus about +1012 (one hundred years before the first drilling in France) gave a description of deep drilling and explained the phenomenon of self-flowing wells by a theory of communicating vessels. So it is possible there was transmission through the Arabs.

I must now turn away from this question to another technological complex, namely, the nature of efficient horse-harness, which is, I think, quite definitely something which came from Central Asia and which has not yet been sufficiently discussed.[1] People ordinarily look at animals drawing vehicles without seeing what the harness is. It never occurs to them to notice it. That was the case with me, at any rate for the first forty-seven years of my life. When I looked at horses and other animals drawing vehicles it never occurred to me to notice the harness. After one's eyes are opened one begins to see that this is an extremely important question because it concerns the efficiency of the animal's tractive force. It raises social implications such as the problem of slavery, because if there is inefficient animal harness it is necessary to use human labour as the tractive force, or conversely if abundant human labour is available, nobody bothers to invent an efficient harness. The man who opened the eyes of European scholars to this question was

[1] [See especially J. Needham & Lu Gwei-Djen, *Physis*, 1960, **2**, 121; 1965, **7**, 70; also *SCC*, vol. 4, pt. 2, pp. 304 ff.]

Lefebvre des Noëttes, a gifted French amateur; not a professional scholar but a soldier who wrote a book entitled *L'Attelage à travers les Ages* after he retired from military service. He asked the simple question: when did modern harness originate? No historian was able to tell him. So Lefebvre des Noëttes collected photographs, reliefs, medallions, and all possible evidence, and having put them together and studied them he arrived at a general theory of the situation, actually now no longer a theory but a cold historical fact, namely, that, believe it or not, for thousands of years in the West the harness used was gravely inefficient. It is known as the 'antique' harness or the 'throat-and-girth' harness, and was used in Sumeria, Chaldea, ancient Egypt, Greece, Rome, and indeed by every people in the Western part of the world until about the + 10th century (Fig. 8 a). Then it died out

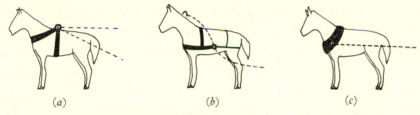

(a)   (b)   (c)

Fig. 8. The three main systems of equine harness: (a) the inefficient throat-and-girth harness, characteristic of occidental antiquity, (b) the breast-strap harness, characteristic of ancient and early medieval China, (c) collar harness, used from late medieval times in the West and from the + 5th century in China.

and by + 1200 one may say that the modern or 'collar' harness was completely adopted in Europe. The throat-and-girth harness suffocated the animal so that its tractive force was greatly impaired. Collar harness doubles or trebles the power which an equine animal can exert.

Now the fascinating point is that the only ancient people who had an efficient horse-harness were the Chinese. In all Han bas-reliefs you will find what some call the 'breast-strap' harness, or what is called in modern Europe the 'postillion' harness, where the strap in front bears right upon the shoulder of the animal so that the tractive force is fully exerted (Fig. 8 b). The trachea is not compressed. A Han chariot was a 'bus compared to a Greek or Roman chariot. The ancient Egyptian chariot had hardly any handrails and at most two people standing. In Han chariots six or eight people are frequently seen riding.

In a drawing of a horse copied from a Han relief we see the various parts of the harness. The breast-strap was called the *yin*, but this and many of the other technical terms used have been obsolete for centuries. The tractive force was conveyed to the chariot by a curiously curved wooden shaft (if two horses there were three such shafts) which was connected with the trace at its central point and therefore took the pull from the shoulder of the animal. But what is the connection between the postillion harness of the ancient Chinese and the modern or collar harness? And secondly, what can be found in Central Asia which bears upon the question of this transmission?

On thinking about it, one can easily see that if the modern collar harness (Fig. 8c) were not stiffened by wood or metal and padded with felt it would in effect approximate to the anterior part of the postillion harness because the pull would tend to deform it, and this is the morphological reason for the connection between collar and postillion harness. All that would be necessary to make it still closer would be to move forward the point of attachment to the chariot, and that seems to have been done, for we find examples of it on mural paintings in the cave-temples at Tunhuang, a place which I need not expatiate on further to the members of this Society. This cannot be intended to represent an 'antique' harness. Another picture from the caves at Tunhuang, of about the same date, also shows the attachment to the lower point of the collar. Moreover, in these mural paintings we find the actual modern collar harness shown in full in a fresco datable at +851, and inferable as early as about +480 [Fig. 9, pl.]. Carvings at the Yünkang cave-temples have since confirmed this late +5th-century date [Fig. 10, pl.]. So the evidence is that somewhere in Central Asia the postillion harness transformed itself by insensible stages into collar harness between the +5th and +9th centuries.

A French scholar, André Haudricourt, has continued the work of Lefebvre des Noëttes in this respect, and has adopted a method not previously much used in such studies, namely, the linguistic method. It may be known to those practically acquainted with harness that the sides of the collar harness in use today are known as *hames* in English. It is a peculiar word. If you look in the Oxford English Dictionary you will find the word was not known in England before +1300, but Haudricourt examined the question of its origin and found it to be Central Asian. I cannot remember all the languages in which the word *hame* occurs, but Mongolian is one; Kirghiz, Tartar, Russian, Finnish, and Lithuanian are others. He listed about twenty-five languages. It is not very far removed from the old Chinese word *'ak* which refers to a similar thing and which in some dialects is aspirated.

Now we come to the most interesting part of Haudricourt's theory, and one which I find quite plausible. He found, when he examined the word, that in some of the Central Asian languages it does not mean the same horse collar as in Europe but rather the horseshoe-shaped padded apparatus which is put round the two humps of a Bactrian camel for the transport of merchandise—that is, the camel pack-saddle. That horseshoe-shaped annulus had only to be applied to a horse's neck at the front end of a postillion harness in order to become the modern harness collar.

The whole story acquires still more plausibility from the fact that felt was a distinctively Mongolian invention. There is a special monograph by Laufer devoted to its origin. Central Asia was just the place where a felt-padded collar was available to be put round the horse's neck by some desert genius and take the strain of the wooden part connected with the shafts or traces. So, we may suppose, it gradually came to Europe.

Exactly what connection this may have with the social history of Europe will remain a matter for argument. Lefebvre des Noëttes was definite in his conviction that it had a relationship with slavery, but others such as Marc Bloch pointed out that in fact slavery in the decline of the Roman Empire actually ended long before it is possible that the collar harness came to Europe. However, one cannot say that there is no connection, because it is regarded as probable that one reason why technical discoveries were not adopted in ancient Mediterranean civilisation was because of the availability of almost unlimited servile man-power. Perhaps when social conditions changed and the labour was no longer available people were driven to have recourse to more efficient machines.[1] Harness must certainly be considered one of the earliest applications of engineering.

Another subject in which I have been taking a good deal of interest lately is that of suspension bridges. I do not know whether one can say that Central Asia acted as a transmission area in that regard also. The facts are that in West China and the Tibetan borderlands suspension bridges have played an important part from an early date. That alone is not perhaps unique because in Peru, for example, and other South American countries, one finds liana bridges used as part of primitive technology. I do not think it was so remarkable a discovery to shoot across a gorge an arrow attached to a string and carry across a bamboo rope after it. What was important was the use of iron chains in suspension bridge construction, and that seems to be considerably older in West China than anywhere else in the world.

I will give you evidence. In a typical West China bamboo bridge you will note the capstans round which the bamboo cables are wound. We ought, of course, to

---

[1] [Cf. M. I. Finley, *Econ. Hist. Rev.* 1965, **18**, 29.]

distinguish between suspension bridges proper and catenary bridges. A suspension bridge has a flat deck suspended from the chains or ropes thrown across the river, whereas a catenary bridge is one across which you walk on a deck which actually follows the curve of the cables hanging freely between the two points of support. Most of the Chinese bridges seem to have been catenary and not flat-deck suspension bridges, but the difference is small.

When a bridge cable is formed of twisted bamboo the tensile strength can be extremely strong. During the war aeroplanes and gliders were made in China of a very strong bamboo-ply which was developed by the Chinese Air Force Research Bureau. Bamboo is certainly usable for bridge spans up to 300 ft. The most famous bamboo bridge in China is that at Kuanhsien on the Min River, and it there forms part of an artificial irrigation system which in itself is extraordinary since it dates in its present form from about − 270. One of the great sights of the place is this six-span suspension bridge across the Min River. Each span is an average of 200 ft.

I could not show a more beautiful or better type of suspension bridge than that over the Mekong, a bridge remarkable for its flat character; it does not sag to any great extent. The danger of all suspension bridges is that they sway; they have little lateral stability. In the gorges over which the bridges are built there is generally no great wind force, and provided a bridge is renewed every fifty years it will be quite satisfactory.

An iron-chain suspension bridge was set up in + 1628 on the borders between Kweichow and Yunnan, but few people now know of it as the modern motor-road does not pass over it or even by it. A special book was written on it in + 1665, and a copy is now in the Library of Congress at Washington; in that book is a rather interesting drawing of the bridge showing the use of chains not only for the basic catenary but also for the guide ropes. This might afford an intermediate stage to the true suspension bridge because all that would be necessary to do would be to flatten the deck and use what were once the guide ropes as the supporting members.

The question arises as to the antiquity of these bridges. I have recently been going into that and the results are quite definite. One finds in the literature a persistent story of the building of a wrought-iron chain suspension bridge at a city called Chingtung, south of Tali in Yunnan, by the emperor Han Ming Ti about + 65. We have traced the origin of this story, and the truth of the matter is that the bridge was repaired in the Yung-Lo reign-period, which is about + 1410, and only local tradition ascribed it to Han Ming Ti's time. But the bridge had been there for a long time in + 1410. I do not see any reason why the Chinese should not have built a chain suspension bridge of wrought-iron chain links in + 65, but it certainly cannot be said to be proved. What is sure is that at another place in Yunnan, Chungtien, near Lichiang, a bridge over the Chin-Sha River was put up with iron chains between + 580 and + 618, and we have the names of those who built it, so that we can certainly go back to the beginning of the + 7th century. Another bridge of this type, which is quite old, is the San Hsia bridge in Szechuan, for which there are records going back to + 1367.

Against this we have to ask, what was the situation in Europe? In Europe it is clear that the first suspension bridge using iron chains was proposed in + 1595 in Italy by Faustus Verantius. It was not until the eighteenth century, from about + 1730 onwards, that iron-chain suspension bridges began to be fairly general in Europe.

There is an architectural book by Fischer von Erlach of Leipzig in + 1725 which depicts the bridge at Chingtung in Yunnan, and the author illustrates it as a practically true suspension bridge where people are not walking on the catenary, but I believe that to be a mistake. In any case, the point is that the bridge was being shown as a great marvel in + 1725, so that it is not surprising that Europeans became interested. Telford built his bridge across the Menai Straits in 1819. The fact seems to be that the Chinese developed the use of iron chains for suspension bridges, and that something like 1,000 years elapsed before they were successfully used in Europe.

In conclusion, it may be said that few have any idea of the extent to which medieval European technology was indebted to the Far East. This has to be brought out because while it remains true that the Chinese had nothing analogous to the great systematisers such as Aristotle, Ptolemy, or Euclid, yet on the other hand medieval Europe was a quite barbarous place compared to medieval China. I should say that for technology before the Renaissance Europe was enormously indebted to the Far East. The best summing-up is one which I came across recently in a book by an Arabic author, an excellent remark made about + 830 by al-Jāhiz in Damascus: 'The curious thing is that the Greeks are interested in theory but do not bother about practice, whereas the Chinese are very interested in practice and do not bother much about the theory.'

4

CHINA, EUROPE, AND THE SEAS BETWEEN[1]

[1966]

THE DEVELOPMENT OF THE SEA ROUTES BETWEEN
THE FAR EAST AND THE FAR WEST

MUTUAL knowledge of Greece and India seems to have begun in the late − 5th century when ambassadors and envoys were exchanged. The greatest Greek traveller in India, about − 300, Megasthenes, brought back a great deal of information. His embassy of course followed the great expeditions and conquests of Alexander, whose army occupied the Indus Valley in − 325 and left behind settlers to form the Greek kingdoms of Bactria and Sogdia. The bilingual coinages and cultures of these regions are well known, but the extent of the mingling of civilisations can be gauged by the fact that the great Indian king Aśoka set up an inscription near Kandahar in − 248 in Greek and Aramaic. Missionaries of Buddhism with 'healing herbs and yet more healing doctrine' visited Hellenistic Syria, and there is much reason for seeing Buddhist influence on early Christian monasticism. Indeed, the culture contacts between the Greek world and India were much wider and deeper than is usually recognised. For instance, Christian writers like Hippolytus knew well the doctrines of the Upanishads, and the pneumatic system in Greek medicine is closely related to the theories about *praña* in older Indian books. Gnosticism itself and Neo-Platonism are full of Indian traits, while conversely the water-mill is supposed to have been introduced to the Indians by a Greek, Metrodorus. In the history of science it is now becoming clear that Babylonian arithmetical-algebraical astronomy took permanent root in Tamil literature; as opposed to post-Hipparchan geometrical astronomy in north Indian Sanskrit sources, penetrating doubtless via Bactria. Greek influence it was, too, which stimulated Buddhism to demand statues and paintings of Gautama and of the Bodhisattvas, so that the original world-denying philosophy was supplemented, or rather overlain, by the system of devotion (*bhakti*) to a personal Saviour.

[1] Material presented at the International Congress of Maritime History, Beirut, September 1966.

There can be no doubt whatever that a great deal of this intellectual and social intercourse went on by sea. The anonymous book entitled *The Periplus of the Erythraean Sea*, written about + 70 by a Graeco-Egyptian seafaring merchant, whose home port was Berenice on the western shore of the Red Sea, is one of the most interesting books of antiquity. As is well known, the development of the sea-routes between the Mediterranean and India depended upon the discovery that use could be made of the monsoon winds to allow of rapid direct passages from the opening of the Red Sea to the ports upon the west coast of the Indian sub-continent. Thus from the + 1st to the middle of the + 3rd century Western ships, Roman in name but actually Graeco-Egyptian, Syrian or Lebanese, were reaching ports all round India; hence towards the latter part of the period a few went as far as Kattigara, which was most probably Hanoi in Vietnam. Certain coastal settlements, such as Arikamedu, modern Virapatnam near Pondicherry in South India, were undoubtedly mercantile centres and emporia for wide-ranging East–West trade. Arretine pottery found its way there from the West, carved ivory objects, like mirror-handles, came from the East. Roman coins have been found in Indo-China. Through these intermediate ports passed lacquer ware from China and beautifully carved bone plaques from India, Chinese silk, of course, and Syrian bronzes and fine glass.

It used to be thought that Chinese long-distance navigation appeared only after the + 3rd century and did not reach full development until the + 14th. It is most probable that Chinese ships were sailing to Penang in Malaya about + 350 and to Ceylon at the end of the + 4th century, while by the + 5th they were probably coming to the mouth of the Euphrates in Iraq and calling at Aden. Ammianus Marcellinus refers to Chinese merchandise at the annual fair of Batanea on the Euphrates round + 360. Prisoners from the Talas River battle returned home to China from the Persian Gulf in + 762 in Chinese junks. In + 850 Sulaimān the Merchant refers to the port of Sīrāf on its north coast as a terminus of Chinese shipping. Chinese texts recently studied, however, seem to indicate that Chinese ships came further westwards in early times than had previously been supposed. The *Chhien Han Shu*, for example, has a passage describing Han trade with the South Seas. This is applicable to the − 1st century, and since the furthest country mentioned in the text is said to require a sea voyage of over twelve months, the whole account being entirely devoid of any legendary quality, eminent scholars such as Pelliot have felt that one should visualise Chinese missions penetrating already at that time as far as the western extremity of the Indian Ocean. It is certainly arresting to think that Chinese merchants walked with Roman citizens from Greece, Syria and Egypt on the quays of Arikamedu. These far-ranging navigations may have anticipated the Han period, because a passage in *Chuang Tzu* (−4th century) describes the sailors of the Yüeh region in South China being absent in the western regions for years. The *Chhien Han Shu* passage also indicates where the trading envoys went in the West in the −1st century. Much depends upon the interpretation of the place-name Huang-chih; the most moderate view is that it was Kanchapura, modern Conjeeveram in Madras, then capital of the Pallava State, but the alternative view is that it may have been the port of Adulis, modern Massawa in the Red Sea, in which case Ethiopia and East Africa would have been known to Chinese trading mariners as early as the −1st century.

Between the +8th and the +14th century the Islamic Arab shipmasters were dominant in the Indian Ocean rather than the Chinese. In +758 they were strong enough to burn and loot Canton, just 100 years after the first Arabic embassy to China. Many Arab colonies or factories were established in the Chinese ports from the +9th century onwards. In these centuries we have important Chinese accounts of maritime commerce and navigation, especially the *Chu Fan Chih* (Records of Foreign Peoples) written by Chao Ju-Kua in +1225. The Arabs had their own special names for the ports on the China coast, all of which are now identifiable. Along with the Arabs came Christian missionaries and perhaps also merchants. The Nestorian mission had been a purely overland phenomenon, so far as we know, in the Thang period (+7th and +8th centuries), but now the Franciscan influence was felt and many Christian tombstones have been found built into the city wall at the port of Chhüanchow in Fukien. One of these may be that of Andrew of Perugia, the first bishop of Zayton, as the Arabs called it (a Franciscan, +1323 to +1332), though this is uncertain. Remarkable tombstones of this period have been found at Yangchow, notably one to a woman, Catherine de Viglione, probably of Genoa, in +1342. She was presumably the wife or daughter of a Christian merchant. Many Muslim families settled in China and their descendants entered the Civil Service; of such origins was Phu Shou-Kêng, Commissioner of Merchant Shipping at Zayton between +1250 and +1275, just before the arrival of Marco Polo. By the +14th century the Chinese sailors were ranging widely again as far as Africa, and in the +15th there came the relatively short period of Chinese maritime supremacy under the Ming dynasty, which brought Chinese sea-going junks to Borneo, the Philippines, Ceylon, Malabar and all East Africa.[1]

[1] [For further information on these subjects, see *SCC*, vols. 1 and 4, pt. 3.]

NAVIGATION IN THE INDIAN OCEAN AND THE CHINA SEAS

Chinese pilots in the period of primitive navigation certainly made use of all the usual aids which were known in ancient times. Like sailors in other civilisations these ancient pilots were observant men, men who took soundings, noted sea-bottom samples, marked the prevailing winds and currents and recorded in their early rutters depths, anchorages, landmarks and tides—nor did they forget to use the services of shore-sighting birds. The stars were extremely important. By night they could tell the time by the circumpolars and the culminations or risings of decan-stars, gaining an idea of their latitude by rough assessment of the height of the pole seen against masts and rigging. By day the varying relations of ecliptic and horizon helped them to construct their wind-rose. Time and distance estimation was still extremely crude, no more than a count of day and night watches with a guess at the way made good.

There is a special reason however for believing that the ancient Chinese mariners were good star-clerks. Chang Hêng, the great astronomer, wrote in his *Ling Hsien* (+118): 'There are in all 2,500 greater stars, not including those which the sea people observe.' There has been doubt as to how the expression *hai jen* should be translated here, but very probably it meant sailors and this interpretation may be applied to a number of books of which we possess now only the titles in the bibliography of the *Chhien Han Shu*. For example there was the *Hai Chung Chan* (Astrology (and Astronomy) of the People in the Midst of the Sea (or of the Sailors)). Or again there was the *Hai Chung Wu Hsing Shun Ni* (The Forward Motions and Retrogradations of the Planets: a Mariner's Manual). There are some seven or eight of these, all finished before the end of the +1st century. Philological researches tend to identify this 'sea-going' corpus as the work of those 'magicians' of the Warring States period and early Han who lived along the coasts of Chhi and Yen, in a word, the 'mathematical practitioners' of the early stages of Chinese navigation. Their skills were doubtless undifferentiated and it would be impossible to disentangle in them the components which today we should call astrology, astronomy, stellar navigation, weather prediction, and the lore of winds, currents and landfalls, all the more so since, as in the work of Dee, Hartgill, Goad, Gadbury and many others, these elements were still wholly confused down to the end of the +17th century in Europe.

The Chinese pilots of ancient times thus knew their stars, but it was they who brought the period of primitive navigation to an end by being the first to employ the magnetic compass at sea.[1] This great revolution in the sailor's art, which ushered in the era of quantitative navigation, is solidly attested for Chinese ships by + 1090, just about a hundred years before its initial appearance in the West. Our first text which shows this also mentions astronomical navigation and soundings, together with the study of sea-bottom samples. Two further accounts in the + 12th century follow before the first European mention, each emphasising the value of the compass on nights of cloud and storm. The exact date at which the magnetic compass first became the mariner's compass after a long career ashore with the geomancers is not known, but some time in the + 10th century would be a very probable guess. Before the end of the + 13th century we have compass bearings recorded in print, and in the following century, before the end of the Yuan dynasty, compilations of these began to be produced.

In all probability from the beginning of its use at sea, the Chinese compass was a magnetised needle floating on water in a small cup. A thousand years earlier, the first and oldest compass had been a spoon-shaped object of lodestone rotating on a bronze plate. At some intervening period the frictional drag of the spoon on the plate had been overcome by inserting the lodestone in a piece of wood with pointed ends, which could be floated, or balanced upon an upward projecting pin. The dry pivot compass had thus been invented. But although these primitive arrangements seem still to have been used as late as the + 13th century, Chinese sailors did not, so far as we know, employ them. At some time between the + 1st and the + 6th centuries the discovery had been made that the directive property of the lodestone could be transferred by induction to the small pieces of iron or steel which the lodestone attracted and that these could also be made to float upon the surface of water by suitable devices. The earliest extant description of a floating compass of this kind dates from + 1044 and involves a thin leaf of magnetised iron with upturned edges cut into the shape of a fish. To floating compasses of this kind Chinese navigators remained faithful for nearly a millennium. We have detailed accounts of their use from the + 15th century, but in the + 16th there came Dutch influence, mediated in part through the Japanese, as the result of which the dry pivoted needle and then the compass-card (doubtless an Italian invention) were adopted on Chinese vessels. The Chinese compass-makers, however, employed a very delicate form of suspension which automatically compensated for variations of dip, and still impressed Western observers as late as the beginning of the 19th century.

[1] [The whole story is in *SCC*, vol. 4, pt. 1, pp. 245 ff.; and, summarised, on pp. 239 ff. below.]

The remarkable series of Chinese maritime expeditions between + 1400 and + 1433 gave rise to a large literature, and much of the navigational material found its way into a + 17th-century compilation, the *Wu Pei Chih* (Records of Armament Technology), the charts preserved in which have attracted a great deal of research.[1] We also find a number of instructive navigational diagrams summarising the star positions to be maintained during a number of regular voyages. Here we find that instead of using astronomical degrees for assessing pole-star altitudes and those of other stars, the sailors preferred another graduation in finger-breadths (*chih*) each of which was divided into four parts (*chio*).

Once we realise that the navigators of the China Seas and the Indian Ocean depended quite as much on polar altitudes as the Portuguese towards the end of the + 15th century, many fascinating impressions arise. Unfortunately we know as yet neither exactly how far back this quantitative oceanic navigation went in Eastern waters nor how far the Europeans of the Atlantic border were influenced by it during the explorations of the West African coast. Certain it is that when the Portuguese showed him their astrolabes and quadrants in the summer of + 1498, Ibn Mājid was not in the least surprised, saying that the Arabs had similar instruments, but the Portuguese were very astonished that he was not surprised. Moreover there are a number of points at which we may suspect East Asian influences on Europe, or where at least we have to grant considerable East Asian priority.

First it is clear that the Chinese navigators of the + 15th century, beside their compass bearings, knew the method of finding and running down the latitude. We are still largely in the dark, however, as to the instruments which the Chinese used. Their astronomers had had armillary spheres quite as long as the Greeks, but no Chinese seamen's quadrants have been found. What they more probably used was some simple type of cross-staff, for it has been shown that Jacob's staff was known in China and used by surveyors three centuries before the description of Levi ben Gerson, i.e. by + 1086 rather than + 1321. This would also be more in line with the practices of the Arab and Indian pilots.

We know that the magnetic compass, the portolan chart, the sand-glass and the marteloio formed a closely connected knot of complementary techniques. The magnetic compass was started by the Chinese, but portolan charts as we know them in the West have not been found in Chinese culture, though their particular

---

[1] [The appearance of the important work of J. V. Mills on +15th-century Chinese navigation is eagerly awaited.]

type can be extrapolated backwards clearly enough from the *Wu Pei Chih* charts. Traverse-tables have not so far been recognised in Chinese rutters, but the use of sand for time measuring opens up curious perspectives. Earlier Chinese historians had concluded that the sand-glass was not known or used on Chinese ships until the end of the + 16th century when they acquired it from the Dutch or the Portuguese. But since that time much new information has come to light about an important development in the history of Chinese mechanical clockwork which occurred about + 1370, namely the substitution of sand for water in clocks of the hydro-mechanical type.[1] These new clocks, unlike the old ones, had a dial and pointer. It is certain that time-keeping by sand-flow was very much in the minds of the Chinese in the + 14th century. It may be necessary therefore to re-examine the Western traditions which make the sand-glass begin with Liutprand of Cremona in the + 10th century (a very doubtful attribution) and to reconsider the suspicion, long ago voiced but since ignored, that the hour-glass came to Europe from the East. An argument of some weight is that since nautical watches (*kêng*) are mentioned or implicit in many descriptions of Chinese navigation from the beginning of the + 12th century onwards, the measurement of such units might have necessitated the sand-glass, since no form of water clepsydra or hydro-mechanical clock would have been imaginable at sea. If those are right who trace the Western nautical sand-glass back to the Venetian glass industry of the late + 12th century, the possibility presents itself that, together with the magnetic compass and the stern-post rudder, it might have formed part of one of those clusters of transmissions from Asia which we find in so many fields of applied science, and which we shall summarise below (p.61). But the time-keeping joss-stick is much the more likely explanation.[2]

A remarkable feature of the polar altitude system is that it was identical with that in use among the Arab shipmasters of the Indian Ocean, who expressed altitude in *isba* (the finger-breadth) and its eighth part, the *zam*. It is thus particularly interesting that when the measurements in the *Muḥīṭ* (The Ocean), a compendium of nautical instructions put together by the scholarly Turkish admiral Sīdī 'alī Re'is ibn Ḥusayn in + 1553, are compared with those in the *Wu Pei Chih*, they are found to be generally speaking in good agreement. The chief difference between the Arab and the Chinese systems seems to be that when a substitute polar mark-point was desired in southerly latitudes, the Arabs chose the classical 'guards'

---

[1] [Cf. *SCC*, vol. 4, pt. 2, pp. 509 ff.]
[2] [Cf. S. Bedini, *Trans. Amer. Philos. Soc.* 1963, **53**, no. 5.]

(β and γ Ursae minoris), which they called *al-farkadain* (the 'calves') while the Chinese chose *Hua Kai*, the declinations being very similar but the right ascensions almost exactly 180° (12 hr) apart.

Western writers such as Marco Polo, Nicolò de Conti and Fra Mauro, were so impressed with the astronomical navigation of the Asian pilots of the Indian Ocean, that they gave the impression that the magnetic compass was not used in those waters. This impression was undoubtedly wrong. The Mediterranean pilot of the + 14th century never took his eyes off the needle and gave orders to the helmsman accordingly while working out his course by bearing and distance. For the Asian pilot the needle was only one of his instruments, and the determination of position by star-sights was at least equally important. This was no doubt because the region sailed by the seamen of the Arabic and Chinese tradition was one of relatively scanty, or at least highly seasonal, rainfall and frequent clear skies, so that orientation by the stars was more inviting and capable of more precision. If Arabic writers did not wax enthusiastic about the leading of the lodestone it was because they had less need of it than others. The Chinese, in their more northerly waters, did wax enthusiastic about it, very much so, but their words were enclosed in the ideographic language not to be understood or appreciated by Westerners until comparatively modern times.

The question arises about the mutual influence of the Arabic and Chinese navigators, but at present we hardly know enough to answer it. They had been in contact, as we have seen, for many centuries before + 1400. Measurements of altitude were particularly prominent in Arabic astronomy, but on the other hand circumpolar markpoints for invisible stars were rather characteristically Chinese. The system of finger-breadth units for altitudes could, of course, easily have arisen independently in the Arabic and Chinese culture-areas. But there is some evidence for it among Chinese military geographers as early as the + 3rd century.

As already noted, the Chinese altitude-measuring device was most probably some kind of cross-staff or Jacob's staff, perhaps like the Arabic 'tablet' or *kamāl*, in which the stock was represented by a knotted string.[1] The fact that Arabic pilots later on called the cross-staff *al-bilisti* must no doubt mean that some of them had received it from the West, but this does not necessarily mean that its origin had been there, or even that their own forefathers had not transmitted it in the opposite direction for, as we have just seen, there is evidence of the cross-staff in

---

[1] [This has very recently been demonstrated by Yen Tun-Chieh, *Kho-Hsüeh Shih Chi-Khan*, 1966, **9**, 77. The Chinese used a set of twelve tablets of different sizes and a cord of constant length.]

China in the +11th century. Evidence is now accumulating that the Chinese were taking altitudes for navigation by the beginning of the +12th. The account of an embassy to Korea in +1124 (the *Hsüan-Ho Fêng-Shih Kao-Li Thu Ching*) says that the pilot steered by the Great Bear and the neighbouring circumpolar stars. Furthermore, a passage recently discovered in the *Sung Hui Yao Kao* (Drafts for the History of the Administrative Statutes of the Sung Dynasty) tells us concerning the fitting out of sea-going warships in +1129 that they should be equipped among other things with polar sighting-tubes (*wang tou*). Now this can only mean sighting-tubes for determining the positions and altitudes of stars in and near the Great Bear, and no doubt it was a quadrant type of sighting-tube (*wang thung*) like that illustrated in the *Ying Tsao Fa Shih* (Treatise on Architectural Methods) of +1103. Perhaps therefore the quantisation of stellar altitudes followed closely on the quantisation of azimuth directions by the Chinese pilots. Perhaps they started with sighting-tubes and went over to instruments like the cross-staff later.

In sum, the development of quantitative navigation in the Eastern Seas began with the introduction of the mariner's compass on Chinese ships some time before +1050, possibly as early as +850. How soon this spread to the Indian Ocean we still do not know. Before +1100 there is hardly any evidence for the taking of star-altitudes at sea by instrument, whether among Arabic, Indian or Chinese navigators. We may not therefore be far off the truth if we say that when Ibn Mājid met Vasco da Gama at Milindi fully quantitative navigation was several centuries old 'East of Suez', but hardly one century old in the West.

Some of the typical navigational compendia in Chinese literature are well worth studying. There is, for example, a MS in the Bodleian Library at Oxford entitled *Shun Fêng Hsiang Sung* (Fair Winds for Escort) written by an anonymous mariner some time before +1430 or at the close of the period of Chinese exploration. Another work is the *Tung Hsi Yang Khao* (Investigation of the Oceans East and West) compiled by Chang Hsieh in +1618, i.e. already in the Jesuit period but showing no evidence of any occidental influences.

THE FIFTEENTH-CENTURY CONFRONTATION

Whoever has had in his life the good fortune both to wander on those Fukienese and Cantonese shores which saw the passage of the great barques of Chêng Ho and to stand also on that hill which overlooks the Tower of Belem and the Praiz de Restelo on the banks of the Tagus cannot but be powerfully impressed by the

strange contemporaneity of the great Portuguese and Chinese voyages of discovery. Indeed it is an extraordinary historical coincidence that Chinese long-
distance navigation from the Far East reached its high-water mark just as the tide
of Portuguese exploration from the Far West was beginning its spectacular flow.
These two great currents almost met, but not quite, and in a single region, the
coasts of the African continent. Their wind-angels, their inspirers, were two
equally extraordinary men active in maritime affairs, on the one side a royal patron
of navigators, on the other an imperial eunuch, ambassador and admiral. The
contrast is inescapable, for this was the apogee of Chinese maritime enterprise.
All the time that the Portuguese were exploring slowly southwards down the
west coast of Africa, the Chinese were trading and visiting all up and down the
east coast, with larger ships and larger fleets; but by the time that the Portuguese
rounded the Cape and made their way to India, a fundamental change of national
policy within China had withdrawn the Chinese fleets from the Indian Ocean so
that the Portuguese found nothing but recent memories of their presence. Such
was the confrontation that failed to occur; it is extremely interesting to speculate
what would have happened if the smaller flotillas full of ambitious, even blood-
thirsty, occidental adventurers had met the larger, peaceful fleets of great junks,
the vessels of a philosophical people which had never had any part in the crusading
strife between Christian and Moor.

It is not necessary to say much about Prince Henry the Navigator (+ 1394 to
+ 1460), a figure so well known to European historians. It is a commonplace to
say that it was under his influence that the west coasts of Africa were first explored.
By + 1444 Nuño Tristão had reached the mouth of the Senegal River, and two
years later Alvaro Fernandes was on the Guinea coast. The year + 1453 was marked
by two events, one colossal blow and one seemingly minor affair: Byzantium fell
to the Turks, as if to show the Portuguese they were none too soon in their
endeavours, while along the African coasts sailed Cid da Sousa, leading the first
expedition to Guinea with the primary object of trade. Just after this Prince Henry
died. By + 1486 Diogo Cão was on the Angola coast and then Bartolomeu Dias
rounded the Cape of Good Hope in + 1488. Thus the way was open for the culminating cruise, that of Vasco da Gama, who left Lisbon in + 1497 and was off
the mouth of the Zambezi early in the following year; then, entering the
'Chinese' area, touched at Malindi in April, just about fifty years after the Ming
navy had ceased to frequent those shores. This was the place where he was fortunate enough to get the services of one of the leading Arab pilots of the age,

Aḥmad ibn Mājid, already mentioned, who brought the Portuguese admiral to Calicut in India the following month. The die was now cast, the Europeans were in the Indian Ocean for good or evil—and of the latter much.

The complementary figure to Prince Henry the Navigator on the Chinese side was Chêng Ho, known in his lifetime and afterwards as San Pao Thai Chien (The Admiral of the Triple Treasure), fl. + 1385 to + 1440. This celebrated eunuch, ambassador, admiral and explorer, 'the Vasco da Gama of China', as Debenham has called him, was born of a Yunnanese Muslim family, the son of a father who had made the pilgrimage to Mecca. Between + 1405 and + 1433 he accomplished seven diplomatic expeditions with large fleets of great junks, the biggest vessels afloat at that time, called 'Treasure-ships' (cf. Figs. 11 and 12, pls.); ranging from Borneo to the Zanguebar coast of East Africa and greatly stimulating Chinese interest in foreign countries. The three voyages between + 1405 and + 1411 went to Champa, Java, Palembang, Siam, Ceylon, Calicut, Cochin, etc. Between + 1413 and + 1415 elaborate visits were paid to Ormuz and the Persian Gulf, as also the Maldive Islands. Then three expeditions between + 1417 and + 1433 covered the whole of the east coast of Africa, including Mogadishiu, Mozambique, etc., as far south as the Straits of Madagascar. The motives of these great voyages, comparable with the Portuguese discoveries and closely parallel with them in time, were (a) to search for the dethroned Chien-Wên Emperor, who was suspected of having fled beyond the seas; (b) to demonstrate the power and glory of China as the leading political and cultural nation in Asia to the kings and sultans of Southern and Western Asian regions; (c) to induce them to acknowledge the nominal suzerainty of the Chinese Emperor and to send tribute missions to the Chinese court; (d) to encourage maritime trade; (e) to collect natural curiosities of all kinds for the imperial cabinets, including strange animals and hitherto unknown drugs; (f) to survey the sea routes and coastal defences; (g) to make reconnaissance of the strength and capacities of neighbouring countries, especially in the South Seas. We may discuss some of these aspects in turn. The half-century of the voyages gave rise to a large literature in Chinese which is only now being fully translated. They certainly led to a great increase in the Chinese knowledge of the rest of the world.

It must not be supposed that Chinese knowledge of Africa began at this time. Already in the middle of the + 9th century a book like the *Yu-Yang Tsa Tsu* contains an interesting account of Berbera on the south coast of the Gulf of Aden. The *Hsin Thang Shu*, written in + 1060, has an account of the country of Malindi on the coast of Kenya. Chao Ju-Kua, in the book already mentioned, has a whole section on Tshêng-Pa, the whole Zanguebar coast between the Juba River in Somaliland and the Mozambique Channel, while in the *Ling Wai Tai Ta* (Information on what is beyond the Passes), written in +1178, Chou Chhü-Fei describes Madagascar at some length. By the +14th century all these regions were well known. Wang Ta-Yuan, who travelled widely between +1330 and +1349, deals with most of them in his *Tao I Chih Lüeh* (Records of the Barbarian Islands), including such places not mentioned before as the Comoro Islands in the Mozambique Channel.

Among the things which the Chinese wanted from Africa were elephant tusks, rhinoceros horns, strings of pearls, aromatic substances, incense gums and the like. Al-Idrīsī, on the other hand, tells us that Aden (and hence the coast) received from China and India iron, damascened sabres, musk and porcelain, saddles, rich textiles, cotton goods, aloes, pepper and South Sea spices. Much of this hardware has actually survived until today. 'I have never seen', wrote Wheeler in 1955, 'so much broken china as in the past fortnight between Dar-es-Salaam and the Kilwa Islands, literally fragments of Chinese porcelain by the shovelful. I think it is fair to say that as far as the Middle Ages is concerned from the +10th century onwards the buried history of Tanganyika is written in Chinese porcelain.' Another kind of Chinese hardware on the East African coast is monetary, coins and coin hoards, the earliest dating from about +620. Early Chinese knowledge of Africa is evinced by considerations of quite a different kind. Rockhill was impressed nearly a century ago by the fact that in Chinese maps of the +16th century South Africa was shown in its right shape, i.e. with its tip pointing to the south. As he well knew, the European cartographical tradition before the Portuguese discoveries was to have it pointing to the east. In the magnificent Korean world map due to the cartographer Yi Hoe and the astronomer Kwŏn Kŭn in +1402, before the first Portuguese caravel had sighted Cape Nun, Africa was made to point south, with a roughly correct triangular shape, and some thirty-five European and African place-names, including Alexandria, were marked upon it. It is interesting that a pagoda-like object represents the Pharos. This map is greatly superior to the Catalan atlas of +1475, and even to the map of Fra Mauro (+1459), presumably because the knowledge of Europe and Africa which the Chinese scholars obtained from their Arab informants was much better and more abundant than all that Marco Polo and the other Western travellers could bring home about East Asia. In fact the Chinese were a good century ahead.

Fascinating comparisons can be made between the Chinese and the Portuguese activities. First let us take the purely maritime point of view and the nautical technology. It would seem on the whole that the Chinese achievement of the + 15th century involved no revolutionary technical break with the past while that of the Portuguese was more original. The Chinese had had their fore-and-aft lug-sails since the + 3rd century at least, and already in the time of Marco Polo and Chao Ju-Kua their ships were many-masted. If they used their mariner's compasses in the Mozambique Channel, they were only doing what their predecessors had done in the Straits of Thaiwan right back to the foundation of the Sung navy at the beginning of the + 12th century. Although their stern-post rudders were attached in weaker fashion to the hulls than those of the Westerners with pintle and gudgeon, they were highly efficient in more ways than one and descended from inventions of the + 1st century (cf. p. 250). The most obvious difference which would have struck everyone if the vessels of da Gama had met those of Chêng Ho lay in the much greater size of the Treasure-ships of the Grand Fleet, for many of these were of 1,500 tons, if not considerably more, while those of Vasco's were of 300 tons and some were much less; but while the Chinese craft were the culmination of a long evolutionary development, those of the Portuguese were relatively new in type. As is well known, the Portuguese had thrown overboard the square-sail rig and had adopted for their famous caravels a fore-and-aft one in the form of the lateen sail, taken from their enemies the Arabs. By + 1436 these ships carried triangular lateen sails on as many as three masts. Then as the century went on, the superior advantages of the square-sail for running before the wind reasserted themselves and ships began to be built which combined both rigs. The originality of the Portuguese, however, seems somewhat qualified when we remember that of the basic inventions they assembled, the mariner's compass and the stern-post rudder were transmissions from much earlier Chinese practice, the principle of multiple masts was characteristically Asian, and the lateen sail was taken directly from the Arabs.

There was another matter, however, in which the Portuguese showed seemingly more originality than the Chinese, namely, the understanding and use of the régime of winds and currents. It might be truer to say that the problems set for them by Nature were much more difficult and that they rose gallantly to the occasion. In the East Asian and South Asian waters monsoon sailing had been and remained traditional. But the more inhospitable Atlantic had never encouraged sailors by the same regularity, and though there had been a number of attempts to sail to the West, that ocean had never been systematically explored. Thus what is called the Sargasso Arc, and later the Brazilian or Cape San Roque Arc, were brilliant applications of what was fundamentally new meteorological and hydrographic knowledge.

In matters of war and conflict the contrast is truly extraordinary. The entire Chinese operations were those of a navy paying friendly visits to foreign ports, while on the other hand the Portuguese East of Suez engaged themselves almost at once in total war. Indeed, the term navy is hardly applicable to the Chinese fleets, which were more like the assemblies of merchant ships of a nationalised trading authority. As long as the Portuguese were working down the West African coast their aggressive activities (apart from slaving) were relatively restrained, and it was only after + 1500, when they were in a position to carry on terrorist warfare against the East African Arabs and then against the Indians and other Asians, that European naval power showed what it could do in earnest. The Portuguese in fact perpetuated the Crusader mentality and applied it in an attempted naval conquest of all South Asia.

The Chinese fleets were certainly armed, and with gunpowder weapons. As a famous Chinese text says: 'They bestowed gifts upon the kings and rulers and those who refused submission they overawed by the show of armed might.' But in fact if the historical records are carefully examined, it is found that the Chinese only got into trouble on three occasions. The first was in + 1406, when they had to repel a surprise attack by a tribal chief of Palembang, who was defeated and sent to Peking. Another small fracas happened in north-western Sumatra, seven or eight years afterwards. The most famous occasion was when in + 1410 the king of Ceylon, Alagakkonara (probably but not certainly Bhuvaneka Bahu V), enticed Chêng Ho's expeditionary guard into the interior and then demanded excessive presents of gold and silk, meanwhile sending troops to burn and sink his ships, which lay in all probability in Galle harbour. But Chêng Ho pushed on to the capital, most probably Kotte, took the king and his court by surprise and then fought his way back to the coast, routing the Sinhalese army on the journey. The prisoners were taken to Peking, where they were kindly treated and sent home again after an arrangement had been made to choose a relative of the king as his successor. Naval armed might thus meant something very different indeed in the Chinese and the Portuguese interpretations. The Chinese set up no forts or strongholds anywhere and founded no colonies of any kind.

As regards trade, we still know relatively little about it, but it was only natural that what was done both by the Chinese and the Portuguese was done under the aegis of their respective economic systems, and these were very different. It seems clear that from the start the Portuguese activities were concerned more with private enterprise than the Chinese. The search for an 'El Dorado' which would make one's personal fortune was after all an integral part of the Conquistador mentality. By contrast, the Chinese expeditions were the well-disciplined operations of an enormous feudal-bureaucratic state, the like of which was not known in Europe. Their impetus was primarily governmental, their trade, though large, was incidental, and the 'irregular' merchant mariners whose trafficking was to be encouraged were mostly men of small means. The slave trade was a particular case. The Chinese and other Asian nations had been using negro slaves for many centuries but the fact that their slavery was basically domestic kept the practice within bounds. Not so the use of African agricultural plantation labourers in the New World, which brought it about that between +1486 and +1641 no less than 1,389,000 slaves were taken by the Portuguese from Angola alone. In general we have the paradox that while the feudal state of Portugal, hardly emerged from the Middle Ages, founded an empire of mercantile capital, on the other hand bureaucratic feudalism, though certainly not the economy of the future, gave to China in the +15th century the lineaments of an empire without imperialism.

But it is important to notice a further difference. The Chinese probably never had to face an adverse balance of trade, for silk, lacquer, porcelain, etc., were everywhere esteemed and quite good in exchange for anything the Chinese ever wanted. On the other hand, the Portuguese were caught in an intractable economic necessity, for Western Europe did not at that time produce anything that Asians wanted to buy. When da Gama first visited Calicut in +1498, a highly significant event occurred. When the Portuguese presented the goods which they had brought, consisting of striped cloths, scarlet hoods, hats, strings of coral, hand washbasins, sugar, oil and honey, the king laughed at them and advised the Admiral rather to offer gold. At the same time the Muslim merchants already on the spot affirmed to the Indians that the Portuguese were essentially pirates, possessed of nothing that the Indians could ever want, and prepared to take what the Indians had by force if they could not get it otherwise. There is something very familiar about this scene; in fact it symbolised perfectly the fundamental pattern of trade imbalance which had been characteristic of relations between Europe and East Asia from the beginning and was destined to continue so until the industrial age of the late 19th century. Broadly speaking, Europeans always wanted Asian products far more than the Easterners wanted Western ones, and the only means of paying for them was in precious metals. This process occurred at many places along the East–West trade routes, but finally, of course, in medieval times at the Levantine borders between Christendom and Islam. Even during the Roman Empire there was a great drain of gold and silver to the East. Nearly two thousand years later the opium traffic with the Manchu dynasty, and hence the Opium Wars, arose because the East India Company, alarmed at the drain of silver from Europe to pay for its silk, tea and lacquer, sought for some substitute commodity. The Portuguese, of course, in the +15th century were after spices. They wanted to break the Muslim monopoly. It was fortunate for them that during the exploration of the West African coast they were able to tap the gold of the Sudan successfully about +1445 and thus in time gained the precious metal necessary for the Eastern trade. It does not seem that the Portuguese defrauded the Africans greatly in this exchange, for, unlike Asians, they were pleased to get horses, wheat, wine and cheese, copperware and other metals, blankets and strong cloth. Unfortunately the gold of Africa was never anything like enough for the ambitions of the Portuguese in the Eastern Seas, and out of the need for bases there arose naturally the temptation to accumulate further treasure by appropriating at sword-point the wealth of coastal port cities such as Malacca. The real criticism of their operations is that they were not content with a reasonable share of Asian trade; what they wanted before long was the complete domination of the trade and the traders. This it was given only to others to achieve and then but for a time. The Chinese, on the other hand, carried on for half a century what was essentially government trading with the emirs and sultans of the western parts of the Indian Ocean. They obtained, as nominal 'tribute', all the strange commodities that they sought for, never having any difficulty in 'paying' for them by means of imperial 'gifts' of silk, lacquer, porcelain, etc.

Equally profound differences between the Chinese and the Portuguese behaviour appear in matters of religion. Missionary activity, well intended no doubt, accompanied the Portuguese explorations from an early time, but before the end of the century the war against all Muslims was being extended to all Hindus and Buddhists too, save those with whom the Portuguese might find it expedient to arrange a temporary alliance. In +1560 the Holy Inquisition was established at Goa, where it soon acquired a reputation even more unsavoury than that which it had in Europe. It subjected the non-Christian as well as the Christian subjects of the Empire to all those forms of secret police terror which have disfigured our own century, yet more abominable here perhaps because enlisted in the interests of higher religion. On board the Chinese ships what a contrast! Without forsaking the basic teachings of the sages Khung and Lao, Chêng Ho and his commanders were 'all things to all men'. In Arabia they conversed in the tongue of the Prophet and recalled the mosques of Yunnan; in India they presented offerings to Hindu temples and venerated the traces of the Buddha in Ceylon. Ceylon indeed provided the occasion for a particularly interesting example of this almost excessive urbanity. In 1911 a stele with an inscription in three languages (Chinese, Tamil and Persian) was unearthed by road engineers within the town of Galle. This act commemorated, as was soon clear, one of the visits of the Ming navy under Chêng Ho, and took the form of an address accompanying religious gifts. Owing to differential weathering, the Chinese version was the first to be deciphered. It says that the Chinese emperor, having heard of the fame of the Buddhist temples of Ceylon, such as the Temple of the Tooth relic, sent envoys to present a number of rich gifts—its date was + 1409. When the Tamil version was deciphered and translated later on, it turned out to say that the Chinese emperor, having heard of the fame of the god Devundara Deviyo, caused the stone to be set up in his praise. Still more remarkable, the Persian version, the most damaged of the three, was found to say that the presentation was to the glory of Allah and some of his (Muslim) saints. But while the texts are thus different, they all agree in one thing—the list of the presents is almost exactly identical. There is thus hardly any escape from the conclusion that three parallel sets of gifts were brought out by sea and handed over to the representatives of the three most important religions practised on the island. Such humanistic catholicity contrasts indeed with the *autos-da-fé* of Goa later on.

We have already mentioned some of the stated motives of the Chinese expeditions, but there may have been others. Purely geographical exploration as such was probably never one of them. What the Chinese sought rather was cultural contacts with foreign peoples, even if quite uncivilised. The Chinese voyages were essentially an urbane but systematic tour of inspection of the known world. The Portuguese motive was not primarily geographical exploration either; their discoveries were really secondary achievements in the great attempt of a nation which believed itself to be the champion of Christendom in an unceasing war to find a way round to the Indies and so to take the Islamic world in the rear. On the Chinese side, probably the search for the dethroned emperor was never more than a pretext, the main motive was surely the demonstration of Chinese prestige by obtaining the nominal allegiance and exchange tribute of far-away princes. If one compares the Chinese and the Portuguese ventures over the whole range of our knowledge of them, it does seem as if the proto-scientific function of collecting natural rarities, strange gems and animals, was more marked in the former. For example the Chinese were extremely interested in giraffes. Undoubtedly one of the departments that concerned Chêng Ho's men was materia medica; the search for new drugs may have been a much more important motive than has usually been thought. There is a great deal of evidence that new drug-plants, and even crop-plants, were sought for intensively by the Chinese and recorded in their biological literature at the time.

The decline and fall of the Chinese navy (or national merchant navy) in the middle of the + 15th century was a rather dramatic process.[1] The great Treasure-ships had always been strongly criticised by the Confucian bureaucrats, whose minds were traditionally agrarian in interest, and to some extent the naval expansion was suspect as being an enterprise of the eunuchs—always in conflict with the Confucian bureaucracy. Military events also intervened, the serious deterioration of the position on the north-western frontiers diverting all attention from the sea, and in + 1450 at the disastrous battle of Thu-mu, that Chinese emperor who had suppressed the Treasure-ship fleets was himself taken into captivity by the Mongol and Tartar armies. At the same time there was a significant shift of population from the south-eastern seaboard provinces, reversing the trend which had been so strong at the beginning of the Southern Sung. The navy simply fell to pieces. By + 1474 only 140 warships of the main fleet of 400 were left. By + 1503 the Têngchow squadron had dropped from 100 vessels to 10. China had to pay heavily for this in the + 16th century when the time came to repel the concerted attacks of Japanese pirates. Gradually Chinese shipping recovered and most of the traditional types of build were preserved, but what a difference it would have made to Lin Tsê-Hsü and his friends at the time of the Opium Wars in the '40s of the 19th century if the short-sighted landsmen of the Ming court had not won the day.

So this was the confrontation. On the one hand there were the voyagers from the East, the Chinese, calm and pacific, unencumbered by a heritage of enmities, generous (up to a point), menacing no man's livelihood, tolerant, if a shade patronising, in panoply of arms yet conquering no colonies and setting up no

[1] [See Lo Jung-Pang, *Oriens Extremus*, 1958, **5**, 149.]

strong points. On the other hand there were the voyagers from the West, the Portuguese Crusader traders out to take hereditary enemies in the rear, wrest a mercantile foothold from unsympathetic soil, hostile to other faiths yet relatively free from racial prejudice, hot in the pursuit of economic power and heralds of the Renaissance. In all the maritime contacts between Europe and Asia in that dramatic age, our forefathers were quite sure who the 'heathen' were; today we suspect that these were not the less civilised of the two.

CULTURAL, SCIENTIFIC AND TECHNOLOGICAL
CONTACTS AND TRANSMISSIONS

The chief stimulus to ancient European knowledge of China was of course the trade overland, or mostly overland, in silk. Some geographical knowledge arose out of this, but the Westerners at that time never really found out much about the Chinese. However, there are interesting rumours of Confucianism in Europe as early as the + 2nd century. For example, Bardasanes recorded that it was considered that the Seres (the Chinese) were very strange people because they were not subject to the dominion of the stars and planets as Europeans were. In China law and custom was much more powerful than astrological influences. A century and a half later Caesarius, the brother of Gregory of Nyssa, even stated that in China custom was followed rather than codified law. Many other writers repeated this and the atheism of the Seres (the Chinese) is mentioned several times by Origen. Bardasanes said that the Chinese were possessed of high morality but had no temples. One may well think that both Bardasanes and Caesarius must somehow have been in contact with the Chinese scholarly tradition, for a merchant would have known that the Chinese did have coded law up to a certain point and also that they did have temples, while only a scholar would have ignored the former in the interests of Confucian *li* (good customs) and minimised the latter as superstitious. We have no idea however by what means such contact was obtained.

There were many embassies during the first ten centuries of the Christian era between the countries of Western Asia and China. Some of these certainly went by sea, at least part of the way. Here we cannot attempt to enumerate them but something may be said of particular aspects of contact during these ages. Several embassies brought jugglers and acrobats with them, a fact which may be more significant than has sometimes been thought, for we know how much attention the Alexandrian mechanicians gave to temple 'miracles' and palace entertainments, stage-play machinery, and the like. Some engineering transmissions may thus be hidden here. Then the Chinese were interested in a number of products of the Syrian regions. One of these was a special gem called the 'night-shining jewel' (*yeh kuang pi*), which so fascinated the Chinese that it was mentioned in all the principal sources. The most likely mineral appears to be chlorophane, which is not an emerald or a beryl but a fluorspar (calcium fluoride), many varieties of which show strong phosphorescence and fluorescence on being heated or scratched in dim light. Other gems from the Roman Empire were regarded, however, by the Chinese as false. This is of particular interest because of the prominent place taken by 'imitative' processes in the beginnings of European chemistry among the Alexandrian 'aurifictors'. Moreover glass was undoubtedly an old Phoenician invention, so it is not surprising that many imitation gems found their way to East Asia. The textual evidence of the Chinese opinion is strikingly verified by the archaeological evidence, because at Arikamedu large numbers of false as well as true gems have been found. Such coloured glass jewellery would date from the first half of the + 1st century.

The Chinese were also interested in the byssus industry. Byssus means the protein threads spun by certain marine lamellibranchiate molluscs which they use to attach themselves to the rocks at the bottom of the shallow waters where they live. In the Hellenistic age it was found that they could be dried and woven into fabrics and at later times there grew up a whole corpus of fables about them. Tarentum in Italy was the main centre of the industry and the Syrian merchants evidently lost no time in exporting its products to China. When the Chinese obtained purses and other goods woven from this curious textile fibre, they thought that they came from the down of a 'water-sheep'. From the + 8th to the + 14th century the fables about the 'water-sheep' became mingled with others which spoke of lambs or sheep engendered in the soil and sprouting on stalks like umbilical cords. This occurs in a number of Chinese texts and may have been the origin of a tale brought back by + 14th-century European travellers in Asia, such as Odoric of Pordenone, about the 'Scythian lamb'. This again was a plant, though it produced something like wool. It was at one time thought that the vegetable lamb fable referred to the cotton plant, but this can hardly be the case because cotton was well known practically everywhere by the + 14th century, and though the Chinese probably had little of it until the Sung, there is nothing reminiscent of it in their presentations of the 'earth-sheep' fables. Another substance which interested the Chinese was storax, a kind of gum which has long been produced in the Middle East. The classical storax was a fragrant resin produced by the tree *Styrax officinalis*, a native of the Levant, but the liquid storax which replaced it in later commerce was an ointment-like product prepared by subjecting the bark of the tree *Liquidambar orientale* to heat and compression. This tree is a native of Asia Minor. Other incense materials from the Middle East were also greatly welcomed in China, possibly in part for use as drugs.

Another object of interest spoken of by the Chinese texts about the Western countries was the striking clepsydra. Hellenistic and Arabic time-keeping technology made great use of anaphoric clocks so arranged as to produce diverse effects when striking the hours. For example, the appearance of jackwork, the dropping of balls with a clang into metal cups and the successive lighting of lamps in a row of niches. There are a number of Chinese descriptions of these and it is quite possible that they exerted some stimulus upon the critical Chinese invention in the +8th century of the escapement. In the development of this, Chinese hydro-mechanical clockwork went far beyond anything which the Hellenistic and Arabic water-clocks had accomplished.

On the medical side, trepanation for blindness was referred to in Chinese texts as a clever Western invention, though here the origin and inspiration may rather have been Indian. The Chinese also knew about theriac, a complicated pharmaceutical preparation supposed to be a universal antidote. This medicine, which had originated from Nicander of Colophon (c. −275), played an important part in medieval Western pharmacy, but the Chinese remained sceptical about it.

Some of these matters may seem rather trivial, and indeed it is a fact that throughout the Middle Ages the really important things were passing from East to West and not from West to East. At the beginning of the period we meet with several inventions which are strangely contemporaneous in time in the Far East and the Far West. For example there is the basic technique of rotary milling and then there was the application of water-power to it. The former was attributed by Varro to the Etruscans of Volsinii and by the Chinese to Kungshu Phan, and thus in both cases the invention would have been of the −3rd or −4th century. Similarly the water-mill belongs both in West and East to the −1st century. The oldest water-mill in the West was in Asia Minor in −65; it belonged to the last Mithridates (that same pharmaceutical monarch who was so interested in theriac); he lost it when he was overthrown by Pompey. From Chinese evidence we know that water-powered trip-hammers were in action during the last half of the −1st century and by + 30 water-power was being effectively used in China, though not so much for the relatively simple job of grinding cereals as for the complicated mechanism of blowing metallurgical bellows in forges and furnaces. Since the time intervening between these two sets of dates is much too short to allow of the belief of transmissions in either direction, it seems more likely that the inventions were both made in some intermediate culture-area, perhaps the Iranian one, spreading afterwards both westwards and eastwards. Unfortunately it has not been possible so far to bring forward any concrete evidence in support of this.

When we come to the later Middle Ages we begin to be able to discern a whole series of transmissions from East to West which seem to come in clusters; one might almost call them 'packaged transmissions'. Thus there is evidence of a wave of adoptions towards the end of the + 12th century, when, within a few decades of the year + 1180, Europeans first began to use the magnetic compass, the stern-post rudder and the windmill. But the greatest cluster was the + 14th century-one. If we choose the date + 1300 to + 1320 as a focal period for the appearance of the first working mechanical clocks in Europe, we may remember that + 1325 was equally focal for gunpowder, the original home of which had been + 9th- or + 10th-century China. Towards + 1380 we find the first blast-furnaces producing cast-iron in Europe—but in China this technique had been developed from the − 4th century onwards and by the Sung period had reached a high level of excellence. Towards + 1375, and also in the Rhineland, comes block-printing, an art which had been current in China since the + 9th century. Still closer to clockwork in time are the great segmental arch bridges of Europe, the first being about + 1340, though in China structures of this kind had first appeared in the + 7th century. The + 14th century thus presents itself as a time of adoption by Europeans of a number of important techniques which had already been known and used for a long time in the Chinese culture-area. One may indeed believe that Europeans did not know exactly where they came from but it is asking too much to maintain that they were all new and independent discoveries. Another important transmission cluster appeared in the + 15th century. The horizontal roasting-spit vane wheel may have been one of these, together with the horizontal windmill, the ball-and-chain flywheel device and perhaps the helicopter top. It may be that both the + 14th-century and the + 15th-century transmission clusters had some connection with a slave trade which brought thousands of Tartar (Mongolian) domestic servants to Italy in medieval times and which reached its height in the first half of the + 15th century. They may have brought all kinds of curious know-how with them and no doubt many voyaged by sea.

The Jesuits in the + 17th century certainly voyaged by sea and hard voyages they had. By this time the scientific revolution was under way in the West and for the first time in two thousand years Europe contributed something technically valuable to China. The Jesuits took with them, for example, the Archimedean screw and worm-gear (as also presumably the simple wood screw and metal screw for fixing objects together), the Ctesibian double force-pump, the crankshaft and the tower-type vertical windmill. The crankshaft was by far the most important of these, though the days of its employment in external- or internal-combustion engines were yet far off. On the other hand, most of the steam-engine's anatomy had already come to the West by the +15th century, for example the double-acting principle of piston and cylinder permitting work on both strokes, used first in bellows for air blast and pumps for water. Another essential component was the standard assembly of eccentric, connecting-rod and piston-rod for the interconversion of rectilinear and rotary motion. This assembly first appeared in medieval China of the Thang or Sung and not in medieval or Renaissance Europe; we do not know exactly when it was transmitted, but most probably it belongs to the + 15th-century cluster. So much for the mutual influences of East and West in historical times.[1]

INVENTION AND TRANSMISSION IN NAUTICAL TECHNOLOGY

Here the first thing to recognise is that Chinese shipbuilding has a pattern entirely of its own. What shipwrights did all over the rest of the Old World quite failed to exhaust the ways in which it is possible to build ships. In Europe and Southern Asia the fundamental beam, the keel, was scarfed at each end to another scarfed beam which turned upwards to form the stem-post and the stern-post respectively. The strakes of the hull which connected them were then held apart in the desired profile by an internal skeleton of bent timbers. But the design of junks exemplified in the oldest and least modified types has a carvel-built hull wanting in all the three components which elsewhere were regarded as essential, keel, stem-post and stern-post. The bottom may be flat or slightly rounded, and the planking does not close in towards the stern but ends abruptly, giving a space which would remain open if it were not filled by a solid transom of straight

---

[1] [For further details, see *SCC, passim.*]

planks. In the most classical types there is no stem either, but a rectangular transom bow. The hull may be compared to the half of a hollow cylinder or parallelepiped bent upwards towards each end and there terminated by final partitions. Moreover there are no true frames or ribs whatever; they are replaced by solid transverse bulkheads of which the stem and stern transoms may be regarded as the terminal units. This is clearly a much more solid method of construction than that found in other civilisations. Fewer bulkheads were required than frames or ribs to give the same degree of strength and rigidity. It is obviously also possible for these bulkheads to be made watertight, and so to give compartments which would preserve most of the buoyancy of the vessel if a leak should occur, or damage below the water-line. In other ways also the bulkhead structure involved corollaries of great importance, for example in providing the essential vertical component necessary for the appearance of the hinged axial rudder. This, together with cognate inventions in the domain of propulsion by sail (of surprisingly early date), we will speak of presently. I need only stop to remark the striking similarity between the bulkhead structure of the Chinese ship and the prominence of the transverse partitions or frameworks so fundamental in Chinese architecture. If the latter prevented a longitudinal vista and permitted the classical curve of the roof, the former provided distinct holds, rendered the vessel extremely strong and gave it the typical bluff bow and stern of large Chinese craft.

One is irresistibly reminded of the fact that the bamboo, that plant so familiar to every Chinese from a thousand uses, has transverse septa. Indeed an important clue may lie herein. The sampan (*shan*) is reminiscent of the bamboo stem just as strongly as the junk. It is an open skiff, bluntly wedge-shaped in plan, shallow, keelless and very broad in the beam aft where the gunwale rail is often continued beyond the stern as an upwardly curved projection endowing the craft with 'cheeks' or 'horns' facing astern. It was the roofing of the space between these projections that led to the overhanging counter of the junk. There is every reason to believe that this particularly characteristic method of ship construction goes back to the earliest times of Chinese civilisation, for in the pictographs of the written language as used during the Shang period of the − 2nd millennium, one finds that the word for ship (*chou*) is represented by a little drawing of what looks like a curved ladder. I know hardly any other Shang pictograph which enshrines so completely such a fundamentally distinctive characteristic of a great branch of Chinese technology. So perhaps the longitudinally sliced bamboo was its paradigm or model.

As for the physical evolution of the junk, the most probable view is that it derives from the bamboo raft. The remarkable sea-going sailing-rafts, still used today in the Chinese and Indo-Chinese areas southwards from the latitude of Thaiwan, would have required only a conversion of the wooden thwart-beams into bulkheads, the substitution of wood planks for bamboo in bottom and sides, and the addition of decking. Chinese and Annamese sailing rafts are of great interest, being equipped with five centre-boards and tall lug-sails, permitting them to sail near the wind.

One of the greatest riddles in the comparative history of civilisations is the possible influence of the ancient countries of the Fertile Crescent on the Far East as well as the Far West. Rather striking examples of Babylonian influence upon China have been found in astronomy, astrology, mathematics and time-keeping, as well as in medico-physiological theory. Influences of Egypt on the Far East have been much less convincing, but it seems rather likely that they did operate in shipbuilding. The most typical and characteristic ancient Egyptian ship was that familiar type which has an extremely long bow and stern sloping and tapering in each direction above the water-line, but these were not the only kind of hull known on the ancient Nile. Especially during the − 3rd millennium a quite different type was prevalent, presenting some extraordinary similarities with certain river-junks still in use in China today. Egyptologists know these two types as the Naqadian and the Horian respectively, for the former can be traced back to the predynastic pottery designs of Naqad and the latter are associated with a conquering people who came from further east and worshipped the god Horus. Both types of ship are sometimes seen in one and the same carving, for example, two funerary reed-bundle Naqadian boats being towed by a Horian ship with square sail set. One has only to compare the latter type with its high stern, stern gallery, low bow and relatively truncated ends with existing Chinese vessels to see the similarity. Since by great good fortune a number of tomb-models of both kinds of boat have survived, it is most striking to find that some of the Horian boats are quite square-ended like all true Chinese ships. Whether or not bulkheads were used is uncertain.

Another common feature is the bipod mast, widely diffused in Southern Chinese shipbuilding. As is well known, such masts were quite characteristic of ancient Egyptian ships of both kinds. This mast type is of course far from being a mere historical curiosity. From the engineering point of view it was excellent and it came back in the days of metal tubing. If there had really been West Asian influence on Chinese naval architecture one might expect to find some exceptions in the Chinese culture-area to the basic principle of boats without keels, stem- or stern-post. There is one, and it is of great interest. The dragon-boat (*lung chhuan*) is used for those races which form such an important feature of the Fifth Month Festival, carrying thirty-six or more paddlers, and it is built with a true keel or kelson. Although bulkheads slotted to the kelson are built in, we have clearly here an archaic element of one of the constituent cultures which fused to give Chinese civilisation. But even more interesting, it is found that in order to prevent 'back-breaking', the drooping of the bow and stern of such a long, narrow craft, a strong bamboo cable is slung from the projecting bow end of the kelson to the stern exactly as was the common practice in ancient Egyptian ships. This is nothing other than the 'anti-hogging truss', and one can hardly fail to see in it an ancient transmission eastwards. The dragon-boat might be described as a canoe-derivate which has survived alone in a world of raft-derivates.

As for a transmission in the opposite direction, the history of the common punt needs further analysis. This is a simple rowed or punted boat of rectangular shape with sloping ends, bulkheads and flat bottom, but its exact origin in Europe is quite obscure. It remains to be seen whether the Chinese build found its way westwards overland early in Bactrian times or through Byzantine contacts, or possibly in the + 13th century due to Chinese technicians following the Mongolian armies.

In south-east Asia, in the regions where Chinese mingled with Indian and indigenous influences, hybrid types of ship developed. A keel was perhaps the first feature to be adopted, then bulkheads were abandoned for rib-frames as in the 'twaqo' of Singapore. The Chinese lug-sails, on account of their great efficiency, are the last components to disappear, and indeed as in the famous Portuguese 'lorcha' of Macao and Hongkong, have co-existed since the + 16th century with a slender hull of normal European type. Probably the centre-board function of the keel helped its adoption. But the most pure and characteristic Chinese designs persisted all over and around Central and North China and even at the present day many ships of highly traditional style are still being built. In 1964, when at Hangchow, I was able to make considerable study of the shipping of the Chhien-thang estuary and also to study the river and lake craft at Chhangsha in Hunan.

It is interesting that small drawings of what are clearly and recognisably junks appear in the Catalan world-map of + 1375 and the world-map of Fra Mauro of

+ 1459, both supposed to be based upon information coming from Marco Polo and his contemporaries. The rectangular build of the junks is not to be mistaken, with their large and prominent rudders, and most of them have four or five masts. They also show towering deckhouses and mat-and-batten sails, and they are much larger than any of the other ships shown in the maps. This is confirmed by the textual evidence of all the Western travellers, whether Europeans or Arabs, in the + 13th and + 14th centuries, who comment without exception on the large size of the Chinese ships. The curious thing is that just as the Europeans were struck by this, the Chinese also were or had been under the impression that the ships of the Far West were larger than their own. In + 1178 Chou Chhü-Fei had written that the ships of the Almoravid dynasty in Spain were the biggest of all in the world; they carried a thousand men, with textile factories and market places, and they did not get back to port for years. However, there was a certain amount of legend in this, for the accompanying accounts describe grains of wheat three inches long, melons six feet round, and sheep the tails of which were so fat and heavy that they had to be supported on trucks. So it seems that people at both ends of the Old World thought that the other end had the largest ships. Objectively, however, the Europeans were right in this opinion while the Chinese were wrong.

In conclusion let us enumerate the ways in which Chinese shipbuilding affected the rest of the world. Building ships on the analogy of the half-bamboo stem with its septa intact led immediately to the absence of stem-post, stern-post and keel, as also to the presence of bulkheads, giving a hull very resistant to deformation. This involved the principle of water-tight compartments with its many advantages, certainly in use in Chinese ships by the + 2nd century, but not adopted in the West until the end of the + 18th, when the provenance was well recognised. Bulkhead construction also offered the possibility of free-flooding compartments which the Chinese found useful both on river rapids and at sea; this was never adopted to any extent in Europe. Finally the bulkhead and the transom construction allowed for the existence of a vertical member to which the axial rudder could be attached in 'line-closure' rather than 'point-closure'. Additional elements not essential to the design but which quickly sprang from it were, first, the approximation to a flat bottom. This obviously had great advantages in tidal waters where shoals and sandbanks abounded, as on the North China coast; but as it offered little resistance to leeward drift it led to the invention of leeboards, earlier here than anywhere else in the world. The flat bottom goes back many centuries in China but it was not adopted in Europe for ships of any size until the

19th century. By the same token there was an approximation to a rectangular cross-section; although old in China, this again was not adopted in Europe until the development of iron and steel. A further peculiarity about the classical Chinese hull was the tendency to place the largest master-couple well aft. This must be at least as old as the Thang period (+8th century) and may well be earlier. Its interest for sailing vessels was not understood in the West until the end of the 19th century.

With regard to propulsion, Chinese seamanship had the lead of Europe for more than a millennium. First, concerning the use of oars and paddles one must note the invention not later than the −1st century of the self-feathering 'propeller' or sculling oar (the *yuloh*); though universally used in China this was never adopted in the West. Secondly, there was the invention of the treadmill-operated paddle-wheel boat[1] in the +8th if not the +5th century, and its great development in the Sung (+12th century) for warships with multiple paddle-wheels (as many as twenty-three on one vessel) and trebuchet artillery. Although proposed in +4th-century Byzantium and discussed in Western Europe in the +14th and +15th centuries, no practical use of the principle was made there until the +16th century in Spain. Thirdly, one must note the complete absence of the multi-oared galley from Chinese civilisation, whether powered by slaves or freemen (apart from the paddled dragon-boats used only for ritual races). This must partly be regarded as the result of the relatively advanced development of sails and rig.

Here several points are important. From at least the +3rd century onwards the ships of the Chinese culture-area were fitted with multiple masts. This may well have been a corollary of the bulkhead build, since these would have invited the placing of several tabernacles along the fore-and-aft mid-ship line (Fig. 12, pl.). Europeans of the +13th century and later were greatly impressed by the large size and many masts of the sea-going junks, as we have seen, and in the +15th century they adopted a system of three which led in due course to the development of the 'full-rigged ship'.

The Chinese also staggered their masts thwartwise in order to avoid the becalming of one sail by another. This is approved by modern sailing-ship designers but was not adopted by Europeans during the period of importance of the sailing-ship. Nor did the Chinese practice of radiating the rakes of the mast like the spines of a fan win acceptance in other parts of the world.

[1] [See further on this pp. 25 and 127.]

One of the earliest solutions of the problem of sailing to windward in large vessels was due to the Chinese of the + 2nd and + 3rd centuries or to their immediate Malayan and Indonesian neighbours at the zone of Sino-Indian culture contact. This involved the development of fore-and-aft sails. The Chinese lug-sail arose in all probability from the Indonesian canted square-sail and hence indirectly from the square-sail of ancient Egypt; perhaps also, as philological evidence suggests, it had something to do with the 'double-mast sprit-sail' now known only in Melanesia, which in its turn had developed from the Indian Ocean 'bifid-mast sprit-sail'. Here again the reference is to the Shang pictograph for sail (*fan*) which shows a trapezoidal, almost square, expanse of cloth or bamboo matting upheld by two parallel masts or spars. The Chinese sprit-sail would have had similar origins. Roman–Indian contacts (see p. 41 above) at the same period (+ 2nd and + 3rd centuries) generated the sprit-sail in Mediterranean waters, but it seems to have fallen out of use there and was introduced a second time from Asia at the beginning of the + 15th century. Meanwhile in the West the lateen sail, characteristic of the Arabic culture-area, dominated in the Mediterranean from about the end of the + 8th century and spread to the ocean-going full-rigged ship in the latter part of the + 15th. The later lug-sails of Europe derived in all probability from Chinese balance lugs.

The earliest aerodynamically correct sails for windward working were the mat-and-batten sails developed in China from the Han period onwards. The system involved many ingenious auxiliary techniques, such as multiple sheeting. Such sails were never used in the West during the period of importance of the sailing-ship but modern research has demonstrated their value, and present-day racing yachts have adopted important elements of Chinese rig, including battens for tautening the sails and a system of multiple sheets (Fig. 13, pl.).

As a corollary from the flat bottom and in view of the development of the fore-and-aft lug-sail, the use of leeboards and centreboards developed in China, probably directly from the sailing rafts of ancient times and certainly by the beginning of the Thang (+ 7th century). They were adopted in Europe a thousand years later, early in the + 17th century. Sailors of the Chinese culture-area raised and lowered them by tackle (especially when the larger rudder served the purpose) or by pivoting or sliding in grooves.

In the domain of vessel control there was a great development of steering-gear which centred on the invention of the axial or median rudder.[1] This device was

---

[1] [On this see further pp. 250 ff. below, and, more fully, in *SCC*, vol. 4, pt. 3.]

fully developed in China, as we know from tomb models, by the end of the + 1st century, and the attachment to the transom stern if not already achieved followed soon afterwards, certainly by the end of the + 4th, but its first appearance in Europe did not occur until the end of the + 12th century. Steering-oars and stern-sweeps, however, although relegated to a secondary position so early in Chinese nautical technology, were never completely abandoned, partly because of their value in negotiating river rapids. For this same purpose bow-sweeps were also often used. Both were useful in harbour and sometimes when changing tack at sea.

Then came the invention of the balanced rudder, hydrodynamically more efficient than the unbalanced type. This was current in China at least as early as the + 11th century but was still regarded as a new and important device at the end of the + 18th in Europe. The further invention of the fenestrated rudder followed, also hydrodynamically advantageous, but this was not adopted in Europe until the era of iron and steel ships.

Among miscellaneous ancillary techniques the following are worthy of remark. Some need was felt for hull sheathing devices. Already in the + 11th century the superimposition of layers of fresh strakes was usual in China; in Europe it became general in the + 16th and later. Sheathing with copper plates was discussed, if not actually performed, in + 4th-century China; lead had been used in Hellenistic Europe. The practical use of copper did not become general either in the East or West until the + 18th century.

Allied to this was the development of armour-plating. The strong predilection of Chinese naval commanders throughout the ages for projectile tactics as opposed to reliance on the close combat of boarding-parties led to the armouring of hulls and upper works above the water-line with iron plates by the end of the + 12th century, and this continued in later times with notable Korean contributions in the + 16th. By that time similar developments were occurring, though rather half-heartedly, in Europe.

Among minor techniques we may mention the Chinese development of non-fouling 'stockless' anchors and the invention of articulated or coupled trains of vessels, probably in the + 16th century. This was not often employed in Europe for shipping but greatly used in other fields of transportation. Besides these there were ingenious dredging practices going back to the Chinese Middle Ages, and advanced techniques of diving generated by the pearling industry in South China. Europeans of the + 16th century also admired what they considered the advanced techniques of bilge clearance by chain-pumps on Chinese ships.

Of course there may have been some degree of independence in the European advances. Even when we have good reason to believe in a transmission from China to the West we know very little of the means by which it took place. But as in all other fields of science and technology the onus of proof lies upon those who wish to maintain fully independent invention, and the longer the period elapsing between the successive appearances of a discovery or invention in two or more cultures concerned, the heavier that onus generally is. The techniques here discussed were, to be sure, often improved upon by the Europeans who later adopted them. All that our analysis indicates is that European seamanship probably owes far more than has generally been supposed to the contributions of the sea-going peoples of East and South Asia.

5

THE CHINESE CONTRIBUTION TO
SCIENCE AND TECHNOLOGY[1]

[1946]

WHEN M. Léveillé said I would tell of the contribution of China to science and technology during the war he was not quite correct. That would have been another subject. I was in China for much of the time during the Second World War, but what I want to speak about here is the contribution of the Chinese people to science and technology as a whole over the centuries. My belief is that that contribution is usually very much underestimated. This arises largely from the fact that knowledge of the Chinese language is not very widespread in the West, and the historians of science who have traced so brilliantly the development of modern science and technology on the shores of the Atlantic have not always been able to see that often the first discoveries were made by our Chinese colleagues right at the other end of Asia.

One example of this is a famous book by Professor J. B. Bury, *The Idea of Progress*. In this book he refers to the fact that at the time of the Renaissance in Europe there were many discussions between scholars, some of whom argued that the 'moderns' were better than the 'ancients', and others that the 'ancients' were better than the 'moderns', meaning the people of the Renaissance. Those who supported the moderns usually argued that they were better because they had made the discoveries of printing, of gunpowder and of the magnetic compass. There is not even one footnote in Professor Bury's book mentioning that all of those discoveries were not made in Europe, but in Asia—we owe them all to the Chinese.

During the time of the Greek city-states and the Roman Republic in Europe the Chou dynasty was reigning in China. In − 221 the first emperor of a united China came to the throne, having forged all the previously isolated proto-feudal states into one. Before that there were only semi-independent feudal princes held

[1] Reprinted from *Reflections on Our Age*, published by Allan Wingate (Publishers) Limited, London, 1948 (ed. D. Hardman & S. Spender). Lectures delivered at the opening session of Unesco at the Sorbonne, Paris, November 1946; tr. from the French 'Conférences de l'Unesco', Fontaine, Paris, 1947.

together by the hegemony of one or other ruler under the nominal suzerainty of the Chou dynasty. Then, about the time of the decay of the Roman Empire, you find China divided into three separate nations (the San Kuo period) until after fifty years there was union again under the Chin dynasty. Eventually, by the + 7th century, we arrive at what is regarded as China's flowering period, the Thang dynasty, when religious thought and practice flourished, and painting, poetry and music were at their height.

But from the point of view of science, however, the Sung period was more interesting because it was then, in the + 11th and + 12th centuries, that Chinese science reached its peak. Later came the Mongol invasions and the foundations of the Yuan dynasty, which lasted about a hundred years. After that a national dynasty again arose, that of the Ming, only to be overthrown in turn by the Manchus (the Chhing dynasty).

I would like to state the general theme of this lecture in the following way: we are interested, of course, in exactly what the Chinese contributed to world science and technology. But what we are also interested in is not only what the Chinese did, but why they did not succeed, as Europe's civilisation did, in giving rise to *modern* science and technology. Why did their science and technology always remain primarily empirical? Why was there no indigenous industrial revolution in China? That, I believe, is one of the greatest problems of all comparative social history, and I believe that we can form some idea of what the inhibiting factors were.

Before I go on to speak of this problem, however, I want to describe certain things in Chinese philosophy and in Chinese science and technology which demonstrate the remarkable triumphs that were achieved in that land.

First of all, my opinion is that Chinese philosophers, both ancient and medieval, could speculate about Nature as well as the Greeks. It must, of course, be admitted that Chinese civilisation did not produce an Aristotle; but we may assume that the factors and circumstances which inhibited the growth of scientific thought in China were already active at a date when an Aristotle could have been produced. But if we look at Chinese methods in ancient philosophy we will see much to make us believe that the Chinese could speculate as well as the Greeks, on whom is usually laid the whole burden of the origins of modern science.

We all know that there were many schools of Chinese philosophy. The Confucian school, which became orthodox, may be said to have always held the preponderance from the Han time onwards. If one studies books like the *Conversa-*

tions and Discourses of Confucius, or the *Discourses* of his great disciple Mencius—if one studies those books one gets quite a clear picture of Confucius. He and his followers were profoundly this-worldly in their outlook, and desired to organise human society in such a way as to obtain the maximum of social justice, as they conceived it. They were extremely social-minded, they emphasised literary as opposed to manual activities, and in fact they built up what one might call a sort of social 'scholasticism'. Although they began as the counsellors of feudal lords, and ended as the personnel of the mandarinate, they also, in many ways, contributed to democratic thought. The sayings of Confucius are famous; they have been widely expounded and are well known. The right of the people to rebel against tyrants was laid down by Confucian scholars in the First Empire and the ethos of the scholar-gentry was fundamental for the nature of the Chinese social system.

But the school of philosophers in whom we are specially interested is the Taoist school. This was a group of philosophers who reacted against the Confucians partly because they felt that however confident Confucius might be that he knew how to organise human society, it would never be possible to do so, and to get real social justice, until we knew more about Nature. So the Taoists retired to the woods and mountains and other remote places and tried to make some examination of Nature. As they never developed an experimental method based on hypotheses, they were not able to proceed much further than Democritus or Lucretius, but in many ways the Taoists are similar to the Epicureans in our European history. The followers of Epicurus were men who believed they had found a satisfactory and substantially true theory of how the universe worked. The Taoists also thought they understood something about the fundamental Order of Nature, which they called the Tao. They believed that the way to be liberated from the terrors of natural forces and primitive gods and demons was to form some rational theory about the nature of the universe; and the peace of mind achieved in that way was very similar among the Epicureans and among the Taoists.

The attitude of the Taoists towards Nature may also be shown in some remarkable passages in some of their greatest books, such as the *Canon of the Virtue of the Tao* in which reference is made to the 'Valley Spirit'. The 'Valley Spirit' was an immortal goddess, not a god. In many other passages of that book the feminine receptivity of the Taoist is emphasised, his lack of prejudice made clear, and his passivity described not as a religious passivity, but a humility in the face of Nature. Man has to come in a humble way to Nature and put his questions without too many preconceived ideas.

If you look in one of the great Taoist books (the *True Classic of Southern Hua* by the incomparable Chuang Tzu) you will find a good story of a king whose butcher was so remarkably skilful that he could cut up a bull with three strokes of his hatchet. The king came and said to him, 'How do you manage to do it?' and the butcher replied, 'Most butchers, it is true, take fifty-five strokes, and even the better ones will do it only in twenty and then blunt their hatchet; but all my life I have been studying the Tao of the bull, so that is why I can do it.' The Taoists believed in going along 'with the grain' of Nature, not in cutting across it.

Another story which helps us to understand what the Taoists meant by the Tao is that in which Chuang Tzu's disciples ask where the Tao is to be found. He answers, 'Everywhere.' They say, 'But surely not in that piece of broken tile?' and he answers, 'Yes, also in that.' 'But not surely in that piece of dung?' 'Yes, in that too.' The Order of Nature runs through everything. At these early stages we are standing at the beginning of religious thought, as well as of science, since the 'One' of religious experience was not yet clearly distinguished from the unity of the natural order.

If I am right about the conceptions of the Taoists, there would have to be some relation between their traditions and activities and the rise of scientific practice. That is precisely the case, as we shall see in a few minutes. Alchemy, which is older in China than in any other civilisation, is found to have grown up in a Taoist milieu; they were trying to find the medicine of immortality, but we will come back to that. You might also like to know that in Chuang Tzu's book dating back to about −290 we find an interesting doctrine of evolution—that animal species are not fixed or immutable but change into each other in the course of time. And while I am speaking of evolution it is very desirable that we should remember another description of evolution which we find in the *Book of Rites*. People who say that Chinese ideas have always been static do not know what they are saying. Primitive barbarism is first described, much as in Lucretius. Next comes the stage of what is called the 'Lesser Tranquillity' in which wars and famines take place, competing national states exist, and the like. Finally, the 'Great Community' is described, as the outcome of social evolution, when the whole world lives in unshakable peace and unity and the people have full social security. Yet this work cannot be later than −300.[1]

Before I leave the Taoists I would like to give you one further quotation from

[1] [For full details, see *SCC*, vol. 2, p. 167, or more particularly J. Needham, 'Time and Eastern Man', Royal Anthropological Institute Occasional Paper, no. 21, the Henry Myers Lecture for 1964.]

their literature. A modern philosopher has said, 'Freedom is the knowledge of necessity.' In order to be free it is necessary to understand the laws of the universe. So in a book of about − 330, attributed to Kuan Tzu, it is said that 'the sage follows after Nature, in order that he may control her'. The political, as well as the Baconian, significance of this dictum is of course not far to seek. And by a remarkable coincidence (if it is only a coincidence) the Taoists were just as revolutionary politically as the Confucians were orthodox and conservative. The Taoists wanted to go back to the pre-feudal collectivist tribal society, before the differentiation of classes. Throughout Chinese history they were present somewhere in the background of every rebellion. And the connection between 'democratic' and scientific or pre-scientific beginnings, which is clear also in Ancient Greece, should not be overlooked.

Another ancient school of thought was that of Mo Tzu. He differed from the Confucians because he rejected the family system, and from the Taoists because he preached a doctrine of universal love. What is not generally realised is that his works and those of his disciples contain a great deal of scientific material; there are certain chapters on optics, and other branches of physics. It is rather curious to find this interest in physics in connection with ethics, because it reminds us of Spinoza's great aim to produce an ethics *more geometrico demonstrata*.

A fourth school was the school of Legalists. There was in ancient China a great controversy between the Confucians, who believed in a kind of paternal justice in which every case at law should be judged on its merits, and another school which said that everything ought to be judged according to a code fixed beforehand. In some of its forms the school of the Legalists anticipated modern authoritarianism. That school is lost today; it did not succeed, and that lack of success is perhaps one of the ideological factors involved in the failure of China to develop modern science and technology. For in European history at any rate there was a close historical connection between the concepts of legal law and of natural law.[1]

I will not follow further the philosophical track, but I want to say that the medieval philosophy of China is just as well worth studying as that of ancient times. In the Han dynasty there arose a very remarkable rationalist scholar, Wang Chhung, who wrote a book called *Ideas Weighed in the Balance*, about the superstitions of his time. In it he said that human beings on the earth's surface were of no more importance to the earth or the stars than parasites on the human body.

[1] [See *SCC*, vol. 2, pp. 518 ff., or more recently J. Needham, 'Human Law and the Laws of Nature' (Hobhouse Lecture ,revised), Hatfield College of Technology, 1961.]

When we come down to the + 11th century we reach the greatest period, with the Neo-Confucians of the Sung dynasty. The greatest of these, Chu Hsi, has been called a + 12th-century Herbert Spencer. The more you read him, the more incredible it is that without the basis of experimental science so realistic and naturalistic a philosophy could have been evolved at that time. I might add that Chu Hsi was the first to identify fossils. He said that the stone animals that were found at the top of mountains proved that the mountains had once been at the bottom of the sea. Thus this was clearly recognised in China about + 1170, while in the West you have to wait for Leonardo da Vinci for a recognition of it.

At the end of the Ming dynasty, about + 1650, there was an upright official who refused to give in to the Manchus. Wang Chhuan-Shan retired to the mountains and wrote many books, including a materialist, almost Marxist, history which deserves much study today, and demonstrates the naturalistic and realistic trend of thought of the Chinese.

The next point which must be made is that in both ancient and medieval China every evidence is shown of experimental manual operations from which valid inductions were drawn. When we say that modern technology did not develop, we mean that the science of the Chinese always remained empirical, and its theories were confined to those of 'primitive' type, such as the Yin and Yang principles and the Five Elements. Theories of the advanced post-Galilean mathematical type were not developed. Alchemy and chemistry illustrate this.

Notable first is that although the Taoists were extremely interested in immortality, they did not want some kind of spiritual immortality in the sky—they wanted to live on here, and they wanted a medicine or plant of immortality which would enable them to do so, any method, in fact, ascetic or otherwise, which would give long life. Material longevity and immortality was what they sought.[1]

In − 133 Li Shao-Chün went to the Emperor Han Wu Ti and said, 'If you will sacrifice to the stove' (i.e. support my researches) 'I will demonstrate to you how to make the yellow gold; out of that gold you may make vessels and drink from them to become immortal.' That is the first reference to alchemy in world history. Later on, in + 142, you find what is without question the first book on alchemy ever written, *The Kinship of the Three*, which describes the use of chemically transformed substances as elixirs of life. We know that the appearance of alchemy in Islam and Europe dates from after that time, because we cannot find

[1] [Cf. p. 337 below.]

it before the + 8th century, perhaps not till the + 10th.[1] The origin of the word 'alchemy' has been much disputed. It has been suggested that it comes from *Khem*, a name for Egypt, said to refer to the black earth of the Nile Valley, but Egyptian alchemy is not ancient. I suggest that the word is really Chinese in origin and comes from the words *lien chin shu*, the art of transmuting gold.[2] This would be pronounced in Cantonese *lien kim shok*.[3] Now it is known that Arabic people and Syrians were trading with China as early as + 200, so the Arabs would naturally put the prefix *al* on to it, and get *al kimm*,[4] 'pertaining to the making of gold'. All the greatest alchemists were Taoists. There are a large number of alchemical books in Chinese, the vast majority of which have never been translated.

If you ask about the theories of these ancient proto-sciences, I may say that the classical theory used in China began very early and lasted very late, in fact, until today. The earliest theory supposed the Universe to be composed of two fundamental principles, Yang and Yin, light and darkness, male and female. This dualism looks as if it might have been Persian in origin; but the fact that good and evil were definitely not part of this antithesis makes such an influence unlikely. Again, the Five Elements were not the same as the four elements of the Greeks—air, water, earth and fire; in ancient China one had metal, wood, water, earth and fire. Pervading everything was *chhi* in various forms, vapour, spirit, subtle influence, something like the *pneuma* of the Greeks. It is certain that the first conception of atoms goes back to them or to the Indians; but I could bring evidence suggesting that the conception of waves really goes back to the Chinese, because whenever the operations of Yin and Yang are described, it is always as a process of maximum and minimum; when one comes up the other goes down, which is a conception of waves. The original idea of the atom, as that which cannot be divided, is no doubt Greek or Indian, whereas the idea of waxing and waning waves may be said to be Chinese.[5]

[1] [There was of course an important development of technical, speculative and mystical chemistry in Hellenistic Europe (first and second centuries), but it was not alchemy for it did not attempt to make true gold from other substances, only to imitate the metal, and it was not primarily concerned with longevity and material immortality. The later ideas of the 'philosopher's stone' and the 'elixir of life', though unattainable phantasms, were those which inspired the invention of most of the practical techniques out of which modern chemistry was born.]

[2] [Unknown to me, a similar suggestion had been put forward by S. Mahdihassan of Karachi earlier in the same year (*Current Sci.* 1946, 15, 136, 234, see also *Journ. Univ. Bombay*, 1951, 20, 107). The equation has now been accepted in principle by a number of scholars (see e.g. H. H. Dubs, *Ambix*, 1961, 9, 23).]

[3] [Actually *lin kêm shut*, but I was speaking from memory. The ancient pronunciation was probably *lien kiem dzhiuet* (Karlgren). Hakka, Korean and Annamese all have *kim* for the main word.]

[4] [Actually *al-kimiya*.]   [5] [Cf. p. 22 above.]

The poor condition of Chinese anatomy in the last century gave people the idea that it had always been backward in China, but that is not the case. If you look at Chinese anatomical pictures dating from the + 7th, + 8th and + 9th centuries you will see on the contrary that they were rather advanced; and the opinion of some anatomists is that Chinese drawings of the human body may well have been the origin of the famous 'series of five pictures' which is one of the most important genres in the history of anatomy in the West. Chinese anatomical pictures occur prominently in many editions of a famous book of forensic medicine, *The Clearing of the Innocent; or, the Washing away of Wrongs*. This treatise, the first on the subject in any civilisation, was written by Sung Tzhu in + 1247.

In certain other ways Chinese scientific effort was very remarkable in the early periods. Systematic measurements of rainfall were being made at about the time of the Norman conquest of England, and at the end of the Roman period, about + 132, the mathematician Chang Hêng invented the first seismograph. Its description is interesting. It was so constructed that if a trembling of the earth occurred, a bronze ball would fall out of the mouth of a bronze animal into a container below. It is stated that by this means knowledge of earthquakes was obtained several days before couriers arrived at the imperial court with the news.[1] About the same time too we get many accounts of other ingenious apparatus. There are records of a carriage which, if set to indicate the south, would continue to do so no matter in what direction it travelled or changed course. This was not the magnetic compass, but a mechanical device, the first of all cybernetic machines.[2] Another vehicle, a 'taximeter', sounded a drum for every mile passed, and this may have helped in the mapping of the empire.[3]

I must pass over many things such as silk technology, and the development of ceramics and porcelain. No doubt the three greatest discoveries of the Chinese were paper and printing, the magnetic compass, and gunpowder.

The question of paper and printing is, of course, of great interest.[4] The Chinese historical records are so good that we know almost to a day when paper was first made. In + 105 Tshai Lun went to the Emperor and said: 'Bamboo tablets are so heavy and silk so expensive that I sought for a way of mixing together fragments of bark, bamboo and fish nets, and I have made a very thin material which is suitable for writing on.' It was not for another six centuries that this was used for printing, but when we come to the + 10th century we find a great desire to

---

[1] [Cf. *SCC*, vol. 3, pp. 626 ff.]   [2] [Cf. *SCC*, vol. 4, pt. 2, pp. 286 ff.]
[3] [Cf. *SCC*, vol. 4, pt. 2, pp. 281 ff.]   [4] [Cf. pp. 22 ff., 32 above.]

get the classics printed instead of engraved on stone. Printing began in about + 700 in West China, and movable block printing three centuries later. Although the latter found its way to Europe just before the time of Gutenberg, it did not make much headway in China because the Chinese characters made it more convenient to go into stereotype straight away, as it were, and to cut the characters of a whole page on one block of wood at the same time. Perhaps the original idea of printing came from the custom of cutting seals, which is very ancient in China.

The Chinese book came to be very different from the Western book because it was printed only on one side of the page, and the pages were folded over and stitched. That was because the original method was to write scrolls on silk and roll them up. When they began to print they took the paper and folded it over and over, so that the book was always printed on one side of the paper only.

It has been remarked that the invention of printing in Europe was one of the great causes which led to the fragmentation of European civilisation after the apparent unity of the Latin-speaking Middle Ages; because if you start printing and circulating very widely in different local dialects you get a tremendous dissemination of language variations. Then they crystallise. This did not happen in China because the Chinese written language is a 'monolithic' unit. It is pronounced differently in different parts of the country, but it cannot be spelt in different ways. The characters are always identical, so it was impossible for printing to have the fragmenting and disintegrating effect on Chinese provinces that it had on European regions in the Renaissance.

Regarding the mariner's compass, there have been many arguments to and fro.[1] We know that the attractive property of the magnetic needle was known to the Romans; it was also known to the Han people in China, but the directive power, the polarity, was known there as well. By the Sung dynasty we find the compass in full use. About + 1085 there was written a book by a very remarkable man, Shen Kua, in which he described the magnetic compass. He said that when magicians want to find the direction of the north, they take a needle, rub it on a lodestone, and hang it up by a thin thread, when it usually points to the south. He adds that there are two kinds of needle, one which points to the north and the other to the south, but that this is not surprising because there are also two kinds of animals which shed their horns respectively in summer and winter. In earlier times the Chinese apparently used to carve their lodestones in the form of spoons. Long before + 1180 (the date of the first knowledge of magnetic polarity

[1] [For the present state of knowledge, see p. 239.]

in Europe) we have records of journeys to Korea, Cambodia, etc., which show definitely that the compass was used to steer the ship's course. I may say that a more complicated compass plate is still used even today in the country because people believe in the importance of having their house face a certain direction.

If we pass on to the question of gunpowder, it is known that crackers were used in Han times, but the evidence is rather against their having anything to do with gunpowder; they were probably pieces of green bamboo. Descriptions of fireworks are found from + 600 to + 900 (the Thang time) which indicate that some sort of inflammable mixtures were known. A clear statement of the compounding of sulphur, saltpetre and carbonaceous material—for the first time in any civilisation—occurs about + 850. The first indication of the use of gunpowder in war is just after + 900. It is not true to say that although the Chinese invented gunpowder they were so humane that they only used it for fireworks. It was first employed in a flame-thrower, using oil ignited by gunpowder, not exploding, but burning as a 'slow match'. Later on we get the rocket (fire-arrow), all sorts of bombs thrown by catapults, and, in the fighting between the Chhitan (Liao) and Jurchen (Chin) Tartars in the north and the Sung people in the south, bombs of a highly destructive character, the proportion of nitrate having been raised. I think there is no doubt that gunpowder can be traced to the alchemy of the Thang Taoists.[1]

But I have still one or two more things to mention. Vaccination is a procedure which is not usually regarded as being of Asian origin, but nevertheless the first form of the process occurs as a dream of a Taoist nun. She took the contents of the smallpox pustule and implanted it into the mucous membrane of the nose, perhaps on some principle of sympathetic magic. The process, 'variolation', is still used among the Mongols; it is a dangerous one because it may cause an epidemic, but it can confer protection on individuals.

I must now make a reference to the literature on pharmaceutical natural history; it is an enormous collection of important works, the first of which appeared in the Han time, and includes descriptions not only of plants, trees and many animals, but of mineral objects of all kinds. In Europe, there was a great controversy in the time of Paracelsus (+ 16th century) who introduced minerals into medicine—mercury, antimony, bismuth, etc., instead of only herbs. At that time the Chinese had been using them for many centuries already.

Another discovery concerned what we call deficiency diseases. It is usually

[1] [See *Legacy of China* (Oxford, 1964), pp. 245 ff.]

supposed that a knowledge of deficiency diseases belongs to our time, together with the recognition of the vitamins that cure them. But if to know that some diseases can be cured by diet alone, without any drugs in the ordinary sense, is an empirical knowledge of deficiency diseases, then the Chinese were well aware of them. We possess a book by Hu Ssu-Hui, of the Yuan dynasty (+ 14th century), which bears the title *Some Diseases can be Cured by Diet Alone*, in which the author gives a description of both the forms of beri-beri and a description of dishes which would restore desperate patients to normal almost in a few hours.[1]

I have talked about the schools of philosophy of the Chinese, the Chinese 'proto-sciences', and some Chinese technological achievements; and lastly I want to come back to the original question: why did not modern technology develop in China? Why did modern science not develop? This must be due to many factors. I shall try to stick to concrete, material things, because one can easily be led astray by emphasising ideas alone. Ideas are of course important, but no more so than the geographical and social factors which conditioned the struggle of the Chinese people through three millennia.

Let us speak first of rainfall in China—China is a monsoon country, and the rainfall is far greater in June and July than in other months; it is also very variable from year to year. On this account the Chinese were faced with the necessity of making large-scale irrigation works and mastering water conservancy at a very early date. Their works in that direction are greater than any others, even those of the Egyptians, and the Grand Canal is one of the greatest hydraulic engineering achievements of the world. It is argued by some Chinese scholars that the necessity of this had two consequences: in the first place, millions of men, of workers, had to be controlled, and if you have to control such a large labour force you have to have a large body of officials. No one who is not acquainted with Chinese civilisation can realise the importance of the Civil Service and the Mandarinate in traditional China. But also the extent of the irrigation to be conducted has to be considered, since if the work was to be effective it had to be done on a large scale. Hence it transcended the boundaries of the fiefs of individual feudal lords. But the more centralised the authority, the less the power of the feudal lords and the more the power of the Emperor.

We must also consider the continental character of China as against the peninsular structure of Europe. The characteristic European unit was the mercantile city-state. The European distribution of land and water led very early to an

---

[1] [For further information on this see Lu Gwei-Djen & J. Needham, *Isis*, 1951, **42**, 12.]

emphasis on maritime navigation and to a mercantile economy. The Chinese solid land-mass, on the contrary, led to a network of towns 'held for the Emperor' by a Governor or Magistrate, and each surrounded by a hundred agricultural villages. One must always contrast the Greek *polis* with the Chinese *hsien*. Now if the Mandarinate was supreme, if the Civil Service was always the great power, there was a bar to the development of any other group in society, so that the merchants were always kept down and unable to rise to a position of power in the State. They had guilds, it is true, but these were never as important as in Europe. Here we might be putting our finger on the main cause of the failure of Chinese civilisation to develop modern technology, because in Europe (as is universally admitted) the development of technology was closely bound up with the rise of the merchant class to power. It is perhaps a question of who is going to put up the money for scientific discovery—it is not the Emperor, it is not the feudal lords; they fear change rather than welcome it. But when you come to the merchants, they are the people who will finance research in order to develop new forms of production and trade; and such was indeed the fact in European history. Chinese society has been called 'bureaucratic feudalism', and that may go a long way to explain why the Chinese, in spite of their brilliant successes in earlier science and technology, were not able, as their colleagues in Europe were, to break through the bonds of medieval ideas, and advance to what we call modern science and technology. I think one of the great reasons is that China was fundamentally an irrigation-agricultural civilisation, as contrasted with the pastoral-navigational civilisation of Europe; with the consequent prevention of the merchants' rise to power.

I feel now that I shall have done my job if I have managed to awaken some sort of interest on your part in the excellence of the contribution made to science and technology by the Chinese in past ages, in the absence of which the whole course of our civilisation in the West would have been impossible, for one cannot imagine the disappearance of feudalism in Europe without gunpowder, paper, printing, and the magnetic needle. And you will see that when full account is taken of the environmental conditions, it is not so much to the credit of Europeans that they developed modern science and technology, nor yet so much of a reproach to our Chinese friends that they failed to do so. The abilities were everywhere, but the favourable conditions were not.

6

THE TRANSLATION OF OLD CHINESE SCIENTIFIC AND TECHNICAL TEXTS[1]

[1957]

MY colleagues and I have for some years been engaged in the preparation of a comprehensive work on the history of science, scientific thought, and technology in China. As historians of science and technology in the Chinese culture-area, our concern is with the content of the writings which we study; and since we have an enormous amount to do, we have not time, I must confess, to go very deeply into the form in which the content is expressed. A professional linguistic sinologist would perhaps, therefore, be much more suitable for this task than anyone like myself. So if this paper does not correspond with your idea of what an interesting account of scientific and technical translation from Chinese should be, I can only humbly apologise. It *is* my idea of what an interesting paper would be, and that is the best I can do for you.

Chinese is generally spoken of as a highly isolating and non-agglutinative language. It has thus always been very difficult for a writer of an Indo-European language (or any other extremely explicit, perhaps excessively explicit, tongue regarding number, tense, mood, gender, etc.). And of course in Chinese the parts of speech are not so rigidly differentiated as they are in other language groups such as the Indo-European; a particular word can function as several different parts of speech, depending to some extent, like all Chinese grammar, on the order of the words in the sentence. One reason, however, why it may be a good thing that a historian of science and technology should deal with this topic is because the sinologists themselves have in the past been rather lax regarding the translation of technical terms. An outstanding example is *kho-lou* 刻鏤. This was translated by one of the greatest sinologists of the past, Alfred Forke, as 'the weaving of stuffs with inserted patterns'; and this was very important (for it was

---

[1] The substance of a lecture first delivered for the Communications Research Centre at University College, London, in April 1957; then published in *Aspects of Translation* (ed. A. H. Smith; Secker & Warburg, London, 1958), p. 65; and finally in *Babel* (International Journal of Translation, published by the International Federation of Translators), **4**, no. 1 (special issue, 'Translation in Asia'), March 1958.

a + 1st-century text) because of the problem of the origin of the drawloom, whereby figures are woven automatically in silk as the weaver proceeds. He presumably confused it with the term *kho-ssu* 刻絲 which means a kind of brocade, and arose very much later. But the proper translation of it is 'the painting, ornamenting and carving of buildings' and it has nothing to do with textile technology at all.

Another example is the word *niang* 釀, which cannot possibly be made to mean 'distillation' though many, even modern, sinologists have tended to translate it so, probably not being very clear in their own minds between 'distillation' and 'fermentation'. In fact, 'fermentation' is what it means. Of course, terms of a more philosophical sort, like the great *Tao* 道—the 'Order of Nature'—have given much trouble. The tendency to translate that by 'law' or 'laws of Nature' has been too much for a great many people.[1] To come to the essence of the subject, Marcel Granet said, I think very well, that the character in Chinese is a more esteemed and significant symbol than any single syllable in an Indo-European language. His actual words were: 'Solidaire d'un signe vocal dans lequel on tient à voir une valeur d'emblème, le signe graphique est lui-même considéré comme une *figuration adéquate*, ou plutôt, si je puis dire, comme une *appellation efficace*.' In other words, it has greater significance as a single unitary emblem or symbol than the syllable of an Indo-European language. And on the other hand, as A. F. Wright has said, it possesses a set of meanings accumulated in the long history of the language, to which one could add a still wider range of allusive undertones derived from rich literary traditions, and finally, still greater flexibility depending on the context of the noun, as in doublet forms and the like.

When one comes to mathematical terms one finds a rather interesting thing; throughout Chinese history algebra and algebraic methods dominated, and geometry was very much in the background. But according to the phrase of Nesselmann, it was 'rhetorical algebra'—that is to say, it was not composed of symbols as we understand algebraic symbols today, but it was in words. In fact, these Chinese words were rather more than words in the alphabetical sense, though undoubtedly less than symbols in the full modern mathematical sense. These are some of them: for example, *khai fang* 開方 for the 'square root', *shih* 實 for the 'dividend', *chhêng* 乘 for 'multiplication', *fa* 法 for the 'divisor'; then, when you come to *ting fa* 定法 you have a term 'the first fixed divisor' in one of the procedures of square root extraction; you have *tzu* 子 'numerator', *mu* 母 'denomin-

[1] [See *SCC*, vol. 2, pp. 36 ff., 573.]

ator', and *chhu* 除 'division', and so on. These could be handled and manipulated in different positions very much as you might put *x* or *ab* in relation to the calculation, so that although they did not appear as single letters, they were units which could be moved about and treated in much the same way as in modern mathematical procedure.

The next point I would like to make is the great continuity of the Chinese tradition. Among all the civilisations of the world we have in Chinese alone a language which has been spoken right through from the middle of the – 2nd millennium. It is not like Sumerian, Hittite or Ancient Egyptian, or even like Greek or Latin, some of which are not spoken at all today, and others did not start very early. Probably Hebrew, as one finds it in Israel still, would be the nearest competitor to Chinese. Now this is very important from the point of view of technical terminology because it means that we have a lexicographic tradition which takes us back at least to the – 3rd century. You may find that tradition in a book like the *Lü Shih Chhun Chhiu* for instance (– 239); and we have great dictionaries at the end of the first century, e.g. the *Shih Ming* and the *Shuo Wên* about + 100. It is of the greatest importance to be able to rely on so continuous a tradition of spoken and written language.

We also get rather interesting results when we look into the mode of origin of some of the basic words used in scientific and technical discourse. The following table of percentages (p. 86) has been drawn up taking a hundred characters at random, all concerned with scientific communication and the expression of facts about Nature (they add up to more than 100 because there are some overlaps). One finds that they grew up, roughly speaking, from three main classes—drawings of non-human natural objects, drawings of the human body and its parts, and diagrams of human actions, especially tools, techniques, and rituals. There are other, smaller classes as well, but these are the three main ones.

It is interesting too that one can find certain technical information in these ancient pictographs. For example, the word *chou* 舟 meaning 'a boat'; if one looks on the oracle-bone writing of the – 2nd millennium, one can see that the actual type of boat involved was not a boat with a stem- and stern-post and a keel, such as Europeans imagine a boat to be, but a boat with transom bulkheads and a square-ended construction. In other words, the ancestor of the junk. We get the same thing in another boat character implying an arrangement for turning and steering, and the old character in the old writing shows the oar and a hand to guide the boat. Another example arises with regard

		%
Pure geometrical symbolism		2·5
Drawings		
non-human natural objects		
inanimate or cosmological	19	
biological	13·5	32·5
human body or its parts (sex 9)		27·5
Diagrams		
human actions		
motions, paths	6	
tools, technique (ritual 6)	29	35·0
social life		7·5
Borrowed homophones		10·0
Symbols of abstract concepts		5·0
Uncertain		1·0
		121·0

to the word *kung* 弓 'a bow'. We talk about self-bows and reflex bows; a self-bow would be like the long-bow of England where you have a single substance (wood) and it is straight, but there is also the reflex bow with a more complicated curve, and, of course, the composite bow as used by all Asian nations, which is made of a number of different things (wood, horn, glue, etc.)

Here again, the old character, the old picture, in the − 2nd millennium (as is also the case with the word *shê* 射 'to shoot'), gives us clearly the reflex bow and the arrow, not the self-bow.

Now suppose we take a much more complicated thing, the cross-bow trigger. This was part of the standard weapon of the army of the Han dynasty, which corresponds in dating with the Roman Republic and Empire. There are many interesting things about the *nu chi* 弩 機—the trigger of the cross-bow—one being that its parts are all named, and we know the names very well from definitions given in the *Shih Ming* dictionary of + 100. For example, the 'stock' is *pi* 臂; the 'hook' which holds the string back is the *ya* 牙; the housing of the trigger mechanism is the *kuo* 郭; then you have the *kuei* 規 meaning the lug; the trigger itself is called the 'hanging knife' (*hsüan tao* 懸 刀); the rocking lever is the *tien chi* 墊 機, and the *chien* 鍵 are the shafts on which the levers pivot. I want just to mention *chi* again because of the interest of the pictograph. The *chi* 機 *par excellence*, the 'machine as such', was the loom; and indeed one can see two hanks of silk in the

86

ancient pictograph of it.[1] I do not say that any character exactly like this has been found on oracle-bones or bronzes, because the wood-radical has been added on the left of a known form, in the same script, but the loom seems to have been the paradigm of all machines in China, and gave its name to all kinds of other machinery.[2]

Then one comes to a still more complicated type of mechanism, the magazine cross-bow, or 'machine-gun' cross-bow, where the arrows fall into place and the weapon automatically sets itself as the string is drawn back. This is called the *Chuko Liang nu* 諸葛京弩 after a +3rd-century general (p. 32), but there is no reason for thinking that it had been invented at that time. The most likely date of its invention was the Middle Ages (in the Sung), when we can probably recognise it under the phrase *kan thung mu nu* 匫筒木弩—the 'tube and box cross-bow'. Then we have the *chê tieh nu* 摺疊弩, the cross-bow with 'things piled up in layers', i.e. the piled-up arrows in the magazine. There is also the expression *chi nu* 積弩, in other words 'piled-up cross-bows', but this more likely means a 'massed formation of archers'. Some have thought that *shu chi* 樞機 meant the magazine cross-bow in the Han, but this is also very unlikely as it probably only refers to the 'shafts' on which the trigger levers pivot.

One of the greatest difficulties about technical terms is that now and again the same word covers two different things; the thing changed while the word remained unchanged. This is a great nuisance from the point of view of the history of technology, because there is nothing for it but to read every possible text you can find in order to throw light on the point when the change in the 'thing' occurred. An outstanding example is the word *thung* 銅 which meant 'copper' long before it meant 'bronze'. So, too, the word *tho* 柂 means 'rudder'; but it did also once mean 'steering-oar'—and of course one of the important things in the history of shipping is to know when the steering-oar gave place to the axial or stern-post rudder. In a case like this, all one can do is to read every passage one can find where these words occur, and see what it says. To put the matter in a nutshell, we find, for example, that a text of the +9th century, the *Kuan shih Ti Li Chih Mêng*, says that if the *tho* goes down too deep into the water, it

[1] [Actually this double component derived not from the silk-radical but from another ancient radical which depicted a germinating seed; the thought-connection, however, between a delicate hypocotyl and a single fibre of new-spun silk was very close in ancient China, and the graphs became interchangeable.]

[2] [Many years later, when I had occasion to live for some time in the city of Senglea in Malta, I used to pass Triq il-Macina (Machine Street) leading down to the quay. But here 'machine as such' meant the crane, not the loom. Hardly anything could symbolise better, I thought, the maritime-mercantile character of Europe in contrast with the agrarian-productive character of China.]

will not do; it will hit upon rocks and on the bottom of the river—now that does not sound like a steering-oar because such oars usually trail out rather a long way after the ship. The *Hua Shu*, about +940, says that it is a remarkable thing how the control of a big junk is assured by a piece of wood no longer than six feet. Six feet is not like a steering-oar which would be at least ten feet, and often much longer. And so one can gradually build up a case in this way for the appearance of a specific 'thing' at a certain time. When we come to +1124, in the *Hsüan-Ho Fêng Shih Kao-Li Thu Ching*, the records of an important embassy to Korea, we find large and small stern rudders fairly clearly described, and the steering-oars are called 'assistant rudders', so that we have probably got it definitely by then. Thus the evidence accumulates for the time of appearance of the 'thing', and as we know that the first axial rudders appeared in Europe about +1180, the argument rather points to China having been the place of their origin. The paradox is that this may be precisely because of the fact that although there were no stem- or stern-posts in Chinese shipbuilding, there were bulkheads forming convenient vertical members to which a vertical rudder-post could be attached.[1]

Another good case of this is found in the terms *hun hsiang* 渾象 and *hun i* 渾儀. There is no doubt at all that *hun i* means, has meant, and always did mean, the observational armillary sphere. In the early stages it may have meant only 'armillary rings', but one single ring could be put in different planes of space, and from the +1st century it certainly meant the 'armillary sphere' in the full sense. But *hun hsiang* is difficult because although after about +430 it certainly means 'celestial globe', i.e. a solid globe with stars marked on it, before that time it seems to have meant an armillary sphere set up for demonstration purposes with a model earth at the centre. One would very much like to know whether that model earth was a little flat plate sticking up on a pin in the polar axis (the earth being sometimes supposed to be flat and square, while the heavens were thought of as round and curved) or whether it was in fact a ball. There were cosmologists in the Han period who said over and over again that 'the earth is floating in the midst of the heavens as round as a cross-bow bullet', and both conceptions therefore would have been possible.

An interesting situation arises where there are orthographic fluctuations. The phrase *kuan li* occurs written in several different ways, either alone 關戾, or with

---

[1] [A year or so after this paragraph was written, a Han ship-model in terracotta with a perfect median axial rudder was excavated from a tomb of the +1st century in Canton, thus validating beyond doubt the elaborate textual argument which Dr Wang Ling and I had built up. Dr Lu Gwei-Djen and I were able to examine it in China in 1958 and 1964. See p. 257 below.]

the hand-radical 關捩, or the wood-radical 關棙, and the contexts in which it appears are quite different. It has to be translated according to them, because the definitions of these technical terms have long been lost. The first of these was probably a trip-lug and it was used by Chang Hêng in the +2nd century for rotating astronomical instruments by water power. He had a lug, a pinion of one, on a shaft, which turned round periodically as the buckets of a water-wheel filled, and moved on a toothed wheel. The form with the hand-radical, however, is quite a different thing; it is a 'stop-valve'. The stop-valve was a little metal fish inside a hollow bamboo tube used by tribal people for ceremonial feasts, so that a man could not drink too fast nor too slowly, but had to keep at just the right rate in the ceremony, otherwise the valve stopped up the tube. Then, thirdly, the 'wood' one is a special kind of pivot, used with the rings in the Cardan suspension, or gimbals, of lanterns used at festivals. So there is orthographic fluctuation: the general sense may be fairly clear, but one has to fit the forms in with the actual context in which they occur. Indeed the word *kuan* in these examples is the word actually used for parts of the escapement of hydro-mechanical clockwork from the +8th century onwards (p. 219), but the dates are much too early for that, from other evidence, so it cannot mean 'escapement' in our forms, and the general sense would be 'pushed up against something' or 'opposing something'.

I have already mentioned the long line of the great general dictionaries and encyclopaedias beginning literally two thousand years ago which one can consult, but I must say something about the tradition of technical glossaries. Everyone knows how confused the terminology is in the field of alchemy and chemistry. The alchemists, of course, both in East and West, were decidedly anxious to cloak their traces, and they did not want to be too explicit. It is therefore remarkable to come across a synonymic dictionary, the *Shih Yao Erh Ya*, written in the year +806 by Mei Piao, constituting a veritable encyclopaedia of drugs and minerals. One gets things like this, of course, in Europe in the eighteenth century, for example Martin Ruhland's *Lexicon Alchemicum*, but that is a thousand years later, or nearly so. Another good example is the Japanese work of Fukane no Sukehito, the *Honzō Wamyō*, written in +918. This is also a most useful compendium.

Again, one must study the usage of words. The phrase *huo yao* 火藥 is the term for gunpowder; and we have never come across it meaning anything else. In other words, 'fire chemical' invariably means mixtures of carbon, sulphur, and nitrate in various proportions, used for rockets, bombs, etc. Other things have quite a different terminology. For example, *pao chu* 爆竹 for 'crackers' arose because in the beginning these 'explosive fireworks' were not fireworks at all but simply pieces of green bamboo which when put into a fire decrepitated with loud bangs. Then fireworks like Bengal lights and things of that sort were known as *huo hsi* 火戲, and smokes as *yen huo* 烟火. We have never met with any case where *huo yao* could mean something other than a 'gunpowder' mixture.[1]

And so, in the same way, one finds great fixity of terminology in metallurgical matters. In this field it is very fortunate that things are so well defined. Thus there is no question that when you find *sêng thieh* 生鐵 or 'raw' iron you have to deal with cast iron; when you find *kang* 鋼 you have a steel, and when you find *shu thieh* 熟鐵, 'ripe' iron, you have wrought iron. There are other rarer words like *hsien* 銑 which is occasionally used for cast iron, *jou* 鍒 for wrought iron, and a very unusual word indeed, *hsieh* 鏶, which is used for a bloom of wrought iron. It is unusual because from a very early period the Chinese had been making cast iron in abundance; certainly more than a thousand years before Europe. And this abundance of cast iron had a remarkable effect: it meant that in China steel was not usually made by the cementation process of adding carbon to wrought iron. Wrought iron, of course, is highly pure and almost no carbon is present. The usual European method was to get wrought iron either direct from a bloomery furnace or by the fining or puddling of cast iron, and then to carbonise it so as to get steel. As far as we can see, the Chinese rarely, if ever, did this. They either de-carbonised their cast iron to steel direct, which was a very skilled thing to do; or else they adopted a process which we call, inadequately, the co-fusion process, about which I shall say something more later on, and which was known as *tsa lien sêng jou* 雜煉生柔—'mixing and heating together the raw and the soft'. In the co-fusion process the cast iron and the wrought iron were heated together so that their carbon content averaged out, and steel was produced. This started at least from the +5th century onwards. An interesting thing is that in the *Mêng Chhi Pi Than* and other Sung books, the writers distinguished between two kinds of steel. One they called *kuan kang* 灌鋼 or 'interfused steel', because the cast iron melts, of course, and bathes the softened lumps of wrought iron; the same co-fusion steel was also called *thuan kang* 團鋼, 'lump steel'. But they also called it *wei kang* 僞鋼, 'false steel', to distinguish it from *chen kang* 眞鋼, true steel, or *shun kang* 純鋼, pure steel, or *lien kang* 煉鋼, refined steel. The reason doubtless was that the carbon was not always very successfully distributed right through

[1] [With one special exception only, 'Fire-element drug' in *nei tan* terminology, cf. p. 272.]

the mass of steel in the co-fusion process, with the result that it did not have such a high reputation, though great amounts of steel were made by this method. The 'pure steel' was made direct from cast iron and must have had a more even carbon content. Both these methods are of the highest interest because they are ancestral to the two great processes still used in the steel industry, the Siemens–Martin process on the one hand, and the Bessemer conversion process on the other.[1]

The whole field of engineering furnishes a great variety of reliable special terms. For example, the square-pallet chain-pump is almost always called the *fan chhê* 翻車, sometimes the *lung ku chhê* 龍骨車, the 'dragon-bone water-raiser', and an interesting thing about it is that the sprocket-wheels used in it have spokes rather than teeth protruding in different directions—hence the name *hsia ma*, the 'spread-eagled toad'. *Fan chhê* are very clearly differentiated from *thung chhê* 筒車, or norias—buckets fixed to the periphery of a wheel to raise water. This *fan chhê* or square-pallet chain-pump dates certainly from the + 2nd and probably from the + 1st century, and it is highly characteristic of Chinese technique.

We now reach the final phases of this discourse. All through the centuries, the Chinese (like medieval Europeans) found great difficulty in coining new technical terms. For example, the clock today is called the *tzu ming chung* 自鳴鐘, 'self-sounding bell', but this term only came in during the + 17th century after the arrival of the Jesuits as a translation of 'clock', 'cloche', 'Glocke', etc. The Jesuits brought to China the newest sorts of clocks made in brass and gold and encrusted with various precious stones, to offer to the Emperor and the high officials; they were doubtless good examples of Renaissance horological technique. As these instruments got the name of 'self-sounding bells' the impression was that the machine itself was something fundamentally new; there were perhaps a few scholars in China at the time who realised that their ancestors had made something similar as far back as the + 8th century, but they carried no weight. Actually what had happened was that there was a complete failure to evolve a new technical term for clockwork when it began in the Thang period in China. I have an interesting extract concerning this; it is a quotation from a memorial made by Su Sung to the Emperor in + 1092 when he was presenting his astronomical clock-tower to the throne. He ends by saying: '...in any case, if we use only one name all the marvellous uses of these three instruments cannot be included in its meaning; yet since our new instrument has three uses, it ought to have a more general name...'

[1] [See, more fully, p. 107 below.]

In other words, with its mechanised armillary sphere, celestial globe, and jack-work, it should not be called just a *hun hsiang* 渾象 or a *hun i* 渾儀: it ought to have some more general name such as *hun thien (chi)* 渾天機 which we might translate as 'cosmic engine'. '...and we are humbly awaiting your Imperial Majesty's opinion and bestowal of a suitable name upon it.' That was in + 1092, but the Emperor had no ideas whatever. As a matter of fact, he was only about seventeen at the time, but in any case he did not produce a name, and nobody else did, so the mechanical clock did not get one; and the arrival of a new term six centuries later gave rise to the idea that there had arrived a new thing.

One sees many examples of the difficulty experienced in making new technical terms long before. For example in the (say) − 4th-century *Kuan Tzu* book there are discussions on the properties of minerals like jade, and the writers talk about the 'benevolence' or 'heroism' of jade, and so on, when what they really want to have are words like 'brittleness' or 'hardness'. In the same way the *Huai Nan Tzu* book of − 120, talking about the Five-Element theory, has technical terms just taken straight over from human relations. And again in the + 12th and + 13th centuries, when the Neo-Confucian philosophers wanted to speak about expansive and contractive forces in Nature, they contented themselves with using the old words *shen* 神 and *kuei* 鬼, i.e. the 'gods and demons'. Thus the common people could continue to talk about gods and demons and spirits, while the Neo-Confucian philosophers could use the same words in a naturalistic and sophisticated sense. It was always very difficult to create new terms.

Of course, this inhibition was in Europe too. If one reads Albertus Magnus on the development of the chick embryo, one can see how desperately he needed new words:

But from the drop of blood out of which the heart is formed, there proceed two vein-like and pulsatile passages, and there is in them a purer blood which forms the chief organs such as the liver and the lungs, and these though very small at first grow and extend at last to the outer membranes which hold the whole material of the egg together. There they ramify in many divisions, but the greater of them appears on the membrane which holds the white of the egg within it...

Here he wanted to say 'allantois' but he could not coin it. Charles Singer has often shown how important was the formation of new technical terms invented by the Arab scientists—a word like 'syrach' for example. The Arab technical terminology was taken over by European men of science later on.

However, the Chinese had especial difficulties in this department; they were faced with a dilemma of translation or transliteration. Of course, this is very well known to those who are familiar with the Chinese world, but it is worth emphasising that the same problem had to be solved by the early translators from Buddhist texts as by those in our own time translating modern scientific terms. Should we transliterate phonetically, with impossibly ugly gibberish resulting? Or should we employ already existing Chinese words and distort the meaning? For example, there was the expression *wei-tha-ming* 維他命 which taken literally would mean 'binding up his fate' or 'destiny'—and so does not say anything; in fact, it simply transliterated the word 'vitamin'. The opposite way of handling this was to write something like *sêng chi su* 生機素—'pure or essential component of the living machinery'—which is quite a good way of talking about it, but does not sound like vitamin. The second method was known by the Buddhist scholars as *ko i* 格義, 'explaining by analogy', and it is the one now generally in use. But it is striking that just the same problems arose from modern scientific and technical terminology as when Buddhism first came to China eighteen centuries ago. In modern times, however, new terms have been formed by inventing new characters much more freely; when new elements are wanted, for instance when we want to talk about lanthanum or palladium or bismuth or argon we use the new characters *lan* 鑭, *pa* 鈀, *pi* 鉍, or *ya* 氬. Similarly, dynamics is 'force science' *li hsüeh* 力學. New characters are made *de novo* using the metal-radical or the vapour-radical and so on.

Lastly, it may be suitable to discuss one or two texts as a whole and see how it feels to get something scientific and technical from them. In order to translate any text, it is absolutely essential to know what the writer is talking about. The great sinologist Friedrich Hirth once wrote a few lines which are worth remembering:

Generally speaking, anyone can translate a chapter of Livy without difficulty with a grammar and dictionary, but you cannot do that with a Chinese text from antiquity or the Middle Ages, because there is so much more than the mere meaning of the words and sentences. The European reader must understand, be familiar with, and know the places, the people, and the things; he must not only translate, he must identify. Only when he has realised what the author is really talking about, can his translation have the breath of life. Even those who know the language extremely well must also be collectors, or as we might say, students of things, if the things are going to be talked about.

Before giving you some examples I should like to make one more point about the ideographic language. Many people insist on saying that it is very vague— and I admit that the classical style is highly laconic, since number, tense, mood, gender, abstraction, and so on, is much less explicit than in European or other languages. At the same time one can often come across what might be called veritable epigrammatic crystals of thought. The following example

CHU HSI (+1130 TO +1200)
ON THE EMERGENT EVOLUTION OF MIND
所覺者，心之理也。
能覺者，氣之靈也。

Chu Tzu Yü Lei, ch. 1, p. 40*b* (ed. *c.* +1270)

is a very beautiful one; it is absolutely transparent. Its meaning is:

Cognition (or apprehension) is the essential pattern of the mind's existence, but that there is (something in the world) which can do this, is (what we may call) the spirituality (inherent in) matter (or, as we might say, emergent from matter).

Now, that is quite a lot to get into so short a phrase, but it is not at all unusual, and it is extremely poetical too, because the words are so evocative. *Li* 理, for example, evokes the natural pattern in jade, or some other exquisite stone; and the word *ling* 靈 is one of the loveliest words in the language, meaning 'spirituality'—indeed 'numinous' might be the best translation of *ling*. Imagine a temple, for instance, in some remote and beautiful place in the mountains, which you are visiting in company with distinguished, elegant and charming people, where the priest-in-charge is intelligent, the associations inspiring, the scenery magnificent and the weather is fine, well, this 'sacredness' of the place, to cap the whole, is summed up by the word 'ling'.

Again, examine this lexicographic crystal from Hsü Shen's *Shuo Wên* (+121). Defining *chi*, the word for 'machine' already discussed, took him only two words —*chu fa*, 'controlled energy-application (or energy-output)', *chu fa wei chih chi*. Could any modern mind have done it more succinctly?

And now having come back to the world of technology for a moment or two, let us have a look at the following classic example

CHHIWU HUAI-WÊN'S METHOD OF MAKING STEEL
(THE CO-FUSION PROCESS) *c.* +545
〔綦毋懷文〕又造宿鐵刀。
其法燒生鐵精以重柔鋌，
數宿則成鋼。
以柔鐵為刀脊。
浴以五牲之溺。
淬以五牲之脂。

Pei Chhi Shu, ch. 49, pp. 8*b* ff. (+636) and *Thai-Phing Yü Lan*, ch. 345, pp. 6*b* ff. (+983)

94

on steel-making. It refers to an interesting man with a strange name, Chhiwu Huai-Wên, who lived in the + 6th century, and made sabres of steel for the last ruler of the Eastern Wei dynasty. The text of the *Pei Chhi Shu* says that 'Chhiwu Huai-Wên also made sabres of overnight iron'. Now what could 'overnight iron' be? The text continues:

His method was to bake the purest cast iron, piling it up in layers with the soft ingots (of wrought iron), until after several (days and) nights, it had all turned to steel.

This is extremely interesting because it clearly describes the melting of the high-carbon cast iron in presence of the low-carbon wrought iron, and the averaging of their carbon-content with the consequent production of steel. Then he goes on to say something else we have not mentioned yet: 'And he used soft iron to make the backbone of the sabre.' This is another way of 'combining the soft and the hard'; it is simply pattern-welding where a soft steel part is placed inside and a hard steel is used for the cutting edge. 'And then', it says, 'for quenching he used the urine of the five animals; and for tempering he used the fat of the five animals.' This is quite interesting too, because all metallurgists are well aware of the difference between rapid and slow quenching. When oils are used, they absorb the heat from the metal object more slowly and the microscopic structure of the steel may be profoundly affected.

My last example concerns the armillary sphere which the Persian astronomer, Jamāl al-Dīn, brought to Peking in 1267. At least we cannot be sure that he actually brought it, but he certainly brought a design.

THE ARMILLARY SPHERE OF JAMĀL AL-DĪN

(+ 1267)

⌊咱禿哈剌吉⌉漢言混天儀也。
其制以銅爲之。
平設單環，刻周天度，畫十二辰位，以準地面。
側立雙環，而結於平環之子午。
半入地下，以分天度。
內第二雙環，亦刻周天度，而參差相交。
以結于側雙環，去地平三十六度，
以爲南北極。
可以旋轉以象天運，爲日行之道。
內第三第四環皆結於第二環。
又去南北極二十四度，亦可以運轉。
凡可運三環，各對綴銅方釘。
有篾以代衡簫之仰窺焉。

Yuan Shih, ch. 48, pp. 10*b*, 11*a* (*c.* + 1370)

95

It was the time of the invention by Kuo Shou-Ching of the equatorial mounting (p.10) which has been used in astronomy ever since. *Tsa-thu ha-la-chi* transliterates Arabic words *dhātu al-halaq-i* (the owner of the rings). Note the purely phonetic syllables, just as in the *wei-tha-ming* we spoke of before (p. 93).

The dhātu al-ḥalaq-i is what is called in Chinese 'armillary sphere'.

It is made of bronze.

Horizontally there is set up a single ring graduated with the degrees of the circumference of the heavens, and the 12 *chhen* positions, for measuring (directions on) the earth's surface.

At right angles to this there is a split (meridian) ring fixed to the horizon ring at the north and south points. Half of this meridian circle goes below the 'earth', and so divides the heavenly circumference.

Inside this is a second split ring, also graduated with the degrees of the heavenly circumference in such a way as to correspond exactly (with the graduations on the meridian ring), and connected with the meridian ring at a point 36° above the earth's surface line (and at a point 36° below it) to mark the north and south poles.

This ring can turn and rotate, representing the revolution of the heavens and marking the path of the sun.

The third and fourth rings are inside and both connected with the second ring. They are placed 24° away from both poles and can also turn and rotate.

So in all three rings rotate, being pivoted on bronze pins. And there are holes which take the place of the sighting-tube for looking up (at the heavenly bodies).

From this we see immediately that Jamāl al-Dīn's sphere was an equatorial one—a fact of considerable interest, for the Chinese had never used the ecliptic co-ordinates of the Greeks and Arabs. As this must have been perfectly well known at Marāghah, it is clear that the design must have been prepared specifically with Chinese practice in mind. The horizon ring and the meridian ring were of course fixed to the base, while the movable equatorial declination split ring for observations rotated in the polar axis. Vanes with holes adjustable on this ring would give the declination of a star, but while this deviated from the Chinese tradition of incorporating a sighting-tube, the splitting of the rings was entirely in the Chinese style. The most unusual feature, however, was the attachment of two further smaller circles at right angles to the declination ring, i.e. parallel to the equator, there being no equator circle either fixed or movable. The right ascension measurement could naturally be obtained by their aid. If the text is taken *au pied de la lettre* we should have to regard them as circles of perpetual visibility and invisibility, but since their placing does not correspond with the polar altitude distinctly stated, it is perhaps more likely that they were 24° removed, not from the south and north polar pivots, but from the equator, in which case they would have been tropic rings. It is interesting that the altitude of 36° fits Teheran or Meshed rather than Marāghah, and suggests that the sphere was designed for the observatory of Phing-yang (mod. Lin-fên) in Shansi, not for Peking. This is by no means all that we can learn from this text, but for the present purpose it must suffice. The outstanding conclusion is that as a contribution to the re-fitting of the Chinese observatories under the Mongol dynasty, the Persian astronomers prepared a design which, though original in conception, was entirely in the Chinese style, departing very widely from their own Ptolemaic and Graeco-Arab traditions.

With this the present survey must come to an end. In the general context of human communication I hope we have shown that across the very great barrier of the ideographic and alphabetic languages, and across the time distance of ten or twenty centuries, minds trained in the observation and experimental study of Nature, and in the techniques which utilise her gifts, can still communicate.

7

THE EARLIEST SNOW CRYSTAL OBSERVATIONS[1]

[1961]

A N interesting question in the history of meteorology asks when and where was the hexagonal system of snow-flake crystals first discovered? Before we came upon the material which we present in this brief paper we would have been inclined to suppose that the observation was made in classical Western antiquity, perhaps even among the pre-Socratic nature-philosophers, but this is far from the truth. The examination of snow-flake forms seems to have been a distinctive achievement of East Asia, for the oldest Chinese statements, going back to the − 2nd century, antedate the first European observations by more than a millennium. Our present contribution thus extends the findings of a former one,[2] and continues the elucidation of ancient and medieval Chinese contributions to meteorological history.

In spite of a fairly careful investigation, nothing appears to have been said about the shape of snow-flake crystals by Aristotle or Seneca in classical times, nor have we been able to find any such observations in other classical authors. Whether or not the Arabs investigated the question remains uncertain, but a starting-point for a European history of the subject is in the writings of Albertus Magnus (c. + 1260). From his meteorological writings we know that he thought the crystals were star-shaped (*figura stellae*) but he seems to have believed that such regular forms fell only in February and March.[3] After that time no further mention occurs in Europe until the book of the great Scandinavian bishop Olaus Magnus in + 1555 (*Historia de Gentibus Septentrionalibus*). In this he devoted a short chapter 'De variis figuris nivium' to the question, and illustrated it by a very bad woodcut,[4] but one which has often been reproduced. Out of twenty-three forms shown in this, there is only one star and three or four star fragments. The rest are of all kinds of queer shapes, such as crescents, arrows, nail-shaped objects, bells, and one like

[1] Reprinted from *Weather* (October 1961), **16** (no. 10), 319.
[2] Ho Ping-Yu & J. Needham, on ancient Chinese observations of solar haloes and parhelia. This and other references will be found in the appended bibliography. [See also *SCC*, vol. 3, p. 474.]
[3] *Meteorol.* **1**, 10.   [4] Ch. 22, p. 37.

a human hand. Olaus Magnus certainly appreciated the great variation of form of snow-flake crystals, but he missed the essential unity of their pattern—the hexagonal symmetry present in all. The recognition of this, and therefore the real beginning of knowledge of snow-flakes in Europe, is undoubtedly due to the great astronomer Johann Kepler, who, at the height of his career, presented to his patron Wackher von Wackenfels a 15-page Latin tractate on snow-flakes as part of the festivities of New Year Day + 1611.[1] This was published later in that year under the title *Strena, seu de Nive Sexangula* (A New Year's Gift of Hexagonal Snow).[2] Kepler's pamphlet was not so much remarkable for its description of the starry form of the snow-flake as for its discussion of the methods by which nature achieved the hexagonal crystal structure. This he attempted to explain on an atomistic basis in relation to mathematical theories of close packing, but he had to fall back upon the thought of a *facultas formatrix* to account for the pheno-menon.[3] Thus the winter of + 1610 saw the true beginning of the investigation of snow-flakes in Europe. We shall say something a little later about its subsequent development in post-Renaissance science, and turn now to consider the medieval Chinese contributions.[4]

The oldest of these is quite venerable. The + 10th-century *Thai-Phing Yü Lan* encyclopaedia preserves[5] for us a passage in a book of the Former Han dynasty written by Han Ying about − 135. This book is entitled *Han Shih Wai Chuan* (Moral Discourses Illustrating the Han Text of the *Book of Odes*). Here we find the statement: 'Flowers of plants and trees are generally five-pointed, but those of snow (*hsüeh*), which are called *ying*, are always six-pointed.' It is obvious that this discovery implied fine observation, and it would be interesting to know at what point in Chinese history any kind of magnifying lens was used to study the snow-flakes. There is here, of course, a direct connection with the ancient doctrine of symbolic correlation in which the Five Elements[6] and other things classifiable in groups of five were associated with particular numbers. Thus in many of the classical Chinese writings we can find that the number six is the symbolic correlation number

[1] A valuable account of this, and the light it throws on Kepler's thinking, has recently been given by Schneer.
[2] Tampach, Frankfurt, 1611. German tr. by R. Klug, 'Des Kaiserlichen Mathematikers Johannes Kepler Neujahrsgeschenk oder über die Sechseckform des Schnees', *Jahresber. d. k. k. Staatsgymnasium zu Linz*, 1907, no. 56. New. ed., annotated and tr. H. Strunz & H. Born, Bosse, Regensburg, 1958.
[3] The merit of Kepler's conviction of the connection between atomism and crystallographic regularity has been acknowledged by E. von Laue. Cf. Hellmann, pp. 12 ff., 49.
[4] Some of these have been collected by Tamura Sennosuke in an interesting recent book (pp. 217 ff.), which stimulated our interest in the matter.
[5] Ch. 12, p. 2*b*.
[6] N.B. not four, as in ancient Greek thought.

for the element Water. The Chhin dynasty ruled, it was said, by virtue of the element Water and had six as its symbolic number. These ideas occur not only in the books of the naturalists like the *Lü Shih Chhun Chhiu* (Master Lü's Spring and Autumn Annals)[1] of −239, but also in the medical writings, such as the famous *Huang Ti Nei Ching, Su Wên* (The Yellow Emperor's Manual of Corporeal [Medicine]; The Pure Questions and Answers)[2] which may be placed in the −2nd century. Numerous other statements[3] could be quoted to show that six was associated with Water and the North, while five was associated with Earth and the Centre.

This contrast between five-pointed plant structures and six-pointed snow-flakes became so well known in subsequent centuries as to be almost a literary common-place. For example, in a poem by Hsiao Thung (+501 to +531), that scholarly Crown Prince of the Liang dynasty who edited the *Wên Hsüan* (Literary Treasury), the greatest anthology of Chinese poetry, there occur the following lines:[4]

> The ruddy clouds float in the four quarters of the caerulean sky
> And the white snow-flakes show forth their six-petalled flowers.

This passage occurs in a series of poems, one for each month of the year, and from his words we visualise a glowing winter sunset with promise of snow.

One of Hsiao Thung's contemporaries was Jen Fang, who early in the +6th century wrote a book entitled *Shu I Chi* (Records of Strange Things). In this he mentioned that there is on Thien-thai Shan (mountain) a peculiar kind of apricot, the flowers of which have 'six petals and five different colours'. In +863 in his *Yu-Yang Tsa Tsu* (Miscellany of the Yu-Yang Mountain Cave), Tuan Chhêng-Shih again referred to the matter, saying that 'among flowers there are few which have six petals, but the *chih tzu* does'. This statement was repeated in an interesting passage by the physician Chang Kao (fl. +1189) in his *I Shuo* (Medical Discourses). There we read:[5]

The physician Li Wei-Hsi of Shuchow was good at discussing natural phenomena. (Among other things, he said) that the reason why double-kernelled peaches and apricots are harmful to people is that the flowers of these trees are properly speaking five-petalled, yet if they develop with sixfold (symmetry) twinning will occur. Plants and trees all have the fivefold pattern; only the yellow-berry (*chih tzu*) and snow-flake crystals are hexagonal. This is one

---

[1] Cf. Wilhelm, p. 463. [2] Ch. 4. [Cf. p. 271 below.]
[3] E.g. the *Kuan Tzu* book, ch. 8; *Huai Nan Tzu*, ch. 4; and *Chhien Han Shu*, ch. 27.
[4] In *Chhüan Shang-Ku San-Tai Chhin Han San-Kuo Liu Chhao Wên*, ed. Yen Kho-Chün (Liang section), ch. 19, p. 10*a*.
[5] Ch. 8, p. 4*b*.

of the principles of Yin and Yang. So if double-kernelled peaches and apricots with an (aberrant) sixfold (symmetry) are harmful it is because these trees have lost their standard rule.

It is not difficult to identify the *chih tzu*,[1] for it is well known in Chinese botany as the plant *Gardenia jasminoides* or *florida*. This belongs to the Rubiaceae and among the dicotyledonous species six-petalled flowers are known though not common. The remarks of this + 11th-century naturalist on the principles of twinning would interest experimental biologists today.

A + 12th-century contemporary of Chang Kao was the great philosopher Chu Hsi, perhaps the greatest in all Chinese history.[2] He was a man of deep insight into all kinds of natural phenomena.[3] Two quotations from him will be of interest to us. In the *Chu Tzu Yü Lei* (Classified Discourses of Master Chu) we find: 'Six generated from Earth is the perfected number of Water, so as snow is water condensed into crystal flowers, these are always six-pointed.' Elsewhere in the *Chu Tzu Chhüan Shu* (Collected Writings of Master Chu) there is a text which says:[4]

The reason why 'flowers' or crystals of snow are six-pointed is because they are only sleet (*hsien*) split open by violent winds (and sleet being half-frozen rain, i.e. water) they must be six-pointed. Just so, if you throw a lump of mud on the ground, it splashes into radiating angular petal-like form. Now six is a Yin number, and *thai-yin hsüan-ching-shih* is also six-pointed, with sharp prismatic angular edges (*lêng*). Everything is due to the numbers inherent in Nature.

The mineral here mentioned is in fact selenite, translucent hexagonal crystals of gypsum or calcium sulphate,[5] so that Chu Hsi specifically made a comparison with the mineral world. This is exceedingly interesting because it prefigures the later development of the cloud-seeding process, to which we shall return below.

All these ideas continued down in the writings of the Chinese naturalists. In the late + 14th or early + 15th century, Wang Khuei, in his *Li Hai Chi* (The Beetle and the Sea), remarked[6] that 'snow is the extreme form of Yin, and so has the Water-number in perfection. Thus it is that snow-flowers are always six-pointed.' Li Shih-Chen, who reproduced this passage in his *Pên Tshao Kang Mu* (The Great Pharmacopoeia) of + 1596, concurred in Wang Khuei's opinion, which indeed had been traditional for many centuries.[7] Li Shih-Chen, however, also quoted[8]

---

[1] Also *chih tzu*, Read, no. 82; Stuart, pp. 183 ff.; Phei Chien & Chou Thai-Yen, vol. 4, no. 195. Why the double-kernelled sports should have been poisonous is not obvious, unless the metabolism of the cyanogenetic glucosides was also affected so that toxic amounts of HCN accumulated in the fruit.

[2] See *SCC*, vol. 2, pp. 455 ff.　　　　　[3] Cf. *SCC*, vol. 3, pp. 598 ff.

[4] Ch. 50, p. 48 *b*.　　　　　[5] Identification in Read & Pak, no. 120.

[6] Ch. 1, p. 2 *a*.　　　　　[7] Ch. 5, p. 8 *a*.　　　　　[8] Ch. 5, p. 9 *a*.

a somewhat different formulation from 'Mr Lu, the Agriculturist' (Lu Nung Shih). In some book, now lost, this writer, whom we may most probably identify with Lu Yung, said:

> The Yin embracing the Yang gives hail (*pao*), the Yang embracing the Yin gives sleet.
> When snow gets six-pointedness it becomes snow crystals.
> When hail gets three-pointedness it becomes solid. This is the sort of difference that arises from Yin and Yang.

Lu Yung is said to have been living about the time when Marco Polo was in China, i.e. *c.* + 1285, and wrote a *Thien Chia Wu Hsing Chih* (Meteorological and Phenological Forecasting for Agriculture according to the Theory of the Five Elements).[1]

A little uncertainty concerning the age-old opinion was introduced by Thang Chin later on in the Ming period in his *Mêng Yü Lu* (Records of my Daydreams).[2] He remarked, indeed, 'that flowers of plants and trees are always five-pointed and snow crystals six-pointed was a saying of the old scholars, for, since six is the true number of Water, when water congeals into flowers they must be six-pointed'. But he added:[3] 'When spring comes the snow crystals are five-pointed.' For this he was reproved to some extent by Hsieh Tsai-Hang, writing in his *Wu Tsa Tsu* (Five Assorted Offering Trays) *c.* +1600. He says: 'There is an old statement that snow crystals can (often) be five-pointed. But every year at the end of winter and the beginning of spring I used to collect snow crystals myself and carefully examine them; all were six-pointed, with five-spike ones fewer than one in ten. So one can see that old sayings are not always quite true.' Upon this final note of personal observation, so strikingly contemporary with Johann Kepler, we shall end our ancient and medieval Chinese quotations. We are convinced that a large number more could be found by further searching in the literature, but this may suffice.

Looking back at the whole story, it is interesting both that the original discovery was made so early in China and also that there was apparently so little development of it during the ensuing centuries. Of course, some good medieval drawings of snow crystals may yet come to light, but our general impression is that the Chinese, having found the hexagonal symmetry, were content to accept it as a fact of Nature and to explain it in accordance with the numerology of

---

[1] See Wang Yü-Hu, pp. 95 ff., who, however, places his *floruit* about a century later.
[2] Ch. 2.
[3] In a strange echo of the idea of Albertus Magnus.

the symbolic correlations. Hellmann may well be right in explaining the absence of Greek knowledge of snow-flake crystal shape by the fact that snow was somewhat rare in the Mediterranean region. No doubt this would have had an inhibiting effect also upon Indian observations of the same kind, but in China it certainly did not operate. Nevertheless one might perhaps have expected some knowledge of the crystals earlier in Northern Europe.

To Europe we now return. After the New Year's letter of Kepler, the next advance was due to Descartes, who in his tractate *Les Météores et la Géométrie*[1] gave drawings of snow-flake crystals based on observations which he had made in + 1635. These, though very diagrammatic, are infinitely better than those of Olaus Magnus nearly a century earlier. A still greater advance in representation was made by Erasmus Bartholinus in his *De Figura Nivis Dissertatio*[2] of + 1660. Here for the first time, the branching of the hexagonal stars was shown, though not quite correctly. Advances now came rapidly. Five years later Robert Hooke printed his observations in *Micrographia*,[3] almost certainly the first which had been made with a microscope; and ten years later that remarkable naturalist, Friedrich Martens, who went to the Arctic as a ship's barber on a whaler, published his classified observations of the crystals in a very meritorious work.[4] Martens was the first to take meteorological observations at the time of the collection of the specimens. If Bartholinus in + 1660 was the first to draw the branching stars, the Italian, Donato Rossetti, a Canon of Livorno, was the first to draw in detail the hexagonal platelet type of crystal. This was just twenty-one years later.[5]

During the eighteenth century progress slowed down again. There were many writers who produced many drawings (often erroneous) and little advance was made. J. K. Wilcke in 1761 seems to have been the first experimentalist to make snow crystals artificially, and this bore fruit in investigations in the following century, involving iodoform and camphor as nucleating agents.[6] Neither those who thus pioneered the seeding of clouds, nor those who ultimately succeeded in producing 'artificial' rain thereby, could have known that Chu Hsi in the + 12th

[1] Leiden, 1637; also in *Oeuvres* (Paris, 1902), vol. 6, p. 298. Hellmann, pp. 13, 50.
[2] First published, Copenhagen, repr. The Hague, 1661. Hellmann, pp. 14, 50 ff.
[3] London, 1665. Hellmann, pp. 14 ff., 51 ff.
[4] *Spitzbergische oder Groenlandische Reise Beschreibung* (Schultzens, Hamburg, 1675). Hellmann, pp. 15 ff., 51.
[5] *La Figura della Neve* (Turin, 1681). Hellmann, pp. 15 ff., 52.
[6] See the reports of Dogiel and Spencer. For the most recent work, which shows that particles of clays and other minerals from the earth's surface are more probably the cause of precipitation than meteoric dust, see Mason & Maybank. Chu Hsi's gypsum is one of the effective ice-nucleating agents.

century had already made the essential comparison between crystals of snow and hexagonal crystals of a salt.

With this we approach the modern period, and it is extremely interesting to find that upon its threshold remarkable work was done in East Asia once again, still almost, if not quite, out of touch with the main scientific tradition proceeding

Fig. 14. Six drawings of snow-flake crystal forms taken from a page of Doi Toshitsuru's *Sekka Zusetsu* (1832).

in the West. The foundations of our modern knowledge were laid by William Scoresby, who as a result of his travels in the Arctic just before 1820 drew up the first systematic classification of the forms of snow-flakes.[1] Scoresby was the first to describe the columnar and complex forms of crystals, such as those in which spicules or prisms have one or both extremities inserted in the centre of a lamellar crystal, forms, in fact, which look like hexagonal platelets threaded upon a hexa-

[1] Hellmann, pp. 18 ff., 54 ff.

gona column. Scoresby was also the first to take careful observations of the temperature at which the snow-flakes formed, and noted the relationship to the shapes produced, a relationship which had already been suggested by Guettard in Warsaw in 1762, but not proved until the work of Fritsch in Prague in 1853 and later writers.[1]

Twelve years after the publication of William Scoresby's work, the feudal lord of Koga, a town in the old Shimōsa province of Japan, Doi Toshitsuru by name, published a most remarkable book entitled *Sekka Zusetsu* (Illustrated Discussion of Snow Blossoms).[2] This included eighty-six excellent sketches and was followed up by a supplement in 1839, *Zoku Sekka Zesetsu*.[3] The drawings of Doi Toshitsuru (1789 to 1848) are very nearly as good as those of James Glaisher in 1855 (twenty-three years later), the last great collection of pictures before the coming of microphotography.[4] Some of Doi's pictures are reproduced in the recent book of Nakaya Ukichiro.[5] Doi Toshitsuru was a learned *daimyō*, who had as his chief retainer a 'Rangaku' scholar (i.e. one who was interested in the so-called Dutch learning, or modern science), Takami Senseki. This man had studied under Kawaguchi Nobutō, one of the pioneer scientists of Japan, and both of them were active between 1812 and 1832. It is highly probable that this group used compound microscopes. One such instrument had been employed a little earlier by another outstanding pioneer of Japanese science, Ono Ranzan (1729 to 1810), probably about 1799, also for the study of snow crystals, but his results were never published. Unfortunately the Japanese work remained unknown to Hellmann,[6] who at the end of the nineteenth century embodied in his book the most complete history of the knowledge of snow crystals which we possess even today.

Perhaps one might be inclined to see in this parallel development of knowledge of snow-flakes in East and West an epitome of the difference between the European and Chinese social environments. The Chinese began very early indeed with a sound observation, but it was allowed to become a commonplace and relatively little development occurred through the centuries. In Europe, on the other hand,

[1] Hellmann, pp. 19 ff., 57. Modern work on this is described by Mason.
[2] Cf. Nakaya, p. 2; *SCC*, vol. 3, p. 472. [See Fig. 15, pl.]
[3] Both works have been reproduced in the *Nihon Kagaku Koten Zensho* series (Collection of Old Japanese Scientific Works), vol. 6.
[4] Hellmann, pp. 20 ff., 57; who himself, with Neuhauss, contributed a number of microphotographs. For subsequent descriptive studies so illustrated, see Dobrowolski; Bentley & Humphreys, and Nakaya Ukichiro.
[5] The last are dated in the winter of the third year of the Tempū reign-period, i.e. 1832.
[6] Still more curious is the fact that Tamura Sennosuke also ignores it.

nothing to speak of was known of this natural phenomenon until the Renaissance, after which, in accordance with the impetus of modern science, knowledge rapidly increased. If Chinese civilisation had been allowed to continue in its traditional form, it is reasonable to suppose that a slow and continuous further development would have occurred. The observations of Doi Toshitsuru form an interesting echo of modern science in the then relatively isolated sphere of Japanese culture. If they had involved only the use of magnifying lenses they could have been made in China almost any time back to the Thang or even the Han, but in fact they were probably accomplished with the aid of a microscope, one of the character-istic products of Renaissance science. In any case it is necessary to seek for a just historical perspective, and in this light the remarkably early Chinese knowledge of the hexagonal symmetry of snow-flake crystals ought to receive its meed of praise.

8

IRON AND STEEL PRODUCTION IN ANCIENT AND MEDIEVAL CHINA[1]

[1956]

THE object of this lecture is to place in the general setting of the comparative history of iron and steel technology the course of events in the Chinese culture-area. I believe that the more we discover about this development, the more we shall find that it differed from the course of history in the rest of the world.

Starting with dates generally accepted, I suppose we may assume that the Iron Age began about − 1200. In a famous cuneiform text of − 1275 Hattušiliš III, the king of the Hittites, referred to the iron which his people were making. By comparison iron came to the Chinese culture-area rather late. The word *thieh*, which means iron in all later centuries, originally meant the colour grey. Then in the − 8th century it was used as a place-name, perhaps referring to the finding of iron ore in such locations. One of the first mentions of iron refers to the making in the year − 512 of cauldrons of that metal upon which were inscribed codes of law. Legends associate the use of iron for swords and weapons with the king of Wu, about − 500, and his famous semi-legendary smiths, such as Kan Chiang. When we come to the − 4th century, there are numerous mentions of iron. They occur in the *Yü Kung*, a kind of catalogue of the products of all the different provinces of China, and in such classical texts as Mencius. The *Kuan Tzu* book also contains a famous passage in which it describes, while dealing with taxation, how every woman must have a needle and knife made of iron, every farmer must have a ploughshare, a hatchet and a hoe, and every wheelwright a saw, a bradawl and shears.

It is a remarkable fact that in the history of iron and steel in China we come so soon upon the preparation of cast iron. Actual cast-iron tools and moulds have

[1] An epitome of the Second Dickinson Biennial Memorial Lecture: Presented at the Science Museum, London, on 9 May 1956. Reprinted from *Transactions of the Newcomen Society*, **30**, 141 (1955–6 and 1956–7). The full text of the Lecture was published as a separate monograph by the Society in 1958, repr. Heffer, Cambridge, 1964. A French translation appeared in *Rev. d'Hist. de la Sidérurgie*, 1961, **2**, 187, 235; 1962, **3**, 1, 62.

been excavated in recent times from tombs of the − 3rd century. The names of eminent iron-masters of that time have come down to us, especially in connection with the early development of industrial production. Kuo Tsung, Cho Shih, and Khung Shih are referred to as such in the *Shih Chi* (the 'Historical Record', the first of the Chinese dynastic histories) written by Ssuma Chhien about − 90. The philosophical writer Hsün Chhing, about − 250, refers to steel, in a passage where he says the people of the state of Chhu use shark skin and rhinoceros hides for armour, as hard as metal or stone, and spears of steel from Wan, as sharp as a bee's sting. Dating from the − 2nd century there are many more references, to deserters from a Chinese Embassy, for instance, who went to teach the making of cast iron to the people of Western Turkistan. After − 100 we have a couple of references to explosions of small blast-furnaces. Actual specimens of cast iron come to us from the + 1st or + 2nd century, such as the cooking-stove described by Laufer; and then, from the early Christian centuries onwards, we have cast-iron statues in considerable numbers [cf. Figs. 16, 17, pls.]. One of the largest ever cast anywhere in the world was made in the + 10th century in China.

One of the most interesting facts about the history of iron in China is that the commonest words for the technique of handling it all mean flowing, liquid metal; for example, *chu*, *yeh*, *shuo* and *hsiao* correspond to the Latin *fundo*, rather than to *excoquo*. These are sharply to be distinguished from the word *tuan*, which means to forge, or *hsieh*, which means a bloom. In other words, iron could be melted and cast in China almost as soon as the metal was known. This is an extraordinary fact, and immediately serves to set on one side the history of iron and steel technology in China from that in the rest of the Old World.

When we seek the causes which brought this about, we may find several. In the first place, there is the high phosphorus content of some iron ores in China. As is well known, this has the effect of reducing the melting point of the ore in the blast furnace. Secondly, it is clear that in some provinces, especially Shansi, there were rather good refractory clays for making crucibles, so that the manufacture of iron by crucible methods was possible from an early date. There is evidence that in some places pieces of bloom were separated by hand, after the cooling, from pieces of higher carbon content which were re-melted to make steel. Thirdly, it is certain that coal was used directly for smelting iron, at least from the + 4th century. If this use was combined with a crucible process, a very hot pile would be produced and the sulphur from the coal would not be likely to enter the contents of the sealed crucibles. A fourth reason why iron-casting developed so early in China was, no doubt, the use of the double-acting piston-bellows, the existence of which is well attested in the + 1st century. This could supply a continuous blast, which undoubtedly would have aided the attainment of higher temperatures than was possible in Europe. We do not know exactly how the metallurgical blowing-engines of the Han dynasty functioned, but we do know how Chinese historians of the + 14th century thought they had functioned. Reasoning from the machinery of their own times, they recorded the existence of horizontal water-wheels driving, by means of a belt, an eccentric lug arrangement which converted the rotary into longitudinal motion, and so permitted the harnessing of water-power to double-acting piston-bellows. That water-power was used for this purpose as early as the + 1st century is clear from texts taken from the dynastic histories.

Coming now to the making of steel, one must first point out that there is no difficulty or uncertainty about the technical names for cast iron (*sêng thieh*), wrought iron (*shu thieh* or *jou thieh*), and steel (*kang* or *kang thieh*). Wrought iron was certainly known throughout Chinese history, and we have illustrations of a kind of puddling process taking place on an open platform, with the addition of silica, and in a position near the blast-furnace. Technological encyclopaedias, such as the *Thien Kung Khai Wu* of + 1637, show the forging of gongs, of anchors and other large wrought-iron objects, and the making of steel needles from steel wire.

Now it is very remarkable that the evidence for the cementation process of steel-making, namely, the packing of wrought-iron ingots with charcoal so as to get the right addition of carbon, seems to be almost lacking in Chinese history. On the contrary, from the + 1st century onwards, we constantly read about what is called the harmony of the hard and the soft (*kang juan chih ho*). Sometimes this means the welding of soft and hard steels together; but elsewhere it undoubtedly refers to a process which one is tempted to call 'co-fusion'. By this I mean the piling up together and heating of billets of wrought iron and cast iron, with the object of obtaining a material, namely steel, which we now know has an intermediate carbon content. It was in fact a method of averaging out the carbon content of wrought iron and cast iron. The term co-fusion is unsatisfactory, because it is fairly sure that at the temperatures attainable the wrought iron did not melt, although it doubtless softened, while the cast iron did melt, and washed or bathed the billets of wrought iron. I have not, so far, succeeded in finding a satisfactory term for this process, but perhaps we might speak of 'co-lavation'. We really need a new term. Now it is certain that this technique, which clearly foreshadowed the Siemens-Martin process of combining cast and wrought iron, dates from at least the + 6th century in China. At the beginning of that century, Thao Hung-Ching refers to the process as the mixing and melting together, and heating, of the raw and the soft (*tsa lien sêng jou*), which is certainly a reference to the process of co-fusion. Then, about + 550 was written a circumstantial account of an ironmaster of one of the northern, foreign dynasties—that is to say, dynasties ruling over parts of China under the aegis of a ruling house which was Turkic or Hunnic—a man whose name was Chhiwu Huai-Wên. According to

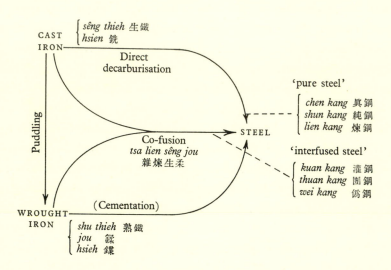

the account (p. 94), he heated the two kinds of iron together continuously for some days, with the result that a transference of the carbon took place, and by repeated forging afterwards, steel was obtained. When we come to the + 11th and + 12th centuries, we find numerous descriptions of this procedure. It is very interesting that experiments recently made in England, at Corby in Northampton-shire, have confirmed that it is indeed possible to average the carbon content of the two kinds of iron, if they are heated together and subsequently forged under appropriate conditions.

Steel made by the co-fusion process was called 'suffused, or interfused (lit. irrigated), steel' (*kuan kang*), or 'lump steel' (*thuan kang*), and sometimes 'false steel' (*wei kang*), probably because of its lesser homogeneity than the kind made by direct decarburisation. An early (+ 7th-century) name was *thiao thieh* ('jumped iron'), probably because of the forging of the billets after the co-fusion. A late name was 'wrought (-iron) steel' (*shu kang*). Steel made by the direct decarburisation of cast iron under cold blast was called 'transformed steel' (*lien kang*), or 'pure steel' (*shun kang*), and by implication 'true steel' (*chen kang*). The process was often known as the 'hundred refinings' (*pai lien*). A late name was 'cast (-iron) steel' (*sêng kang*). It seems as if the direct decarburisation must have been something like the 'mazéage' of the Belgian iron workers of the late Middle Ages, and something like the process by which the nail-makers of Birmingham converted cast iron directly into steel. In the +17th century a strange account by a Dutch traveller in Japan tells of blowing a cold blast into heated cast iron, almost as if he was describing a precursor of the Bessemer process. This description is not yet fully understood, but we hope that light will be thrown upon it by further study of the texts.

A feature of Chinese and Japanese iron and steel technology, which is much better known than those to which I have hitherto referred, is the welding together of hard and soft steels to produce sword blades of most remarkable properties. This has been described by many writers (such as Chikashige, Joly & Inada, Inami, etc.). Like most other Japanese techniques, it is possible to trace it back to its origins in China, probably as far back as the +3rd century. Chhiwu Huai-Wên certainly knew about it, and probably practised it, but by the time of Shen Kua, in the +11th century, it had already become a speciality of the Japanese. Evidently the procedure was not so well carried out in China; at any rate, the Japanese sabres were then preferred by connoisseurs. This question of welding naturally raises great problems concerned with origins and transmission, since, as we know from many investigators, particularly from the group at Nancy, this welding of steels was practised by the late Celtic and Merovingian swordsmiths in the +5th and +8th centuries. It would be very interesting to know from what part of the Old World this procedure radiated, and no doubt further research will throw more light on the subject.

I am afraid that there is no space in this résumé to say more about such interesting questions as the Chinese methods of quenching and tempering; nor can I deal with the question of the use of the trip-hammer in Chinese forges, and the application of water-power to it. There has not been space to say anything about the history of wire-drawing; or about the methods of testing the iron and steel products; or on the importation of wootz steel from India to China, for which we have evidence as early as the +5th or +6th century.

To sum up: I think it is clear that systematic work on the history of science and technology in China, carried out by the thorough examination of texts that have,

III

so far, either been unknown or misunderstood by sinologists and scholars of literary culture, will show that the evolution of iron and steel technology in the Chinese culture-area followed a course rather different from that of the Far West. In many ways, it was much more advanced in the earlier periods. It has often been said that the Chinese farmer was ploughing with a cast-iron ploughshare long before the farmers of the West knew anything else but wood; but that he continued to plough with a cast-iron ploughshare long after the farmers of the Far West had ploughshares of steel. One finds a very early appearance of the mastery of cast-iron technology. From the + 6th century onwards clear descriptions exist of the making of steel by co-fusion of cast and wrought iron. It seems, too, that there was a tendency to replace cementation by the direct decarburisation of cast iron. There is also the very important technique, carried to great perfection by the Japanese, of the welding of hard and soft steels to produce remarkable sword blades. Perhaps, after all, in view of the circumstances, the early appearance of cast iron is not so difficult to understand. In the − 2nd millennium there were extremely brilliant Chinese bronze-founders, and it is natural that they should have been on the look-out for fusible metal of another kind. Nevertheless, their masterly development of steel-making by advanced methods, so long before the rest of the world, is an achievement of very great interest to the history of technology in general.

9

CLASSICAL CHINESE CONTRIBUTIONS TO MECHANICAL ENGINEERING[1]

[1961]

I

SOMEBODY has said that the history of science and technology is the palladium of our freedom. As I understand this, it means that when we know something of the origins and way of development of our assured knowledge of the natural world, we are delivered from the bondage of certain preconceived opinions otherwise very easy to acquire. We understand better the limited value, and limited life, of all hypotheses; we see how misunderstandings arose and were resolved; and we become able to put the achievements of our own time in a more just and proper historical perspective. Among the preconceived ideas which can thus be rectified is one very persistent belief, namely, that Western civilisation has a penchant for mechanical invention not shared by any other culture. On the contrary, the more we know of the other great classical civilisations of the Old World, the more we realise that they had a vital part to play in the matter, and that some of them were indeed much more advanced than Europe between the − 2nd and the + 15th centuries. Thus it is that I am to talk to you this evening about the contributions of classical China to mechanical engineering.

Before the river of Chinese science flowed, like all other such rivers, into the sea of modern science, there had been remarkable achievements in mathematics.[2] Decimal place-value and a blank space for the zero had begun in the land of the Yellow River earlier than anywhere else, and decimal metrology had gone along with it. By the + 1st century Chinese artisans were checking their work with sliding calipers decimally graduated. Chinese mathematical thought was always profoundly algebraic, not geometrical, and in the Sung and Yuan periods (+ 12th to + 14th centuries) the Chinese school led the world in the solution of equations; so that the triangle called by Pascal's name was already old in China in + 1300. As for astronomy, I need only say that the Chinese were the most persistent,

---

[1] Forty-first Earl Grey Memorial Lecture, delivered at King's College, Newcastle upon Tyne, 1961.
[2] SCC, vol. 3, pp. 1–168.

113

successful and accurate observers of celestial phenomena before the Renaissance.[1] Although geometrical planetary theory did not develop among them they conceived an enlightened cosmology, mapped the heavens using our modern coordinates, and kept records of eclipses, comets, novae and meteors, still useful, to radio-astronomers, for example, today. A brilliant development of astronomical instruments also occurred, including the invention of the equatorial mounting and the clock-drive; and this development was in close dependence upon the contemporary capabilities of the Chinese engineers. Towards the end of this lecture we shall have something to say about this field. Such skill affected also other sciences such as seismology,[2] for it was a Chinese man of science, Chang Hêng, who built the first practical seismograph in +132.

Three branches of physics were particularly well developed in ancient and medieval China—optics, acoustics and magnetism. This was in striking contrast with the West, where mechanics and dynamics were relatively advanced but magnetic phenomena almost unknown. Yet China and Europe differed most profoundly perhaps in the great debate between continuity and discontinuity, for just as Chinese mathematics was essentially algebraic rather than geometrical, so Chinese physics was faithful to a prototypic wave theory and perennially averse to atoms.[3] One can even find such contrasts in minor preferences in the field of engineering, for whenever an engineer in classical China could mount a wheel horizontally he would do so, while our forefathers preferred vertical mountings, as witness typically the water-mill and the windmill.

Always one has to consider the Chinese contributions in relation to the achievements of other civilisations. It is already quite clear that in the history of science and technology, the Old World must be thought of as a whole. But when this is done, a great paradox presents itself. Why did modern science, the mathematisation of hypotheses about Nature, with all its implications for contemporary technology, take its meteoric rise only in the West, at the time of Galileo? This is the most obvious question, which many have asked but few have answered. But there is another of quite equal importance. Why was it that between the −2nd and the +15th centuries East Asian culture was much more efficient than that of the European West in applying human knowledge of Nature to useful purposes? Only an analysis of the social and economic structures of Eastern and

---

[1] *SCC*, vol. 3, pp. 171 ff.   [2] *SCC*, vol. 3, pp. 624 ff.
[3] *SCC*, vol. 4, pt. 1, pp. 3 ff., 126 ff., 229 ff. Also J. Needham & K. Robinson, 'Ondes et Particules dans la Pensée Scientifique Chinoise', *Sciences*, 1960, **1** (no. 4), 65.

Western cultures, not forgetting the great role of systems of ideas, will in the end suggest an explanation of both these patterns.

Here we cannot occupy ourselves with such high matters, but rather with more concrete things, inventions and devices not unworthy of discussion, I hope, in this university and city. I propose to begin with the water-wheel and the water-mill, out of which will come some unexpected conclusions about the lineage of the steam-engine; for in China water-mills were older and more important for blowing metallurgical bellows than for grinding corn. We must make a distinction, at the outset, between ex-aqueous wheels, those which were turned by the force of water's flow and fall, and ad-aqueous wheels, those which imparted motion to water, or brought about motion over water, by the application of some other force. Mill-wheels were still ex-aqueous when they were mounted on ships moored in a current, but their position was now ambiguous in that it invited the transition to the ad-aqueous arrangement and the consequent birth of the automotive paddle-wheel ship. But besides mills that were mounted on boats there were also in medieval China mills that told the time. For the missing link between the clepsydras or leaking-vessel water-clocks general in antiquity and the fully mechanical clocks of + 14th-century Europe arose in China, when an escapement mechanism was first applied to a water-wheel with buckets fed from a constant-level tank of water or mercury. To explain these remarkable developments will take all the time at our disposal.

II

A word or two, to begin with, about the first origins of the water-wheel. Nobody knows what they were, but in East Asia there have survived until the present day certain curious pieces of apparatus which we may call spoon tilt-hammers, and these might be an ancestral form. Here the spoon filled with a continuous stream of water empties periodically, operating as it does so a tilt-hammer. This is the *tshao-tui*, and the oldest Chinese picture of it dates from + 1313. It comes from a compendium of agriculture and rural engineering which I shall refer to henceforward simply as the *Nung Shu*, compiled by Wang Chên; and we may think of its contents as at least of + 1300, for what was described in some detail thirteen years later was probably valid for at least a century earlier. This exceedingly simple apparatus is archaic and probably very old. It survives down to the present day, and Troup at the beginning of this century found many examples in Japan, where the number of spoons had increased to two, and then to four, and to six or eight, forming already a bucket-wheel. The simple device of the tipping bucket runs through all medieval engineering and was especially popular among the Arabs. In the eighteenth century (+ 1736) a forge-blower was built in Sweden by Martin Triewald using the same arrangement of buckets, in this case to depress alternately two bells in water which forced the air into the blast.

Now in the medieval West the typical stamp-mill was a vertical apparatus, as we know from drawings such as those in the manuscripts of the anonymous Hussite engineer (+ 1430), but in China water-power was very early applied to a quite different type of stamp-mill with recumbent hammers, known as a *shui-tui*. The hammers are lifted one after the other as the lugs on the driving-shaft come round. The first reference to this simple machine occurs in + 20 in an essay by Huan Than, which demonstrates that already by that time water-power had been applied to such pounding. Most probably a vertical water-wheel was here used from the start on account of its greater simplicity, though in fact the most widely used type of water-mill in China later on had, and has, horizontally mounted water-wheels. I became very familiar with the *shui-tui* during my travels in the Chinese countryside, and many photographs are available to illustrate living examples. One can see at once how these ancient machines are the lineal ancestors of the martinet forge-hammer of modern times by comparing them with the famous picture in Diderot's encyclopaedia (+ 1765). In the + 12th century Lou Shou wrote a poem about these machines, then used for hulling rice, which contrasts rather charmingly with the + 18th-century industrial picture. He wrote:

> The graceful moon rides over the wall,
> The leaves make a noise 'sho, sho, sho' in the breeze;
> All over the country villages at this time
> The sound of pounding echoes like mutual question and answer;
> You may enjoy at your will the jade fragrance of cooking rice
> Or watch the water flowing in and out of the slippery spoon
> Or listen to the water-worn wheel industriously turning.

We noted just now that the typical water-mill in China remains to this day the horizontal form, quite similar to that known in Europe as the Norse or Shetlands mill.[1] The blades of the paddle-wheel were generally inserted slanting in Europe, but not in China, where, however, they are 'shrouded', i.e. enclosed

---

[1] P. N. Wilson, *Water-mills with Horizontal Wheels*. Society for the Protection of Ancient Buildings, London, 1960 (Water-mills Committee Booklets, no. 7).

by a rim, and not free-standing. This most typical water-mill of China is of course depicted in the *Nung Shu* about + 1300. When I was in China again two years ago in Szechuan and Kansu, I was able to study a number of such mills, climbing down underneath to examine the construction. Flumes or nozzles are always provided for horizontal water-wheels. This type of mill-work had a rather important future before it, because there is good reason for thinking that the turbine developed from the horizontal rather than from the vertical wheel, through Besson's tub turbine of + 1563. In the West, the Vitruvian vertical water-wheel, with its right-angle millgear (so called because of its description in Vitruvius' great book written about − 25, only a few years before the essay of Huan Than mentioned just now), was always more widespread and more familiar. Nevertheless all through the Middle Ages the Chinese also used vertical wheels, and about + 1300 the *Nung Shu* shows an example of nine sets of mill-stones connected by gearing and worked from a single vertical wheel. Something which surprises the economic historians is the application of water-power at this time in China, indeed probably throughout the thirteenth century, to textile machinery. The *Nung Shu* shows a silk-throwing mill or a spinning-mill for hemp or ramie powered by a vertical water-wheel, and the text assures us that it was common practice. China may have been a land of abundant labour but Chinese culture never seems to have rejected labour-saving inventions.

III

I turn now to a different subject altogether, the use of water-power in China for metallurgical blowing-engines. Few Westerners realise it, but in fact before the last four or five centuries China was the iron-age culture, not the West.[1] No-one in Europe could get a small pig of cast iron for love or money before about + 1380, but in China iron was being systematically cast on an industrial scale (if one can use the word for those days), in the − 4th century. Quite apart from impressive textual evidence, many objects of cast iron have been recovered during the past few decades from tombs of this period as well as those of the Han and later, mostly agricultural implements and moulds for implements. We do not know whether these were for bronze or some other alloy, or for casting cast iron itself, as is quite possible.

When one reaches the Middle Ages, one begins to find good pictures of small

---

[1] J. Needham, *The Development of Iron and Steel Technology in China* (Second Biennial Dickinson Lecture), Newcomen Society, London, 1956.

blast-furnaces. The oldest known to me comes from a book entitled *Ao Pho Thu Yung* (The Boiling down of the Sea), written in + 1334 by Chhen Chhun, and illustrated by Chhü Shou-Jen and his brother. It refers to the salt industry, but salt and iron always went together because the salt people needed great pans for the evaporation of the brine, and these were made of cast iron. Alongside this + 14th-century Chinese blast-furnace, one can see the blowing equipment in the form of a fan- or piston-bellows. All through the Middle Ages China's command of iron manifests itself—whether in the statue of the Goddess of Mercy, dated

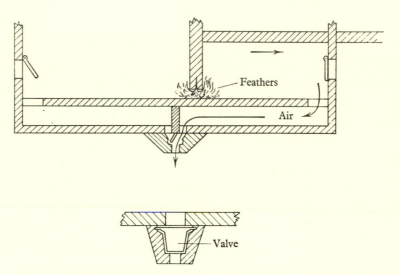

Fig. 18. Longitudinal section to show the action of the Chinese double-acting piston-bellows (after Hommel).

+ 550, under the Northern Chhi dynasty, or the Great Lion of Tshangchou, cast in + 954 to commemorate the victories of the Later Chou dynasty against the Liao Tartars. In + 1061, just about the time of our William the Conqueror, the Chinese were actually making pagodas of cast iron, such as that which still exists at Tang-yang in Hupei. Moreover, iron was so plentiful that it was used not only for pagoda building but also for roofing temples. On the top of the sacred mountain Thai-shan, where winds are very severe, a number of the temple halls were roofed with cast-iron tiles early in the + 15th century in order to avoid the trouble of constantly repairing their roofs.

Now one of the factors (certainly not the only one, but an important one), which made possible the early development of iron-casting, was undoubtedly the double-acting piston-bellows which the Chinese developed. A diagram of it is shown in Fig. 18. By a simple but ingenious arrangement of valves a continuous blast can be obtained. This form of bellows is still universally used by smiths of all kinds in China and one becomes very familiar with it there. If the piston was hinged and followed an arc in its travel, as in some of the 'fan-bellows' of later China and Japan, the basic principle was in no way affected.

The most ancient textual reference to the casting of iron by the use of water-powered bellows occurs in +31. The *Hou Han Shu* (History of the Later Han Dynasty) says:

In the second year of the Chien-Wu reign-period Tu Shih was posted to be Prefect of Nan-yang. He was a generous man and his policies were peaceful. He destroyed evil-doers and established the dignity of his office. Good at planning, he loved the common people and wished to save their labour. He invented a water-power blowing-engine (*shui-phai*), for the casting of metal (iron) agricultural implements. Thus the people got great benefit for little labour and it was widely adopted and used.

Exactly what apparatus was used by Tu Shih for this purpose we do not know, but it may have been an arrangement of lugs on a shaft exactly as for the trip-hammers, only acting upon a set of crescentic buffers, and operating bellows which were returned to position by a series of strong bamboo springs. If so, it probably depended on a vertical water-wheel. Such a design was still described by Wang Chên about +1300; but his other design, apparently in common use by that time, was much more important and of more interest for us. As it is good to quote the actual words of his text, I shall read it as follows. He said:

According to modern study, leather bag bellows were used in the olden times, but now they always use a wooden fan (or piston). A place beside a rushing torrent is selected, and a wooden shaft set up in a frame with two horizontal wheels, the lower one being rotated by the force of the water. The upper one is connected to a 'winding-wheel' in front by a driving-belt. When the main wheel moves, all the parts follow it, including the 'moving spindle' carrying the eccentric lug. Thus the 'horizontal axle' (i.e. the connecting-rod) is pushed and pulled to left and right by the 'grasping ear' (i.e. the lug), and the straight piston-rod comes and goes back and forth, operating the furnace bellows much more quickly (and violently) than would be possible with man-power.

The construction of this machine can be understood quite clearly, especially when the picture in the *Nung Shu* (Fig. 34, p. 163) is compared with similar representations in later Chinese books (Fig. 19). A diagrammatic reconstruction may also be found useful (Fig. 40, p. 179). The horizontal water-wheel drives

Fig. 19. The traditional Chinese blowing-engine for forge and furnace (*shui-phai*, water-powered pusher, or reciprocator), from the *Thu Shu Chi Chhêng* encyclopaedia (+1726).

an upper wheel connected by a crossed driving-belt with a small pulley carrying an eccentric lug; this is attached to a connecting-rod, and that in turn works the piston-rod of the bellows through a rocking roller or bell-crank arrangement. Here then we have that whole system of eccentric, connecting-rod and piston-rod, which constitutes what might be called the standard method of converting rotary to longitudinal motion. The oldest European picture of forge or blast bellows worked by water-power occurs in the book *De Gentibus Septentrionalibus* written by Olaus Magnus in +1565 (Fig. 51), and here the simple trip-lug type of bellows is still at work. However, by +1588 the picture in Ramelli shows bell-crank rocking rollers and eccentrics (cranks) essentially similar to those in the Chinese design (Fig. 52, pl.). The basic principle remains the same in John Wilkinson's blowing-engine design of +1757, the only difference being that there is now a true crank-shaft.

When we carefully consider this combination of crank or eccentric, connecting-rod and piston-rod, which developed in China by +1200, certainly by +1300,[1] we realise with astonishment that it was really the 'morphological' equivalent of the reciprocating steam-engine. But it was working the opposite way, for water-power was bringing about rotary motion and that was being transferred into rectilinear motion for the piston-bellows. In the steam-engine the problem was to transmit the power of steam applied to the piston through the piston-rod and connecting-rod so as to give continuous motion to flywheel and shafting. The pattern thus being identical, one might almost say that the great 'physiological' triumphs of 'ex-pistonian' Europe were built upon a 'morphological' foundation which had been developed five or six centuries earlier in 'ad-pistonian' China. If there was any genetic connection this is the place to look for it rather than in the use of steam jets or aeolipiles. It is true that the Chinese blowing-engine has a bell-crank rocking roller intervening between the connecting-rod and the piston-rod, but this does not affect the logical situation, and the device may indeed be considered an interim solution in the search for stability ultimately attained in the slide guides and cross-head. One may also reflect that the beam of the beam-engine was simply a straightened bell-crank, so that as long as beam-engines lasted (and that was almost down to our own time) Wang Chên's design was perpetuated in every single detail. The only thing lacking in +13th-century China was the true crankshaft, and although that was a European invention of the +15th century, it did not appear in a blowing-engine anywhere until the time of John Wilkinson four centuries later.

Let us now have a new look at the general history of the interconversion of continuous rotary and longitudinal or rectilinear reciprocating motion. The oldest examples of this achievement are no doubt the bow-drill, the pump-drill and the pole-lathe, but they all involved non-continuous belting, and they did not lead very far. Then there were the lugs on the rotating shaft, with springs to ensure

[1] [As we now know, by +900; cf. p. 186 below.]

return travel. But Bertrand Gille is perhaps only stating the half of a half-truth when he says that the sole way of effecting this conversion known to the Middle Ages involved springs.[1] He was of course thinking of Villard de Honnecourt's

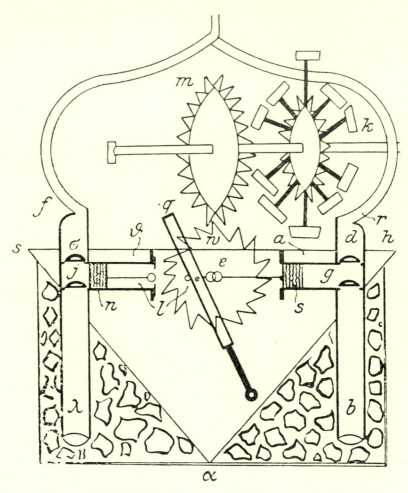

Fig. 20. Al-Jazarī's slot-rod pump worked by a water-wheel (+1206).

saw-mill (c. +1250), where the lugs on a water-powered shaft operated the saw, and a spring returned it to its starting position at each stroke. There are, however, other medieval machines which are really more interesting in this connection, though not mentioned by Gille. First I should like to refer to the slot-rod water-

[1] B. Gille, 'Machines (in the Mediterranean Civilisations and the Middle Ages)', in *A History of Technology*, ed. C. Singer *et al.*, vol. 2, pp. 629 ff. (p. 652), Oxford, 1956.

pump described by the great Arabic engineer Ismā'īl ibn al-Razzāz al-Jazarī in his book *Kitāb fī Ma'rifat al-Ḥiyal al-Handasīya* (The Knowledge of Ingenious Mechanical Contrivances) finished in +1206. A re-drawing from one of the manuscripts[1] is given in Fig. 20. Here the vertical water-wheel (*k*) rotates a large gear-wheel on the same shaft (*m*), engaging with another gear-wheel (*n*). The shaft of this is fitted not centrally but eccentrically, being carried on a kind of universal joint at one end and running round in an annular slot at the other.[2] And this unusual eccentric, instead of being attached to a connecting-rod, slides up and down within a slotted rod (*q*), attached to a fixed pivot below and having at each side midway along its length connections with two piston-rods. Thus as the lower wheel rotated, the slot-rod was forced alternately to left and right, moving the pistons successively back and forth. This was a fine piece of mechanism for the early +13th century, but it was not as direct an ancestor of the reciprocating steam-engine's connecting-rod system as the Chinese machines, which were certainly in practical use at the same time, +1200.

Another machine which demands consideration among the ancestors of the steam-engine is the Chinese silk-winding or reeling apparatus. We know that this goes back to the +11th century because it is fully described in a text of that date, the *Tshan Shu* (Book of Sericulture) by Chhin Kuan (d. +1101). Let us look at a picture of the *sao-chhê* (Fig. 21), taken from one of the traditional Chinese eighteenth- and early nineteenth-century treatises on the silk industry such as the *Tshan Sang Ho Pien* (Collected Notes on Sericulture), by Sha Shih-An, Lu I-Mei and Wei Mo-Shen. Here we have the crank or eccentric and the connecting-rod, but no piston-rod. The silk-worm cocoons are unrolled in the hot-water pan on the left, and the fresh fibres come up through 'eyes' on a frame and over a set of rollers towards the large reel on which they are to be wound. In order to get them laid down regularly on to this reel they are moved laterally back and forth by a 'ramping-arm', the ancestor of the 'flyer', which is operated automatically. The pedal arrangement rotates the reel by means of a crank, and the reel's axis is connected by a driving-belt with a small pulley bearing an eccentric lug, which oscillates the ramping-arm back and forth.

Here then we have two of the components of the steam-engine's pattern, but

---

[1] E. Wiedemann & F. Hauser, 'Über Vorrichtungen zum Heben von Wasser in d. Islamischen Welt', *Beitrage z. Geschichte d. Technik u. Industrie*, 1918, **8**, 121.

[2] [Such, at least, is the reconstruction of Professor Aubrey Burstall, which certainly works, but other friends such as Dr A. G. Drachmann and Mr T. Schiøler prefer the simpler eccentric lug. The text is not entirely clear. See *SCC*, vol. 4, pt. 2, p. 382.]

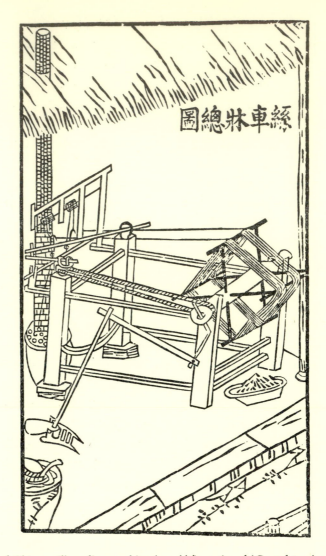

Fig. 21. The classical Chinese silk-reeling machine (*sao-chhê*) captioned 'Complete view of the frame of the silk machine'. Though taken from the sericultural treatise *Tshan Sang Ho Pien* of Sha Shih-An, published in 1843, this illustration corresponds in every detail with the description given in Chhin Kuan's *Tshan Shu* of *c.* +1090. An older picture, not quite so clear, appears in the first edition of Sung Ying-Hsing's *Thien Kung Khai Wu* (+1637). The silk is being wound off from the cocoons in hot water in the basin under the chimney on the left. A 'sewing-machine' treadle crank drive rotates not only the main reel but also an early form of flyer, the motion being communicated to the pulley and its eccentric by means of a driving-belt. The system therefore lacks only the piston-rod component of the standard inter-conversion assembly. But as it dates back in all probability a long time before +1090 it is a predecessor of the water-powered reciprocator of Figs. 19 and 34, perhaps even also of that in Fig. 45 (pl.), and hence of the definitive form of the steam-engine. In view of the high antiquity of the silk industry in China it would not be wise to exclude its possible dating to Han rather than Thang, Sung and Yuan times. The machine shown used coins with holes in their centres as the guides for the silk fibres on the flyer, but from the *Erh Ya* encyclopaedia (−3rd century) we know that the leaves of certain ferns were then used for guiding the fibres. How far the flyer of the Han *sao-chhê* was automated, however, we do not as yet know.

124

not all three. As we shall see in a moment, this limitation was still not transcended in +15th-century Europe. Yet the silk-reeling machine was fully developed by the +11th century in China. Moreover, in view of the great age of the Chinese silk industry, which was already well advanced in the Shang period (*c.* −14th century),[1] it is overwhelmingly probable that the silk-reeling machine is several centuries older than the time of the first textual description of it which happens to have come down to us. Of Villard de Honnecourt's +13th-century hydraulic saw-mill Lynn White has said that it was the earliest instance of a fully automatic industrial machine involving two separate but correlated motions (that of the saw and that of the feed).[2] It would thus mark an epoch in the development of mechanical devices. But I am not at all sure that the *sao-chhê* is not a better candidate for this honour, for it certainly had an automatic secondary motion. As we have seen, water-power was in wide application in China for textile machinery by +1300, and therefore with high probability in Villard de Honnecourt's time fifty years before. It is true that we do not know whether such a drive was applied at that time to this particular machine, but power of some sort it had, and the criterion of separate but correlated motions was fully satisfied.

When Leonardo da Vinci faces the problem of interconversion late in the +15th century he does everything he can to get out of doing it in the way which would naturally occur to us. The great German and Italian engineering manuscripts of the previous two centuries sometimes show eccentrics and connecting-rods, but they never have a piston-rod on the end. Leonardo knew that this combination could be constructed but he did everything to avoid it.[3] The only example of his use of the complete set of three component pieces is in a design for a saw-mill. His subterfuges were numerous. In one case he had a set of reciprocating alternating ratchets imparting a continuous winding motion to a rope or chain. Elsewhere he used a most ingenious method, a rod sliding in spiral grooves on what one might call a rotating cylindrical cam; and he found he could also gain his objective by link-work alone which was pushed back and forth by an arrangement of cams in rotary motion—finally he had recourse, as did so many +16th-century engineers, to the half-gear. I need not enlarge upon this; it was a way of converting continuous to alternating rotary motion and then converting that to

---

[1] Chêng Tê-Khun, *Archaeology in China*, vol. 2, *Shang China*. Heffer, Cambridge, 1960, pp. 198, 241.
[2] Lynn White, Review of the second edition of A. P. Usher's *History of Mechanical Inventions*. *Isis*, 1955, **46**, 290.
[3] T. Beck, *Beiträge z. Geschichte d. Maschinenbaues*. Springer, Berlin, 1900, esp. pp. 323, 417, 419, 421.

alternating longitudinal motion by means of windlass chains. One must emphasise again that the +15th-century engineers of Europe seem not to have known of the triple 'steam-engine' combination; Leonardo (*c.* +1480) did, but he used it very sparingly. Why this was so remains a puzzle, but quite probably the trouble was the excessive wear of the bearings. Cast iron was relatively new in Europe in Konrad Kyeser's time (+1400) but when Chhin Kuan was writing in China (+1090) it was already very old. Steel pins working in cast-iron bushes would have been just the thing for Wang Chên's *shui-phai* blower, and it may well have been lack of experience with the forms of iron which inhibited the development of the 'steam-engine' pattern in Europe. Tenth-century China was where it began.

IV

The steam-engine itself, a typically European post-Renaissance development, came to China in the 1840s. The first Chinese picture of a locomotive was published in Ting Kung-Chhen's *Yen Phao Thu Shuo* (1843), and about the same time Chêng Fu-Kuang, in his *Huo Lun Chhuan Thu Shuo*, also drew the first picture of a steam paddle-boat such as was then appearing in Chinese waters.[1] It will be remembered that at that time the British were propagating the narcotics trade against the objections of the Chinese government, and during the Opium War they used steam paddle-boats in the battles at the mouth of the Yangtze round the Wusung Forts. The Chinese actually brought against them (unsuccessfully) paddle-wheel war-junks operated by man-power tread-mills, and these, when captured and examined by British officers, were the subject of great interest. Everyone of course believed that they had been copied from the paddle-wheel steam-boats, and the Chinese were patronisingly congratulated on their imitative skill. Nevertheless this was a complete misunderstanding.

In beginning this chapter, I said a word about ship-mills, i.e. the ex-aqueous wheels of mills mounted on boats and moored in a current. This goes back to the historian Procopius, who tells us that when the Byzantine general Belisarius was besieged in Rome by the Goths in +536, they cut off the water-supply of the city's corn-mills. But he saved the situation by setting mill-wheels on boats and using the Tiber's current to work them. Such mills continued in use long afterwards, as we may see from the pictures of Giuliano di San Gallo (+1445 to

---

[1] Chhen Chhi-Thien, *Lin Tsê-Hsü, Pioneer Promoter of the Adoption of Western Means of Maritime Defence in China.* Dept. of Economics, Yenching Univ., and French Bookstore, Peiping, 1934.

+ 1516). The Danube and the Po, among other European rivers, were familiar with such craft down to our own time. But on Chinese rivers they were no less well known. I have not actually seen them there myself, but they are particularly prominent above the gorges of the Yangtze and especially in the neighbourhood of Fou-chou; Worcester has given us a scale drawing which shows the four mill-wheels and their machinery mounted in each ship.[1] There was of course another idea of mounting a vertical ex-aqueous paddle-wheel on a ship, and that concerned the log. After talking about the hodometer, Vitruvius in − 25 has a passage about mounting a wheel on a ship in order to record the distance sailed. This was probably tried from time to time, and Renaissance editions of Vitruvius have pictures of it, but it never worked well. Sir Isaac Newton himself made a report on the paddle-wheel log of Saumarez which was not very complimentary,[2] and even Smeaton tried it, without success. They had to wait for the advent of the screw. But let us return to the main thread of this discourse.

The earliest idea of an ad-aqueous wheel on a ship occurs in a Byzantine manuscript called the Anonymus' *De Rebus Bellicis*,[3] which is now dated *c.* + 370. The oldest drawing of it which we have comes from + 1440, but there is no doubt that the description was in the original manuscript and that its date is of that order. However, there is no evidence at all that the ship was ever constructed, and it seems that the manuscript was pigeon-holed in the archives of the Byzantine War Office and never put to any practical use. Not until + 1543 did Blasco de Garay construct in the Catalan port of Barcelona a treadmill-operated paddle-wheel tug, and contemporary illustrations show that the idea was in the air. But the idea had been in the air a great deal earlier in China, for references, more or less clear, to paddle-wheel boats actually constructed occur in Chinese texts from *c.* + 494 onwards. By the Thang dynasty the evidence for practical paddle-wheel boats becomes indubitable. Allow me to quote the actual words concerning Li Kao, Prince of Tshao, Governor of Hung-chou in + 783. The *Thang Shu* (History of the Thang Dynasty) says:

Always eager about ingenious machines, Li Kao caused naval vessels to be constructed, each of which had two wheels attached to the side of the boat, and made to revolve by tread-mills. These ships moved like the wind, raising waves as if sails were set. As for the method of construction, it was simple and robust so that the boats did not wear out.

---

[1] G. R. G. Worcester, *Junks and Sampans of the Upper Yangtze*. Shanghai, 1940 (Chinese Maritime Customs Publications, Series III (Misc.), no. 51), pp. 24 ff.

[2] H. R. Spencer, 'Sir Isaac Newton on Saumarez' patent log', *Amer. Neptune*, 1954, 14, 214.

[3] E. A. Thompson & B. Flower, *A Roman Reformer and Inventor*. Oxford, 1952.

Something very like Li Kao's patrol-boats is illustrated in the *Thu Shu Chi Chhêng* (Imperial Encyclopaedia) of +1726. European scholars who saw this picture all agreed that it must have been derived from the Jesuit mission in the +17th century or later. But they were quite mistaken, for we now have a mass of evidence about the use of treadmill-operated paddle-wheel warships in the Sung navy[1] starting about +1130.

At that time a serious rebellion under Yang Yao was being suppressed in the Southern Sung, and we learn from the texts that

one of the soldiers engaged in these operations, Kao Hsüan, who had formerly been Chief Carpenter of the Yellow River Naval Guard Force, submitted a specification for wheeled ships which he claimed could cope with the enemy. He first built an eight-wheeled ship as a model, completing it in a few days. Men were ordered to pedal the wheels of this boat up and down the river; it proved speedy and easy to handle whether going forward or backwards. It had planks on both sides to protect the wheels so that they themselves were not visible. Seeing the boat move by itself like a dragon, onlookers thought it quite miraculous. Gradually the number and size of the wheels were increased until large ships were built which had 20 to 23 wheels and could carry 200 or 300 men. The pirate boats, being small, could not withstand them.

However, very soon afterwards, the rebels captured a fleet of these and got hold of Kao Hsüan himself, so that for the rest of the operations he was building paddle-wheel battleships for them. In the end the rebellion was put down, but the fate of this remarkable naval engineer is not recorded. The odd numbers of wheels mentioned in many of the sources would be puzzling if we did not know of a Governor of Nanking, Shih Chêng-Chih, who constructed in +1168 a useful class of warship with one single twelve-bladed wheel, evidently a stern-wheeler. Thus Kao Hsüan's 23-wheelers doubtless had 11 wheels a side plus a stern-wheel. The reason for this multiplicity of wheels is fairly obvious: as long as they were not constructed of iron their size was severely limited, and one could only increase their number. The largest crew of pedallers mentioned for a single warship is about 200. No doubt these vessels were equipped with trebuchet catapults for throwing bombs and grenades, as well as a complement of crossbowmen and the usual marines for boarding-parties.

It is not generally known that these Chinese paddle-wheel boats lasted on down into modern times, and only a few years ago you could take passage on a

---

[1] Lo Jung-Pang, 'China's paddle-wheel boats; the mechanised craft used in the opium wars and their historical background', *Tsinghua Journ. Chinese Studies*, 1960, **2** (no. 1), 189.

pedalled paddle-wheel boat which ran from Shanghai to Suchow overnight, about 100 miles, with rather roomy sleeping accommodation.[1] Extant models show how three sets of men could pedal at the same time, the axles being connected with the stern-wheel shaft by a simple coupling-rod as on a locomotive.[2] Other tread-mill paddle-wheel boats were common on the Pearl River near Canton in recent times, where they were photographed thirty years since by Mr P. Paris. When I was in Canton two years ago I did my best to find some of these remaining antique paddle-wheelers but in spite of unstinting assistance by the local branch of the National Academy and the Cantonese Port Authority we could unfortunately find none of them left.

I would only like to add to the epic story of the Chinese paddle-boat the conclusion that it did very good service indeed in the century-long wars of the Sung against the Chin (Jurchen) Tartars (+1100 to +1230). Once the northerners had been driven back across the Yangtze River in +1130 they never succeeded in coming south of it again. The only serious danger was at the time of the Battle of Tshai-shih in +1161, which was a great victory for the Sung navy and its fleet of paddle-wheel warships. If the Sung had relied upon sail alone, the Chin forces could perhaps have got across in small boats on a calm windless night, but with the automotive paddle-wheel boats full of archers, trebuchet artillerists and marines, patrolling the Yangtze as they did, nothing could be done. As I have already indicated, gunpowder weapons were at that time already in use, especially in the form of bombs and rocket compositions,[3] so that the historical role of the Sung paddle-wheel warships was quite considerable. Nor is it hard to see the reasons for their rise and fall. As soon as the Yangtze became the frontier between the imperial State, the home of traditional Chinese culture, and the nomad-dominated north, the shipbuilders came into their own at once and a brilliant naval development followed. But when by +1280 the Sung empire was conquered and China unified by another nomadic people, the Mongols, under the Yuan dynasty, river and lake combat lost importance, while attention shifted to the sea. Here the paddle-wheel boat, in its pre-steam incarnation, was unsuitable, and sail once again came to the fore.

[1] [My friend the late Dr Victor Purcell remembered these very clearly.]
[2] [I thought at first that this must have been an introduction from Western engineering, but later found out that a coupling-rod is used on the crank-handles of the Chinese rope-maker's frame (Fig. 22, pl.). The device is therefore probably traditional and indigenous.]
[3] Wang Ling, 'On the invention and use of gunpowder and firearms in China', *Isis*, 1947, **37**, 160. Also J. R. Partington, *A History of Greek Fire and Gunpowder*. Heffer, Cambridge, 1960.

V

We reach now the last subject which I should like to discuss, namely the question of time-keeping. I have dealt with mills mounted on boats, and the results of them, and must now deal with mills that told the time. We are all familiar enough with the first mechanical clocks of the West, in the early + 14th century, with their verge-and-foliot escapement, pallets and crown-wheel. The problem which had to be solved was how to slow down a set of wheels sufficiently to allow them to keep pace with the apparent diurnal revolution of the heavens, humanity's primary clock. I do not need to remind you about the application of the pendulum to the same machinery by Galileo in + 1641 and Huygens in + 1673, nor to spend words on the equally familiar anchor escapement, probably introduced by William Clement about + 1680. What I have to describe is something quite different, and much older—the water-wheel link-work escapement developed in China from the early + 8th century onwards.[1] The great astronomical clocks built in China between + 700 and + 1400 had a complicated form of escapement which involved two trip-levers or weigh-bridges, the second one of which released a gate at the top of the wheel and allowed the next buckets to come forward.

Our greatest literary monument of Chinese horological engineering is the *Hsin I Hsiang Fa Yao* (New Design for an Armillary Clock) by the great scientist-statesman Su Sung in + 1090. It records the construction of an astronomical clock-tower for the Sung emperor at Khaifêng in the year + 1088, in which Su Sung was assisted by a brilliant engineer and mathematician Han Kung-Lien. The illustrations in this wonderful book show what the clock was like. There was an arrangement by which puppets came round on a series of levels, dressed in different coloured clothes, to announce the time by means of placards. On the first floor of the tower was a celestial globe continuously rotated by clockwork. On the roof there was an armillary sphere used for astronomical observations, made of bronze and about 20 tons in weight, to which a clock-drive was applied so as to form a kind of coarse adjustment. From the main reservoir, with its constant-level tank, water (or mercury) poured continuously into the driving-wheel buckets. The great wheel operated, by right-angle gearing, a driving-shaft which moved all the puppet-wheels on a separate shaft, at the top of which there was an oblique drive for the celestial globe; meanwhile at the top of the main shaft there was another oblique drive for the armillary sphere. One of the most

[1] J. Needham, Wang Ling & D. J. de S. Price, *Heavenly Clockwork, the Great Astronomical Clocks of Medieval China*. Cambridge, 1960. (Antiquarian Horological Society Monographs, no. 1.)

striking things about this whole instrument was the use of a chain-drive. The very long main shaft soon proved unsatisfactory, as could have been predicted, since it was certainly made of wood; so the Chinese engineers substituted for it an iron chain-drive, probably the first power-transmitting chain-drive in history. They used first a long one and then a shorter one, so that 'Mark III' was a great improvement on the original. The chain-drive was called a *thien-thi* or 'heavenly ladder', and the text describes the interlocking iron links. At the top it had a little gear-box. Historical evidence demonstrates that this great clock ran from + 1090 until + 1126, when the Chin army captured it, after which it was taken away to Peking. The Tartars, however, failed to capture the engineers, who escaped to the south, and got some second-rate technicians who did not succeed in making the clock work more than a couple of decades longer. However, that was a fairly long run for William the Conqueror's time.

As for 'winding up' Su Sung's clock, a couple of hand-operated norias raised the water into the reservoir every 24 hours, about a ton and a half of it, ready for the next day; and we suppose that a squad from the Ministry of Works came round to the Palace every night to carry out this task. It has been suggested that clock-towers of this kind may have been responsible for the very idea of perpetual motion, for Indian ambassadors or Arab merchants may well have seen these devices working several centuries before Su Sung's time, and not knowing how it was done (since the use of a free-running mill-stream of water does not occur in any description we have come across), may have supposed that it was accomplished somehow by magic or magnetism to correspond with the apparent daily rotation of the heavens. In any case, the water-wheel escapement clocks of China occupy a position of great historical importance as the missing link between the most ancient water-clocks of Babylonia and Egypt, and the spring-driven wrist-watches we are all wearing today. For their operation was neither wholly dependent upon liquid flow nor upon purely mechanical (oscillating) devices. The history of horology has thus at last become one continuous story.

As I have mentioned already, the original development of the water-wheel link-work escapement occurred long before Su Sung's time, indeed during the Thang dynasty, about + 724. That was the year when a most gifted Buddhist astronomer, a Tantric monk named I-Hsing, together with another engineer Liang Ling-Tsan, first introduced the device and used it for rotating a kind of planetarium. Both of these brilliant associate engineers and mathematicians (Han Kung-Lien and Liang Ling-Tsan) were minor officials in government offices until their gifts were discovered, after which they were taken from their desks and given work for which they were really suited. It was possible to find out about these earlier achievements because once we had acquainted ourselves with the technical terminology used in Su Sung's book, we could follow up his historical account and identify in texts which otherwise would have been quite incomprehensible what had been going on as far back as the beginning of the + 8th century.

The question finally arises whether there was any influence of these water-wheel clocks upon the development of clockwork in Europe. We do not really know. We may date the beginnings of the Western mechanical clock and the verge-and-foliot escapement fairly safely at about + 1320. Unlike the Chinese clocks this had, of course, a weight drive. But during about a century before that time one finds a good many references to 'horologia' which were certainly not mechanical clocks of the later type, but which do not sound like sun-dials or leaking-vessel water-clocks either. There is mention, not only of water, but also of various kinds of wheels and weights and pipes with orifices; so there is something still unexplained about the period + 1220 to + 1320, and it may well be that water-wheel clocks were then known and used in Europe. There is for example the story about the monks of Bury St Edmunds; when in + 1198 the abbey was about to go up in flames they rushed to the 'clock' to get water. A French manuscript in the Bodleian Library, dating from + 1285, has a picture showing the prophet Isaiah putting the clock back ten degrees for King Hezekiah during his illness.[1] This shows something which looks quite like a bucket water-wheel, with a nozzle delivering water into a bowl like a sump, a row of bells above, and in one or two places what appear to be cords passing back and forth. The whole mechanism is hard to make out, but it certainly is not just a leaking-vessel water-clock. A row of bells in a curiously similar position occurs in another representation of a clock, carved in a relief of Tubal Cain on the façade of the cathedral at Orvieto about + 1320, where it is combined with a set of right-angle gear-wheels and what might conceivably be a weight drive.[2] We must therefore retain the possibility that the Chinese water-wheel link-work escapement system was known and used in + 12th- and + 13th-century Europe. If it should turn out that this was not so, then the idea of the escapement as such was a striking case of stimulus diffusion.

[1] C. B. Drover, 'A Medieval Monastic Water-Clock', *Antiq. Horol.* 1954, **1**, 54. Cf. Fig. 78.
[2] J. White, 'The Reliefs on the Façade of the Duomo at Orvieto', *Journ. Warburg and Courtauld Institutes,* 1959, **22**, 254.

VI

This lecture has been as full of dates as 'the palms of the heathen Chinee'. I trust that so many dates will be forgiven me by my hearers. But after all from the point of view of the history of technology and the history of science one has to be quantitative; it is as if one were titrating one civilisation against another, and one does want to know where the end-point is.

In conclusion I have only two things to say. The first concerns the form and nature of these transmissions of stimuli which we may imagine took place between East and West. If we can take + 1320 as a focal point for the appearance of the first working mechanical clocks in Europe, we ought to remember that + 1327 was also equally focal for gunpowder, the original home of which had undoubtedly been + 9th- or + 10th-century China. Towards + 1380, again in the same century, we find the first blast-furnace producing cast iron in Europe, though this technique goes back (as we have seen) far earlier in East Asia, the other end of the Old World. Towards + 1375 (and also in the Rhineland) comes the first European block-printing, an art which had been current in China since the + 8th century; and still closer to clockwork in time are the great segmental-arch bridges of Europe, the first being about + 1340, though in China structures of this kind had appeared long before in the work of that brilliant bridge-builder Li Chhun (+ 610). The + 14th century thus presents itself as a time of adoption of a number of important techniques which had already been known and used for centuries in the Chinese culture-area. One may indeed believe that people in Europe did not know exactly where they had come from, but it is probably significant that these adoptions occurred just after the Mongol period, the Pax Tartarica, when intercourse had been so easy between the western and the eastern ends of the Old World. After all, Marco Polo had been in China from about + 1275 to + 1292, and he was only the most famous of many European merchant travellers. We have evidence, moreover, for a somewhat similar wave or cluster of adoptions towards the end of the + 12th century, when within a few decades of the year + 1180 Europeans came to know and use the magnetic compass, the stern-post rudder and the windmill. It is curious that these things should have come in 'packaged' clusters, one towards the end of the + 12th century and another during the + 14th. Further research will, we hope, uncover more facts about the history of science and technology at both ends of the Old World, and also more about the transmissions between them.

The second subject, the last on which I want to touch, concerns the way in which our colleagues and friends in China are thinking of these matters. In China today there is a vigorous recovery of past achievements, and a great deal of work in progress on science and technology in the classical culture. The National Academy (Academia Sinica) has had for some time past a special research department on the history of science and technology, and this is headed by an outstanding historian of mathematics, Dr Li Nien.[1] Moreover there is in China a great enthusiasm for science as the indispensable means of the raising of the general standard of life to equality with the rest of the world. Science is powerfully supported, new generations of brilliant young scientists are growing up, and the Chinese people are proud of the great discoveries, observations and inventions made by their ancestors. They are getting to know about facts which the sandstorms of history have hidden for centuries, and which some occidental historians have not always been happy to uncover. Too long have the disinherited thinkers and technicians of Asia been intimidated by the unduly high prestige of European culture. Is it not important for them to realise that although (for example) the first complete Western descriptions of mock sun displays and parhelic phenomena, haloes and strange illuminations caused by ice-crystals in the upper atmosphere, were given in the +17th century, every single component of the complex effects had been observed and named by Chinese astronomers just a thousand years earlier? And not in the +7th, but already in the −4th century, had Chinese naturalists recognised the hexagonal system of snowflake crystals.[2] Must it not encourage China's students of science today to know that the celestial co-ordinates of modern astronomy are neither Greek nor Arabic, but Chinese? Should they not take legitimate pride in the fact which we have been discussing this evening, namely that the combination of crank, connecting-rod and piston-rod first appears, not in Leonardo da Vinci or in the German engineers of the +15th century in the West, certainly not among the geometrical Greeks or the Alexandrian mechanicians, but in the Chinese water-power blowing-engines of the +13th and +14th? Nowadays in China one even finds little books of pictures for school-children explaining to them about Chang Hêng and the seismograph, or about Tshai Lun's invention of paper in the +2nd century, or Pi Shêng's creation of movable type in the +11th. All these achievements are now well established. Science and its applications need therefore no longer be regarded in China or by

---

[1] [Now, alas, no longer with us. His successor is also an eminent historian of mathematics, Dr Chhien Pao-Tsung.]   [2] [Cf. p. 98 above.]

other Asian people as something for which they should feel themselves beholden to the generosity of the West, something with no roots in their own culture. On the contrary, it had many great and illustrious roots, roots which helped to sustain the scientific Renaissance itself, and it is right that the Chinese should become more and more conscious of them. Though *modern* science originated only in Europe, to modern science everyone in the last resort contributed. So in the end I think it is fair to say that our studies may turn out to be a contribution not only to objective history, but also to the cause of international understanding and friendship. When all the debts are acknowledged (and certainly no-one can ever pay them) Asians and Europeans will be able to go forward together without hesitation, on a just and mutually appreciative basis, 'neither afore or after other', truly 'without any difference or inequality'.

10

THE PRE-NATAL HISTORY OF THE STEAM-ENGINE[1]

INTRODUCTION   [1963]

A timely book, published during this centenary year of our eponymous hero Thomas Newcomen, is entitled *The Prehistory of the Steam-Engine*, by L.T.C. Rolt. When I was asked to contribute to the celebrations something on the hidden ancestry of the component parts of the steam-engine, I knew that I must deal, as it were, with the 'palaeolithic' rather than the 'neolithic' aspect of the problem, and thus the title took shape as 'The Pre-Natal History of the Steam-Engine'. My object is to trace back into the remote past the origins of the main constituents of the definitive form of the steam-engine attained about 1800. The general result is rather paradoxical, for it demonstrates that some of the earliest prototypes exerted their effects at a relatively late stage, but it throws greater lustre upon Thomas Newcomen than he has generally received. Taking his invention as the focal point, it is possible to show that the entire morphology (and some of the physiology) of the reciprocating steam-engine of the early nineteenth century was prefigured in Asian, especially Chinese, machinery, widely used at the beginning of the thirteenth. In order to pass beyond the single-stroke atmospheric stage (Fig. 23) and the non-rotary rocking beam stage, as Watt, Trevithick and their contemporaries did, currents of design were required which went back centuries earlier than Newcomen. Moreover, these currents were characteristically East Asian, in contrast to the line of development from Torricelli and von Guericke to Newcomen, which was based almost (though not quite) completely on Greek antecedents. Newcomen himself therefore, as a man of the Scientific Revolution, appears more original, and also at the same time more European, than could be realised before the conclusion of the analysis here attempted.

Let us look first at the Chart[2] on which I have sought to give a diagrammatic view of the as yet unrecognised, as well as the recognised, elements in the development of the reciprocating steam-engine. If I pass very rapidly over the well-known

---

[1] Newcomen Centenary Lecture, London, 1963, delivered at the Institution of Mechanical Engineers; reprinted from *Transactions of the Newcomen Society*, 1962–3, **35**, 3.   [2] Opp. p. 202.

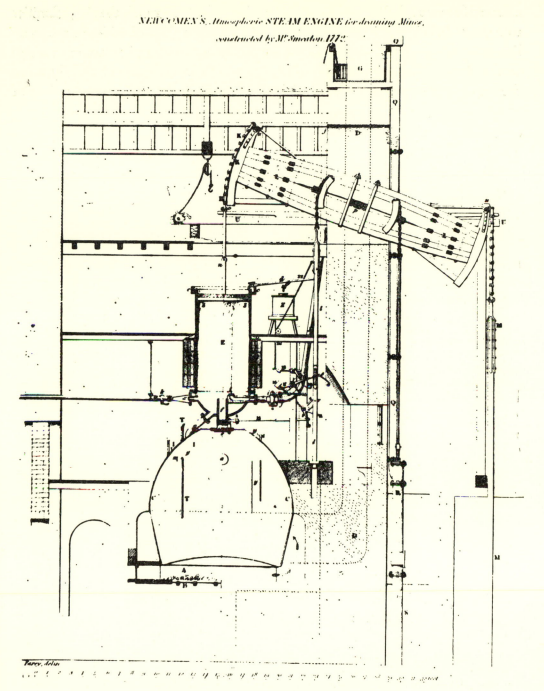

Fig. 23. 'Newcomen's atmospheric steam-engine for draining mines, constructed by Mr Smeaton, 1772', an illustration from Farey.

137

course of events I shall surely be forgiven since I have so much that is unfamiliar to report. One is struck at once by the two manifest discontinuities in the main line of development, first when the steam displacement pumping systems yielded to the piston-and-steam machines of Papin and Newcomen, and secondly when the atmospheric engines of Watt yielded to the high-pressure steam-engines of

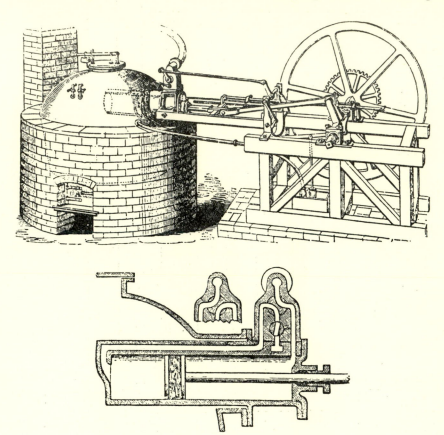

Fig. 24. Trevithick's high-pressure engine, *c.* 1800
(Dickinson & Titley, after Farey).

Trevithick (Fig. 24). As we shall see, the main effect of Asian precursors, though not all of it, was exerted at this second discontinuity. I am not among those who refuse[1] to the Alexandrians any claims of paternity to the harnessing of steam,[2]

[1] As, for example, Cardwell.

[2] In order to keep this contribution within some reasonable bounds of space, I propose to neglect, so far as possible, all inventions of the steam-turbine type, from Heron's 'aeolipile' through Branca to de Laval and Parsons. They may be regarded as constituting a quite separate chapter in the history of power engineering, easily separable anatomically from that of the reciprocating steam-engine. Cf. Burstall, pp. 339 ff.

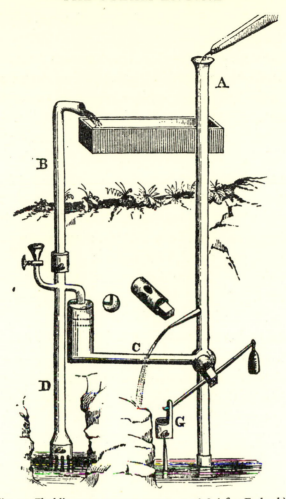

Fig. 25. Fludd's water-pressure pump, +1618 (after Ewbank).

for it is clear that Heron (fl. +62) used it to displace water and wine from a closed vessel,[1] and thus became the ancestor of the seventeenth-century pioneers of steam displacement (J. B. della Porta,[2] Solomon de Caus,[3] and others) who undoubtedly knew of his devices. Out of the steam displacement ideas came the steam displacement pumping systems, some shadowy like those of Somerset (the 2nd Marquis of Worcester)[4] or d'Acres,[5] but others quite solidly historical and

[1] See Drachmann KPH, p. 129, discussing *Pneumatica*, ch. 2, no. 21; cf. Woodcroft, ch. 60, p. 83.

[2] See Dickinson SE, p. 5; Thurston, p. 14, etc., cf. immediately below. *Pneumaticorum Libri III* (Naples, 1601); *I Tre Libri de' Spiritali* (Naples, 1606).

[3] See Dickinson SE, p. 12; Thurston, pp. 15 ff. *Les Raisons des Forces Mouvantes...* (Paris, 1615, 1624).

[4] See Dircks, pp. 540 ff.; Dickinson SE, pp. 13 ff.; Galloway, p. 57; Thurston, pp. 19 ff.

[5] Ed. Jenkins.

effective up to a point in their water-raising duties, notably the 'fire-engines' set up by Captain Savery.[1]

In any evolutionary story it is always interesting to trace the 'missing links' between particular stable stages, and these can certainly be found at the two manifest discontinuities of which we have spoken above. Intermediate forms which appeared at the second of these we shall meet with later (pp. 154, 201); here we may glance for a moment at a transitory device which belongs to the first—the 'free' or 'floating' piston. In his 'fire-engine' of 1707 Papin provided a floating piston with no piston-rod between the water and the steam in the displacement vessel, his object being to protect the latter from too sudden condensation on contact with the cold water.[2] Though not in strict chronological sequence, this was certainly a 'missing link' between the two great phases in the development of the steam-engine, Savery's and Newcomen's. But if the floating piston had no great future it had had an interesting past. It had, in fact, formed part of the water-pressure pump described in +1617 by that strange character Robert Fludd, mystical philosopher, Paracelsian physician and Rosicrucian alchemist (+1574 to +1637),[3] who seems to have encountered it in his travels through the mining districts of Germany and Central Europe. Its duty was to raise fresh water to a height by the pressure of a supply of waste water conveyed from a still higher source. The two kinds of water were separated in a cylinder by a floating piston; when this was forced upwards by the waste water it raised the fresh through a

[1] Dickinson SE, pp. 18 ff.; Galloway, pp. 56 ff.; Thurston, pp. 31 ff.; Farey, pp. 99 ff. Savery's patent was of 1698, and several pumps of his design were erected during the subsequent dozen years. Cardwell again is prone to minimise the role of the steam-displacement pumping systems in the steam-engine's history. But they embodied two essential features, the injection of steam from a boiler into a space, and the subsequent condensation of that steam. They were in a way ahead of their time because of their tendency to use high-pressure steam, although no suitable containers for it existed. In so far as Savery failed, this was why Newcomen succeeded. Besides, the steam-displacement pumping systems lasted on for more than a century after atmospheric engines had proved their worth, and indeed they live on still, in the form of the pulsometer pumps used for draining excavations and for handling water-containing solids (cf. Burstall, pp. 190 ff.). A new steam pump of Saverian character was described with some pride by the Marquis de Manoury in 1821, and as late as 1835 such installations were still considered capable of performing useful services by Colladon & Championnière. In 1909 the displacement principle entered a new phase with the invention of Humphrey's gas explosion pump, which raises a column of water by the direct impulsion of a series of internal combustion cycles (Burstall, pp. 407, 413). In recent years this form of pump has aroused much interest among Chinese engineers, and simplified versions of it have been produced by Tai Kuei-Jui, Phêng Ting-I and other inventors (cf. Ho Shan).

[2] See Galloway, p. 75; Thurston, p. 53.

[3] *Utriusque Cosmi Majoris scilicet et Minoris Metaphysica, Physica atque Technica Historia*, Galler, Oppenheim, 2 vols., 1617–19. This work, a veritable treatise on the universe and its creator, appeared in a number of parts, some of which have the title *Naturae Simia seu Technica Macrocosmi Historia*, and one is dated 1618. The engraving of the water-pressure pump occurs on p. 467 in Tract. II, Pt. 7, bk. 2, ch. 10 (see Fig. 25).

rising pipe aloft, and when the waste water was allowed to flow out, the piston, descending, sucked in a further supply of fresh. The two-way cock which admitted the waste water to the cylinder was automated by an ingenious device in which the cock was attached to a lever with a weight at one end and bearing a small vessel at the other. This vessel filled gradually from an auxiliary orifice in the side of the down waste-water pipe, and when it descended turned the cock so that the pressure head was cut off and the waste water in the cylinder could escape. At the lowest point of the vessel's travel a valve in its floor was opened in good fifteenth-century fashion by a suitably placed projecting pin, the water flowed out, the counterweight restored the vessel to its previous position, and with the corresponding turn of the cock the cycle recommenced. The whole principle was very similar to that of the 'gaining and losing bucket' machines much discussed in the seventeenth and eighteenth centuries,[1] and Fludd's water-pressure pump, the ancestor of many later water-pressure engines,[2] must have been in use towards the end of the sixteenth, perhaps even in Agricola's time, though obviously limited in applicability. There was nothing exotic about it, for the floating piston could easily have originated from the float of the Hellenistic anaphoric clock,[3] but even though it separated two different kinds of water rather than water and steam it is worth remembrance in the steam-engine's ancestral hall. At this point we come to the first irruption of orientalism into our engineering story.

HISTORY AND TRAVELS OF THE STEAM JET

Most people would naturally be inclined to say that in the history of the steam jet Asia could not have been involved. But the matter cannot be dismissed quite so easily. From ancient times onwards a jet of steam was used in many parts of the world for a purpose both domestic and industrial, i.e. for blowing up fires. The story of these steam fire-blowers or 'sufflatores' has been fully related by Hildburgh, who points out that the famous aeolipile of Heron, rotating by jet propulsion on account of its L-shaped tubes, was only a special form of aeolipile.[4] The commoner sort was simply a kettle or boiler with a pinhole orifice so arranged that a jet of steam could be directed upon a fire. This has several effects: it accelerates combustion because of the draught of air which accompanies the jet, products of combustion are removed, and at the same time the steam is decomposed by

---

[1] See Ewbank, pp. 64 ff.   [2] Cf. pp. 151, 153 below, and Ewbank, pp. 352 ff.
[3] See Needham, Wang & Price, with references.
[4] The word derives from the name of the god of the winds, and *pilos* a ball.

the glowing coal to carbon monoxide and hydrogen (water-gas) which immediately burns away.[1] To Philon of Byzantium (*c.* −210) is ascribed[2] an
incense-burner kept glowing by a steam jet, and similar arrangements are mentioned by Vitruvius[3] and Heron of Alexandria.[4] Then in the +13th century
Albertus Magnus[5] describes a *sufflator* of this kind in detail, and in the later
technological MSS there are several pictures of bronze busts or heads which
direct steam jets from their mouths (Kyeser, 1405; Leonardo da Vinci, *c.* 1490).[6]
The jet 'turbine' follows shortly afterwards (Branca, 1629;[7] Wilkins, for rotating
a spit,[8] 1648). In 1545 a hollow bronze statue of a man was found in some ruins
at Sondershausen in Germany, and gradually other similar objects appeared
(Fig. 26, pl.); their use was the subject of much dispute until Feldhaus established
that these 'Püstriche', or blowers, were fire-blowers corresponding to the description of Albertus Magnus.[9] Examples have been found as far east as south Russia
(Ekaterinoslav).

The position of Leonardo da Vinci demands a few words to itself. He was
particularly interested in the expansion of volume which occurs when water is
converted into steam, and he described a couple of experiments which he designed
to measure it.[10] Both used square-section boiler-cylinders[11] with boards as pistons,
each connected to a counterweight, in one case by means of a cord and pulley,
in the other by a rigid piston-rod attached to one end of a lever with a central
suspension.[12] How near Leonardo came to the 1700-fold factor is not recorded.[13]
His interest was quite practical in its way, for, as is well known, he designed a

---

[1] Water-gas has quite a wide use in modern technique, and steam jet blowers are used for industrial boiler
furnaces. By varying the mixture with air, a thermostatic control is achieved.

[2] *Pneumatica*, no. 57; Feldhaus TDV, col. 179; Drachmann KPH, p. 125, who, however, regards this as an
interpolation later than Heron's steam displacement device.

[3] I, 6.2.

[4] *Pneumatica*, ch. 2, nos. 34 and 35, the *miliarium* discussed by Drachmann KPH, pp. 130 ff.; Beck BGM,
pp. 22 ff.; cf. Woodcroft, chs. 74, 75, pp. 100, 103.

[5] *De Meteoris*, bk. 3, tr. 2, ch. 17; Paris ed. 1890, vol. 4, p. 634; the passage is translated in Feldhaus TDV,
col. 844.

[6] See Feldhaus TDV, col. 843 ff.; Hart, pl. 101.

[7] Wolf XVII, p. 546; Dickinson SE, p. 193; Thurston, p. 17, and very frequently discussed and figured.

[8] *Mathematical Magic*, p. 149.   [9] Feldhaus ZEP.

[10] See Hart, pp. 249 ff. More than 250 years before the work of Watt, this is an extraordinary anticipation.

[11] Here we have the same undifferentiated container used 200 years later by Papin (cf. p. 148 below).

[12] This was surely the most extraordinary anticipation of all, since it prefigured the arrangement of Newcomen's engine, though without the separation of boiler and cylinder. Yet the likeness was mainly formal,
since Leonardo had nothing but a counterweight at the other end of the 'beam', and presumably it never
occurred to him to throw cold water over the 'cylinder', when by the condensation of the steam the motion
of the lever would be reversed.

[13] [But in one place he says: 'one ounce of water gives a skinful of steam', and this volume measure has been
interpreted as 1,500 litres.]

steam cannon, the famous Architronito,[1] in which a jet of high-pressure steam was suddenly admitted behind a ball[2] to shoot it forth through a long barrel.[3]

With the *sufflatores* we are now recognisably in presence of the pre-natal form of the steam-engine's boiler, so there is significance in the gradual abandonment of the 'wind-cherub' theme still beloved of cartographers long afterwards. When Cesariano translated Vitruvius into Italian early in the sixteenth century he depicted the boilers in various elegant, but now no longer anthropomorphic, forms;[4] they were turning, so to say, into the celebrated (if legendary) kettle of James Watt. So also when Ercker illustrated his 'Treatise on Ores and Assaying' later in the century the steam blowers were just alembics placed alongside the furnaces.[5] But Branca still used an Aeolian head for his steam jet turbine in 1629, and Athanasius Kircher followed suit in 1641.[6] The real question was, on or into what should the jet be directed? When della Porta in 1601 demonstrated the emptying of water from a closed vessel by the injection of a jet of steam into it, and the drawing up of water by the vacuum induced in a vessel by the emptying of water from another vessel in hermetical connection with it, he was much more truly on the rails which were to lead to the primary break-through of steam power, i.e. the reciprocating steam-engine rather than the steam turbine.[7] This was the green light for Ramsay (1631), Dobrzensky (1657), d'Acres (1659), the Marquis of Worcester (1663) and all their successors.[8] Dircks can surely not be

---

[1] See Hart, pp. 295 ff., pl. 100, etc.

[2] Might one not think of this as a free piston, and the piston as a tethered cannon-ball?

[3] The position of Leonardo in the pre-natal history of the steam-engine has been most recently considered by Reti (*RDI*, 1957, 21), whose ideas much influenced Hart. Reti sought to demonstrate that Leonardo had a fairly clear idea of the prime mover implicit in steam power, and that his work directly influenced later pioneers such as della Porta and de Caus. [He certainly described a steam turbine long before Branca. More important, he also clearly recognised the formation of a vacuum by the condensation of steam, and actually used the partial vacuum formed by the explosion of gunpowder in a closed space to do work (cf. p. 147). It would of course have been very difficult for anyone in Leonardo's time to foresee that the way forward to the successful prime mover lay not through high-pressure steam but through the properties of the vacuum. Europeans were naturally Faustian, yet they accomplished this truly Taoist feat.]

[4] Como, 1521; see Feldhaus TDV, col. 26, fig. 10; Lynn White, p. 91.

[5] Prague, 1574, and many later editions; see Sisco & Smith, frontispiece and pp. 219, 326 ff.; cf. Lynn White, p. 91.   [6] *Magnes...* (Rome, 1641), p. 616; cf. Lynn White, p. 92.

[7] *Pneumaticorum libri III* (Naples, 1601), *I Tre Libri de' Spiritali* (Naples, 1606), ch. 7. See Beck BGM, pp. 258 ff.; Dickinson SE, pp. 4 ff.; Wolf XVII, p. 544 (confused). The vacuum left by condensed steam had been utilised thaumaturgically in ancient China (cf. p. 145 below). Della Porta now showed that it would suck up water. Cf. his *Magiae Naturalis Libri XX* (Naples, 1589), Eng. ed. of 1658, p. 386.

[8] See Dircks, pp. 540 ff. Lynn White, p. 162, provides evidence that something similar was brewing in Bohemia before the end of the sixteenth century, though I cannot find anything more than heated air in Dobrzensky. If I am right in this, he had not got beyond the stage of the Banū Mūsā brothers (the three sons of Mūsā ibn Shākir, who were at work between +840 and +860). They proposed a hermetically sealed space at the top of a well, in which twenty lamps of naphtha should burn all night, so that

wrong in his conviction that the *sufflator* was the chief ancestor of the steam-engine.[1]

But here comes in the unexpected. The heads and busts were a product of European fancy,[2] but for the steam-jet fire-blowers there is another focal area, namely the Himalayan region, especially Tibet and Nepal. There they take the form of bottle-shaped conical copper kettles surmounted by birds' heads, the beaks of which, sometimes quite elongated, point downwards and have the pin-hole at the tip (Fig. 27, pl.).[3] The fact that an ordinary kettle, which emits steam under very little pressure, does not have the same effect, might plead for a single point of origin of the discovery; if so, the device can hardly have reached the Bactrian region from the West in Alexander the Great's time (as Hildburgh suggests), but could have been used there by the later Bactrian Greeks. It would certainly have been useful for the fires of *argali* dung used by travellers on the Old Silk Road and other desert and mountain regions of Central Asia where wood is scarce and altitude considerable. So far we have no evidence that the Chinese made much use of it, and until this is found we may like to accept the distribution as being primarily Western and Bactrian. There is no proof as yet, however, that the device was not original and ancient in Central Asia, travelling westwards with Alexander's returning veterans in time to stimulate Philon and Heron.[4]

THE VACUUM IN EAST AND WEST

We must now take up the thread of another Alexandrian initiative. Ctesibius was responsible about −230 for a simple and fundamental machine, the piston

water would be sucked up by the partial vacuum produced by the removal of the oxygen (cf. Hauser, and especially Schmeller, pp. 26 ff.). And this itself was Alexandrian, for Philon had seen (ch. 8) water sucked into a closed space with a burning candle inside it; as also (ch. 7) the expulsion of water from a closed space in which air was heated (cf. Drachmann KPH, pp. 119 ff.). [Of course, the effect of the burning candle is but small, only 12 per cent in favour, since carbon dioxide is produced; unless something is introduced to absorb this. A similar proposal to that of the Banū Mūsā was made by Giovanni Fontana, about +1420, with arrangements to trap the water raised, so he may be considered a link between them and Savery.]

[1] Here the accent is on the word steam. One can also put the accent on the word engine, and, as we shall presently show, the Chinese culture-area had a part to play in the development of this much more important than steam jets or vaned wheels. It is strange that the judicious Dickinson failed to take account of the *sufflatores*, all the more so as they had been very fully discussed by Ewbank, pp. 395 ff.

[2] A rather curious fancy, too, for many of the figures are prominently phallic, e.g. the celebrated Jack of Hilton described by Robert Plot in 1592, p. 433, and still connected with a jocular tenure.

[3] There are examples in the Cambridge Ethnological Museum, one of which was kindly tested by Dr Bushnell in the writer's presence, with most striking effect.

[4] If this were to be verified it would be very suggestive, since the Central Asian region of Mongol–Tibetan–Persian culture was that in which the windmill seems to have its oldest home. The windmill was first

air-pump, known from the descriptions of later mechanicians.[1] It powered what is confusingly known as the 'water-organ' because the air was pumped into a reservoir where it was held at approximately constant pressure by means of a water-seal. Though the Vitruvian text describes two cylinders, the Philonic and the Heronic speak only of one, and the apparatus has nearly always been so reconstructed—in any case the pump or pumps were invariably single-acting, 'inhaling' and 'exhaling' on alternate strokes. This simplest of pumps entered upon a new reincarnation in the seventeenth century when the virtuosi began to explore with excitement the properties of vacuous spaces, for what had been invented originally as a bellows for pumping air into something now found fresh employ-ment as *the* 'air-pump' for getting as much air as possible out of it. There is no need for me to emphasise how this famous development helped the steam-engine to emerge from the chrysalis of the Saverian 'water-commanding' machine. Suffice it to say that after the Italians, especially Torricelli, had opened men's minds to the conception of the weight of the air and the physical reality of the vacuum,[2] von Guericke about 1650 embarked upon his crucial experiments—the illustrious Magdeburg hemispheres, the crumpling of evacuated copper globes, the single-acting single-cylinder piston air-pumps, and the pistons which raised weights or men into the air when sucked down by the vacuum.[3] It will hardly be believed that there was an ancient Chinese experiment which anticipated Magdeburgian effects, but this is so. In the −2nd-century *Huai Nan Wan Pi Shu* (Ten Thousand Infallible Arts of the Prince of Huai-Nan) we find the following:[4] 'To make a sound like thunder in a copper vessel. Put boiling water into such a

---

harnessed to practical use, so far as we know, in Seistan, a region of Parthia close to Greek Bactria (though some seven centuries after Greek Bactria had ceased to exist as such). We might have to look for the ancestor of Branca's proposals in the Arabic–Iranian direction, a first combination of steam jet and vaned wheel occurring somewhere in western central Asia. Cf. Lynn White, pp. 85 ff.

[1] See Heron, *Pneumatica*, ch. 1, no. 42; Philon, *Pneumatica*, App. 1; Vitruvius, x, 8. Discussion in Beck BGM, pp. 24 ff., Drachmann KPH, pp. 7 ff., 100, MTGR, p. 206. Cf. Woodcroft, chs. 76, 77, pp. 105, 108.

[2] The events are well and succinctly described in Dickinson SE, pp. 5 ff., and at greater length, of course, in all the histories of physics. The monograph of de Waard may also be mentioned. [Cf. W. Gerlach, *Hoch-schuhl-Dienst*, 1967 (no. 2).]

[3] Von Guericke demonstrated his experiments at the Diet of Ratisbon in 1654 and a preliminary account of them was published by the Jesuit Caspar Schott in his *Mechanica Hydraulico-Pneumatica* (1657). Von Guericke's definitive account did not appear till 1672 under the title *Experimenta Nova (ut vocantur) Magdeburgica de Vacuo Spatio*. There is a German translation in Ostwald's series, Klassiker d. exakten Wissenschaften, no. 59. For brief accounts see Wolf XVII, pp. 99 ff.; Dickinson SE, pp. 7 ff.; and preferably Gerland & Traumüller, pp. 129 ff.

[4] Preserved in *Thai-Phing Yü Lan*, ch. 736, p. 8b; ch. 758, p. 3b, under another heading, repeats the passage, adding that the orifice of the vessel has to be closed extremely tightly. Cf. SCC, vol. 4, pt. 1, pp. 69 ff., for the context. On this notable +10th-century encyclopaedia, see below, p. 157.

vessel and then sink it into a well. It will make a noise which will be heard several dozen miles away.' If it is assumed that the vessel was full of steam when the lid was tightly closed, a vacuum would have followed the sudden condensation of the steam; and if the vessel were thin-walled an implosion would result, just as van Guericke found. The noise of this would doubtless be intensified by the echo of the sides of the well, and might have been heard at some considerable distance if not as far as the ancient author boasted. The device was certainly intended for military or thaumaturgical purposes, some of which indeed it could well have fulfilled, but its chief interest perhaps for us is the evidence which it adds to so much other evidence that the Chinese mechanicians were experimenting about − 120 just like their Hellenistic counterparts.[1] Now, seventeen-and-a-half centuries later, it was revealed that all these effects were due to an omni-present force—'the spring and weight of the air', from which man might draw an infinite profit if he could set it to work.[2]

On the uses of steam in traditional Chinese culture we cannot here expatiate, but it may be noted that bread was (and is) generally steamed rather than baked. In his *Travels in China*[3] John Barrow wrote: 'In like manner, they (the Chinese) are well acquainted with the effect of steam upon certain bodies that are immersed in it; that its heat is much greater than that of boiling water. Yet although for ages they have been in the habit of confining it in close vessels, something like Papin's digester, for the purpose of softening horn, from which their thin, transparent and capacious lanterns are made, they seem not to have discovered its extraordinary force when thus pent up; at least, they have never thought of applying that power to purposes which animal strength has not been adequate to effect.' One would like to know more about traditional horn-working in China, but what Barrow failed perhaps to appreciate was that the way to the steam-engine historically lay not directly through pressure-cookers, but indirectly through evacuated vessels, and the understanding of the vacuum was a characteristic

[1] See throughout *SCC*, vol. 4, pt. 2.
[2] In this connection Cardwell has written memorable words. 'To the newly discovered agent', he says (p. 9), 'so powerful that it could overcome the strongest horses and evidently rivalled the largest water-wheels and windmills, science and common sense could set no obvious limit. The wind might blow where, and when, it listed, but the atmosphere *always* exerted a pressure of about fifteen pounds per square inch *every-where* on the face of the earth. The point was, could one utilise this immense force, this "head" of power? Could one, in effect, invent an atmospheric water-wheel or windmill, driven by the deadweight pressure of the atmosphere rather than by the pressure of moving air or water against sails or blades? To the speculative and enterprising of the seventeenth century the possibilities may have seemed as revolutionary as those of nuclear power in our generation, probably with greater reason, certainly with fewer moral reservations.'
[3] 1804, p. 298.

result of the methods of modern science born in the Scientific Revolution. In other words, the way to high-pressure steam, and all that it could do, lay dialectically through its precise opposite; and the whole historical process was an extraordinary justification of the classical idea of Taoist philosophy that emptiness would be the gateway to all power.

THE COMING OF THE VACUUM OR ATMOSPHERIC ENGINE

By 1670 an acutely interesting question had arisen. What could one do to create a vacuum underneath a piston (and so make it do work), otherwise than by the previous use of another piston in an exhaustive air-pump according to the Magdeburgian art? When young Denis Papin from Blois took up his post as assistant curator of experiments under the great Christian Huygens at the Royal Academy in Paris in 1671 this was certainly one of the things they discussed, and the explosion of gunpowder was certainly one of the methods proposed.[1] By 1678 Huygens had succeeded in obtaining erratic piston stroke motion by exploding successive small quantities of gunpowder at the base of a cylinder,[2] and in the same year de Hautefeuille [is said to have] proposed the same source of rapid gas production both for Saverian water-displacement systems and for the cylinder and piston.[3] Papin continued working on the idea with persistence, and his publication of 1687 described the best form which it was capable of taking in the state of technique at that time, but this was simply not good enough for practical use.[4] The gases of the explosion certainly drove out most of the air,

---

[1] Cf. Galloway, pp. 14 ff., for more detailed references. It is not generally known that the gunpowder engine goes back to Leonardo da Vinci (Hart, pp. 295 ff.). He called it 'a way of lifting a weight with fire, like a cupping glass'. A weight was suspended from a piston with leather packing, and gunpowder was exploded above it, the touch-hole and any other openings in the cylinder being then closed. As the gases cooled a semi-vacuum was created and the piston raised the weight, which was secured in its new position by the insertion of blocks and wedges. Here Leonardo came nearer to the ultimate gateway of success, the vacuum, and certainly anticipated the seventeenth-century physicists by 150 years or more; but without a deeper analysis of the phenomenon of the cupping glass he could not go further. At this stage, too, the method of internal combustion was essentially unpractical.

[2] Galloway, pp. 21 ff.; Wolf XVII, p. 548; Thurston, pp. 25 ff.; best in Gerland & Traumüller, pp. 226 ff. I forbear from emphasising the important circumstance, as striking perhaps as any of the other facts reported in this paper, that gunpowder itself was of strictly Chinese provenance. Though first known in Europe in the late +13th century, it had been a discovery of +8th- or +9th-century China, and had been used there for military purposes from the early +10th century onwards.

[3] Galloway, pp. 18 ff.; Thurston, pp. 24 ff. [Actually he used the gunpowder as an expansion mechanism, not for creating a partial vacuum.] De Hautefeuille [is also said to have] suggested the evaporation of alcohol as a method of clearing the cylinder of air (cf. Stuart, vol. 1, p. 70). [Actually the alcohol was part of an attempt to rotate a shaft by a triangular or three-bulbed distillation apparatus mounted at one end of it. I am indebted to my friend Dr Ladislao Reti for several of these corrections.]

[4] Galloway, pp. 41 ff., 44 ff.; Gerland & Traumüller, pp. 227 ff.

but when they cooled about a fifth of the air was always left, and this reduced the piston's down-stroke due to atmospheric pressure by as much as a half.[1] So Papin gave it up, and turned instead in 1690 to take the decisive step of having nothing but water and steam underneath the piston.[2] Here at last was an effective cycle, the removal of the air and the condensation of the steam. Though Denis Papin never harnessed his piston-rod to anything,[3] his historical position in the development of the steam-engine is a central one, and Thomas Newcomen himself would surely never grudge him his statue among the vegetable-stalls that overlook the Loire on the great flight of steps at Blois.

I find it almost impossible to believe that Newcomen did not know of Papin's steam cylinder.[4] The over-confident statement of Robison nearly a century later[5] led perhaps to excessive scepticism in our own time, for Newcomen could have seen not only the paper of Papin in the *Acta Eruditorum* of 1690, as Dickinson admitted, but also the small treatise which Papin published, in French as well as in Latin, at Marburg and Cassel in 1695.[6] In any case, Newcomen set about the matter with all the craftsmanly ability and determined practicality of a West-country artisan of genius, so that in 1712 the cylinder found itself set directly upon a boiler, with its piston-rod harnessed to a huge pump-handle in the shape of a rocking beam,[7] and arrangements for the injection of a jet to bring about rapid

[1] The gunpowder engine was of course one of the first steps on the way to the internal-combustion engine, but that is another story. One cannot refrain, however, from noting the interest of another later step, the 'pyréolophore' of the Niepce brothers invented in 1806, an extremely complicated machine which used lycopodium powder as the fuel combusted within the cylinder (cf. Daumas BPN). It is also fascinating to observe how completely the morphology of the definitive form of the steam-engine imposed itself upon the early stages of gas and petrol engines in Lenoir's time (1860); cf. Burstall, pp. 332 ff.

[2] Dickinson SE, pp. 10 ff.; Galloway, pp. 47 ff.; Wolf XVII, p. 550; Thurston, pp. 50 ff.; best in Gerland & Traumüller, pp. 228 ff.

[3] At any rate, not successfully. There are accounts that towards the end of his life, in 1707, Papin experimented with a small steam-driven paddle-wheel boat on the Fulda near Cassel, but it is not easy to see how it could have worked even with a Newcomen beam-engine as the stroke frequency would have been so slow. Our information remains obscure and somewhat contradictory on this last phase. See Galloway, pp. 76 ff.; Thurston, pp. 224 ff., where more detailed references will be found.

[4] At least by hearsay.

[5] Cf. Dickinson SE, pp. 32 ff.; Galloway, pp. 78, 85; Wolf XVIII, p. 611; Gerland & Traumüller, p. 240.

[6] *Fasciculus Dissertationum de Novis quibusdam Machinis...* (Marburg, 1695); *Recueil de Diverses Pieces touchant quelques Nouvelles Machines* (Cassel, 1695). There was also the *Traité de plusiers Nouvelles Machines et Inventions Extraordinaires sur Différents Sujets* (Paris, 1698). Surely there was someone at Dartmouth, one of the gentry perhaps, who could read French and gave Newcomen access to Papin's work.

[7] It is obvious that the model for the beam and column must have been the arm and post of the swape, or counter-weighted bailing bucket, but I failed to see this until it was pointed out by Cdr. John Mosse at the centenary meeting when this paper was read. Though the swape died out long ago in the West Country, and indeed all over Britain, it is still widely used in many parts of Europe from Touraine to Transylvania, and Newcomen was undoubtedly familiar with it. Its history is of course age-old, for it was used in Meso-

steam condensation in each cycle (Fig. 23).[1] From this point onwards the road towards the nineteenth-century developments was an arterial one, steadily leading through stages of increasing efficiency—the separate condenser[2] of James Watt, imagined in 1765 and at work by 1776,[3] then the application of the double-acting principle so that the engine was no longer idle on alternate strokes, and the simultaneous conversion to rotary motion, in 1783,[4] finally the introduction of high-pressure steam in the Cornish engines by Richard Trevithick from 1811 onwards.[5] Such was the biography of the 'atmospheric engine'. Its close was marked by the second discontinuity which we have already mentioned, for when the high-pressure steam permitted by improved boilers and fire-boxes took over in the very last years of the eighteenth century it was made by Trevithick to act upon both sides of the piston alternately, and the 14 lb/sq. in. of the atmosphere had no further part to play (Fig. 24). As we shall presently see, an ancient Asian machine (the Chinese double-acting piston-bellows) exerted profound influence in just those two late decades. That it did not exert its influence upon the seventeenth- and eighteenth-century air-pumps is what I would now like to point out.

potamia and Ancient Egypt as far back as the −2nd millennium. [In China it certainly goes back to the −1st.] The fact that it was always primarily a water-raising device links it directly and inescapably with the column and beam of the atmospheric steam-engine.

[1] See particularly Dickinson SE, pp. 29 ff.; Galloway, pp. 78 ff.; Thurston, pp. 57 ff.; Wolf XVIII, pp. 612 ff. Newcomen engines spread rather rapidly through Europe. In 1722 Joseph Emmanuel Fischer von Erlach (1680 to c. 1740) built one at Baňská Štiavnica in Slovakia for mine-drainage and, in the following year, another at Vienna for the water-supply to a palace and its park. The original model of this latter engine is still to be seen in the Technical Museum at Vienna (cf. Nagler). This engineer was the son of a famous architect, Jean Bernard Fischer von Erlach (1650 to c. 1740) who played a part in the transmission of the idea of iron-chain suspension bridges from China to Europe [see p. 39 above and SCC, vol. 4, pt. 3]. The condensing jet was not peculiar to Newcomen; it was used also in Desaguliers' modification of the Savery steam-pumping system in 1718 (Farey, p. 111).

[2] Dickinson's biological analogy is of some interest here; with Newcomen, boiler and cylinder were differentiated, with Watt, condenser also.

[3] Dickinson SE, pp. 66 ff., JW, pp. 32 ff.; Galloway, pp. 142 ff.; Thurston, pp. 88 ff. (apparently confused); Wolf XVIII, pp. 618 ff. As has often been pointed out, Black's scientific discovery of latent heat played a similar role in this technical advance to Boyle's exploration of the vacuum which suggested the atmospheric cylinder.

[4] Dickinson SE, pp. 79 ff., JW, pp. 124 ff., 134 ff.; Galloway, pp. 162 ff.; Thurston, pp. 103 ff.; Wolf XVIII, pp. 621 ff. It will be remembered that this was the point at which Watt was obliged to use the sun-and-planet gear for the conversion of rectilinear into rotary motion because the crank, connecting-rod and piston-rod combination had, it was thought, been patented by Pickard. The full significance of this extraordinary fact will be apparent later (p. 199).

[5] Dickinson & Titley, pp. 127 ff., 144; Pole, p. 51; Galloway, p. 192.

THE GREEK ANCESTRY OF THE VACUUM PUMP

Everyone is familiar with the exploration of the properties of the vacuum carried out by the scientific men of Robert Boyle's time. But the engineering rather than the physical aspect of the air-pump is not so well known. To find a good treatment of the successive types of air-pump from von Guericke onwards is much more difficult than one would expect,[1] but it seems quite clear that all the pumps of this period were single-acting, whether they consisted of one cylinder only (von Guericke, 1654; Boyle & Hooke, 1659; Huygens & Papin, 1674) or of two cylinders working with an alternate motion (Boyle & Papin, 1676; Hawksbee, 1709; 's Gravesande, 1720), etc.[2] We are therefore bound to regard them as essentially Hellenistic in style, based either upon the Heronic organ air-pump already described, or, in the latter case, upon the Ctesibian double-cylinder force-pump for liquids, which raised water from a tank in which it stood.[3] This also bears sometimes a confusing name, the 'fire-engine' or 'fire-pump', because the Heronic description refers to its use against conflagrations, though it had nothing else to do with fire.[4] I think that the strictly Hellenistic character of the seventeenth- and eighteenth-century air-pumps has not previously been recognised.

THE ARABIC ANCESTRY OF CATARACT AUTOMATION

One Asian influence there was, however, upon the mechanism of the atmospheric engine. It played a part in the early forms of automatic mechanisation of the cycle. There is good reason to believe that the engines of Newcomen were 'self-acting' from the first, and indeed this feature in his designs might almost be considered his most ingenious achievement, especially as one looks in vain for any comparable cyclical automation in earlier times.[5] On a 'plug-rod' hanging from

---

[1] Daumas LIS, to which I was first directed, pp. 83 ff., 115 ff., 285 ff., I found disappointing, but the photographs of the different types are good. Poggendorff, pp. 423, 471, 473, etc., is concise and authoritative; well supplemented by the illustrations and descriptions in Gerland & Traumüller, pp. 129 ff., 157, 193 ff., 304 ff., 312 ff. Cf. Konen, p. 39, fig. 2; Wolf XVII, pp. 99 ff.; Andrade; Wilson. None of these is at all exhaustive, for example, none mentions the air-pump of John Christopher Sturm, *Collegium Experimentale...* (1676); and it seems to me that a monograph on the detailed evolution of the air-pump would be a useful contribution. [Cf. H. D. Turner, *Nature*, 1959, **184**, 395.]

[2] Sturm's pump has been called 'double-acting', but I cannot see that it was.

[3] See Heron, *Pneumatica*, ch. 1, no. 28; Philon, *Pneumatica*, App. 1; Vitruvius, x. 7. All sources ascribe it to Ctesibius. Discussion in Beck BGM, p. 14; Drachmann KPH, pp. 4 ff., MTGR, pp. 155 ff. Cf. Woodcroft, ch. 27, p. 44. Several actual examples of Roman pumps of this kind still exist in museums. Cf. p. 168 below.

[4] The Vitruvian text does not mention this fire-brigade purpose but simply speaks of raising water on high.

[5] Dickinson SE, pp. 40 ff.; Wolf XVIII, pp. 612 ff.; Becker & Titley. This is not to say, of course, that the idea of a 'self-acting' machine was in itself unprecedented. Leonardo, for example, sketched a double water-

the beams, pins were so placed as to actuate the steam valve and the injection cock respectively at the right moments in the up and down strokes, or else a float responsive to the boiler pressure operated the mechanism as soon as a further supply of steam was available. Here we cannot describe the varying systems of levers and detents which all the atmospheric engines had; I wish only to mention the 'cataract', as it was called, a device which Watt found already in use in Cornwall when his first engines were installed there.[1] This was a stroke frequency governor in the form of a tipping bucket; a small balanced tank into which water flowed at constant speed until at a predetermined time the bucket tipped and simultaneously opened the injection cock (Fig. 28). Now the tipping bucket was not prominent in the designs of the Alexandrians, but it was an especial favourite among the medieval Arabs, and is constantly found in the jackwork of their great anaphoric clocks.[2] The works of al-Jazarī, for example, around 1206, are full of them,[3] and the principle was well known in the seventeenth century as we see from the book of machines of Isaac de

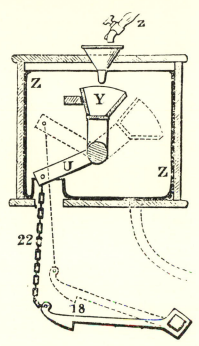

Fig. 28. The 'cataract' or tipping-bucket stroke-frequency regulator used in many forms of atmospheric steam-engines (Farey).

pump of concertina air-bellows type, which was to supply a fountain and to work automatically given only an incoming feed of water at suitable rate. See Beck BGM, p. 471. A similar automation had been achieved in the water-pressure floating-piston pump of Fludd (p. 139 above), and in the 'gaining and losing bucket' water-raising machines of the +17th century there also referred to. It may even be considered to go back to the Alexandrians, for Heron made some use of suspended counterweight buckets alternately filled and emptied by siphons so as to give continuously repeating automatic effects; cf. Woodcroft, p .31, ch. 15, p. 57, ch. 37; Drachmann KPH, pp. 114 ff., 127–8. I am thinking here naturally only of cyclical automations dependent upon the continuous supply of a liquid or a gas, and not of those which may be derived from continuous mechanical rotations by the use of half-gears and the like (cf. p. 197 below). 'Homoeostatic' or cybernetic controls are something else again; for example, the fantail gear of windmills or the governor-balls of the steam-engine. In a sense, the hydro-mechanical clocks of medieval China were both cyclical and self-acting given only the constant-rate water-supply, but Leonardo's pump, because of its two 'cylinders', is a much nearer anticipation of Newcomen's systems. On the clocks, see Needham, Wang & Price.

[1] Dickinson SE, p. 77; Farey, pp. 188 ff.; Stuart, vol. 2, pp. 298 ff. A form of cataract had already been employed in 1744 by de Genssane (MEE) in his automatisation of the Savery steam displacement pumping system, cf. Stuart, vol. 1, p. 199. This was almost identical with a design of Leonardo's (Beck BGM, p. 478).

[2] See Wiedemann & Hauser UBI, *in extenso*; more conveniently Needham, Wang & Price, pp. 112, 198, 190, figs. 44, 63.

[3] Cf. Coomaraswamy, pls. 4, 8, with accompanying descriptions.

Caus in 1644 (Fig. 29, pl.).[1] Thus a simple device, which had originated, perhaps, in the horological sphere for cutting time into predetermined intervals and operating jackwork, found a new application in helping to fix the duration of the steam-engine's cycle of events. It must have been part of the stock-in-trade of mechanical devices which late-medieval Europe inherited from the 'Saracens'.

ROMAN AND CHINESE PRECURSORS OF TUBULAR BOILERS

Our traverse of familiar ground is now nearly over. But first we must just remind ourselves of what happened when the atmospheric engine had had its day.[2] Its severe limitations of working speed could only be overcome by the use of high-pressure steam, a break-through which was primarily pioneered by Trevithick and primarily dependent upon a rapid evolution of strong fire-tube and water-tube boilers out of the simple steam-kettles which had sufficed for working at atmospheric pressure. The idea of a high-pressure steam-engine had been in the air as early as 1725, when Leupold had proposed a double-cylinder engine which exhausted the steam into the air after each stroke (a 'puffer').[3] As this was single-acting it was the perfect morphological converse of the double-cylinder exhaustive air-pump, and hence ultimately of the Ctesibian double-cylinder impulsive force-pump for liquids, owing much less (in fact only its association with steam, and the fact that it was designed to do work) to Papin's steam-and-vacuum piston and cylinder. It was clear that if such an idea could be realised it would enormously reduce the size of the steam-engine plant, indeed it would open the doors for it to run about in all directions as a universal haulier. Perhaps Trevithick saw Leupold's diagram, certainly he was busy with double-acting water-pressure

---

[1] *Nouvelle Invention de Lever l'Eau plus Hault que sa Source...* (London, 1644; Eng. tr. by J. Leak, London, 1659). His plate depicting a clock, the hands of which are moved on by a tipping bucket periodically operating a ratchet mechanism, is also reproduced by Needham, Wang & Price, fig. 45. The device was almost identical with one of al-Jazarī's (Fig. 30).

[2] This is of course a wholly schematic expression. Just as the steam-displacement pumping systems lasted on more than a century after the atmospheric engines had begun to work satisfactorily, so these in turn remained in use for more than another century after the first introduction of high-pressure steam-engines of definitive type. Thus atmospheric engines functioned long after Watt's time. It is clear from what Lardner says in 1840 (p. 72) that they were often then preferred to the more advanced types in mining areas where cheap coal was available (slack), partly on account of their cheaper initial cost of construction. It is hardly to be believed that Newcomen engines were at work well into the present century. One was used near Glasgow till 1915 (Dickinson SE, p. 65), and another near Sheffield till 1926, and again during an emergency in 1928 (Newbould).

[3] *Theatrum Machinarum Hydraulicarum* (Leipzig, 1725).

pumping engines[1] working at pressures up to 65 lb/sq. in. in the last years of the eighteenth century. It was in + 1798 that he built the first successful high-pressure engine in model form,[2] and by 1802 he had produced a design in which we can clearly see the definitive form of the nineteenth-century reciprocating prime mover (cf. Fig. 24).[3] Following close upon his heels was Oliver Evans at Philadelphia, whose engine of 1804 shows equally clearly the attainment of the same definitive form.[4] The vacuum had gone for good, slide-valves had come in, and boiler-construction was now the limiting factor for the pressures which could be applied to each side of the piston.[5]

In contemplating the efforts that were now made to increase the heating surface and strengthen the structure[6] it is worth remembering that they had curious antecedents in previous ages. Dickinson illustrated a remarkable Roman water-heater found at Pompeii and therefore earlier than + 79 in date, which has grate-bars in the form of water-tubes.[7] But there are other precursors, from Asia, notably the unexpectedly complicated water-jacket devices, sometimes amounting to a piped central cooling system, used by the alchemists of medieval China for keeping under control the temperatures in their reaction-vessels.[8] The most extensive description which we have of these occurs in a book of + 1225, the *Chin Hua Chhung Pi Tan Ching Pi Chih* (Confidential Hints on the Manual of the Heaven-Piercing Golden Flower Elixir), written by Phêng Ssu. But, of course, neither the ancient Roman nor the medieval Chinese tubing systems were

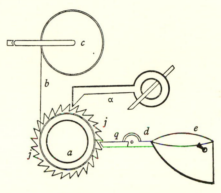

Fig. 30. Arabic tipping-bucket mechanism for rotating a disc or globe by water-power, from al-Jazarī's treatise on striking clepsydras, + 1206 (Wiedemann & Hauser).

---

[1] Dickinson & Titley, pp. 39 ff. These engines were almost exactly the converse of the double-acting water-pumps of de la Hire (see p. 175), motive power being derived from water pressure on the piston, not applied externally to the piston-rod. On the earlier history of water-pressure engines see pp. 140, 151 above, and Ewbank, pp. 355 ff. [See now also D. S. L. Cardwell, *TCULT*, 1965, **6**, 189, 195.]

[2] Dickinson SE, pp. 93 ff.; Dickinson & Titley, pp. 43 ff.

[3] Dickinson & Titley, pp. 53 ff., 60 ff., 269 ff. and especially fig. 7, as also figs. 39, 40 and 41 dissecting pl. VII.

[4] Dickinson SE, pp. 94 ff. and pl. IV.

[5] By 1802 Trevithick was working with a pressure of 145 lb sq. in. (Dickinson SE, p. 53).

[6] See Dickinson SE, pp. 117 ff.

[7] Figured by Dickinson SE, p. 124, this is now in the British Museum.

[8] See Ho Ping-Yü & Needham, pp. 77 ff.

intended to withstand any considerable pressures. Thus the problem before the builders of boilers from about + 1770 onwards was one on a new level of difficulty and indeed many years elapsed before its full resolution.[1]

THE CHINESE ANCESTRY OF THE DOUBLE-ACTING PRINCIPLE

We are now free to concentrate our attention upon the double-acting principle, i.e. upon that way of construction which ensures that the piston shall do effective work on every one of its strokes back and forth. This principle was, we know, embodied in the atmospheric engines with separate condensers built by Watt in and after + 1783;[2] it was necessarily accompanied by his trapezoidal pantograph link-work connection between the piston-rod and the beam, and went along with his conversion to rotary motion by the sun-and-planet gear. Though Leupold had followed Ctesibian tradition, Trevithick, Evans and all subsequent high-pressure engineers invariably made their systems double-acting, slide-valves being available after Murdock in + 1799 and Murray in 1801.[3] We have shown

[1] A number of interesting hybrid designs enjoyed brief life at the time when atmospheric engines were giving place to high-pressure engines. In + 1790 Adam Heslop, a Coalbrookdale man, combined the two principles by attaching to one end of the beam a 'hot' cylinder which used steam a little above atmospheric pressure, and to the other a 'cold' one in which the steam was then condensed, the condenser acting as a Newcomen cylinder (Raistrick, pp. 156 ff. and opp. p. 151; Farey, p. 671). Eight years later, James Sadler, an engineer also associated with Coalbrookdale, patented a similar arrangement, both pistons acting on the same end of the beam, and the 'hot' one fitted with a valve through which the steam passed on to the 'cold' condensing atmospheric cylinder (cf. Farey, pp. 669 ff.). The idea of having valves in the pistons themselves now led to developments in which two pistons were set upon a single piston-rod, developments which thus echoed long antecedent practices. For we find two pistons in a single cylinder and on a single rod in + 11th-century China, as the pumping mechanism of a military naphtha flame-thrower (cf. SCC, vol. 4, pt. 2, p. 148). And also in the ethnological field air-pumps with two pistons on a single rod in one long cylinder occur in Madagascar (Ewbank, p. 252, see cut opp.); they are evidently related to the Chinese double-acting piston-bellows (see pp. 118 ff., 155 ff.) but lack the air conduits to the common valve. Their presence in Madagascar indicates a more ancient Asian (Malayan) origin. Now, in the steam-engine's period of fluidity, the Rev. Edmund Cartwright (1797) mounted a Wattian air-pump piston on the same rod as his main one, a valve in which permitted the steam to pass through at the right moment to the condenser below (Farey, pp. 665 ff.). But this design had been employed in a more interesting way four years earlier when the same James Sadler had used it for combined high-pressure and low-pressure working in a double-acting tandem compound arrangement where the two cylinders were separated by a diaphragm (Raistrick, pp. 158 ff. and opp. p. 176). Steam above atmospheric pressure was admitted to the small h.p. cylinder below, and pressed down the piston till a valve in the diaphragm was automatically opened and the steam expanded against the larger-diameter piston in the l.p. cylinder above, forcing it upwards. At the end of the stroke it was evacuated to a condenser, and fresh steam supplied to the h.p. cylinder. In this system the two cylinders had to be of different diameters, as the Chinese and Madagascan ones had not been, but the basic morphology was similar. Here it is hard to imagine any direct genetic influence, but strange are the capillary channels of history through which such influences have flowed.

[2] The first successful experimental engine was executed at the Soho works in 1782. See Dickinson SE, pp. 79 ff., JW, pp. 123 ff; Farey, pp. 426 ff.; also in Singer et al., vol. 4, p. 187; Wailes (model). The vacuum was formed alternately above and below the piston.   [3] Dickinson SE, pp. 113 ff.

above that there was no model for this in the air-pump, otherwise so influential in the steam-engine's history. Where then did it come from? I shall now suggest that it came from far away, from the other extremity of the Old World.

BELLOWS AND BLOWING-ENGINES IN EAST AND WEST

Universal in China today for all artisanal purposes, and even on a larger scale for minor industries, is the box-bellows (*fêng hsiang*). Hommel says rightly that it surpasses in efficiency any other air-pump made before the advent of modern machinery.[1] From the longitudinal section (Fig. 18, p. 118) it can be seen that the box-bellows is a double-acting force and suction pump;[2] at each stroke, while actively expelling the air on one side of the piston, it draws in an equal amount of air on the other side. Whenever this bellows first came into general use it provided that fundamental metallurgical necessity, a continuous blast of air.[3] No less than twelve of the illustrations in the *Thien Kung Khai Wu* (+1637)[4] show its use by metal workers (cf. Fig. 31, a bronze foundry). In the ordinary Chinese box-bellows intake valves (*huo mên*) for air are provided at each end of the box, and a single double-acting valve underneath or on one side at the junction of the two outlet channels.[5] The

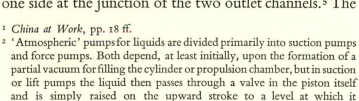

Madagascan air-pump
(see opp.).

[1] *China at Work*, pp. 18 ff.
[2] 'Atmospheric' pumps for liquids are divided primarily into suction pumps and force pumps. Both depend, at least initially, upon the formation of a partial vacuum for filling the cylinder or propulsion chamber, but in suction or lift pumps the liquid then passes through a valve in the piston itself and is simply raised on the upward stroke to a level at which it can discharge through an outlet; while in force pumps the unpierced piston drives out the cylinder's contents on the downward stroke into a rising pipe against as much pressure as the system will stand. Cf. Ewbank, 2nd ed. (1847), pp. 213, 222. If the upper end of a lift-pump is closed, a rising pipe may, of course, be fitted to it, and if this is paired with a force-pump, work will be done on both up and down strokes. Such an arrangement was used with Newcomen engines for town water supplies (Farey, pp. 213, 243 ff., 251 ff.). The Chinese air-pump, with its solid piston, is so arranged that it sucks and expels on both strokes.
[3] My attention was drawn to this long ago by my eminent metallurgical friend Dr Yeh Chu-Phei. Sir William Chambers, who figured it in 1757, called it a 'perpetual bellows'. There are classical descriptions of it in Lockhart, p. 87, and Ewbank, pp. 247 ff. Cf. Dinwiddie in Proudfoot, p. 74.
[4] *The Exploitation of the Works of Nature*, a technical encyclopaedia written by China's Diderot, Sung Ying-Hsing.   [5] It is interesting that Leonardo also sketched such a valve, for a tuyère (Beck BGM, p. 341).

155

Fig. 31. Chinese double-acting piston-bellows in characteristic use; a scene in a bronze foundry, from the *Thien Kung Khai Wu*, 1637 (Chhing drawing). Caption: 'Casting a tripod cauldron.'

cross-section is usually rectangular,[1] allowing for easy construction from wood, and the piston (*huo sai*) is packed with feathers (the ancestors of piston-rings). The common Japanese bellows, though similar, is less ingenious, since the piston

[1] It is not infrequently circular, however.

carries a valve, and the blast takes place only on the push and not on the pull.[1]

Ewbank much admired the Chinese box-bellows.[2] He regarded it as essentially equivalent to the Ctesibian double-cylinder force-pump for liquids, but with the two cylinders elegantly combined into one.[3] If pipes were connected to the intakes it would, he said, become the pump of de la Hire (+ 1716);[4] and he saw its reverse anticipation of the principle of the steam-engine cylinder of Boulton and Watt (+ 1783), the first to do work on both strokes. 'What may be surprising to some persons,' he went on, 'its construction is identical with that of the steam-engine; for let it be furnished with a crank and flywheel to regulate the movements of its piston, and with apparatus to open and close its valves, then admit steam through its nozzle, and it becomes the double-acting engine' commonly in use (he was writing in 1842). Besides its close formal resemblance to the steam-engine, the medieval Chinese piston-bellows could also have served in principle, said Ewbank, as a Boylian air-pump if both the intakes were derived from a sealed chamber. 'The most perfect blowing-machine, and the *chef d'oeuvre* of modern modifications of the pump', he concluded, 'are also its facsimiles.'

It is difficult, unfortunately, to bring forward much evidence about the precise antiquity of this machine, for little research has been devoted to it, and the obvious sources—encyclopaedias such as the *Thai-Phing Yü Lan*—[5] are at first sight unhelpful.[6] But pending the appearance of a history of bellows something like the following may be said. Bellows for metal-working played a very important part in ancient Chinese thought and mythology.[7] One of the legendary rebels against the Confucian sage-kings was named Huan-Tou, 'Peaceable Bellows'.[8] The oldest name for bellows was *tho*, the character for which is closely related to *nang*, a skin bag.[9] The most ancient type of bellows in China was therefore no doubt the

[1] Hommel, p. 20. For Japanese metallurgical piston-bellows see Gowland CAP, p. 17, and many other illustrations in his papers. [2] Pp. 247 ff., 251.

[3] Ewbank hinted that the Chinese form might have been the origin of the Alexandrian one, but what we know of the possibilities of transmission at that time does not [greatly] encourage such an idea. Cf. *SCC*, vol. 1, *passim*.

[4] Ewbank, p. 271. We shall return presently (p. 175) to the consideration of this device, the historical importance of which has not perhaps before been noticed.

[5] 'The Thai-Phing reign-period Imperial Encyclopaedia (lit. Daily Readings for the Emperor)', an invaluable source edited by Li Fang in +983.

[6] Liu Hsien-Chou could find nothing at all, but we shall be able to offer a few hints. Recently an attempt at a history of blowing machinery has been made by Yang Khuan TIC; this is valuable though very brief. [7] Cf. Granet under 'outre'. [8] Cf. *SCC*, vol. 2, p. 117.

[9] Nevertheless one may note the presence of the wood radical in *tho*, and the stone one too, suggesting that leather was not the only component. The word *tho* (bellows) occurs also in a doublet as *tho-tho*, the oldest

skin of a whole animal, with the legs tied off, some kind of valve at one end, and a delivery tuyère fastened into the other; working perhaps in pairs. The next development would have been to make the walls of the bellows partly of pottery or wood, with holes in the skin coverings of the pots upon which the feet of the operators took the place of valves[1]—such as are seen in well-known ancient Egyptian representations,[2] or used by contemporary African peoples.[3] The later term *ko pai* probably refers to this type.[4] Since these skin-covered pots resemble drums it is not surprising that one of the earliest verbs for the act of plying bellows was *ku* 'to drum', and this is very often found in Chou and Han texts. Reference has elsewhere been made[5] to the iron cauldrons which were cast with legal statutes upon them in − 512; here the *Tso Chuan* text has the expression *i ku thieh*. While some commentators[6] took this to mean a measure of weight which the inhabitants were taxed, others have interpreted it as 'iron blown by the bellows', i.e. cast iron.[7]

Some light on the bellows of the Warring States period is forthcoming from an unexpected source, namely the chapters in the *Mo Tzu* book on military technology. From these it is clear that in the late − 4th century it was customary to use toxic smokes made by burning balls of dried mustard[8] and other plants in stoves; the smoke being directed by bellows against troops attacking cities, or blown into the openings of enemy sap tunnels. We learn[9] that 'the bellows are made of ox hide, with two pots to each furnace, and they are worked by a swape lever tens and hundreds of times (up and down) (*tho i niu phi, lu yu liang fou, i chhiao ku chih pai shih*)'. Or 'each stove has four bellows, and when the enemy's tunnel is about to be penetrated, then the oscillating swape levers are furiously worked to blow the bellows fast and fumigate the tunnel (*tsao yung ssu tho, hsüeh*

---

name of the camel. We cannot follow Yang Khuan in his argument that this implies that the earliest bellows were camel-back-shaped. As Schafer shows, the borrowing was rather in the opposite direction, and for *tho-tho*, probably a loan-word from some Central Asian language, characters were chosen which implied 'the sack-carrier', because of its humps.

[1] It is interesting to reflect that descendants of these pot pumps are still in use, in the form of diaphragm pumps such as those for the petrol feed in automobiles, and those which with sleepy rhythm keep street excavations dry during the hours of night. [See p. 162 and the diagram on p. 164 below.]

[2] Westcott, pl. XIX; Gowland CAP, p. 15; Neuburger, pp. 25, 49; Feldhaus TDV, col. 369; Ewbank, p. 238; Wilkinson, vol. 2, p. 316; Blümner, vol. 4, pp. 140 ff.

[3] Cline. For ancient European references see Feldhaus TDV, col. 367 ff.; Blümner, vol. 4.

[4] *Pai* is often misprinted in Chinese texts by *pi*, which is a technical term relating to harness.

[5] SCC, vol. 2, p. 522. The text is Duke Chao, 29th year (Couvreur, vol. 3, p. 456).

[6] As Fu Chhien in the Han.   [7] So Tu Yü in the Chin, and to some extent Khung Ying-Ta in the Thang.

[8] The volatile oil of this plant (as also of the onion and horse-radish) is highly irritant; its active principle is allyl iso-sulphocyanide (cf. Sollmann, p. 693, or any pharmacological treatise).

[9] Ch. 62, p. 20*b* (ch. 62, p. 26*a* in Wu Yü-Chiang's reconstructed edition).

chhieh yü i chieh-kao chhung chih, chi ku tho hsün chih)'.[1] The interest of these passages lies in the fact that already by this date a mechanisation of the push-and-pull motion alternating between two cylinders (or pots) seems to have been introduced, exactly as in the Alexandrian double-cylinder force-pump[2] and so many subsequent 'fire-engines' in later Europe. The use of bellows with toxic

Fig. 32. Annamese single-acting double-cylinder piston air-pump (bellows), also used in Laos and other parts of South-east Asia (Frémont, after Schroeder). Caption: 'Drumming blast boxes for smelting copper.'

smoke projectors must go back to the beginning of the −4th century, for the early chapters of the *Mo Tzu* book also mention them,[3] but without reference to the mechanical swape lever. This invention did away with the necessity of having one person to work each single-acting barrel, such as we see in Schroeder's drawing (Fig. 32) of an Annamese piston-bellows, the continuous blast being obtained by the fact that one piston was rising while the other was descending. From this point it was not really a far cry to the conflation of the pistons and cylinders into an efficient unity (once you thought of it), though unfortunately we know nothing of the ingenious conflator.

---

[1] Ch. 52, pp. 9b, 10a (ch. 62, p. 25a in Wu Yü-Chiang). Cf. chs. 52, p. 11a, 61, p. 17b, and 62, p. 20a. A reference of nearly the same date, perhaps fifty years later, is in *Han Fei Tzu*, ch. 47, p. 4b (Book of Master Han Fei), Liao-Wên-Kuei tr., vol. 2, p. 252.
[2] Cf. the Heronic reconstruction of Woodcroft, p. 44. But Drachmann MTGR, p. 155, doubts whether only a single lever was used.   [3] Ch. 20, p. 2b (Book of Master Mo).

There is one curious hint that this may have occurred as early as the fourth century. The *Tao Tê Ching* (conservatively of this date) says:[1]

> Heaven and Earth and all that lies between,
> Is like a bellows with its tuyère (*tho yo*);
> Although it is empty it does not collapse (*hsü erh pu chhü*),
> And the more it is worked the more it gives forth (*tung erh yü chhu*).

Fig. 33. Japanese *tatara* forge and furnace bellows of all-wood cuneate fan or 'hinged piston' type (Ledebur). Caption: 'The foot-bellows for iron.'

The statement in the third line could hardly have been made of any kind of collapsible leather bellows, but would clearly apply to the piston variety whether the latter moved in an arc or longitudinally, i.e. whether it was hinged or straight-sliding. Commentators from Wang Pi[2] to Huang I-Chou[3] say that what Lao Tzu

---

[1] 'Canon of the Tao and its Virtue', the greatest Taoist scripture, ch. 5, tr. Hughes, p. 147, mod. The version of Waley fails to convey the technical point present in the third line.
[2] Of the San Kuo period (mid+3rd century).
[3] In his *Shih Nang Tho* (Philological Study of Bags and Bellows), an eighteenth-century monograph.

was referring to was the *phai tho*, i.e. the 'push-and-pull' bellows. One of them[1] explains further that the *tho* is the outer box or case (*tu*) into which the tuyère (*yo*) is fitted, and that the latter is a tube through which passes the air forced by the 'drumming'. An interesting point is that in the Han period the bellows were worked by hand, if we may judge from the word *chhien*, which the *Shuo Wên*[2] defines as 'the handle of the bellows' but which originally meant the ears of a jar.[3] Nevertheless some foot-operated types long persisted, such as the large hinged swinging-flap bellows used in the Japanese *tatara* method of iron-smelting (Fig. 33).[4]

Evidence for the existence of the piston air-pump in the Han may perhaps be derived from an interesting passage in the *Huai Nan Tzu* book.[5] Complaining about the decline of primitive simplicity, the writer says that the demands of the metal-workers for charcoal have even led to the destruction of forests. Among the extravagances of the age, 'bellows are violently worked to send the blast through the tuyères in order to melt the bronze and the iron (*ku tho chhui tuo i hsiao thung thieh*)'. *Ku tho* could be taken as 'drum-bellows' but the commentator, Kao Yu, who lived about +200, explains that the word is a verb and means to strike or beat (*chi*). He says further that the bellows (*tho*) is a 'push-and-pull' bellows for the smelting furnace (*yeh lu phai tho yeh*). And he elucidates the unusual word *tuo* by saying that it is a tube of iron conducting the blast of the bellows into the fire (*tho khou thieh tung, ju huo chung chhui huo yeh*). His use of the word *phai* is all the more significant because it was the traditional term for the metallurgical blowing-engines operated by water power which came into use early in the +1st century.[6]

The expression 'bellows with hinged pistons' used just now may sound curious and needs a little explanation. We are probably safe in assuming that the bellows of wood and leather shaped like the frustum of a pyramidal wedge is one of the oldest forms of blower used in the West, though still to be found beside our domestic fireplaces.[7] Medieval and Renaissance European metallurgical industry depended

---

[1] Probably Wu Chhêng of the Yuan (early +14th century).
[2] Hsü Shen's great analytical dictionary of +121.
[3] Later dictionaries such as the *Kuang Yün* say 'the handle of a push-and-pull bellows' (+10th century). For a further discussion on this see Yang Khuan TIC.
[4] Cf. Gowland EMC, pp. 306 ff., and Muramatsu Teijirō.
[5] Ch. 8, p. 10 (Book of the Prince of Huai-Nan), one of the greatest compendia of natural philosophy in ancient Chinese literature.
[6] The translation of this important passage by Morgan, p. 95, was grossly astray.
[7] On the evidence adduced by classical archaeologists, if the whole-skin bag was Homeric and Hellenic, the cuneate wood and leather bellows was at least Hellenistic, for we find it in model form on bronze lamps. See

on it almost exclusively. We can see hundreds of such bellows in Mariano Taccola (+ 1440), in the great works of Agricola and Ercker, and abundantly in the *Pirotechnia* of Biringuccio about + 1540, where batteries of them are worked in various ways by trip-lugs on water-driven shafts, or by systems of cranks, levers and weights.[1] According to the researches of Beckmann[2] ironfounders about this time grew tired of the expense and trouble of oiling and maintaining bellows with flexible leather parts, and blowers made wholly of wood and metal began to take their place. The introduction is ascribed to Hans Lobsinger of Nuremberg, the Schelhorns and others, about + 1550, and the new forms became fairly widespread during the first half of the + 17th century.[3] Available drawings[4] show that the pyramidal cuneate form was retained, one half being slightly smaller than the other, which slid up and down over it, thus compressing and expelling the air. Arrangements of springs at the sides secured tightness. This swinging system lasted on until very recent times.[5]

Exactly at what period it was widely outmoded by reciprocating piston-and-cylinder air blowers is difficult to say in a sentence, because different industries changed at different times. In the iron and steel industry, where the strength of the blast was particularly important, the process was taking place in the middle of the eighteenth century.[6] In + 1757 John Wilkinson patented a blowing-engine of Ctesibian type (i.e. with two single-acting cylinders) powered by a water-wheel,[7] and by about + 1762 John Smeaton was building very effective blowers of this

Daremberg & Saglio, under *ferrum, flabellum, follis* and *uter* (vol. 2, pp. 1087, 1149 ff., 1227; vol. 5, pp. 613 ff. and figs. 3122, 3133, 3134); Blümner, vol. 2, pp. 190 ff. The first literary reference appears to be in Ausonius (+ 4th century), who refers to the valve. The water-powered bellows used in German mining centres from + 1200 onwards, as described by Johannsen, were certainly of this type.

[1] See Beck BGM, pp. 116 ff., 154, 289, 470, and also Agricola; Hoover & Hoover ed., p. 365. [Cf. Fig. 52, pl.]

[2] 4th ed., vol. 1, pp. 63 ff.; Singer *et al.*, vol. 3, p. 32.

[3] Leonardo saw at Brescia late in the + 15th century cuneate bellows wholly made of wood but open at the bottom and dipping into water, which formed the seal. He also described the 'Harzer Wettersatz', in which the bellows consist simply of two barrels open at the bottom and thrust down into water alternately (Beck BGM, pp. 340, 341). This was perhaps the first way in which leather was dispensed with. [Cf. p. 116.]

[4] E.g. in Schlutter (Schlüter), vol. 1, pl. III*b* and p. 325; vol. 2, pl. VI*g, h, i*, and p. 55; de Genssane; Ure, 1st and 3rd eds., pp. 1127, 1128; Paulinyi, p. 167. Cf. Singer *et al.*, vol. 4, p. 125.

[5] In its final eighteenth- and nineteenth-century stages the pyramidal form was lost and the bellows became purely cuneate (cf. Lindroth), thus assimilating very closely to the Chinese + 14th-century construction.

[6] Leonardo, as one might expect, had been thinking in this direction long before, but, so far as we know, without practical effect. He imagined a very curious single-acting cylinder bellows in which the loose-fitting piston alternately everted and invaginated a 'glove-finger' of leather; perhaps this was a device to solve the packing problem at a time when cylinders were so hard to make accurately (cf. Beck BGM, pp. 339 ff.). This was to some extent the ancestor of Samuel Morland's plunger pump of 1674, with its two 'hat-leathers', for raising water (Wolf XVII, p. 541).

[7] See Dickinson JWE. Transmission was by means of a two-throw crankshaft.

kind.[1] In +1776 James Watt harnessed one of his single-acting steam-engines to a single-acting blowing-cylinder of large bore, with a water blast regulator, at an ironworks in Shropshire.[2] This service was therefore one of the very first that

 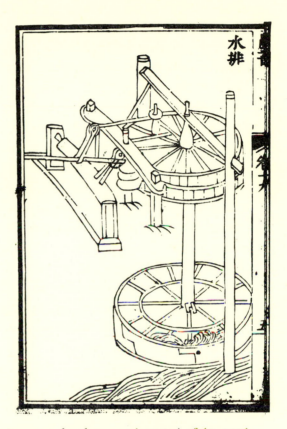

Fig. 34. Water-powered blowing-engine (*shui phai*, water-powered pusher, or reciprocator) of the +13th century, for blast-furnaces and forges; with horizontal water-wheel, flywheel, belt-drive and fan or hinged piston-bellows; the conversion of rotary to longitudinal motion being effected by the standard crank, connecting-rod and piston-rod combination. From the *Nung Shu* (+1313).

the Watt separate-condenser engines were called upon to perform.[3] In the following year Watt built a steam blowing-engine, also for Wilkinson, in Lancashire, so arranged that the blowing-cylinder was double-acting;[4] 'this', says Farey, 'was

[1] See Farey, pp. 273 ff.

[2] Dickinson JW, p. 90; Farey, pp. 320, 724.

[3] During the following dozen years Newcomen engines were also put to this work in various places (Farey, pp. 281 ff., 285 ff.), generally with weighted floating pistons in a third or 'regulating' cylinder, rather than with water regulators.

[4] Farey, pp. 328, 724.

the first application of the double-acting air-pump, or blowing-cylinder, for furnaces; it is now the universal practice.'[1] After the appearance of Watt's double-acting engines in +1783, these also were applied to work the double-acting blowers;[2] but it is clear—and this is of some importance for our general argument—that the double-acting principle was embodied first in 'piston-bellows' worked by steam prime movers, and then in the prime movers themselves, the one use leading to the other by a natural transition.[3] These dates are highly significant, as we shall later see (p. 174).

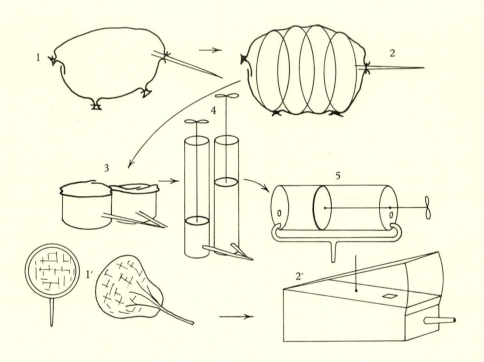

A similar process seems at first sight to have taken place in China and Japan, leading to the 'swinging fan' or 'hinged-piston' bellows always shown in the Chinese drawings of water-powered blowing-engines (Fig. 34), and in those of

[1] This engine had a weighted floating piston in the third (regulating) cylinder, not a water blast regulator, but after about 1794 the latter device became practically universal (Farey, pp. 286 ff.). Its direct descent from the pressure regulation device of Heron's 'water-organ' (p. 145 above) is interesting to note.

[2] See Raistrick, opp. p. 241; Singer *et al.*, vol. 4, pp. 103, 104, 111 and fig. 58. This is redrawn from a sketch due to William Minor and dated 1798. Farey, pp. 723 ff. and pl. xxv, illustrates one of 1807.

[3] It is interesting to find in Tredgold (1827), pl. xiii, a drawing of a double-acting Watt engine working a double-acting de la Hire water-pump, but the date when this was first practised is not clear. This combination does not seem to have been so much in the direct line of evolution as was the piston-bellows.

the Japanese *tatara* bellows[1] (Fig. 33).[2] But the manually operated double-acting piston-bellows so completely ousted the earlier leather forms in the Chinese Middle Ages that we have very little idea of what they were like. I have already suggested that in the Chou period (− 1st millennium) they were originally large leather bags formed from the skins of whole animals, then later they were probably leather-covered pots like drums, and later still tree-trunks hollowed out to form single-acting cylinders with reinforced leather pistons. Eventually came our main subject of discussion, the *fêng hsiang*, consisting entirely of wood, with no leather at all. Thought of in this way, one can perhaps regard the piston as having arisen (in China at any rate) by a gradual reduction of the leather component of the bellows, and the gradual augmentation of the non-flexible tuyère component, along one continuous line of evolution (see diagram). Notwithstanding the evidence just adduced concerning the bellows of the − 4th and − 2nd centuries, it is quite likely that the whole-skin type persisted into the − 1st century,[3] the time when, as we shall soon see, water-power was applied on a considerable scale to iron and steel manufacture. This type seems by then to have been made non-collapsible by the use of interior rings of bamboo.[4] But strange to say the one thing that never occurs in traditional Chinese culture is precisely the pyramidal cuneate form of wood and leather so characteristic of Europe, nor is there any evidence, textual or epigraphical, that it ever did. One has to look therefore for some other origin for the hinged-piston or swinging-flap form of medieval China and Japan. There can be no doubt at all that it was derived from that very ancient Chinese amenity, the fan—not the folding fan of today, which was a Japanese or Korean invention of the late + 10th century, but the leaf-shaped *flabellum*[5]

[1] Cf. Needham DITC, figs. 32, 33, from Ledebur. See also Sévoz.

[2] The oldest representation so far discovered of the East Asian fan bellows, or hinged piston working in a geometrical cylinder segment, is found in a Hsi-Hsia fresco at the Wan-fo-hsia (Yü-lin-khu) cave-temples in Kansu province (Fig. 35, pl.); it must date from between the + 10th and + 13th centuries. [The apparatus occurs again in the oldest Chinese picture of a blast-furnace, that in the *Ao Phu Thu Yung* of + 1334; see Needham DITC, fig. 25.]

[3] In many parts of the world, whole-skin bellows can still be found today, e.g. the sheepskins of the smaller Persian forges (pers. obs. Mr Hans Wulff).

[4] Something like the cylindrical leather bellows with rings of metal used by the Portuguese tin-smelters in Agricola's time (cf. Hoover & Hoover ed., p. 419, and Beck BGM, p. 156, figs. 173, 174). Leonardo made a drawing of a similar one (Beck BGM, p. 340, fig. 466). One might call this the 'concertina' as opposed to the 'cuneate' type. The concertina form has persisted into modern technology, often made wholly of metal, as e.g. in aneroid barometers, thermostats for automobile cooling systems, and 'iron lungs' in hospitals (cf. Burstall, pp. 390 ff., 405). [Cf. Figs. 41, 44, 50.]

[5] See the article already quoted in Daremberg & Saglio, vol. 2, pp. 1149 ff. The most notable contemporary appearance of the occidental *flabellum* perhaps (apart from the simple hand-fans still used in Mediterranean countries) is as one of the liturgical objects carried by clerks of the Orthodox churches.

still used in untold millions during every Asian summer, and indeed through-out the year by every Chinese housewife who still cooks so cleverly on the age-old charcoal stove.[1] Moreover there are plenty of examples of such single-leaf fans used industrially or agriculturally in China; one need only instance those elongated laterally-swinging types depicted in scenes of winnowing in Han reliefs.[2]

THE RELEVANCE OF THE MALAYSIAN FIRE-PISTON

That the double-acting piston-bellows was well known even to literary scholars and philosophers in the Sung dynasty (+11th and +12th centuries) it is easy to demonstrate, as in a moment shall be done. For the general argument of this lecture, I admit, it matters little whether the *fêng hsiang* was an achievement of the +1st or of the +11th century but the question of its date is an intriguing one, and a point of much engineering interest arises. The probability of its de-velopment in relatively early times in China is strengthened, I believe, by certain remarkable facts of an anthropological nature. Perhaps the leather skin of the pot or drum bellows was induced to turn into a piston at a surprisingly early stage because of the presence in one strand of Chinese culture of a simple piston and cylinder which was not designed to push either air or water from place to place, but to heat air to an ignition-point by a sort of adiabatic compression, and so to make fire. It is a familiar observation among us that when one inflates a bicycle tyre the lower end of the pump becomes hot. And indeed the rapid reduction of air to one-fifth of its original volume gives a temperature sufficient to ignite tinder.[3] This fact was discovered by the primitive peoples of south-east Asia (especially the Malayan–Indonesian region), who made use of it in one of the most remarkable of all eotechnic devices, the piston fire-lighter (Fig. 36, pl.).[4] At the bottom of the instrument the tinder is held in a small cavity. The ethnography of the device was sketched in a classical memoir by Balfour, and its presence in

[1] On the history of the fan in China see Forke, vol. 2, App. III, pp. 490 ff.

[2] Cf. *SCC*, vol. 4, pt. 2, fig. 436.

[3] This fact is at the basis of the most modern theories of the nature of explosion (Bowden & Yoffe). Initiation of explosion by mechanical means is due to the formation of minute hot spots, either because of the adia-batic compression of microscopic pockets of trapped vapour, or, more rarely, friction and viscous heating. Thus the physical chemists became interested in the ancient device of the fire-piston.

[4] Hough, pp. 109 ff.; Leroi-Gourhan, vol. 1, p. 68. We may well believe that it was associated with that other ancestor of all piston-engines and projectile-guns, the blowpipe gun, which also belongs to the culture of the area. Was the piston a tethered projectile, or the projectile a liberated piston? Cf. Lynn White, pp. 93 ff.

Madagascar was one of the pieces of evidence which decisively showed the Malay origin of the indigenous Madagascans.[1] Just how far back in time this application of the piston goes is an extremely difficult question to answer, for there is no way of telling to what extent the primitive populations of such regions have been technologically static over the centuries, but there is certainly every reason for thinking that the invention was autochthonous in south-east Asia.[2] It is moreover indigenous in Yunnan province.[3] If then it could be considered part of the stock-in-trade of the ancient Malay–Indonesian–Oceanic component of Chinese culture, the Chinese piston-bellows might well be regarded, at any rate on a working hypothesis, as partly derived from it. Piston-bellows are rather widespread in the more primitive cultures of East Asia, e.g. the double form used by the Khas of Laos.[4] If this is accepted, we may have to suppose that the invention of clack valves occurred twice, once in the Chinese area, and once in the Mediterranean region when the Hellenistic water-pumps were derived from the ancient Egyptian syringe.[5]

In the fullness of time the fire-piston proved capable of exerting a seminal influence far away from its south-east Asian home. About 1877 Carl Linde, the pioneer of artificial refrigeration, gave a lecture at Munich in the course of which he demonstrated a cigar-lighter made on the fire-piston principle. Among his hearers was Rudolph Diesel, who said in later years that this experience was one of those which had most stimulated him to the invention of the high-compression internal-combustion engine now universally known by his name.

[1] Madagascan single-acting piston-bellows used in pairs are figured by Ewbank, p. 246. They are similar to the Annamese type in Fig. 32. Some Madagascan forms are apparently like suction-pumps, with valves in the pistons. Cf. p. 155 above.

[2] It was actually patented in early nineteenth-century Europe, but this development is often considered to have been derivative from south-east Asia. [However, R. Fox, in an article in *TCULT* 1969, **10**, 355, shows fairly conclusively that the fire-piston of nineteenth-century Europe was an independent invention. In seeking to derive the south-east Asian device from this, however, he is much less convincing, and for us the European invention remains a re-invention.]

[3] Medhurst.

[4] Illustrated by Sarraut & Robequin. My late friend Mr Hans Wulff, who knew Laos well, confirms this from personal observation—two single-acting cylinders like those of the Annamese.

[5] Frémont OES suggested that the valve derived in the West from the round Greek and Roman shield which, when suspended, was used to open or shut ventilator holes in roofs or the tops of cupolas. See also Montandon, p. 275.

THE RELEVANCE OF WATER-PUMPS IN EAST AND WEST

The history of piston-bellows is closely related to the history of water-pumps.[1] Here the limiting factors were primarily the pressures which any palaeotechnic system might be likely to withstand. That the Alexandrians developed, and the Romans used, simple bucket suction pumps in which the water is lifted by the piston during its upward stroke, passing through it by a valve from below during its downward stroke, was supposed by Ewbank[2] (particularly for clearing ships' bilges), but Usher[3] regarded this as very doubtful, and he was right.[4] By the time of Agricola, however (mid + 16th century), these pumps were in wide use.[5] The Ctesibian force-pump (in which the liquid does not pass through the piston, but is forced out by an exit pipe) was, on the other hand, as we have already seen, well understood in Hellenistic times, as is shown by the discussion in Vitruvius,[6] who speaks of the cylinder and the piston as *modiolus* and *embolus* respectively. The machine must have been used quite considerably, for a number of such pumps have been found, from the Roman example contrived in a solid wood block and brought to light in a famous excavation at Silchester by Hope & Fox, to the remarkable bronze castings of Bolsena.[7] Westcott believes that this type of pump was little used in subsequent centuries, presumably owing to its greater complexity, and it hardly appears again until the time of Cardan (+ 1550) and Zeising (+ 1613).[8]

Generally speaking, pumps for liquids were not a feature of the Chinese eotechnic tradition, and their illustrations in the *Chhi Chhi Thu Shuo*[9] of + 1627

---

[1] Just how closely may be appreciated from the many Renaissance designs for water-pumps having precisely the form of leather air bellows. Leonardo sketched one of the cuneate type which was to be compressed by worm-gear (Beck BGM, p. 470). Both he and Ramelli also planned to use bellows of the concertina type as water-pumps (see Beck BGM, pp. 227, 470, 471). To the last of these we have already referred. I doubt very much whether any of these designs were made use of in practice. But they go back long before the Renaissance, for one can find them in the works of the Banū Mūsā brothers (fl. + 850), both cuneate and concertina (Schmeller, pp. 24 ff.).

[2] P. 214.

[3] P. 85.

[4] [It has now been demonstrated that the first depiction of a suction-lift pump can be dated to + 1433, a drawing of Jacopo Mariano Taccola (S. Shapiro, *TCULT*, 1964, **5**, 566). This raises rather acutely the possibility that the idea of the valved piston was transmitted from the ancient Chinese brine borehole tube-brackets described just below.]

[5] Hoover & Hoover ed., pp. 177 ff.

[6] x. 7; illustrated again in Neuburger, p. 229; Usher, p. 86; Perrault ed., p. 321.

[7] Smith, p. 120. Some of these have poppet valves like a modern internal-combustion engine.

[8] Westcott, p. 37; Beck BGM, pp. 176, 214 ff., 396 ff. Beckmann, vol. 2, pp. 245 ff., engaged in an elaborate discussion of the extent to which the Ctesibian force-pump was used as a fire-engine in Hellenistic antiquity and the early Middle Ages, but the question remains obscure.

[9] Ch. 3, pp. 20b, 56a ff. (Diagrams and Explanations of Wonderful Machines) by Têng Yü-Han (Johann Schreck, S.J.) & Wang Chêng, 1627.

汲鹵

Fig. 37. Szechuanese brine-field deep-drilled borehole derrick and valved bamboo bucket, from the *Thien Kung Khai Wu*, +1637 (Chhing drawing). Caption: 'Sucking up the brine'.

were no doubt a novelty at the time.[1] Yet there had been one element of traditional art which involved a principle near to that of the suction-lift pump, namely the long bamboo tube-buckets (*chi shui thung*) which were being sent down, from Han times onward, to the brine at the bottom of the boreholes of the Szechuan salt-field (Fig. 37). These buckets carried a valve at the base by which they were filled, and would have constituted suction-lift pumps if they had fitted tightly to the walls of the boreholes.[2] But the Chinese aim was different; it must be remembered that the contents had to be raised from 1,000 to 2,000 ft, not spilled out after a short haul within the limits of a vacuum which atmospheric pressure could fill. According to Esterer's observations forty years ago, the filling time at the brine was 180 seconds, the emptying time at the bore-head 300, the raising time $25\frac{1}{2}$ minutes for each load, the dimensions of the buckets 25 m. (about 75 ft) long by 7·66 cm. (3 in.) internal diameter, and the contents 132 kg. (about 28 gallons). This was a considerable engineering operation.

The relationship of the brine bucket valves to valves in air-pumps or bellows was perfectly appreciated by the great poet Su Tung-Pho in a passage written about + 1060. In his description of the Szechuan salt industy, he says:[3]

Szechuan is far from the sea, so people get salt from brine boreholes. Those at Lingchow are the oldest, but the brine at Yü-ching and Fu-shun has also long been known. At Phu-chiang Hsien in Chiungchow the (brine) wells were bored in the Hsiang-fu reign-period (+1008 to +1016) by a man of the people, Wang Luan, and they profited the populace exceedingly. In the Chhing-Li (+1041 to +1048) and Huang-Yu (+1049 to +1054) reign-periods and subsequently the people in Szechuan began to use 'tube' wells. They employ a round boring tool the size of a bowl, and drill to great depths of several hundreds of feet, then they use large bamboo stems with the nodes removed and fitted together by male–female joints to form the borehole and act as a tube (down below). Then waste fresh water is poured in at the side at the top, whereupon the strong brine comes up by itself.[4]

They also use smaller bamboo tubes which travel up and down in the wells; these cylinders have no (fixed) bottom, and possess an orifice in the top. Pieces of leather several inches in size are attached (to the bottom, forming a valve). As the buckets go in and out of the brine, the

---

[1] I say 'generally speaking' for there are certain accounts such as that of the water-raising machinery built by Pi Lan in +186 which are hard to explain if piston-pumps were not used. [See p. 361 below, and *SCC*, vol. 4, pt. 2, p. 345.]

[2] See Esterer, p. 148. The *Thien Kung Khai Wu* (ch. 5, p. 4a) says: 'When the borehole has reached the brine level, good bamboo culms (several) tens of feet long are selected. The intermediate septa are bored clean through but the lowest one is allowed to remain in place (except for a hole which carries the valve). This 'throat' at the bottom assures the entry of the brine into the (elongated) bucket, sucking and exhaling like a cycle of breathing. A long cable is fastened to the bucket, which is lowered until it becomes full of brine...' The valve may sometimes have been at the side of the tube rather than through the basal septum, but the latter was the commonest arrangement.

[3] *Tung-Pho Chih Lin*, ch. 6, p. 8b. [4] [Paragraph newly added.]

air by pushing and sucking makes (the valve) open and close automatically. Each such cylinder brings up several *tou* (pecks) of brine. All these bore-holes use machinery (hoists). Where profit is to be had, no one fails to know about it.

The *Hou Han Shu*[1] speaks of 'water (-driven) bellows' (*shui pai*).[2] This is applied to iron-working in Szechuan, and large ones are used.[3] It seems to me to be the same kind of method as that used in these salt-wells. Prince Hsien[4] did not understand this, and his ideas on the subject were wrong.[5]

This is valuable evidence on a number of points. Since Su Tung-Pho identifies valves in bellows working like those of the buckets in the shafts, the piston-bellows in some form or other must have been fairly familiar to him, and though water-driven blowing-engines were probably less common, he speaks as if he had himself seen them.[6] He reproaches Li Hsien for wanting to substitute words meaning 'leather bellows' for the 'push-and-pull' of the text. Then, just over a century later, we find a further reference to piston-bellows in one of the works of the great Neo-Confucian philosopher, Chu Hsi. About +1180 he wrote a commentary on the Han alchemical book of Wei Po-Yang, entitling it *Chou I Tshan Thung Chhi Khao I*. Wei had said that four male and female *kua* (the hexagram symbols of the 'Book of Changes') functioned like the bellows and the tuyère, to which Chu Hsi added the following remark:[7]

These *kua* are those in which the Yin and the Yang are combined, namely Chen (no. 51), Tui (no. 58), Sun (no. 57) and Kên (no. 52).[8] The bellows (*tho*), the piston (*pai*), the bellows-bag (*nang*) and the tuyère (*yao*) are the tubular spaces (through which they work)...The bellows should sometimes be worked slowly and sometimes rapidly (according to the degree of heating desired), just as the moon waxes and wanes.[9]

[1] Ch. 61, p. 3*b* (History of the Later Han Dynasty).

[2] The expression it actually uses is *shui phai*.

[3] One cannot tell from the text whether this sentence should be translated in the past or the present tense; the latter is equally likely.

[4] Li Hsien (fl. *c.* +670); he made a commentary on the *Hou Han Shu* in the Thang dynasty.

[5] Li Hsien said that *phai* ought to be written *pai* (equiv. *pai*), these two latter characters suggesting bellows of flexible leather. This *may* mean that in his time the fan or piston types had not yet been applied to water-powered blowing-engines.

[6] He came from Szechuan province himself. His evidence is all the more valuable as he was a poet, not a technician.　　　　　　　　　　　　　　　　　　　　　[7] P. 3*a*.

[8] Cf. *SCC*, vol. 2, table 14.

[9] Chu Hsi must be speaking of at least two different types of bellows here. *Nang* presumably refers, perhaps archaically, to the ancient 'concertina' whole-skin type, such as was used in the metallurgical blowers of the Han period (cf. p. 187 below) and may possibly still have existed here and there in the Sung, when Chu was writing. Since, however, we know from the work of Wang Chên (p. 180 below) that the usual practice in the Sung was to use the more powerful bellows of the fan type (or perhaps also the piston type proper), *tho* and *pai* must refer to them. The Japanese term for the man-powered rocking foot fan type, *tatara*, is equivalent to *tao pai* in Chinese; and we have just seen that Su Tung-Pho, in the century before Chu Hsi,

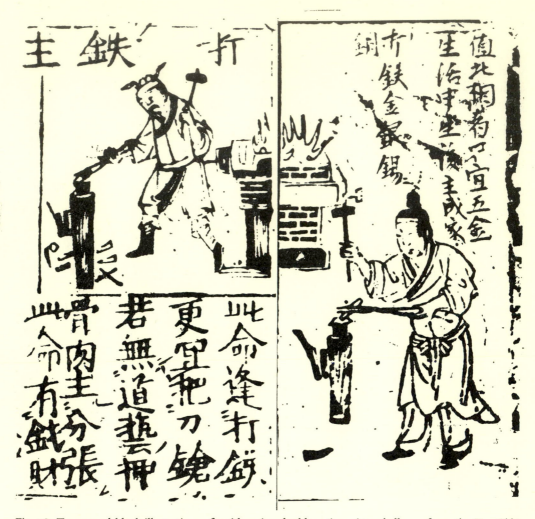

Fig. 38. Two wood-block illustrations of smiths using double-acting piston-bellows, from the *Yen Chhin Tou Shu San Shih Hsiang Shu*, a book on divination printed about +1270. Left, the blacksmith. The text says: 'When a blacksmith has this destiny it is especially fitting that he forge sabres and spears. If he has not got the knack and the skill, his family will have to endure separation.' Right, the silversmith. The text says: 'He who meets with this prognostication will best make a livelihood from the five metals; he will become the head of the family. He will work both copper and iron, gold, silver and tin.'

wrote *shui pai* to refer to water-powered blowing-engines; moreover, he took it as equivalent to the *shui phai* of the text from which he was quoting, i.e. to 'water-driven pullers-and-pushers'. Since Chu Hsi mentions *tho* as well as *pai*, we translate the latter as 'piston', recognising, however, that it may at least as well mean the fan or 'hinged piston'. The word *pai* has fallen out of use in modern Chinese, and the term for piston is now *huo sai*, 'the live, i.e. mobile, stopper-up'. Further evidence which clinches this interpretation will be found in C, p. 125, D, p. 885 and *Chu Chhi Thu Shuo*, p. 1*b*.

And finally one can actually illustrate piston-bellows from the century following Chu Hsi, for a book printed about + 1270 gives two small pictures of smiths working at their anvils with unmistakable piston-bellows by their side (Fig. 38). This is the *Yen Chhin Tou Shu San Shih Hsiang Shu* (Book of Physiognomical, Astrological and Ornithomantic Divination according to the Three Schools),[1] attributed to Yuan Thien-Kang. We can thus quite safely conclude that the piston-bellows was well known in the Sung. Reasons have already been given for thinking that it probably goes back much further, perhaps long before the Thang. The fact that Yuan Thien-Kang was a diviner of that period (d. *c*. + 635) adds further reinforcement to this view.

Looking at these pictures, and knowing the simplicity and elegance of the mechanism, we can understand the exuberance of Ewbank, who prefaced his remarks on the *fêng hsiang* with the following exordium:[2]

The cylindrical forcing-pump [he said] is the bellows of the most numerous and most singular of all existing peoples—a people the wisdom of whose government has preserved them as a nation through periods of time unexampled in the history of the world, and which still preserves them amidst the prostration by European cupidity of nearly all the nations around them; a people, too, who notwithstanding all that our vanity may suggest to depreciate, have furnished evidence of an excellence in some of the arts that has never been surpassed. The Chinese, like the ancient Egyptians, whom they greatly resemble, have been the instructors of Europeans in several of the useful arts, but the pupils, like the Greeks of old, have often refused to acknowledge the source whence many inventions possessed by them were derived, claiming them as their own. Of the truth of this remark, we need only mention printing,[3] the mariner's compass[4] and gunpowder[5] as examples. In the bellows of the Chinese, too, we perceive the characteristic ingenuity and originality of that people's inventions.

This brings us to the end of a chapter, but at the same time to a vital question: how did the double-acting piston-bellows become known to Europe in the eighteenth-century steam-engine age?[6]

---

[1] Photolithographically reproduced, Tokyo, 1933, ch. 2, pp. 35*a*, 36*a*. I am grateful to Dr Piet van der Loon for bringing the reprint of this rare Sung book to my notice. It contains, I found, other representations of technical interest besides the bellows.

[2] P. 247. It will be noted that Ewbank was an American, writing in 1842. At that time Americans disapproved of the imperialism of the 'European expansion in Asia', as the Cambridge History Tripos calls it.

[3] Cf. Carter.   [4] Cf. *SCC*, vol. 4, pt. 1.   [5] Cf. Wang Ling; Goodrich & Fêng.

[6] [The argument of the following three pages suggests that it was a + 16th- or + 17th-century introduction from China. While it remains true that de la Hire's water-pump was the first European machine to use the exact design of the Chinese single-piston single-cylinder bellows, L. Reti (*TCULT*, in the press) has now shown that Ramelli (+ 1588) came very close to it. Furthermore, he has found evidence of the double-acting principle as such in six European designs from + 1475 onwards, either with two pistons in a single cylinder (like the earlier Chinese flame-thrower, cf. p. 154), or with hinged pistons (like the earlier Chinese cuneate fan bellows, Figs. 34; 35, pl.). If therefore transmission took place, it must rather belong to the + 15th-century cluster (cf. p. 201).]

WESTWARD TRANSMISSION OF THE DOUBLE-ACTING PRINCIPLE AND
ITS INCORPORATION IN THE STEAM-ENGINE

There was plenty of time for it to come. Chinese sea-captains and their crews had been met with first by the Portuguese at Malacca from + 1498 onwards, and + 1517 was the year of the ill-fated embassy of Tomé Pires through Canton. Between + 1522 and + 1550 the Portuguese traded up and down the coast in an illicit manner, i.e. contrary to Chinese government policy but with the sympathy of the local people, and the accounts of some who were imprisoned on this account towards the end of the period, such as Galeote Pereira, have come down to us. The settlement of Macao was founded unofficially in + 1557, and in + 1569 the first European book on China, the *Tractado...as Cousas de China* by the Dominican Gaspar da Cruz, was printed at Evora in Portugal.[1] Before the end of the century (+ 1585) that best-seller, the *Historia* of Juan Gonzalez de Mendoza, appeared, soon translated into half-a-dozen European languages; three years earlier the Jesuits had established themselves in Macao, with momentous results. Before the end of the century, too, the Dutch were writing about China, as e.g. Jan Huyghen van Linschoten in + 1596. In + 1600 the East India Company was founded in England, and eleven years later Captain John Saris sailed for the Far East, whence he brought back the maps of China that Samuel Purchas published. In + 1637 Captain John Weddell explored the possibility of trade at Canton, but without success, and the Hon. Company's factory there was not established until + 1715. Nevertheless, the intensity of Sino-European relations throughout the + 17th century was very great, and a formidable literature exists upon them.[2] Thus by the time that another Dominican, Domingo de Navarrete, referred to the double-acting piston-bellows, in + 1676, as 'much more convenient than, and just as useful as, those of Europe',[3] there had been plenty of time, not only for other literary references, but for the westward transmission of a number of actual examples (which would have been bought for a song), by traders, sea-captains and missionaries.

The first effect of this in Europe was, as might be expected, upon the design of the water-pump. In 1716 Jean Nicolas de la Hire described a pump of beautiful

[1] On this whole period, as well as the work of Gaspar da Cruz, see Boxer.

[2] Here, to avoid enumerating a shelfful of books, I will refer only to the long-expected treatise of Professor Donald Lach of Chicago [*Asia in the Making of Europe*, 2 vols., Univ. Chicago Press, Chicago, 1965].

[3] This passage is contained, not in the autobiographical portion so excellently edited by Cummins recently, but in another chapter. Apart from the original Spanish, this can only be found in the translation of Churchill & Churchill (1704), vol. 1, p. 58.

simplicity in which the intake pipe was led to both ends of a cylinder while the rising pipe took off also at both ends on the opposite side (Fig. 39).[1] The piston thus did work on both strokes. From what source de la Hire drew his inspiration is clear from his words: 'de même qu'un soufflet double fait un vent continu.' If

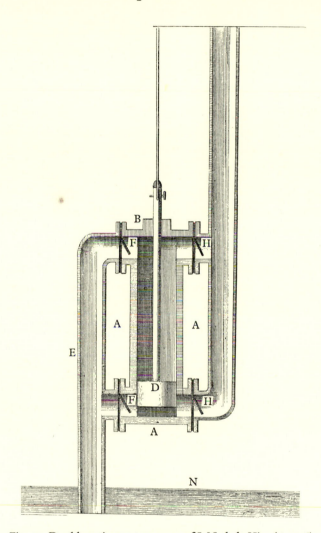

Fig. 39. Double-acting water-pump of J. N. de la Hire (+1716).

[1] It is interesting that this invention was not made by his father or his elder brother, both more eminent, and both surveyors and engineers, but by Jean Nicolas, the physician and botanist. It does not appear in the series *Machines et Inventions approuvées par l'Académie* (des Sciences, de France), 1666 to 1754, where one is tempted to look for it, but in the *Mém. de l'Acad. Roy. des Sciences*, 1716, p. 322. Cf. Westcott, p. 38. We have noted Ewbank's mention of it above (p. 157).

this was not the leading influence on the mind of James Watt in + 1777 and + 1782 we may take it with certainty that he was acquainted with the elegant album which the architect Sir William Chambers had published in + 1757 after a personal visit to Canton. Entitled 'Designs of Chinese Buildings...', it included a clear description and diagram of what he called the 'perpetual bellows'.[1] The pump of de la Hire explains perhaps why it was only natural that Trevithick should make his water-pressure engines double-acting in + 1798, but if anything further was wanting to ensure that the high-pressure steam-engines of the first years of the nineteenth century should also be so, descriptions of the *fêng hsiang* kept on coming out.[2] The French minister, H. L. J. B. Bertin (1719 to 1792), who had been much occupied with Franco-Chinese relations, amassed a large cabinet of Chinese curiosities, and from the description of this by Breton de la Martinière we know that double-acting piston-bellows were among them, for he figured and explained them.[3] Much more could be said, but surely all this is sufficient to prove that from the beginning of the eighteenth century onwards European engineers were very well acquainted with the *fêng hsiang*, whether or not they consciously had it in mind when designing prime movers. That they could have thought of the double-acting principle for themselves may well be true, but the business of the historian of technology is to trace ideas to their origins, and not to assume independent invention unless it can be shown to have occurred.

THE CHINESE ANCESTRY OF THE STANDARD METHOD OF INTER-CONVERSION OF ROTARY AND RECTILINEAR MOTION

Having dealt with cylinders, pistons, and valves I now propose to show that the other kinematic half of Newcomen's child, the reciprocating steam-engine, was

[1] P. 13 and pl. XVIII, fig. 1.

[2] On his embassy to Peking in 1793 Lord Macartney was accompanied by at least two scientific men, Dr James Dinwiddie and Dr Hugh Gillan, both of whom wrote elaborate accounts of scientific and technological observations in China, which circulated in MS form. Besides these, the Ambassador himself, an exceptionally intelligent man interested in these matters, wrote a third account of what was seen. Dinwiddie's description of the bellows appeared in print much later, in papers published by a relative (Proudfoot, p. 74). For the opinions of Macartney and Gillan, with a critical study of them, see the recent book edited by Cranmer-Byng. In the late eighteenth and early nineteenth centuries many portfolios of paintings of traditional Chinese trades and crafts circulated in Europe; there is a drawing of a smith with his piston-bellows in this style in Frémont LFM. Other mentions of about this time occur in van Braam Houckgeest's *Authentic Account of the Embassy of the Dutch East-India Company to the Court of the Emperor of China in the Years 1794 and 1795* (French ed., Philadelphia, 1797), vol. 1, pp. 275 ff.; English ed., Phillips, London, 1798, vol. 2, p. 78; and in J. Barrow's *Travels in China* (London, 1804), p. 312.

[3] Eng. tr., 3rd ed., London, 1812-13, vol. 3, pp. 70 ff. *China, its Customs, Arts, Manufactures, etc.*, edited principally from the Originals in the Cabinet of the late Mons. Bertin, with Observations Explanatory, Historical

anticipated by, and even indirectly derived from, a pattern which had arisen in medieval China for an exactly inverse purpose.

When we ask about the earliest water-wheel in Chinese history[1] we come upon the paradox that it was not used for turning simple cereal grindstones, but for the more complicated job of blowing metallurgical bellows. This must mean that there was a tradition of millwrights going back some considerable time before, even though we cannot trace it in literary references. The essential texts run as follows: first, the *Hou Han Shu*:[2]

In the seventh year of the Chien-Wu reign-period (+31) Tu Shih was posted to be Prefect of Nanyang. He was a generous man and his policies were peaceful; he destroyed evil-doers and established the dignity (of his office). Good at planning, he loved the common people and wished to save their labour.[3] He invented a water-power reciprocator (blowing-engine, *shui phai*) for the casting of (iron) agricultural implements.

(Comm: Those who smelted and cast already had the push-bellows (*phai*) to blow up their charcoal fires, and now they were able to use the rushing of the water (*chi shui*) to operate it...)[4]

Thus the people got great benefit for little labour. They found the 'water(-powered) bellows' convenient and adopted it widely.

This advanced mechanism comes, therefore, between the dates of Vitruvius and Pliny. The tradition of Tu Shih and his engineers must have persisted in Nanyang, for it was a Nanyang man who became prominent as an official two centuries later, and spread the knowledge of the technique. This we know from the *San Kuo Chih* which says:[5]

Han Chi, when Prefect of Lo-ling,[6] was made Superintendent of Metallurgical Production. The old method was to use horse-power for the blowing-engines, and each picul[7] of refined

---

and Literary... The engravings in the Cambridge University Library copy are hand-painted. Later there was a description in the *Chinese Repository* (see Anon.).

[1] I refer to the first specific description. A general statement of +20 shows that water-power was then well known for trip-hammer pounding, and implies that water-wheels had been in use for several decades of the −1st century. See *SCC*, vol. 4, pt. 2, p. 392.

[2] Ch. 61, p. 3*b* (History of the Later Han Dynasty).

[3] [Or, perhaps better, 'he loved to lighten the labour of the common people'.] I see no reason for taking a superior attitude about this. Well-authenticated cases are known where humanitarian feelings have motivated invention, e.g. Jouffroy's paddle-boat of 1783, for Jouffroy had seen the sufferings of the galley-slaves (Schuhl, p. 53).

[4] This commentary is in itself a piece of information for it was written by Li Hsien (cf. p. 171 above) about +670, and shows that at that time people were quite familiar with the idea of water-powered blowing-engines.

[5] Ch. 24, p. 1*b* (*Wei Shu*; History of the Three Kingdoms Period, Wei section). Both these passages are quoted in *Thai-Phing Yü Lan*, ch. 833, and *Nung Shu* (Treatise on Agriculture), ch. 19, p. 6*b*.

[6] On the North China plain, north of the Yellow River, but just within Shantung province.

[7] The *tan* or *shih*, 120 lb, occasionally 133 lb.

wrought (iron) took the work of a hundred horses. Man-power was also used, which was exceedingly expensive. So Han Chi adapted the furnace bellows to the use of ever-flowing water, and an efficiency three times greater than before was reached. During his seven years of office, (iron) implements became very abundant. Upon receiving his report, the emperor rewarded him and gave him the title of Commander of the Metal-Workers (*Ssu Chin Tu Wei*).

The period referred to must have been a little before +238. The story continues through the +5th century, for Phi Ling wrote in his *Wu Chhang Chi*:[1]

The origin of the Northern Chi Lake is that it was (artificially constructed) for the Hsin-Hsing (iron-) smelting and casting works. At the beginning of the Yuan-Chia reign-period (+424 to +453) there was a great development of water-power for blowing bellows for metallurgical purposes. But later, Yen Mao, finding that the earthworks of the lake leaked and that it was not much good, destroyed them, substituting man-power bellows (*jen ku phai*) so that it was called the 'treadmill bellows' (*pu yeh*) lake. Now it has got so much out of repair that it cannot be used for smelting and dries up altogether in winter.

Moreover the *Shui Ching Chu* quotes[2] Tai Tsu's *Hsi Chêng Chi*, a record of the campaign against Yao Hsing, which was written about +410, as saying that the Ku Shui River[3] had been used for the iron industry. At a certain place there were staithes or quays with machinery for smelting by water-power (*shui yeh*) and an office of the government Iron Authority. Thus we see, says Li Tao-Yuan, that the Ku Shui was used for driving metallurgical blowers. Later on, further mentions may be found in the Thang geography, *Yuan-Ho Chün Hsien Thu Chih*[4] of +814, and probably every century would yield them on investigation.

BLAST-FURNACE BLOWING-ENGINES OF MEDIEVAL CHINA

Machines of this kind are illustrated in nearly all the relevant Chinese books from the time of the *Nung Shu*[5] onwards (+1313), not only for metallurgical bellows, but also for operating flour-sifters[6] and any other machinery requiring longitudinal reciprocating motion. They have been seen in their traditional form by modern travellers (Hosie;[7] Rocher;[8] A. Williamson; Hommel[9]). Their great importance

[1] (Record of Wu-Chhang) cit. *Thai-Phing Yü Lan*, ch. 833, p. 3b.
[2] Ch. 16, p. 2b (Commentary on the 'Waterways Classic', quoting the 'Record of the Western Expedition'). We owe this reference to Yang Khuan TIC.
[3] It runs from the west into the Lo River some distance south-east of the city of Lo-yang.
[4] (Maps and Descriptions of all Prefectures and Counties, of the Yuan-Ho reign-period.)
[5] Wang Chên's great treatise on agriculture and rural engineering.
[6] [See fig. 461 (p. 208) in *SCC*, vol. 4, pt. 2.]
[7] P. 96.
[8] Vol. 2, pp. 196 ff., whose description almost exactly parallels that in the *Nung Shu*.
[9] In part, p. 86.

is due to the fact that at some time or other between Han and Sung they began to embody a conversion of rotary to longitudinal motion in heavy-duty machines. For this reason they may be considered the morphological precursors of the reciprocating steam-engine, which effects the opposite conversion.

From the time of the first illustrations[1] at any rate (Fig. 34 shows that given by the *Nung Shu*), a conversion of rotary to rectilinear motion is apparent. It is, however, at first sight, owing to the many mistakes made by the artist, almost impossible to make out from this how the machine worked.[2] But by the aid of Wang Chên's *Nung Shu* text, together with the illustrations redrawn in later books, especially the *Thu Shu Chi Chhêng* encyclopaedia (+1726),[3] Fig. 19, a

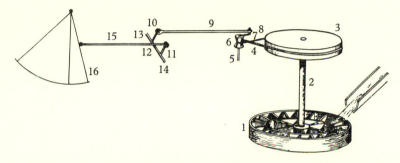

Fig. 40. Diagram to show the operation of the water-powered reciprocator (*shui phai*) depicted in Fig. 34.

clear understanding of the mechanism can be reached (Fig. 40). We may summarise it as follows. The motive power came from a horizontal water-wheel, the shaft of which bore at the top a driving-wheel of similar size.[4] Upon this an eccentric could have been mounted directly if the bearings of the wheel had not

---

[1] [The reference here is to printed illustrations; see p. 186 below.]

[2] Feldhaus GBC had to stop at this point, but he deserves credit for drawing attention to the importance of the machine, which he knew only from the *San Tshai Thu Hui* (Encyclopaedia of the Three Powers, 1609), *Chhi yung* section, ch. 10, p. 30 *a*, *b*. This has the worst picture of all.

[3] *I shu tien*, ch. 6, *hui khao* 4, pp. 2 *a* and 8 *b*; the latter for a flour-sifter. *Shou Shih Thung Khao*, ch. 40, p. 31 *a*, also has this (Comprehensive Study of the Works and Days, 1742). We can just catch a glimpse of one of these in the upper right-hand part of the elaborate flour-mill on a mountain river shown in a magnificent scroll-painting of about +1300 recently published (Yang Jen-Khai & Tung Yen-Ming, pl. 111). This gives valuable confirmation of many statements in the *Nung Shu*. It is reproduced in *SCC*, vol. 4, pt. 2, fig. 627 *b*, opp. p. 405.

[4] This driving-wheel performed the function, it is worth noting, of a flywheel, smoothing any sudden variations in the flow of water. The importance of a flywheel was quickly appreciated by engineering inventors in the period when rotary motion was first being derived from the atmospheric engine (cf. p. 201 below) and the steam-engine. On this see Dickinson SE, pp. 80 ff.; Galloway, p. 160; Thurston, pp. 105, 109, 118. Trevithick's designs were notable for their very large flywheels.

been built above it. Since upper bearings were preferred, a smaller wheel or pulley[1] was mounted alongside in a second frame, and driven by a driving-belt working off the large drive-wheel.[2] This secondary shaft bore, above its bearings, an eccentric lug[3] which connected by means of a rod and working joint with a bell-crank rocking roller,[4] this in its turn operating the pole or piston-rod of the bellows itself through another working joint. Evidence has been given that at least during the Sung, the blowing apparatus in question was some large type of piston-bellows, probably the fan type. Considerable mechanical advantage was of course gained by having the eccentric mounted on the small wheel instead of the large one,[5] and this was no doubt purposely intended, in which case the securing of the main drive-wheel by upper instead of lower bearings would have been deliberate.[6] Now let us compare this with what Wang Chên himself says in a most interesting passage.[7]

According to modern study [1313!], leather bag bellows (*wei nang*) were used in olden times, but now they always use wooden fan (bellows) (*mu shan*).[8] ...A place beside a rushing torrent is selected, and a vertical shaft (*li chu*) is set up in a framework with two horizontal wheels (*wo lun*) so that the lower one is rotated by the force of the water (*yung shui chi chuan*). The upper one is connected by a driving-belt (*hsien so*) to a (smaller) wheel (*hsüan ku*) in front of it, which bears an eccentric lug (lit. oscillating rod, *tiao chih*). Then all as one, following the turning (of the driving-wheel), the connecting-rod (*hsing kuang*) attached to the eccentric lug pushes and pulls the rocking roller (*wo chu*), the levers (*phan erh*) to left and right of which

---

[1] In all the pictures of this iconographic family the artists disguised this pulley as something which looks like superimposed bubbles.

[2] My friend Professor Aubrey Burstall, who has constructed a working model of the water-powered blowing-engine as described by Wang Chên, believes that the power transmission was effected here by a form of frictional drive rather than by a driving-belt, but I regret that I cannot follow him in this view.

[3] In Chinese textile technology we find other important examples of just this device, together with clear evidence that it goes back at least to the eleventh century. See p. 194 below.

[4] The *Nung Shu* artist failed to bring out the connection of the eccentric with the lug, and inexcusably made the attachment of the connecting-rod's pin as if it were with the piston-rod itself, instead of fitting it into the rocking roller. The *Thu Shu Chi Chhêng* artist left the eccentric lug floating alone in mid-air, and the *Shou Shih Thung Khao* artist, who managed the crank part very well, forgot to show the upper part of its pulley.

[5] The number of strokes of the piston-rod for each revolution of the water-wheel would be much augmented.

[6] We were glad to find, long after writing this, that the interpretation of Wang Chên's machinery by Yang Khuan OIT agrees fully with our own. He adds the interesting point that in the similar crossed-belt horse-whim mill described elsewhere in the *Nung Shu* (ch. 16, pp. 6a, b) the smaller wheel is said to rotate fifteen times for each revolution of the driving-wheel.   [7] *Nung Shu*, ch. 19, p. 6b.

[8] I think it is certain that Wang Chên is not referring to any kind of rotary fan (as in the winnowing-fan), but is thinking of the piston itself. This may well have been a hinged one, as shown in the traditional drawing, but in any case valves on the intake and output sides of the bellows would have been necessary. The bellows in his illustration, and in all those which derive from it, are quite like the cuneate wood bellows of the European eighteenth and early nineteenth centuries, shaped similarly to the wood-and-leather bellows of earlier times but no longer pyramidal. Unfortunately, we have found no textual description of them in Chinese. Of course, descriptions of the straight piston-bellows are also extremely rare.

assure the transmission of the motion to the piston-rod (*chih mu*). Thus this is pushed back and forth, operating the furnace-bellows far more quickly than would be possible with man-power.

Another method is also used. The wooden (piston-) rod which comes out from the front of the bellows is connected to a horizontal bar (*hsün*)[1] about 3 ft long, which has at its head a crescent-shaped board set with its rounded edge upwards, and suspended from above on ropes like a swing (*chhiu chhien*).[2] Then in front of the bellows there are set strong bamboo (*ching chu*) (springs) connected with it by ropes; this is what controls the motion of the fan of the bellows (*phai shan*). Then in accordance with the turning of the (vertical) water-wheel, the lug (*kuai mu*) fixed on the driving shaft (*wo chu*) automatically (*tzu jan*) presses upon and pushes the curved board (attached to the piston-rod), which correspondingly moves down (lit. inwards). When the lug has fully come down, the bamboo (springs) act on the bellows and restore it to its original position.

Moreover, using one main shaft it is possible to actuate several bellows (by lugs on the shaft), on the same principle as the water-powered trip-hammers (*shui tui*). This is also very convenient and quick, so I think it worth recording here.

As for metallurgical works in general, they are most profitable for the country. When Metallurgical Bureaux are established, they often spend a great deal of money and hire much labour to work the bellows, which is very expensive indeed. But by these methods (using water-power) great savings can be made. Now it is a long time since the inventions were first devised, and some of them have been lost, so I travelled to many places to explore and recover the techniques involved. And I have drawn the accompanying diagrams according to what I found, for the enrichment of the country by the official metallurgists and the greater convenience of private smelters. This is really one of the secret arts of benefiting the world, and I only hope that those who can understand it will hand it down.

As the poet says:[3]

> I often heard that the good officials of old
> Honoured the forging of tools for the farmers' trade
> And wishing to save the sweat of the smiths, they made
> Engines for blowing, and wheels by the water roll'd.
> This blast, this breath, is part of the tides of the world[4]
> As we know by the Symbols anciently display'd.[5]
> Come, see the fires whipp'd up and the metal flow
> Into its channels; merry the work shall go,
> Speeding the plough and the stalk of the living jade.[6]

The first paragraph requires no comment, and can be easily understood from the diagram in Fig. 40. The second system, however, though not illustrated in

---

[1] This term is one generally used for the cross-bar of a frame supporting bells or gongs.
[2] On the swing, see Laufer; he concluded that it was introduced to China from the northern barbarians between the Han and the Sui.
[3] Perhaps himself.
[4] The allusion is to the *Tao Tê Chīng* (cf. p. 160 above) where the universe is compared to a bellows.
[5] The hexagrams Sun (no. 57) and Li (no. 30) (cf. *SCC*, vol. 2, table 14).
[6] The translation of this poem is somewhat free.

any of the books, is also quite significant, for it describes a type of blowing-engine which used only cams for the conversion of rotary to longitudinal motion. Since mechanisms of this kind were much too difficult to apply to the opposite conversion in the early days of the steam-engine,[1] this second type leads us into a digression from the main line of our argument, but it is much too interesting to pass over. Li Chhung-Chou was the first to see that it depended upon a vertically mounted water-wheel[2] and to realise that Wang Chên was using the term *wo*

Fig. 41. Li Chhung-Chou's reconstruction of Wang Chên's 'second system' of making a blowing-engine (+1313), with vertical water-wheel and trip-lug shaft.

---

[1] Here one thinks of the Michell crankless engine devised by A. G. M. Michell during the first world war. In this 'slant-slipper' mechanism, as it was called, reciprocating is converted into rotary motion according to a very simple kinematic principle. The cylinders, four or more, are arranged uniformly round the main shaft with their axes parallel to it, and the pistons make contact with a 'slant' or swash-plate which is keyed to the main shaft at an inclination of about $22\frac{1}{2}°$ to the normal plane (see Cherry). In spite of its great ingenuity the Michell crankless engine has never come into wide use.

[2] In all probability, that is. The machine could have been composed of a horizontal water-wheel and right-angle gearing, but this would have been unnecessarily complicated. It is true, however, that the text does not specify any departure from the horizontal driving-wheel of the first model. Moreover, such gearing is depicted quite often in the Chinese illustrations, and the fondness of medieval Chinese engineers for horizontally mounted water-wheels must be remembered. An actual example of a picture of a horizontal water-wheel working a square-pallet chain-pump by means of right-angle gearing occurs in the *Thien Kung Khai Wu* of 1637 (see *SCC*, vol. 4, pt. 2, fig. 582). I admit that I have never encountered this arrange-

chu in two quite different senses, first for a horizontal rocking roller like a bell-crank pin, and secondly for a horizontal main driving-shaft fitted with lugs. His reconstruction of the second design (Fig. 41) made the lugs push back the suspended crescent-shaped boards as they passed them, but the action of the mechanism is not at all clear from his drawing, nor does his text explain it.[1] During our work on the preparation of the section on mechanical engineering for Volume 4, pt. 2, of *Science and Civilisation in China*, my collaborators and I therefore had recourse to a modification in which the lugs, rounded like cams, and doubtless greased,

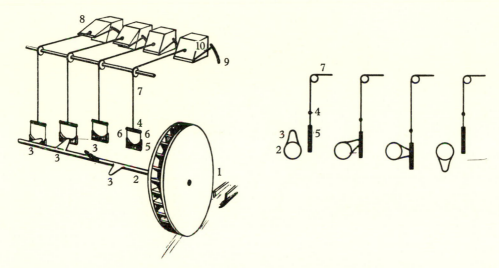

Fig. 42. Reconstruction of Wang Chên's 'second system' of making a blowing-engine (+1313), with vertical water-wheel and trip-lug shaft, by J. Needham & Wang Ling.

bore down upon the crescent-shaped boards, depressing and releasing them alternately, as shown in Fig. 42. This, however, necessitated that the rope of each bellows should be carried over a pulley above, no mention of which occurs in Wang Chên's text, and also perhaps that the hanging cages with the crescent-shaped boards should be secured at ground level by a further rope connection (not shown in the diagram) so that the cages could not slip away from the lugs. Meanwhile Yang Khuan had proposed another solution.[2] He felt that greater significance

ment in my travels in China, nor can I adduce a description by any other traveller of contemporary times, but it may have been more common in the Ming and early Chhing periods, when millwrights had greater encouragement and no competition from modern machinery.

[1] Note his representation of the bellows as of the whole-skin reinforced concertina type, to indicate that a blowing-engine of the Han or Liu Chhao period is in question.

[2] Yang Khuan FPHB.

183

ought to be attached to the concavity of the crescent-shaped boards; since the lugs themselves move in arcs the concavity was surely functional. For him, therefore, the boards, made much wider, were attached directly to the ends of the piston-rods facing sideways, not upwards, and suspended by ropes as stated, each being pushed back with every passage of one of the driving-shaft lugs and (as in all the other reconstructions)[1] returned to position by the bamboo springs (Fig. 43). We ourselves finally adopted this reconstruction in our treatise,[2] and the translation of the relevant paragraph therein will therefore be found to differ slightly from that which has been given here, but we did so not without some

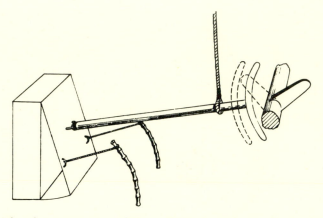

Fig. 43. Yang Khuan's reconstruction of Wang Chên's 'second system' of making a blowing-engine (+1313), with vertical water-wheel and trip-lug shaft (our drawing).

philological misgivings. Difficulties of interpretation arise not only from uncertainty about the meaning of the technical terms used, but also from the fact that in Chinese texts of the +14th century there is no punctuation in the modern sense. The solution of Yang Khuan necessitates taking *hsün* to mean the piston-rod instead of the upper bar of the hanging cage, and has it suspended by a single rope. But (*a*) the crescent-shaped board seems to be specified as 'recumbent' (*yen mu hsing ju chhu yüeh*), 'a recumbent piece of wood shaped like the crescent of the new moon', though it is true that *yen* can also mean just curved or bent, (*b*) this

---

[1] In having the return excursion effected by the bamboo springs, this design was strikingly reminiscent of that of Villard de Honnecourt's water-powered saw-mill (Usher, p. 144, cf. Lynn White, p. 118 and often figured). Though the date (+13th century) is more or less contemporary with him, the Chinese design must be much older because it is far the simpler and more archaic of the two which Wang Chên expounded. Besides, it is closely related to the water-powered trip-hammers which were, as we know, widespread in the Han (+1st century).

[2] In consultation with one of our kind advisers, Professor Aubrey Burstall.

board is said to hang like a swing, and a swing normally implies a pair of ropes,[1] (c) the piston-rod is not termed *hsün* in the preceding paragraph, and in any case the word means primarily a crossbar, (d) 3 ft seems rather too short for such a piston-rod, and (e) if a horizontal oscillating motion alone was required the piston-rod could have been mounted on or between greased rollers, without need for any form of suspension at all. On the other hand, certain other nuances in the text may be held to favour Yang Khuan's solution, so as the alternative reconstructions are less workmanlike from the engineering point of view this is perhaps the best one.

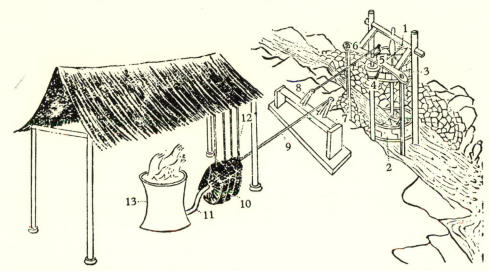

Fig. 44. Li Chhung-Chou's reconstruction of the water-powered pusher, or reciprocator (*shui phai*), used for a blowing-engine (+1313), embodying whole-skin concertina-type bellows, though these were more characteristic of the Han than the Yuan.

All this, as Wang Chên indicates, was a development of the water-powered multiple trip-hammer batteries, those simple machines for decorticating rice, pounding minerals, etc., so common in Han China.[2] Here a series of tilt-hammers are alternately raised and allowed to fall by lugs or cams rotating with the main shaft. This was the ancestor of the martinet hammer of the eighteenth-century West, indeed identical with it in design.[3] One is strongly tempted to believe that

---

[1] This requirement is not met by Li Chhung-Chou's reconstruction, where the suspended cages are rigid.

[2] See Needham DITC, fig. 14, and, for textual evidence on dating, *SCC*, vol. 4, pt. 2, p. 392.

[3] With the sole difference that the martinet was usually fitted parallel to the lug-shaft, while Chinese hammers or pestles were generally at right-angles to it. To what extent such tilt-hammers were used for forges in ancient and medieval China I cannot say. At present I am not able to adduce any textual evidence in favour of such use, but all that we know of Chinese siderurgy encourages the view that it would have been natural.

the blowing-engines of Tu Shih and Han Chi (+ 1st and + 3rd centuries) must have been of this simpler description, but there remains always the difficulty (unless a persistent impression is wrong) that horizontally mounted mill-wheels seem in general to have been more characteristically Chinese than the vertical 'Vitruvian' ones. Yang Khuan adds the further argument that Han Chi's blowing-engines are distinctly said to have taken the place of horse-power whims, which would almost certainly have had horizontal driving-wheels.[1]

Another consideration which would plead for an earlier rather than a later appearance of the crank, connecting-rod and piston-rod arrangement in which we are more interested, is the view now rapidly gaining ground that the crank itself was a Chinese invention.[2] The best evidence at present available is that (apart from ancient Egyptian crank drills of which the interpretation is dubious) we find it first in the handles of rotary-fan winnowing-machines represented in Han tomb farmyard models.[3] Thus as in the case of the fire-piston, ancient Chinese culture embodied certain devices which could have been ancestral to the whole course of technological development which we are here examining.[4] However,

[1] Yang Khuan OIT, p. 58. Atkinson has given us the latest study of the whim.
[2] Not the crankshaft, which belongs to fifteenth-century Germany.
[3] The matter is discussed in *SCC*, vol. 4, pt. 2, p. 118, fig. 415; see Lynn White, p. 104.
[4] [At the time of the delivery of this lecture, and the publication of *SCC*, vol. 4, pt. 2, it seemed safe to assert (cf. p. 200 below) that the standard method of interconversion of rotary and longitudinal motion (the crank, connecting-rod and piston-rod arrangement) could be found in China by + 1200, since it was obviously not at all new when Wang Chên published his description of it in + 1313. We are now in a position to affirm it for the + 10th century. Chêng Wei (in *Wên Wu*, 1966 (no. 2), 17) has given a description of a scroll-painting of undoubted authenticity by Wei Hsien (fl. + 940 to + 980) entitled 'Cha Khou Phan Chhê Thu' (Horizontal Water-Wheels by the Canal Sluice-Gate), here reproduced in Fig. 45 (pl.). Wei Hsien was one of the court painters of Li Hou Chu, the last ruler of the Southern Thang dynasty (r. + 961 to + 975), so the picture was probably painted about + 965. The large central horizontal water-wheel works a mill in the ordinary way (and this picture is also the oldest illustration of such a *shui mo* that we have, textual references and tomb-models alone witnessing for the previous thousand years). On its right is a smaller water-wheel working something more complicated (Fig. 46a, pl.), which Chêng Wei has convincingly reconstructed in Fig. 46b (pl.). There can be no question that the machine is a bolter or flour-sifter, and it is only regrettable that Wei Hsien had to obscure by the carved woodwork of the building some of the more important parts of the power transmission system. The design has a clearly archaic character, for the return excursion of the sifter must have been effected by springs, and the jointing below the rocking roller looks very awkward. The crank, connecting-rod and piston-rod assembly is thus established for the neighbourhood of + 900, since there is no indication that Wei Hsien was illustrating a new invention.

As in all other pictures of water-powered reciprocators (*shui phai*), a driving-belt is shown or implied. When *SCC*, vol. 4, pt. 2, was published, a painting of *c.* + 1270 was given in fig. 405 as the oldest representation of the spinning-wheel yet known in any culture. This is no longer so, for the Liaoning Museum in Shenyang possesses a copy of a scroll-painting by Wang Chü-Chêng, datable in the neighbourhood of + 1035, which depicts a spinning-wheel in action (Fig. 47, pl.). A seated girl holding a baby turns the wheel by a crank-handle in the form of an eccentric peg, while the spinning is done by an old woman who moves away from the wheel as the yarn is formed. The driving-belt, which passes over two spindles, is very clearly

as yet we cannot say with certainty just what construction was used by Tu Shih and Han Chi for their metallurgical blowing-engines.[1]

On the other hand we may feel fairly sure that the bellows of those early times

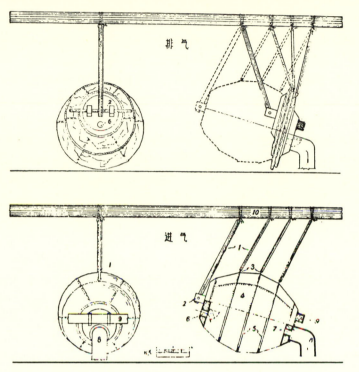

Fig. 50. Wang Chen-To's reconstruction of the whole-skin concertina-type bellows probably identifiable in the +1st-century Thêng-hsien relief.

drawn, and the wheel is mounted on a vertical post with a tripod stand. Here again is an archaic type, not only older than the pedal system which freed both hands, but also older than the forms which allow the operator to turn the wheel with one hand and spin with the other. For further information see Thien Hsiu (in *Wên Wu*, 1961 (no. 2), 44). But the use of the driving-belt in the spinning-wheel was not the limiting factor for its application in the water-powered reciprocator, for stone reliefs of Han date (−1st or +1st century) attest the driving-belt in quilling-wheels already then (cf. *SCC*, vol. 4, pt. 2, p. 269).

Another fragment of the 'standard assembly' would be the quern connecting-rod, used as a convenient handle by farmers (see *SCC*, fig. 413). It seemed likely to be old, but we could not date it from book illustrations earlier than +1210. The Chiangsu Historical Museum at Nanking, however, has a fine model of the quern conrod and handle taken recently from a tomb of the Nan Chhao period (between +420 and +589) at Têng-fu Shan (Fig. 48, pl.). Thus two of the components had married by about +500. In this way evidence is building up to reveal all the successive stages in the evolution of the standard interconversion method—an evolution which took place however in the Eastern rather than the Western regions of the Old World.]

[1] Let it not be said that though Chinese invention may be ancient, practical application had to await the mechanical genius of the West. The rotary-fan winnowing-machine and the water-powered blowing-engine were in very widespread use throughout the Chinese medieval centuries.

were leather bags in some form or other. Our Chinese colleagues tend to see them as ring-reinforced whole-skin bellows of the concertina type; Li Chhung-Chou so depicts them not only in his reconstruction of the early lug-shaft system (Wang Chên's second method) (Fig. 41), but also in a sketch of the crank and connecting-rod system (Wang Chên's first method) (Fig. 44), where it is less convincing because all surviving medieval illustrations show the fan (hinged piston) type of bellows. Some authority for the concertina bellows is to be derived from a re-markable relief of Han date depicting an iron foundry or else a forge, found at Thêng-hsien in Shantung province a few years ago and now in the Archaeological Museum at Chinan (Fig. 49, pl.).[1] In the centre and on the right there are scenes of forging, while on the left there is an object which might be either a furnace or a bellows. On the former interpretation (to which we at first inclined) the curving line is molten iron pouring from it; a reading entirely justifiable on historical grounds.[2] But Wang Chen-To may well be right in preferring the latter interpre-tation, in which the curving line represents the pipe to the tuyère. His ingenious reconstruction of the bellows is seen in Fig. 50. As for the fan types, they were later, though almost certainly pre-Sung, introductions. Whether the double-acting non-hinged piston-bellows was ever incorporated in water-power blowing-engines we still do not know, but the fact that it is so often depicted in scenes of metallurgical operations makes this probable enough.

Chinese literature contains (or rather contained) at least one monograph devoted to the construction of metallurgical blowing-engines. The following curious passage was noticed by Yang Khuan[3] in the *Anyang Hsien Chih* (Topography of Anyang District):

Forty *li* north-west of Yeh-chhêng there is the place called Copper Mountain (Thung-Shan) where formerly copper was produced. According to the *Shui Yeh Chiu Ching* (Old Manual of Metallurgical Water-Power Technology) water was canalised to blow furnace bellows here in the Later Wei period (+5th and +6th centuries). This was called 'water-power smelting' (*shui yeh*). It was said to have been established (here) by the Director of the Ministries Depart-ment Kao Lung-Chih. (The water-wheels were) 1 ft deep (i.e. broad) and a pace and a half (i.e. 7½ ft) in diameter.

The appearance of Kao here is not at all surprising. Born in +494, he was an eminent architect, engineer and city builder of the Northern Chhi dynasty, well

---

[1] My collaborator Dr Lu Gwei-Djen and I had the pleasure of examining a large rubbing of this relief in the Chinan Museum with the curator, Dr Li Chi-Thao, in 1958, who then already preferred the bellows interpretation.

[2] For abundance of evidence concerning early iron-casting in China see *SCC*, vol. 5, pt. 1, and meanwhile Needham DITC.   [3] OIT, p. 65.

known for the many hydraulic engineering works which he carried out before his death in + 554. He was particularly remembered for the construction of water-mills working many different kinds of machinery.[1] One would give a good deal to recover the manual mentioned by the author of the *Anyang Hsien Chih*.

WATER-POWER AND THE IRON AND STEEL INDUSTRY IN EAST AND WEST

It is needless to emphasise the outstanding importance of power-driven bellows and forges for the metallurgy of iron, and the remarkably early successes of the Chinese in cast-iron technology cannot be unconnected with the machinery here described. The water-powered blowers of Tu Shih and his successors, to which there was nothing comparable in the Graeco-Roman world, had indeed a glorious future before them.[2] It is now clear that the use of water-power for metallurgical purposes began in Europe, notably in Germany, Denmark and France, very much later than in China. Forge-hammers were the first to be mechanised, about the beginning of the + 12th century, and the application of water-power to the bellows for the air-blast followed early in the thirteenth, i.e. a century or so before Wang Chên made his researches in the history of technology.[3] Of the specific origin of the plans for the European machines we know nothing, but if anything came westwards overland it would have been the trip-hammer lug first, and then long afterwards the eccentric drive. For the designs in the + 15th-century MSS show

---

[1] Cf. *Pei Chhi Shu*, ch. 18, p. 5*b* (History of the Northern Chhi Dynasty).

[2] If in the Chinese story water-powered forge-hammers are so much less conspicuous than water-powered blowing-engines, this may be due precisely to the fact that bloomeries were replaced by blast-furnaces at such an early time. How far hydraulic or martinet hammers were used in e.g. the forging of co-fusion steel in the Chinese Middle Ages is not yet clear (cf. Needham, DITC, pp. 26 ff.).

[3] This statement depends on records which are very difficult to interpret (cf. Johannsen AWH). Bellows worked by water-power are well established for + 1214 and + 1219 (in the Tyrolese silver and Harz copper districts). Vertical stamp-mills for ore-crushing are clear for + 1135 and + 1175 in Styrian sources, which speak, for instance, of 'unum molendinum et unum stampf'. Mechanised hammers for forging the iron itself first appear probably in + 1116, certainly by + 1249, both sources being French. If 'molendinum' here means only the water-wheel then a series of references to 'mills' in connection with iron-working may take the vertical ore stamp-mill back to + 1010; if it means the bellows as well as the water-wheel then these same records may attest water-powered blowers so operated back to that date. Unfortunately its meaning probably varied at different times and places. Johannsen GDE, pp. 92 ff., followed by Dickmann, pp. 29 ff., gives examples of water-blowers going as far back as + 738, but does not support them with precise evidence; it is most probable that these were trombes, i.e. devices like filter-pumps which needed no moving components (cf. Ewbank, p. 476). They were cold damp analogues of the *sufflatores* (p. 142 above). Trombes probably partnered vertical ore stamp-mills in many places as late as the + 15th century. The first blast-furnace of Europe did not appear until about + 1380. It is quite astonishing to read in Johannsen GDE, p. 21, that water-power was never used in Chinese metallurgy until modern times.

only cuneate wood-and-leather bellows worked by lugs on the shafts of vertical overshot wheels,[1] and this practice continued through the time of Olaus Magnus (+1565) (Fig. 51) and Biringuccio (+1540) who figures them in great variety,[2] to the eighteenth century, when John Wilkinson actually patented in +1757 a hydraulic blowing-engine essentially similar to that described in the

Fig. 51. Water-powered furnace-bellows and martinet forge-hammers depicted in a woodcut in Olaus Magnus, *De Gentibus Septentrionalibus* (+1565). Though not the oldest European illustration of these machines, for the recumbent water-powered trip-hammer appears in the Anonymous Hussite (+1430) and the *Flores Musicae* of Hugo Spechtsart von Reutlingen (+1488), it is one of the most attractive.

Nung Shu of +1313.[3] The only difference (apart from the pistons) was that while some of Biringuccio's machines had included crank motions instead of lugs, Wilkinson's embodied a two-throw crankshaft.[4] The designs of Ramelli especially (+1558) were strangely close to Sung and Yuan Chinese patterns on account of

[1] For instance, the Anonymous Hussite (+1430), Mariano Taccola (+1440), the *Mittelalterliche Hausbuch* of +1480. Cf. Berthelot; Frémont LFM, fig. 30; Beck BGM, p. 289; Gille, p. 643; Forbes MMA, p. 68; PMA, pp. 612 ff.

[2] Cf. Beck BGM, pp. 116 ff.

[3] See Dickinson JWE. [A schematic diagram is given in *SCC*, vol. 4, pt. 2, fig. 687.] Cf. de Bélidor, vol. 1, pls. 34, 35.

[4] As already noted, the true crankshaft appears in the German engineering MSS of the +15th century as well as in Agricola. But a 'piston-rod' as well as a connecting-rod is never shown with it. Even in Agricola (Hoover & Hoover ed., p. 180) the two are combined.

his extensive use of rocking rollers (Fig. 52, pl.).[1] So also in the previous century we find an intimate resemblance in the sketches of Antonio Filarete (c. + 1462),[2] and in the subsequent one the plates of Böckler.[3] One cannot escape the conviction that there was a genetic connection.[4] In any case the Chinese engineers seem to have had a precedence of some ten centuries for the practical trip-hammer principle, and three or four for the combination of eccentric, connecting-rod and piston-rod.[5]

ARABIC SLOT-ROD PUMPS AND CHINESE SILK-REELING MACHINERY

All this brings up in rather acute form the history of the interconversion of rotary and longitudinal reciprocating motion. The oldest examples of this achievement (the bow-drill, the pump-drill, and the pole-lathe)[6] all involved non-continuous belting, and in the development of machinery they did not lead far. Next came the use of lugs on a rotating shaft. But Gille was stating the half of a half-truth when he said that 'the only way of effecting these conversions known to the Middle Ages involved springs'.[7] He was thinking, of course, of the hydraulic saw of Villard de Honnecourt (c. + 1237), already mentioned,[8] which used the trip-hammer principle but assured the return excursion of the saw by means of a spring pole.[9] More commonly gravity had been used rather than springs—as in

[1] Esp. pl. 137, reproduced in Frémont EFC, fig. 142 and OES, fig. 88; Forbes PMA, p. 613, and elsewhere. [I always feel that the ancient blowing-engine or water-powered reciprocator must still be at work somewhere in China. Although I have not so far found it, I was delighted to encounter at the Commune of Chia-chia-chuang near Fênyang in Shansi in 1964 a flour-sifter built with a rocking-roller connection just as in Fig. 34, and in full use with an electric motor under the floor as the power source (Fig. 53, pl.). Thus has the millwrights' know-how come down from the + 14th century.]

[2] Reproduced and discussed by Johannsen FAE. See particularly the discussion which took place in the pages of *Technology and Culture* following the paper of Spencer recently [1963, **4**, 201; 1964, **5**, 386, 391, 398, 404, 406].

[3] Esp. pl. 78. In spite of what has been said above (p. 162), it is rather striking that down to the end of the + 17th century the illustrations of water-powered blowing-engines almost always show wood-and-leather bellows of the pyramidal cuneate form, rather than all-wood bellows of this shape; the purely cuneate ones appeared in Europe only much later.

[4] We still know almost nothing of the bellows of the intermediate regions of the Old World. Afet Inan, p. 41, quotes an interesting account from the travels of Evliya Chelebi (1611 to 1682) of a bellows at the iron mines of Samakov in Turkey, now Bulgaria, which ten men could not have moved.

[5] Assuming that this was not Han and had been developing later through Thang and Sung before Wang Chên's time. [We can now say definitely five centuries.]

[6] Cf. Childe; and Gille, p. 645.   [7] Gille, p. 652.   [8] Pp. 125, 184.

[9] As we have just seen, springs were also necessary in the second (vertical water-wheel) design of Wang Chên. They are in no way to be despised as primitive, for they occur in some of the most ingenious modern reciprocating mechanisms. Artificial respiration pumps, for instance, require a reciprocation cycle of constant velocity. Since this cannot be obtained from the conventional eccentric, connecting-rod and piston-rod

the vertical stamp-mills of Western medieval times, or the bellows with counter-poises,[1] or the ancient Chinese trip-hammers and the later European fulling-mills and martinets which so closely resembled them.[2] Gille's formulation, moreover, omitted the two machines of the Middle Ages which lie most directly in the line of ancestry of the steam-engine and the locomotive, namely the connecting-rod system of the Chinese blowing-engines which we have just been examining, and the slot-rod force-pump described by al-Jazarī a century earlier than Wang Chên.

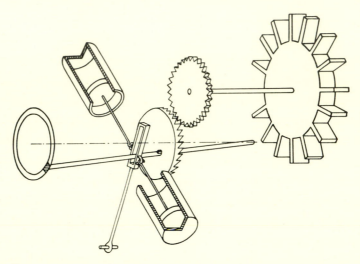

Fig. 54. A. Burstall's reconstruction of al-Jazarī's slot-rod pump; the lower toothed wheel is mounted on a shaft which is loosely pivoted at the right, and free to rotate in an annular groove at the left, so that while the wheel turns upon its geometrical centre the shaft describes a conical path and acts as an eccentric lug, swaying the slot-rod to right and left alternately as it rotates.

combination, a continuously rotating spring arm is made to drop into two notches on a flat wheel connected by steel tape with a bellows-actuating member. The arm is lifted out of the notches by two cam-faced dogs twice in each cycle, and the two return strokes are effected by means of a spring (cf. Jones & Horton, vol. 3, pp. 183 ff.). It is interesting that Wang Chên's second type of blower would have had a stroke of approximately constant velocity.

[1] Cf. e.g. Beck BGM, p. 119.

[2] Gille, p. 635, attributes a 'camshaft' and trip-hammer principle to Heron of Alexandria. He may have been thinking of the puppet effects produced by lugs and little levers in the *Automatic Theatre* (cf. Beck HAA, p. 192; Drachmann MTGR, p. 197), but as Usher points out (1st ed., p. 92, 2nd ed., p. 140) these were essentially for producing motion rather than power. Alternatively, he may have been thinking of the system of lugs depressing a lever attached to the piston-rod of an organ air-pump. This occurs only in the windmill organ-blower (*Pneumatica*, I, 43; cf. Woodcroft, ch. 77, p. 108; Drachmann HWM). This is slightly nearer power production, but the whole design is so hypothetical and unpractical that it cannot convincingly represent the origin of all the trip-lugs of the European Middle Ages. What we know about the tilt-hammer and trip-hammer in ancient China shows that it was essentially practical and in constant heavy-duty use. It therefore has a better claim to have been the ancestral form of the practical pounding and blowing machinery of Europe.

This last machine, described and illustrated (Fig. 20) in his 'Book of the Knowledge of Ingenious Mechanical Contrivances' of + 1206, has been discussed by several modern writers.[1] A vertical water-wheel rotates a large gear-wheel on the same shaft, and this engages with another below (exactly in what relative plane is not quite clear from text or illustration) which is mounted very strangely. It is set upon a shaft which is held in loose bearings at one end and free to rotate in an annular groove at the other, so that while the wheel turns upon its geometrical centre the shaft describes a conical path. It thus acts as an eccentric lug or crank, an arrangement which to us would seem much more natural.[2] Instead of being linked to any connecting-rod, however, it slides up and down within a slotted rod attached to a fixed pivot below and to two piston-rods one on each side. Thus a continuous flow of water assured by the double Ctesibian single-acting force-pump cylinders and valves rises up the discharge pipe, as the slot-rod and the pistons move back and forth alternately to left and right (Fig. 54). This was a fine piece of mechanism for the early thirteenth century.[3] Although al-Jazarī's device was not as direct an ancestor of the reciprocating steam-engine as the Chinese system which Wang Chên described, the slot-rod had a cardinal part to play in the last stages of achievement of its definitive pattern in the nineteenth century. When William Howe in 1842 invented his link-motion reversing-gear[4]

[1] E.g. Wiedemann & Hauser VHW. Coomaraswamy gave the fine reproduction of the drawing which we reproduce, but, being an art historian and no engineer, his explanatory text (p. 17) contains a number of puzzling misnomers.

[2] This reconstruction is due to Professor Aubrey Burstall, who has made a beautiful working model on the basis of it, and to whom I am grateful for permission to reproduce a drawing. [See however the note on p. 123 above.]

[3] The slotted member is an ingenious device to which engineers of later times often had recourse, and it is still an important component of many machines. In the form of the slotted cross-head or Scots yoke (cf. Jones & Horton, vol. 1, pp. 250 ff.) it has played a considerable part in steam-engine and pump design by eliminating the connecting-rod and giving the piston-rod a uniform harmonic motion. It is also valuable for quick-return effects, as in shapers (cf. vol. 1, pp. 300 ff.; vol. 3, pp. 188 ff.), and it occurs in an ingenious arrangement for giving positive accurately timed reciprocation with a firm lock during the dwell at the end of each stroke (vol. 3, pp. 192 ff.). Slotted rods are also sometimes furnished with an internal rack and periodically thrown by cams from side to side so as to engage with a pinion on a driving-shaft in alternate senses, thus bringing about a reciprocating motion in the direction of their elongated axis. This gives a particularly long stroke (as in windmill pumps; Jones & Horton, vol. 1, pp. 260 ff., cf. also the converse case of the Napier motion, where a single rack with two faces perambulates in alternate upper and lower contact with a continuously rotating pinion, pp. 263 ff.). The windmill pump type of internally racked slot-rod was used as early as + 1615 by Solomon de Caus for a similar drive (cf. Beck BGM, p. 510). Curved slots with racks give the various varieties of the well-known mangle gear (Jones & Horton, vol. 2, pp. 245 ff.). A theoretical treatment of slot systems, both plain and racked, was given by Willis, pp. 287 ff., 294, 323, and, of course, in many later books.

[4] See Dickinson SE, pp. 114 ff.; Thurston, pp. 205 ff. Howe's reversing link-motion was immediately applied to the locomotive by Robert Stephenson.

it is highly improbable that he knew anything of al-Jazarī, but nevertheless some day it may be possible to trace back a continuous line of connection between them. The resemblance is clear, for the right-hand piston-rod is replaced by the slide-valve rod, and the left-hand one by the two eccentrics and connecting-rods of the valve motion; yet the profit of the motion is different, for the slotted member does not move from left to right and back, it is raised or lowered by auxiliary gear so that one eccentric may operate and one be idle, assuring thus forward or backward motion of the engine respectively.

Another Chinese machine which is worth considering in close connection with the hydraulic blowing-engine is the silk-winding or reeling apparatus, which draws off the raw silk from the cocoons. Though our best illustrations of this come from the early nineteenth century they agree not only with + 17th-century descriptions but also quite precisely with the text of the *Tshan Shu* (Book of Sericulture) written by Chhin Kuan about + 1090. In this machine (Fig. 21) the main winding barrels or reels are worked by a crank and pedal motion, but the ramping-arm (the forerunner of the 'flyer')[1] is also operated from the same power-source by means of a driving-belt connecting the main shaft with a pulley at the other end of the frame. This subsidiary wheel then moves the ramping-arm back and forth by means of a lug eccentrically placed. Here then we have the driving-belt just as in Wang Chên's hydraulic blower,[2] as also the smaller wheel with its eccentric, so that the ramping-arm corresponds to the connecting-rod. There is nothing, however, corresponding to the piston-rod, and instead of having any link connection at the further end, the arm is simply held in a ring through which it slides back and forth.[3] Thus the silk-reeler did not have quite all the components of the hydraulic blower. But the fact that it was already a standard piece of mechanism at the end of the + 11th century, and indeed owing to the antiquity of the silk industry may well have been established practice long before, considerably strengthens the probability that the water-powered blower, with its

[1] The function of this, of course, is to assure the even laying down of the fibre or yarn on the reel or bobbin.

[2] It may seem surprising that for these connections the Chinese preferred to use driving-belts rather than gearing, especially so where heavy-duty machines such as the hydraulic blowers were concerned. We know that gearing was in no way strange to them. The reason escapes us, but we may be sure that the mechanisms such as the silk-reeler and the hydraulic blower worked well enough, for the Chinese were always an intensely practical people, and the texts in question were primarily intended for the dissemination of practical knowledge.

[3] Thus this system was still at the stage of the European + 15th-century military engineers, whose crankshafts are so often depicted with connecting-rods but without piston-rods. It resembles too the age-old connecting-rods fitted to hand-driven mills for cereals in the Chinese culture-area; and in S.E. Asia too (personal obs. Mr H. Wulff). [See the note on p. 187.]

full 'steam-engine' arrangement for converting rotary to reciprocating longitudinal motion, developed during the Thang and Sung, i.e. during the four or five centuries preceding Wang Chên's description of it. This is almost certainly older, therefore, than al-Jazarī's swaying slot-rod.

WESTWARD TRANSMISSION OF THE STANDARD METHOD AND ITS INCORPORATION IN THE STEAM-ENGINE

It is necessary to emphasise that the system of three parts (eccentric, connecting-rod and piston-rod) has not so far been found in any fourteenth-century European illustration, and occurs only rarely in the fifteenth century. Lynn White, best of guides, gives us nothing earlier than a drawing, now in the Louvre, by Antonio Pisanello, which depicts in very workmanlike fashion a pair of piston-pumps operated by rocking levers raised and lowered by connecting-rods from two cranks fitted 180° apart on the two sides of an overshot water-wheel (Fig. 55, pl.).[1] It would be reasonable to date this about $+1445$ for Pisanello died in $+1456$. The same arrangement appears again in the sawmill shown in one of the seventy-two reliefs of machines by Ambrogio di Milano after the drawings of Francesco di Giorgio Martini ($+1474$), which are now preserved in the corridors of the palace at Urbino.[2] Finally the full assembly is to be found again in the *Mittel-alterliche Hausbuch* about $+1480$ applied to what looks like a single vertical stamp-mill, but here with only a single crank, connecting-rod and rocking lever.[3] By this time we are fully within the period of activity of Leonardo da Vinci.

[1] No. 2286; see Degenhart, fig. 147. The same drawing includes a *sāqīya* driven by an overshot water-wheel through two-stage reduction gearing. I regret that I must exclude what Lynn White, p. 113 and fig. 7, proposes as 'the earliest evidence of compound crank and connecting-rod', a MS drawing of Mariano di Jacopo Taccola which dates between $+1441$ and $+1458$ (Bayerische Staatsbibliothek, München, Cod. Lat. 197, fol. 82v). I do so not because the two-throw crankshaft is doubly mis-drawn, but because it effects the excursion of a lift-pump by what is quite evidently a rope instead of a connecting-rod, gravity alone ensuring the return. This device certainly constitutes an interesting premonition of the three-part system, but no more. By an extraordinary inversion, Lynn White, p. 81, misinterprets the construction of the hydraulic blowing-engine of the *Nung Shu* in exactly the same sense. The water-wheel, he says, 'turned a vertical shaft carrying an upper wheel which, by means of an eccentric peg and a cord, worked the bellows of a furnace for smelting iron'. But no such system was used; text and drawing fully concur in making the connecting-rod a rigid bar. Nor have I encountered any Chinese machine which used a cord with an eccentric peg. I am very grateful, however, to Professor Lynn White for his kindness in bringing these two $+15$th-century machines to my notice, and showing me an advance set of the plates of his book.

[2] See Feldhaus MLV, p. 245, fig. 167.

[3] Essenwein ed., pls. 24*b*, 25*b*; Bossert & Storck ed., p. 32. Here again there is a mis-drawing, for the rocking lever is made to pass through a horizontal instead of a vertical slot. The arrangement does not seem suitable for a stamp-mill, and perhaps a pump is intended, for in the 'Liebesgarten' on the left there is a fountain playing. No text or explanation has survived to help.

When Leonardo faces the problem of interconversion in the late fifteenth century, nearly 200 years after Wang Chên, he shows, as Gille has acutely pointed out,[1] a most curious disinclination to use the eccentric (or crank), connecting-rod and piston-rod combination. In fact, he does so only for a mechanical saw (Fig. 56).[2]

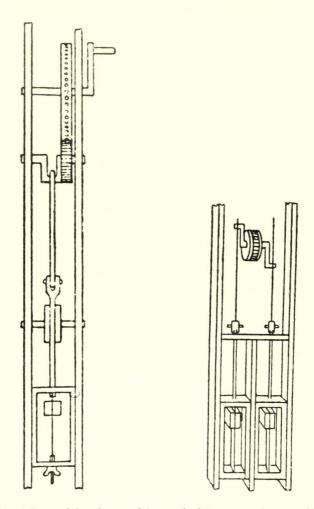

Fig. 56. Leonardo's only use of the standard interconversion assembly, a saw-mill, *c.* +1490 (Beck).

[1] P. 654. It will be remembered that the full crankshaft had appeared in Europe during the first half of the +14th century, and, in the form of the carpenter's brace, during the first half of the +15th.

[2] See Beck B G M, p. 323. The treadle lathe drive, so familiar in the sewing-machine, is a not far distant relation, Leonardo certainly knew this, for he sketched it with crankshaft and flywheel (cf. Burstall, pp. 122, 141 ff.). But he had been anticipated by the treadle drive of the Chinese silk-reeling machine (at least +11th century), and probably also by that of the Chinese cotton-gin.

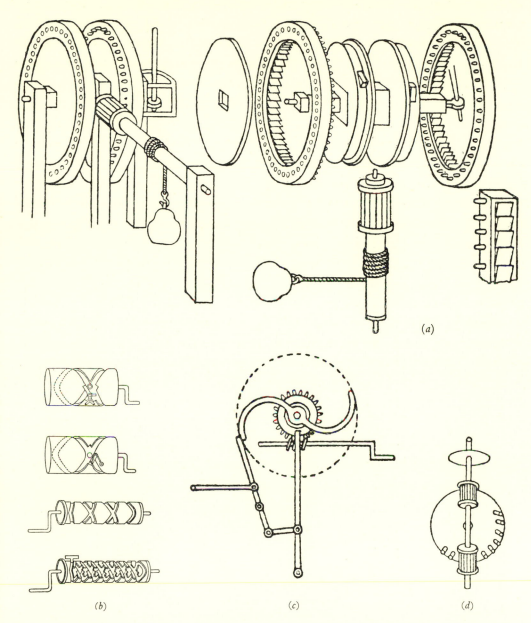

(a)

(b)   (c)   (d)

Fig. 57. Leonardo's other methods of interconversion of rotary and rectilinear motion: (a) the ratchet-windlass, (b) cylindrical cams with double helical grooves in which a follower pin moves back and forth along the left- and right-hand threads alternately, guided in some cases by automatic gates, (c) rotating cams or lugs and linkwork, (d) the half-gear (after Beck).

In order to avoid it he has recourse time after time to the most complicated and improbable devices. One celebrated arrangement (Fig. 57, *a*) has a lantern pinion on a windlass engaging on each side with right-angle gear in the form of pin-wheels, the motion of which is restricted by internal ratchets so that they can turn only in opposite directions. A lever working back and forth moves them on by means of internal drums with projecting movable teeth, thus turning the windlass continuously in one direction.[1] A much more brilliant solution was found when Leonardo devised a cylinder with a double helical groove into which a peg on the end of a piston-rod fitted, so that it was driven back and forth as the cylinder continuously rotated. With extraordinary ingenuity he then added a variety of automatically acting gates which prevented misrouting of the peg at the places where the grooves crossed each other[2] (Fig. 57, *b*). A third design used cam-shaped lugs on a rotating shaft acting upon a system of levers hinged by linkwork in such a way that each cam effected successively two excursions of the piston-rod[3] (Fig. 57, *c*). Lastly, Leonardo initiated that scheme which was to be such a favourite of the later Renaissance engineers, the half-gear.[4] In this system two lantern pinions on the same shaft engage at right angles with a single gear-wheel which has pins set round only half of the circumference of its disc. Thus with the successive contacts the shaft is driven round alternately in each direction. The resulting alternating rotary motion is then easily converted to alternating rectilinear motion by two

[1] Cf. Beck BGM, p. 421; Ucelli di Nemi, no. 16. This was the principle utilised by Keane Fitzgerald (p. 199 below) for a steam-engine beam. Double-acting ratchet gearing pushed on by both excursions of a lever is still used in machines of various kinds (cf. Jones & Horton, vol. 1, pp. 29, 31; vol. 3, pp. 71 ff.). Wheels with internal ratchets are employed, for instance, in the drives of belt-conveyors (cf. Jones & Horton, vol. 1, pp. 53 ff.; vol. 3, p. 70).

[2] Cf. Beck BGM, pp. 417 ff.; Ucelli di Nemi, no. 122; Willis, pp. 157, 321. These cylindrical cams with follower rollers sliding back and forth in the grooves are widely used in contemporary machine construction; cf. Jones & Horton, vol. 1, pp. 4, 8; vol. 2, pp. 11 ff., 19 ff., 294; vol. 3, pp. 178, 182. Automatic gates like Leonardo's can be found in wire-making machines and gas-engines; cf. Jones & Horton, vol. 1, p. 19; vol. 2, pp. 44 ff. The cylindrical cam finds particular application in the textile industry, as for instance in the 'rotoconer', a machine for winding woollen and worsted yarn on to cones and 'cheeses'. Here the spiral groove acts as a traversing device. But no doubt the most familiar example of the use of the grooved cylinder is in the spring screwdrivers sold in every modern tool-shop. The simplest form of the groove system is, of course, found in the 'face cam', where a pin and roller follows a groove in a rotating plate. Examples of the current use of this may be seen in Jones & Horton, vol. 1, pp. 3 ff.; vol. 3, pp. 191 ff. It was already used in the +16th century by Ramelli; cf. Beck BGM, p. 219. In his design levers were worked up and down by a single channel mounted eccentrically upon a rotating horizontal wheel. I am much indebted to Dr Norman Heatley and Dr A. J. W. Haigh of Oxford for discussions on this subject.

[3] Cf. Beck BGM, p. 419.

[4] And not only with them. Intermittent gears are quite common in modern machinery; cf. Jones & Horton, vol. 1, pp. 68 ff., 93 ff., 98 ff.; vol. 2, p. 71; vol. 3, pp. 24, 46 ff., 176. The principle is used in all kinds of ways, with right-angle bevel gears, epicycle gears, reciprocating racks, and so on.

windlass chains or other means[1] (Fig. 57, *d*). All these complicated arrangements proved useful in one way or another in the machinery of later centuries,[2] but the simpler and fundamentally important steam-engine system derives from Wang Chên rather than from Leonardo. Why there was this aversion from it in Europe is not at all clear. Serious difficulties were perhaps encountered in assembling the moving parts so that friction and wear were sufficiently overcome,[3] but in this case we should like to know just how and why the technique of the medieval Chinese engineers was more advanced. Better steel bearings may well be the answer, and particularly a combination of cast-iron and steel surfaces.

Perhaps the most extraordinary part of the whole story is that James Watt was driven to the invention of the sun-and-planet gear[4] by the fact that the basic method of converting rotary to rectilinear motion by crank and connecting-rod had been patented by James Pickard in 1780.[5] Watt had not patented it himself because he knew that it was old, but probably none of those involved knew anything of the +15th-century German engineers, and certainly no one at that time could have had any suspicion that the Chinese of the Sung period had been intimately and practically acquainted with it. Indeed, on our present information, they were its real inventors.[6]

In considering the pre-natal history of all reciprocating steam-engines and derivative prime movers there are thus two outstanding conclusions. Not only, as we have just found, was the combination of eccentric, connecting-rod and

---

[1] Cf. Ucelli di Nemi, no. 71, where Strobino's reconstruction applies the device to the co-axial textile flyer, and Beck BGM, p. 321. For the later applications in Ramelli see Beck BGM, pp. 213 ff., 223 ff., and in de Caus, Beck BGM, p. 508. Theoretical treatment in Willis, p. 293.

[2] There are, of course, a number of other devices too which Leonardo did not mention, e.g. the swash-plate (cf. Willis, p. 319). Cf. p. 182 above.

[3] As suggested by Gille, p. 653.

[4] Cf. Reuleaux, p. 245; Willis, p. 373.

[5] Or so it was thought, and Watt had his reasons for not challenging patents. The story is complex and need not be repeated here (see Dickinson SE, pp. 80 ff., JW, pp. 126 ff.; Farey, pp. 423 ff.; Lardner, pp. 182 ff.; Smiles, p. 227; Feldhaus TDV, col. 593 ff.; Wolf XVIII, pp. 621 ff., and other obvious sources). Earlier attempts to obtain rotary motion from atmospheric engines had generally if not always invoked different kinds of ratchet gear, as e.g. that of Keane Fitzgerald in 1758 and Matthew Wasborough in 1779 (cf. Farey, pp. 406 ff.; Dickinson SE, p. 80). The 'fire-mills' of the unfortunate abbé Etienne (Scipion) d'Arnal at Nîmes in 1780 were (it has been stated) also of this kind, but his work is so obscure that he is not even mentioned by Arago, whose concern it was to insist, quite rightly, upon the part played by French inventors in the history of the steam-engine. Farey, however, who should have known, states (p. 655) that d'Arnal attempted no rotary conversion, simply working ten water-wheels from a reservoir kept full by two atmospheric engines. This was a practice quite common in eighteenth-century England also (cf. Farey, pp. 296 ff., 413). Since the same water could be used over and over again, such engines were called 'returning engines'. Pickard's patent expired in 1794.

[6] Unless, of course, the system goes back to Tu Shih and his Han engineers in the +1st century, a possibility which is not as yet quite excluded.

piston-rod apparently first worked out in Sung or Thang China, achieving there rather than in Europe the most effective of all inventions for the interconversion of rotary and rectilinear motion;[1] but also the double-acting piston and cylinder principle made its first appearance in the Chinese air-pump, sucking and expelling on both strokes, which again was fully developed certainly in the Sung, probably in the Thang, and possibly in the Han.[2] In these inventions the Hellenistic age does not compete, and it is noteworthy that all their datings long precede not only the Renaissance, but also the times of Leonardo da Vinci and even Guido da Vigevano (+ 14th century). What constituted the fundamental revolution of the European + 17th and + 18th centuries was the inversion of the direction of motion so that force was transmitted not to but from the piston. One may thus justly conclude (if it is not putting too much strain on our adopted terminology) that the great 'physiological' triumphs of 'ex-pistonian' Europe were built upon a foundation of formal or 'morphological' identity laid by 'ad-pistonian' China. If there was, as seems most probable, a direct genetic connection, this is the place to look for it—not in that strange episode when Jesuit mechanicians put a model steam turbine locomotive through its paces in the palace gardens of the Khang-Hsi emperor about + 1671.[3] The transmission of the three-component assembly would have taken place long before. In the late Middle Ages and the early Renaissance the relations between Europe and China were quite close; it was the period of the Pax Tartarica. Marco Polo was in China between + 1275 and + 1292, just about

[1] [It is true that in the Chinese blowing-engine a kind of 'bell-crank' rocking roller intervenes between the connecting-rod and the piston-rod, but this does not affect the analogy of the structures, and the device may best be regarded as an interim solution in the search for stability ultimately attained in the slides and cross-head. One may also reflect that the beam of the beam-engine was simply a straightened bell-crank, intervening between the piston-rod and the pump-rod, so that Wang Chên's rocking rollers were re-incarnated for a whole century in that form; and after + 1783 when rotary motion was combined with the beam by various means (p. 201), the similarity became extremely close. About the same time another turning-point was reached when Watt made his atmospheric beam-engines double-acting (p. 149), for as the piston now pushed as well as pulled, the flexible connection with the beam end would no longer work. Since planing machines had not been invented, guide slides were as much out of the question for Watt as they would have been for Wang—hence Watt's elegant solution of the pantographic 'parallel linkage' (cf. Dickinson SE, p. 82, JW, pp. 136 ff.), which however was indispensable only for a couple of decades. The only desirable component lacking in the Chinese blowing-engine was the crankshaft, and though of European origin (cf. *SCC*, vol. 4, pt. 2, p. 113), that did not appear in such a machine until the time of Wilkinson four centuries later than Wang and eight centuries later than Wei.]

[2] In traditional China probably no one noticed the characteristics of the stroke; the irregularity of speed, greatest at mid-stroke, and the momentary dwells in the neighbourhood of each dead centre. Only with the construction of precision steam-driven metal machinery in the European nineteenth century did such features become evident. Since then much ingenuity has been expended in devising means of obtaining strokes with any desired velocity characteristics (cf. Jones & Horton, vol. 1, pp. 249 ff.; vol. 2, pp. 260 ff.; vol. 3, pp. 206 ff.).

[3] See *SCC*, vol. 4, pt. 2, p. 225.

the time when Wang Chên was writing his great treatise on traditional Chinese agriculture and industry. Marco Polo, if the greatest, was by no means the only, traveller of those times across the length of the Old World, and there are many indications of East Asian influences conveyed through the medium of other merchant adventurers contemporary with Marco Milione. Towards the end of the + 13th and the beginning of the + 14th centuries there appears in the West a whole cluster of such inventions—gunpowder, silk machinery, the mechanical clock, and the segmental arch bridge. Nor did the movement come to an end with the decline of merchant travel, for during the whole + 14th century and the first half of the + 15th, there was a little-known but very important trade in Tartar (i.e. Mongol and Chinese) slaves, who were much sought after as servants in Italy. We have no specific evidence of what techniques they brought with them, but we do know that the second half of the + 14th century saw the arrival of the blast-furnace for iron smelting and casting, and the coming of block-printing, soon followed by movable-type printing, together with the use of lock-gates on summit canals—all demonstrably prior Chinese inventions. If the three-component assembly for inter-converting rotary and rectilinear motion first appears in Europe about + 1445, when we know that it was being widely used in China in Marco Polo's time, a century and a half before, then with all due respect to Leonardo's undeniable genius, independent invention is what stands in need of proof.

How the double-acting principle incorporated itself in the development of steam prime movers before the end of the atmospheric engine with Watt's design of + 1783 we have already observed (see Chart). How the crank, connecting-rod and piston-rod assembly joined with it in moulding all steam-engine design after + 1798 will by now be evident. But strange to say, the conversion assembly also influenced atmospheric engine practice, since after about + 1780 these engines also were often made rotary with the addition of a flywheel; known as 'whimsey' engines, they were widely used for winding from shallow pits (Fig. 58, pl.).[1] It is

[1] Dickinson SE, pp. 64 ff.; Galloway, pp. 160 ff.; Farey, pp. 409, 410 ff., 422. One of the most informative treatments is that of Davey. We have already remarked upon the late date to which these engines persisted, p. 152 above. Among the more interesting Newcomen 'whimsey' engines were those which combined two atmospheric cylinders working in opposite senses so as to render the motion more continuous. In 1793 Francis Thompson patented a system in which two Newcomen pistons were placed on the same piston-rod (cf. p. 154 above), the upper one working in an inverted atmospheric cylinder. These two single-acting cylinders operated a connecting-rod, crank and flywheel through two intervening pinions, by means of the usual beam (Farey, pp. 427, 658). Several of these engines were used in mills. Alternatively the Falck system (proposed as early as 1779, and used in England from 1794 onwards) had two Newcomen cylinders, one attached to each end of the beam, and giving a working stroke alternately (Farey, pp. 426, 662); this again was capable of doing good service, and a few of such engines were rebuilt with separate Watt

hard to believe that they could have worked with any degree of smoothness, but they were cheap, and one or two lasted on in use until our own time.[1]

ENVOI

To conclude this offering to the memory of Thomas Newcomen, and to sum up all the material, some of it no doubt rather unfamiliar, which has been presented herein, one may say paradoxically that if Newcomen's work is taken as the turning-point, the hidden pre-natal history of the reciprocating steam-engine played a remarkably large part in its 'post-natal' history. For in order to go beyond the single-stroke stage and the non-rotary beam stage, currents of design were required which went back many centuries earlier than Newcomen and Watt. That the reciprocating steam-engine in its definitive form, attained about + 1800, consisted essentially of two structural patterns which had been working widely and effectively in China about + 1200 can be conclusively demonstrated; that these patterns passed into it by a long chain of direct genetic connections is over-whelmingly probable, however conscious or semi-conscious may have been the indebtedness of engineers of succeeding generations. No single man was 'the father of the steam-engine'; no single civilisation either—the case was more like the 'Quatre-vingts Chasseurs' of the old French song. Yet no one comes nearer to deserving the title than Thomas Newcomen. In the light of the foregoing analysis he stands out as a typical figure of that modern science and technology which grew up in Europe only, while his successors, great as they were, drew upon older Asian inventions more than has hitherto been recognised.

condensers at the close of the century. Perhaps all these exceptional machines show the difficulty of acceptance of the double-acting principle as well as the search for low costs and the exigencies of the Watt patents.

[1] In fact fully into the era of the internal-combustion engine, when the single-acting principle had come into its own again. We are now, however, witnessing a revival of high-pressure steam-engines in ocean-going trawlers, but with single-acting pistons in rows like Diesel engines.

11

THE MISSING LINK IN HOROLOGICAL HISTORY: A CHINESE CONTRIBUTION[1]

[1958]

IT seems that first of all this Lecturer is in private duty bound to celebrate the name of that revolutionary bishop and oecumenical philosopher John Wilkins (+ 1614 to + 1672). He was probably the only man who ever became both Warden of Wadham and Master of Trinity—certainly the only one who married the sister of the Lord Protector and yet was raised to the episcopate under King Charles the Second. The first of all our Secretaries, he had been prominent among the members of the Invisible College in + 1645, and occupied the chair at that meeting of + 1660 which brought the Society towards its definitive form. 'At Cambridge', wrote Burnet, 'he joined with others who studied to propagate better thoughts, to take men off from being in parties or from narrow notions, from superstitious conceits, and fierceness about opinions. He was a great preserver and promoter of experimental philosophy, a lover of mankind, and had a delight in doing good.'

The subject of this afternoon's discourse would, I believe, have had the interest of John Wilkins, and his blessing. For his first work, published in + 1638 and + 1640, bore the title: *The Discovery of a World in the Moone; or, a Discourse tending to prove that 'tis probable that there may be another Habitable World in that Planet: together with a Discourse concerning the Possibility of a Passage thither*. I cite this title not so much for the strange contemporaneity of its second half, but because the book formed part of that great movement of thought, led by such men as Bruno and Gilbert, which destroyed the solid crystalline spheres of Aristotelian–Ptolemaic tradition. And elsewhere it has been possible to show that one of the influences here at work was the new knowledge of Europeans that the astronomers of China had always believed in the floating of the heavenly bodies in infinite space.[2]

In his greatest and largest work, however, the relations of Wilkins with Chinese culture were even closer. For the *Essay towards a Real Character and a Philosophical*

[1] Wilkins Lecture, Royal Society; London, 1958. Reprinted from *Proc. Roy. Soc. Lond. A*, 1959, **250**, 147.
[2] *SCC*, vol. 3, pp. 438 ff.

203

Language of +1668 set out, among other things, to elaborate a '*common* character' or ideographic script, a 'real character which shall be legible in all languages'. This work, like that other foundation-stone of mathematical logic, the *De Arte Combinatoria* of Leibniz, written two years earlier, was partly inspired by 'what is commonly reported of the men of China, who do now, and have for many Ages used such a general Character, by which the Inhabitants of that large Kingdom, many of them of different Tongues, do communicate with one another', everyone understanding the common script but reading it in his own dialect.[1] Indeed, Chinese influences were manifold at this time. Timothy Bright, to whom Wilkins indirectly but clearly refers, had based his system of shorthand, introduced in +1587, on ideograms in vertical columns like Chinese characters.[2] Robert Hooke himself wrote on the Chinese language[3] and took tea with Chinese merchant friends in the city of London.[4] Thus there is reason for thinking that the early Fellows of the Society would have appreciated warmly the story which I now desire to put before you.

The clock is the earliest and most important of complex scientific machines. No one can doubt that the invention of the mechanical clock was one of the greatest achievements in the history of all science and technology. 'The fundamental solution', wrote von Bertele, 'of the problem of securing steady motion by intersecting the progress of a weight-driven (or any powered) train into intervals of equal duration, must be considered as the work of the brain of a genius.' The essential task was to devise means of slowing down the rotation of a wheel so that it would keep a constant speed continuously in time with the apparent diurnal revolution of the heavens. The essential invention was the escapement. In what follows I hope to show that the escapement arose in China in the middle of a very long line of development of mechanisms for the slow rotation of astronomical models (demonstrational armillary spheres or celestial globes), the primary aim of which was computational rather than for time-keeping as such. I also hope to show that its first application was to a water-wheel like that of a vertically mounted water-mill, so that although in later ages mechanical clocks were mostly driven by falling weights or expanding springs, their earliest representatives depended on water-power. The mechanical clock may thus be said to owe its existence largely to the art of Chinese millwrights. The story takes some

[1] For fuller information on Wilkins's work the interesting essay of Hogben may be consulted. The first of the present series of Lectures, by their founder, J. D. Griffith Davies, was on Wilkins, but seems never to have been published.   [2] See Carlton.   [3] *Phil. Trans.* 1686, **16**, 35.
[4] Gunther, vols. 6, 7, pp. 681, 694; vol. 10, pp. 253, 263. The year was +1693.

telling. Obviously it differs widely from the account accepted hitherto. How is it that the Chinese contributions to clock-making have been hidden from world history?[1]

It will readily be allowed that few historical events were so rich in consequences as the decision taken by certain southern Chinese officials in +1583 to invite into China some of the Jesuit missionaries who were waiting in Macao. It was the first decisive step in the long process of unification of world science in Eastern Asia, and the better mutual understanding of the great cultures of China and Europe. The two men chiefly concerned were Chhen Jui (+1513 to *c.* +1585) who was for a short time Viceroy of the two Kuang provinces, and Wang Phan (+1539 to *c.* +1600) who was Governor of the city of Chao-chhing. They were particularly interested in reports that the Jesuits had, or knew how to make, chiming clocks of modern type, i.e. of metal with spring or weight drives and striking mechanisms. These became known as 'self-sounding bells' (*tzu ming chung*), by a direct translation of the word 'clock' or *cloche* (*glocke*). This is important, for an entirely new name, as this one was, naturally suggested an entirely new thing. The mechanical clocks of the Chinese Middle Ages had been, as we shall see, extremely cumbrous and probably never very widespread; moreover, no special name had distinguished them from non-mechanised astronomical instruments. It was therefore not surprising that the majority of Chinese, even scholars in official positions, now got the impression that the mechanical clock was a new invention of dazzling ingenuity which European intelligence alone could have brought into being. And of course the missionaries (as men of the Renaissance) quite sincerely believed in this higher European science, seeking by analogy to commend the religion of the Europeans as something equally on a higher plane than any indigenous faith.

There can be no doubt that Ricci and his companions regarded mechanical clocks as something absolutely new and unheard of in China. He says this in his memoirs on several occasions. It is true that Ricci and Trigault had something to tell of Chinese clocks with driving-wheels which they found on their travels, though they laid little emphasis on them and their descriptions are obscure. At the same time a number of contemporary Chinese scholars recognised that the clockwork of the Jesuit 'self-sounding bells' was not something fundamentally new in Chinese culture. The former were not anxious (in China) to exalt the achievements of the indigenous past. The latter were insufficiently learned to

[1] In *SCC*, vol. 4, pt. 2, a more complete and fully documented treatment of the material presented in this Lecture will be found, not however superseding the monograph of Needham, Wang & Price.

expound that past as it deserved. This complicated situation had its effects upon the thinking of Europeans later on, and in particular on that of European historians of science. If the Jesuits so firmly believed in the novelty of the mechanical clocks which they introduced to China, who were the later historians of science, penned within the ring fence of the alphabetic languages, to gainsay them?

According to the view which became generally accepted, the first successful achievement of slow, regular, and continuous rotation in time with the diurnal revolution of the stars, by means of an escapement acting upon the driving-wheel of a train of gearing, occurred in Europe very shortly after the beginning of the + 14th century. One may take an authoritative statement from von Basserman-Jordan, who wrote: 'We must place the birth of the wheel clock around + 1300. When clocks are mentioned before that time, either sundials or clepsydras are meant, or else there is always something doubtful about the evidence...The soul of the wheel clock is the escapement, which hinders the rapid revolution of the wheels. This invention is one of the greatest and cleverest ever made by man, yet the inventor remains unknown and forgotten, commemorated neither by stone nor monument.' It has indeed been lately shown that these first mechanical time-keepers were not so much of an innovation as used to be supposed. They descended in fact from a long series of complicated astronomical 'pre-clocks', planetary models, mechanically rotated star-maps, and similar devices designed primarily for exhibition and demonstration rather than for accurate time-measurement. Traces of these remain from Greek, Hellenistic and Arabic times, but the remains are fragmentary and the texts tantalisingly incomplete. We shall mention some of them from time to time.

First, let us gain a clear idea of the + 14th-century mechanisms, and then vivify this skeleton by a few textual references. The simplest form of the early European mechanical clock drew its power from the rotation of a drum brought about by the fall of a suspended weight. This was connected with trains of gearing in great variety, but the movement of the whole was slowed to the required extent by the escapement device known as the verge-and-foliot. This can best be appreciated by the accompanying illustrations. Fig. 59 (pl.) shows this type of clock in its simplest form (a Nuremberg bracket-clock described by Zinner). The essential parts of the device were the crown-wheel, a toothed wheel with projections like right-angled triangles set perpendicularly to its main plane; the verge or rod standing across this wheel and bearing (at right angles to each other) its two pallets or little plates so as to engage with the crown-wheel; and finally the foliot (crazy dancer), i.e. two weights carried one at each end of a bar set at the top of the verge. The method of action was very simple. The torque of the crown-wheel pushed one of the pallets out of the way, giving a swing to the foliot, but this only led to the coming into action of the other pallet and then a swing of the inert weights in the opposite direction. In this way the motion of the wheel was arrested alternately by the two

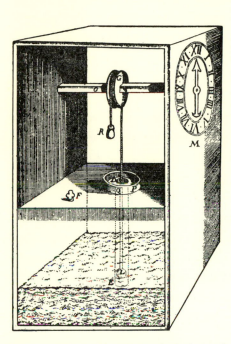

Fig. 60. Hellenistic anaphoric clock depicted by Isaac de Caus, +1644.

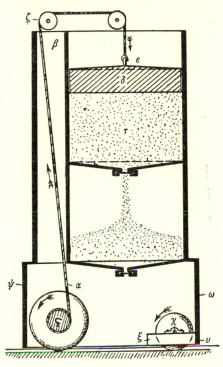

Fig. 61. Automobile puppet theatre of Heron of Alexandria, +1st century (Beck).

pallets. Thus an oscillatory component received its impulses from the weight drive and at the same time imposed a step-by-step or ticking movement upon the train.

There can be no doubt that many of the component parts of these clocks were Hellenistic in origin. The falling weight had originally, no doubt, been a falling float, such as we see in the Roman anaphoric clocks, where a dial bearing astronomical markings was made to rotate slowly by a cord attached to a float sinking in a clepsydra (cf. Fig. 60). The idea of a dial would have come from the same source. In the West there was also the automobile puppet theatre minutely described by Heron of Alexandria and reconstructed by Beck, where slow motion was obtained by the descent of a heavy weight as grains of sand or cereal escaped through an hour-glass hole at the bottom of the container (cf. Fig. 61). Thorndike has drawn attention to a very interesting passage in the commentary of Robertus Anglicus on the *Sphere* of Sacrobosco. Writing in + 1271, he said:

Now it is hardly possible for any time-keeping device (*horologium*) to follow the indications of astronomy with absolute accuracy. Yet clock-makers (*artifices horologiorum*) are trying to make a wheel which will accomplish complete revolution for each one of the equinoctial circle, but they cannot quite perfect their work. If they could, it would be a really accurate clock and worth more than any astrolabe or other astronomical instrument for reckoning the hours, if one knew how to do this according to the aforesaid method. The way would be this, that a man make a disc of uniform weight in every part so far as could possibly be done. Then a lead weight would be hung from the axis of that wheel, moving it so that it would complete one revolution from sunrise to sunrise...

These words strongly suggest that at this time attempts were being made to construct a practical weight-driven escapement clock, but that success had not yet been achieved. However, the mercury clock described in the *Libros del Saber de Astronomia*, compiled for King Alphonso X at Toledo about + 1276, could have worked satisfactorily. Here the weight-drive was combined with a hollow drum containing twelve compartments and half filled with mercury, the escapement effect being obtained by the resistance offered by the mercury in passing through small holes in the walls of the compartments.[1] Why this system did not spread remains puzzling. Of much significance, however, is the fact that it was used to rotate an astrolabic or anaphoric dial.

As for the verge-and-foliot escapement, Frémont must surely be right in his suggestion that it derived from the radial bob type of flywheel. This had been associated since Hellenistic times with the upper ends of the worms of screw-presses, whether used for making wine or oil, and later on, in the + 15th century, for printing books. In the + 16th and probably earlier, it was used to assist crank action. The originality lay in its combination with the pallets and crown-wheel so that it oscillated back and forth rather than continuously turning. Now one of the greatest mysteries of the early European clocks was the origin of the escapement principle. For a long time it was thought to appear in a strange design found in the notebook of Villard de Honnecourt about + 1237, where a cord carrying weights at each end is wound round two axles, one vertical and one horizontal, finally passing between the spokes of a large wheel on the second axle. It was supposed that the motion was periodically checked, and then released on

---

[1] [The original illustration is reproduced in *SCC*, vol. 4, pt. 2, fig. 647, opp. p. 443.]

the recoil. The object of the device was to make a figure of an angel turn and point its finger at the sun. Another design was intended to make an eagle turn its head towards the place where the priest and clerks stood to read the Gospel. But it is now agreed that these mechanisms cannot have been escapements, but simply a means of turning the figures by hand. If so, no predecessor for the first European escapement remains—except the Chinese type shortly to be described.

For three hundred years the verge-and-foliot clock remained unchanged, save for increasing elaboration of striking trains with complex systems of levers and detents. But towards the end of the +16th century the pendulum was coming into wider technological use, and its property of isochronicity began to attract attention. The first application of it to the clock escapement may have been made by Jobst Burgi of Prague in +1612, but more credit is due to Galileo and most of all to Huygens. In +1641, after he became blind, a pendulum clock was built for Galileo by his son Vincenzio. But the main features of the pendulum clock, including even the cycloidal arc of swing, were due to Huygens, who constructed his first successful apparatus in +1657, and whose book *Horologium Oscillatorium* achieved its final form in +1673. At first the pendulum was combined with the verge and pallets. But about +1680 Wm Clement devised the familiar anchor escapement, in which the crown-wheel was replaced by a scape-wheel having teeth in the plane of its rotation. This device has persisted till the present day in many modified forms. Probably its most important modification was the dead-beat escapement introduced by George Graham in +1715; by adjusting the shape of the teeth and pallets this eliminated all that recoil which had been one of the most wasteful features of the early clocks. Then, with progressive solutions of problems such as temperature compensation in the eighteenth and nineteenth centuries, we are fully in the modern period, into which it is unnecessary to go further for our present purpose (cf. Fig. 79).

In the meantime there had been one other invention of major importance, the application of the spring-drive instead of the falling weight. This permitted the making of portable watches as well as stationary clocks. But it introduced a new difficulty, that of compensating for the variable force exerted as the steel spring ran down; this was overcome by various devices, first an auxiliary spring known as the stack-freed, and then the conical drum known as the fusee. This was a driving barrel of varying diameter, so arranged that the maximal leverage of the cord or chain acting on its largest diameter came into play at the end of the spring's activity when its pull was weakest.

All these conclusions naturally depend upon the evidence of texts as well as of remaining clocks or their parts. It may be considered as quite certain that the earliest type was in use by about + 1310 and that all the characteristic features were assembled by + 1335. Yet no + 14th-century clock has survived immune from later reconstructions so extensive as to render difficult the restoration of the original condition. One of the earliest literary references is due to Dante, who describes quite clearly in a text of + 1319 the gear-work of a striking clock. Authentic accounts of clocks occur in the *Chronicles* of G. Fiamma for + 1335 and + 1344; the former one erected in the tower of a palace chapel at Milan, the latter in another tower at Padua. This was due to Jacopo di Dondi, whose son Giovanni, besides constructing a great clock at Pavia in + 1364, wrote a splendid horological treatise which has been carefully studied and summarised by Lloyd (1954). His clock was an astronomical masterpiece embodying gear-trains of great complexity to portray accurately the motions of the planets, and to show the ecclesiastical festivals, both fixed and movable. The first mention of the verge-and-foliot, by Froissart, occurs about the same time, in + 1368, and this is also the year of the first escapement clock in England.

Such was the picture of the development of the mechanical clock as it stood on the basis of researches in European history alone. The invention of the escapement seemed to have occurred at the beginning of the + 14th century with no recognisable antecedents. As Bolton wrote: 'Weight-driven clocks come suddenly into notice at this period in a very advanced stage as regards design, though their workmanship was rough. Their previous evolution must have taken a long time, but there is no reliable record of its stages nor of the men responsible for it.' In the latter part of 1955, however, a way of solving this problem opened out before us. For the *Hsin I Hsiang Fa Yao*, written in + 1092 by a distinguished scientific scholar and civil servant of the Northern Sung dynasty, Su Sung, describes the erection in + 1088 of elaborate machinery for effecting the measured slow rotation of an armillary sphere and celestial globe, together with a profusion of time-keeping jack-work. The book's pithy title might be translated 'New Design for an Astronomical Clock' (lit. 'Essentials of a New Method for [Mechanising the Rotation of] an [Armillary] Sphere and a [Celestial] Globe').[1] The whole 'Combined Tower' (*Ho Thai*) could in fact have been nothing more

---

[1] Square brackets indicate insertions in translations necessitated by the more elaborate grammar and less laconic style of the English language as compared with the Chinese, or indispensable explanations, or parts of proper names omitted in the original. Round brackets in translations are reserved for parenthetical words actually in the texts. This system is different from that used in *Science and Civilisation in China*.

nor less than a great astronomical clock necessitating some form of escapement. And indeed the full translation and study of the elaborate and detailed text not only showed that this was the case, but revealed the considerably earlier origins and development of time-keeping machinery recorded for posterity in Su Sung's

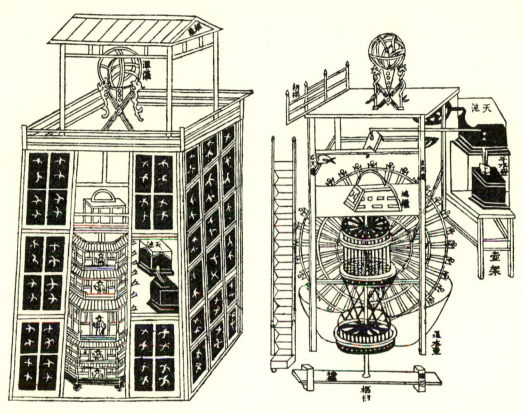

Fig. 62 a. General view of Su Sung's clock tower, from *Hsin I Hsiang Fa Yao* (+1094).

Fig. 62 b. General view of the works of Su Sung's clock tower.

remarkable historical introduction. In this way six centuries of Chinese horological engineering, previously hidden, came to light.

First, a word or two on the transmission of the text to us. It was printed in the south by Shih Yuan-Chih in +1172. A copy of this edition was owned by the late Ming scholar Chhien Tsêng (+1629 to +1699), who reproduced it in a new edition with extreme care. It was printed again later on by Chang Hai-Phêng (+1755 to 1816) and more numerously by Chhien Hsi-Tso (+1799 to 1844) in the latter year. With a solicitude for the history of science somewhat unexpected in the imperial editors of +1781, they wrote:[1] 'The dynasty of your Imperial Majesty now has instruments which in excellence and precision far exceed all those made during the past thousand years. Of course the invention of Su Sung is not to be compared with them. However, we may have something to learn

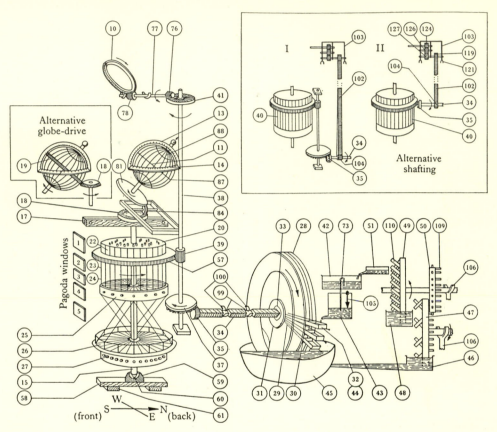

Fig. 63. Reconstruction of the mechanism of the 'Water-powered Sphere and Globe Tower (*Shui Yün I Hsiang Thai*)' of Su Sung, begun in +1088. The numbers identify technical terms in the text, fully listed and explained in Needham, Wang & Price (1960). [See also *SCC*, vol. 4, pt. 2, fig. 652.]

by paying attention to these old matters, for they show that the people of that time were also interested in new inventions...His book should be considered as something indeed valuable and precious.'

Another point of great interest is that it was by no means the only book on astronomical clockwork written during the Sung dynasty. The bibliographical chapters of the *Sung Shih* also record a *Shui Yün Hun Thien Chi Yao* (Essentials

[1] *Ssu Khu Chhüan Shu Tsung Mu Thi Yao*, ch. 106.

of the [Technique of] making Astronomical Apparatus revolve by Water-Power) written by Juan Thai-Fa. But nothing can be ascertained about this author, or his work, or date.

We are now in a position to study the illustrations in Su Sung's book and the working drawings of the reconstruction made by Needham, Wang & Price (1956).

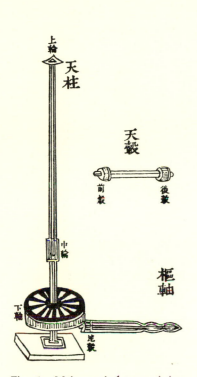

Fig. 65. Main vertical transmission shaft in Su Sung's clockwork.

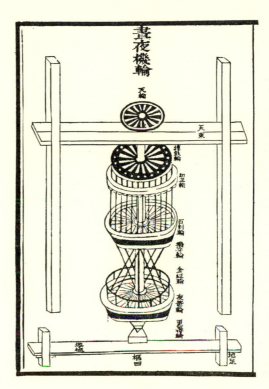

Fig. 66. Time-keeping shaft and annunciator jack-wheels in Su Sung's clockwork.

Fig. 62*a* shows the general external appearance of the 'Combined Tower' or Tower for the Water-Powered Sphere and Globe (*Shui Yün I Hsiang Thai*). The armillary sphere (*hun i*) is on the platform at the top, the celestial globe (*hun hsiang*) is in the upper chamber of the tower, half sunk in its wooden casing; and below this stands the pagoda-like façade with its five superimposed storeys and doors at which the time-announcing figures (jacks) appeared. On the right the housing is partly removed to show the water-storage tanks. The scale of the whole is clearly deducible from the internal evidence of the text; the height must have been between 30 and 40 ft in all. [Cf. Fig. 64, pl.]

Su Sung's general diagram of the works appears in Fig. 62 *b*, but its explanation may best be followed in the modern drawing of Fig. 63. The former sees the structure from the south or front, the latter from the south-east. The great driving-wheel, 11 ft in diameter (Fig. 63, no. 28), carries thirty-six scoops on its circumference, into each of which in turn water pours at uniform rate from the

Fig. 67. One of the time-keeping jack-wheels in Su Sung's clockwork.

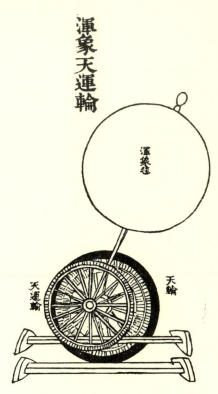

Fig. 68. Oblique polar axis globe-drive in Su Sung's clockwork.

constant-level tank. The main driving-shaft of iron (no. 34), with its cylindrical necks (*yuan hsiang*) supported on iron crescent-shaped bearings (*thieh yang yüeh*), ends in a pinion which engages with a gear-wheel at the lower end of the main vertical transmission shaft. This (Fig. 65) drives two components. A suitably placed pinion (no. 39) connects it with the time-keeping gear-wheel which rotates the whole of the jack-work borne on the time-keeping shaft (Fig. 66). This consists of half-a-dozen superimposed horizontal wheels carrying round figures or jacks. Since each of these wheels is from 6 to 8 ft in diameter, the total weight involved must have been very considerable, so the base of their shaft is fitted with a pointed cap (*tsuan*) and supported in an iron mortar-shaped end-bearing (*thieh shu chiu*). The jack-work wheels performed a variety of functions, their figures either appearing with placards on which the time was marked, or ringing bells, striking gongs or beating drums as they made their appearances in the pagoda

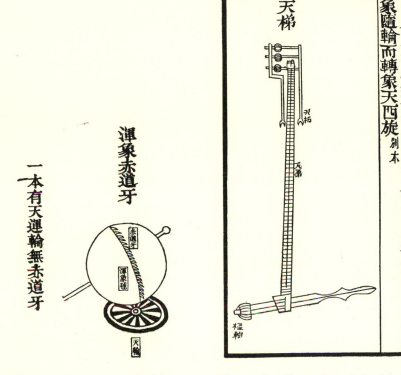

Fig. 69. Direct equatorial globe-drive in Su Sung's clockwork.

Fig. 70. Armillary sphere chain-drive and gear-box in Su Sung's clockwork.

doorways. Su Sung's picture of one of these time-keeping wheels is shown in Fig. 67. Their rotation, however, was not the only duty of the time-keeping shaft, for at its upper end it engages by means of oblique gearing and an intermediate idling pinion with a gear-wheel on the polar axis of the celestial globe (Fig. 68). The angle of these gears corresponded of course to the polar altitude at Khaifêng. Now the text contains a number of notes which record improvements in the clock, probably dating from the last years of the +11th century, and in these an alternative globe-drive is given, the uppermost gear-wheel directly enmeshing with an equatorial gear-ring (*chhih tao ya*) on the globe (Fig. 69). Presumably the original gearing proved difficult to maintain.

We must now return to the main vertical transmission shaft and the second component which it drives. Its uppermost end provides the power for the rotation of the armillary sphere. This is effected by right-angled gears (no. 41) and oblique gears connected by a short idling shaft. The oblique engagement is made with a toothed ring called the diurnal motion gear-ring (*thien yün huan*) fitted round the intermediate nest or shell of the armillary sphere not equatorially but along a declination parallel near the southern pole. In this case also the original model proved unsatisfactory and improvements were made as time went on. We know that the main vertical transmission shaft was made of wood and nearly 20 ft long. This must soon have shown itself to be mechanically unsound, and in the later variants (probably *c.* +1100) it was first shortened and finally abolished altogether. These designs are shown in the inset in Fig. 63. In the first modification the main vertical transmission shaft had no other duty than to turn the chief time-keeping gear-wheel, while in the second, the 'earth-wheel' pinion (*ti ku*) connected the main driving-shaft directly with the time-keeping gear-wheel itself, so that no transmission shaft was necessary. But in both cases the motive power was conveyed to the armillary sphere on the upper platform by means of an endless chain-drive (*thien thi*) rotating three small pinions, presumably of different sizes, in a gear-box (*thien tho*); see Fig. 70. In the final design the chain-drive achieved shorter and therefore more efficient form.

This feature of the clock may perhaps be considered the most remarkable of all for its time (+11th century), for although an endless belt of a kind had been incorporated in the magazine arcuballista of Philon of Byzantium (−2nd century) there is really no evidence that this was ever built, and it certainly did not transmit power continuously. A likelier source for Su Sung's chain-drive may be found in the square-pallet chain-pump [Fig. 71*a*, *b*, pls.] so widespread in the Chinese culture area, a device the origin of which may be traced back at least to the +2nd century, and probably to the +1st. Of course this also was for conveying material and not for transmitting power from one shaft to another—hence the originality of Su Sung and his assistants, to whom perhaps indeed all true chain-drives are owing. The point is so important that it is worth pursuing. Historians of engineering, for example Uccelli (1945, p. 75), Feldhaus (1914, cols. 562, 444, 445), Matschoss & Kutzbach (1940), mention no chain-drive in the true sense in Europe until the nineteenth century. Endless chains used as conveyor belts are of course far older; apart from Philon's magazine arcuballista and the Chinese square-pallet chain-pump, there was also the even better known chain of pots, the Hellenistic and typically Arabic *sāqīya*.[1] But the transmission of power from one axle to another by an endless chain seems to have come much more slowly. About + 1438 Jacopo Mariano Taccola figures an endless hanging chain for manual use like

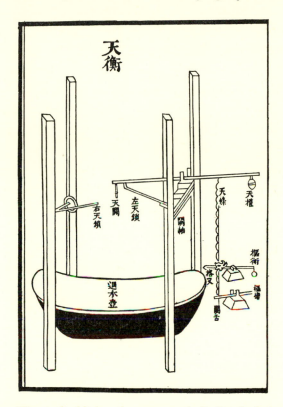

Fig. 72. Water-wheel linkwork escapement in Su Sung's clockwork.

those for small hoists in engineering workshops today. About + 1490 Leonardo da Vinci made elaborate sketches of hinged-link chains, and used them for purposes such as turning the wheel-lock of a gun. This transmitted the power of a spiral spring, but the chain was not endless.[2] In + 1588 Ramelli depicted a chain (again not endless) in oscillatory motion over the geared driving-wheel of a double-barrel pump. Not until 1832 did Galle invent a type of hinged-link chain suitable for a chain-drive, and it was put to use in 1863 by Aveling for cars and in 1869 by J. F. Trefz for bicycles. Equally surprising at the other end of the story,

[1] [See *SCC*, vol. 4, pt. 2, fig. 587.]   [2] [He knew the true endless chain, however, by + 1494 (L. Reti).]

Su Sung and Han Kung-Lien were not the first to use the chain-drive in an astronomical clock. As we shall shortly see, it probably goes back to one of their predecessors, Chang Ssu-Hsün, about +978.

A brief description of the water-power parts must follow here. Water stored in the upper reservoir is delivered into the constant-level tank by a siphon and so

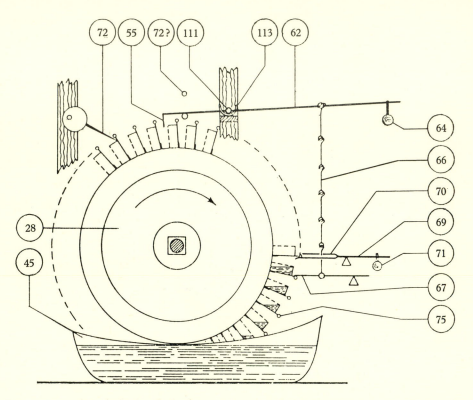

Fig. 73. Reconstruction of the mechanism of the water-wheel linkwork escapement. The numbers identify technical terms in the text (see Needham, Wang & Price, 1960).[1]

passes to the scoops of the driving-wheel, each of which has a capacity of 0·2 cu. ft. As each scoop in turn descends the water is delivered into a sump. Apparently the clock was never so located as to be able to take advantage of a continuous water supply; instead of this, the water was raised by hand-operated norias (*shêng shui lun*) in two stages to the upper reservoir. The bearings of these norias were supported on crutched columns (*chha shou chu*, no. 106).

[1] [This diagram no longer fully represents our knowledge of the Chinese hydro-mechanical escapement, and the reader is referred to *SCC*, vol. 4, pt. 2, fig. 658, for an adequate scale-drawing.]

We can now examine what von Basserman-Jordan calls the soul of any time-keeping machine, namely, the escapement. All that Su Sung's draftsmen could depict of it for his book is seen in Fig. 72, but fortunately the text is elaborate and for the most part clear, enabling the reconstruction of Fig. 73 to be made with certainty. It is seen from the south or front. The whole mechanism was called the 'celestial balance' (*thien hêng*) and it did indeed depend upon two steel-yards or weighbridges upon which each of the scoops acted in turn. The first of these (no. 69), the 'lower balancing lever' (*shu hêng*), prevents the fall of each scoop until full, by means of a 'checking fork' (*ko chha*). The specifications given suggest that release could hardly have occurred more often than every ten minutes, probably every quarter of an hour.[1] But then the descending scoop has to trip by means of its pin another lever (no. 67), the 'stopping tongue' (*kuan shê*),[2] which is connected by means of a chain forming a parallel linkage system (the 'iron crane-bird's knee', *thieh ho hsi*)[3] with another weighbridge (no. 62), the 'upper balancing lever' (*thien hêng*). This lever is fitted at its fulcrum with a crosswise axle moving in a special concave bearing (a 'camel back', *tho fêng*; with two 'iron cheeks', *thieh hsia*), and ends above the driving-wheel in an upper stop (*thien kuan*).[4] As the scoop falls, its pin depresses the stopping tongue, thus pulling down the connecting chain (*thien thiao*) and the right-hand end of the upper balancing lever, thereby raising the left-hand end and withdrawing the upper stop from between the empty scoops and pins at the top of the wheel. At the same time as this gate is opened, recoil is prevented by upper locks (*thien so*)[5] which insert themselves in ratchet manner behind each passing scoop. These motions have the effect of bringing the next scoop to rest under the constant stream of water. It will at once be seen that the whole design is reminiscent of the anchor escapement

---

[1] [As the result of important experimental work carried out by Mr J. H. Combridge in close co-operation with us, we now know that the time interval or 'tick' was only 24 seconds. His findings down to 1965 are incorporated in the account of Chinese horological engineering in *SCC*, vol. 4, pt. 2, pp. 446 ff. See pp. xviii and 11 above.]

[2] [We now translate 'coupling tongue'.]

[3] This interesting technical term originated from a phrase applied to a weapon, the war-flail. In Fig. 74 are shown side by side the well-known farmer's jointed flail (from the *Nung Shu* of +1313, ch. 14, p. 28 b), and the war-flail in which a piece of iron is connected to the handle by a chain (from the *Wu Ching Tsung Yao* of +1044, ch. 13, p. 14 a). Since the latter was called the 'crane-bird's knee' as early as the +3rd century, it was natural enough that the engineers should have borrowed the name for any arrangement of rods linked by chains. For fuller details see Needham, Wang & Price (1960).

[4] [We now understand this differently. The *thien kuan* or 'upper link' was a short length of chain joining the left end of the upper balancing lever to the end of a hinged right 'upper lock' (*thien so*) just below it. The upper balancing lever could thus act as an energy-accumulator for the critical operation of dis-engaging the right-hand lock abruptly from the heavily loaded wheel.]

[5] [Properly, a left upper lock, which inserts itself . . .]

of the late seventeenth century, since the driving-wheel is a scape-wheel and the 'pallets' are inserted alternately at two points on its circumference separated by 90° or less, rather than the 180° of the crown-wheel. But of course the solution of the problem by chain and linkwork naturally has a certain medieval awkwardness, and the action of the release is brought about not by any mechanical oscillation but by gravity exerted periodically as a continuous steady flow of liquid fills containers of limited size. This type of escapement had remained quite unknown to historians of technology until the elucidation of Su Sung's text. Its peculiar interest lies in the fact that it constitutes an intermediate stage or 'missing link' between the time-measuring properties of liquid flow and those of mechanical oscillation. It thus unites, under the significant sign of the millwright's art, the clepsydra and the mechanical clock in one continuous line of evolution.

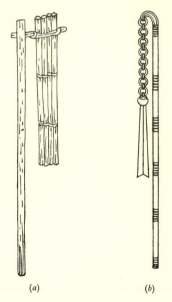

(a)   (b)

Fig. 74. The 'iron crane-bird's knee'; Chinese flails, (a) farmers' flail from the *Nung Shu* (Treatise on Agriculture) of +1313, (b) war-flail from the *Wu Ching Tsung Yao* (Compendium of the most important Military Techniques) of +1044.

At one point in the text it is said that one of the jack-wheels intercepts a gong to strike the night watches as it turns. All the auditory performances of the jacks must have involved simple contrivances of springs, probably of bamboo. Hence the interest of the statement made in the following century by Hsüeh Chi-Hsüan:[1] 'Nowadays time-keeping devices (*kuei lou*) are of four different kinds. There is the clepsydra (lit. the bronze vessels, *thung hu*), the [burning] incense stick (*hsiang chuan*), the sun-dial (*kuei piao*), and the revolving and snapping springs (*kun than*).'

This last expression seems to be rather a rare one, absent both from the copious engineering vocabulary of Su Sung himself and from the mass of other texts which concern the development of clockwork in medieval China. Nevertheless, it can only refer to the springs which worked the bells and drums as the figures on Su Sung's jack-wheels made their daily round. The text is notable, for it was adduced in the time of the Jesuits by the few Chinese scholars who knew enough

---

[1] About +1160, quoted by Wang Ying-Lin in his *Hsiao Hsüeh Kan Chu* (Useful Observations on Elementary Knowledge) of about +1270, first printed +1299, ch. 1, p. 42b.

on such subjects in those days to point out that their Renaissance clocks were not the first which had been known in China.

This completes the account of the mechanism of the great astronomical clock of Su Sung, set up in the form of a working wooden pilot model in the imperial palace at Khaifêng in + 1088. It was the time of our Doomsday Book and the youth of Abelard. Two years later the metal parts, i.e. the armillary sphere and celestial globe, were duly cast in bronze. The writing of the explanatory monograph must have been well under way in + 1092, and it was finally presented to the throne in + 1094. Prefixed to it is a remarkable memorial in which Su Sung not only describes the principles of the clock itself, but gives a historical disquisition on all instruments of a similar kind which had existed in previous centuries. This it was which illuminated many other texts not previously comprehensible, permitting the establishment of a history of Chinese clockwork, the outline of which will be found in the following pages. First, however, we must pause a moment to read a little in Su Sung's memorial, for it contains many matters of interest. He wrote:

When formerly [i.e. after the edict of + 1086 ordering the construction of a new clock] I was seeking for help, I met Han Kung-Lien, a minor official in the Ministry of Personnel, who having mastered the *Chiu Chang Suan Shu* (Nine Chapters of Mathematical Art), often used geometry [lit. the methods of right-angled triangles] to investigate the degrees of [motion of the] celestial bodies. Thinking it over, I also became convinced that the ancients used the techniques of the *Chou Pei* [*Suan Ching*] in studying the heavens...

I therefore told [Han Kung-Lien] about the apparatus of Chang Hêng, I-Hsing and Liang Ling-Tsan, and the designs of Chang Ssu-Hsün, and asked him whether he could study the matter and prepare similar plans. Han Kung-Lien said that they could be successfully completed, if mathematical rules were followed and the [remains of] the former machines taken as a basis. Afterwards he wrote a memorandum in one chapter, entitled 'Verification of the Armillary Clock by the Right-Angled Triangle Method' (*Chiu Chang Kou Ku Tshê Yen Hun Thien Shu*), and he also made a wooden model of the mechanism with time-keeping wheels. After studying this model I formed the opinion that although it was not in complete agreement with ancient principles, yet it showed great ingenuity, especially with regard to the water-powered driving-wheel, and that it would be desirable to entrust him with the building of it. I therefore recommended to your Imperial Majesty that a [complete] wooden pilot model should first be made and presented to you, and that some officials should be ordered to test its use. If the time-recording proved to be correct, then instruments of bronze could be made. On the 16th day of the eighth month in the 2nd year [of the Yuan-Yu reign-period, i.e. + 1087] your Imperial Majesty gave an order that my suggestion should be carried out, and that a [special] bureau should be set up, officials appointed and the necessary materials prepared.

I therefore recommended that Wang Yuan-Chih, Professor at the Public College of Shouchow, formerly Acting Registrar of Yuan-wu in Chêngchow prefecture, should be in charge of the construction and receipt of public materials; while Chou Jih-Yen, Director of Astronomical Observations (Southern Region) of the Bureau of Astronomy and Calendar, Yü Thai-Ku, Director of Astronomical Observations (Western Region) of the same Bureau, Chang Chung-Hsüan, Director of Astronomical Observations (Northern Region), and Han Kung-Lien, should be appointed to supervise the construction. [I further recommended] the Assistants in the Bureau, Yuan Wei-Chi, Miao Ching-Chang, Tuan Chieh-Chi, and Liu Chung-Ching; and the Students, Hou Yung-Ho and Yü Thang-Chhen, as investigators of the sun's shadow, the clepsydras, and so on. [Lastly, I recommended] the Bureau of Works Foreman Yin Chhing to be Clerk of the Works.

In the fifth month of the 3rd year of the Yuan-Yu reign-period [+ 1088] a small pilot model was finished, and at your Imperial Majesty's order presented for testing. Afterwards the full-scale machinery was built in wood and completed by the twelfth month. I [then] begged your Imperial Majesty to send a court official to the Bureau [of Astronomy and Calendar] to explain [the parts to the workmen] in preparation for moving the clock to the palace for presentation ...In the tenth month we had sent in a request for instructions regarding the installation, and the Palace Guard Superintendent detailed the Aide-de-Camp Huang Chhing-Tsun [to look after the matter]. On the 2nd day of the twelfth month, a letter was sent up asking exactly where [the clock] was to be assembled, and your Imperial Majesty's order came to erect it in the Chi Ying Hall [of the Palace].

With all its vividness of detail, this passage concerning the organisation of one of the greatest technical achievements of the medieval time in any civilisation is certainly worthy of appreciation. Moreover, reading rightly 'between the lines' brings out several significant points. Han Kung-Lien, that man of brilliant mathematical and mechanical talent, had no post in which he could use it but was found by Su Sung in the minor ranks of his own administrative Ministry. Contrary to common conceptions of medieval working, the new armillary clock was not put together haphazardly by trial and error, but planned in a special memorandum with all the geometrical knowledge that Han could put into it. This certainly makes it easier to understand how the gearing, chain-drives, and other devices were made to carry out successfully their duty of rotating steadily an armillary sphere weighing some 10–20 tons as well as a bronze celestial globe $4\frac{1}{2}$ ft in diameter. It is also noteworthy that a small wooden model was made first, then a full-scale one was tested against four types of clepsydra as well as star transits, and only after four years were the parts destined for bronze duly cast.

In the last paragraph of his memorial Su Sung wrote:

Thus, as we have seen, the [demonstrational] armillary sphere, the bronze observational armillary sphere, and the celestial globe, are three things different from one another...In any case if we use only one name, all the marvellous uses of [the three] instruments cannot be included in its meaning. Yet since our newly-built machine embodies two instruments but has three uses, it ought to have some [more general] name such as 'Hun Thien' (Cosmic [Engine]). We are humbly awaiting your Imperial Majesty's opinion and bestowal of a suitable name upon it.

And he signed with all his ranks and titles, Imperial Tutor to the Crown Prince, Grand Protector of the Army, Khai-Kuo Marquis of Wukung, etc. Now the two instruments were, of course, the mechanised observational armillary sphere and the mechanised celestial globe. Clearly the three uses were: (a) astronomical observations with the armillary sphere; (b) indication of time, both visual and auditory, by globe, sphere and jackwork; and (c) indication on the globe of the positions of all constellations whatever the weather, and their relations to models of the sun, moon, and planets attached to the globe, for calendrical verifications. Thus Su Sung's request for a new name was of great historical significance. The mechanised astronomical instrument was trembling on the verge of becoming a purely time-keeping machine. Inaudible echo must have answered 'A clock'. But history does not record that the young emperor had any good ideas on nomenclature, and the time-measuring function continued to go unnamed until five hundred years later the Jesuits came with their 'self-sounding bells' to ring in the age of unified world science with its unlimited expansion of appropriate technical terms.

The derivation of mechanical clocks from clepsydras has now been made clear. But the story of the evolution remains to be told. In his memorial Su Sung wrote:

According to your servant's opinion there have been many systems and designs for astronomical instruments during past dynasties all differing from one another in minor respects. But the principle of the use of water-power for the driving mechanism has always been the same. The heavens move without ceasing but so also does water flow [and fall]. Thus if the water is made to pour with perfect evenness, then the comparison of the rotary movements [of the heavens and the machine] will show no discrepancy or contradiction; for the unresting follows the unceasing.

This was a nice appreciation of what Europeans were later to think of as the universal writ of the 'law' of gravitation. But Su Sung goes on to give brief descriptions of the predecessors of his own clock, beginning with Chang Hêng's device of the +2nd century.

The most important clock in the Sung dynasty prior to that of Su Sung himself was built by Chang Ssu-Hsün, towards the end of the +10th century. It included

223

sphere and globe, powered by a scoop-bearing driving-wheel and gearing, together with jack figures to report and sound the hours. Eleven technical terms occur in the description with exactly the same meanings as in Su Sung's text. Chang's clock was a particularly fine and interesting work as it used mercury in the closed circuit instead of water, thus assuring time-keeping in frosty winters. But it must have been somewhat ahead of its age, for Su Sung tells us that after Chang's death it soon went out of order and there was no one able to keep it going.

It is interesting that Chang Ssu-Hsün was a Szechuanese, for that populous province in the west had been the scene, just previously during the late Thang and Five Dynasties periods, of the earliest expansion of another admirable invention, that of block printing.[1] Liu Phien tells us how in +883 on his holiday outings he used to examine the printed books being sold outside the city walls of Chhêngtu (cf. Carter, 1955, p. 60). These were mostly on proto-scientific subjects (oneiromancy, geomancy, astrology, planetary astronomy, and speculations of the Yin-Yang school). The province thus seems to have been rather fertile in technological advances at this time. One remembers that it had long been the home of the proto-industrial area of the Tzu-liu-ching brine-field.[2]

Chang's clock was very like Su Sung's, with similar drive and similar escapement. The use of mercury was particularly ingenious, however, and it seems certain that a set of planetary models was rotated automatically. Though Su returned to the classical method of moving them by hand (as in the anaphoric clock), later Sung specifications (cf. p. 227 below) also had them on geared wheels, like the orreries and planetaria of later Europe. Perhaps most interesting is the mention of the gear-box, which rather implies the use of a chain-drive like Su Sung's; if so, Chang Ssu-Hsün was an anticipator of Leonardo by five hundred years.

From this we can pass directly to the Thang clock-makers. Who were these men who made, in the +8th century, the most venerable of all escapement clocks? One was a Tantric Buddhist monk, perhaps the most learned and skilled astronomer and mathematician of his time, I-Hsing, the other a scholar, Liang Ling-Tsan, who, like Han Kung-Lien later, occupied a minor administrative post. The technical terms employed in the relevant passages again reveal the essential similarity of the machine to the clock of Su Sung.

These passages are to be found in the official histories of the Thang dynasty[3]

[1] [Cf. p. 22 above.]   [2] [Cf. p. 33 above.]
[3] *Hsin Thang Shu*, ch. 31, pp. 1*b* ff.; *Chiu Thang Shu*, ch. 35, pp. 1*a* ff.

and in the *Chi Hsien Chu Chi* (Records of the College of All Sages)[1] written about +750 by Wei Shu. The context of I-Hsing's astronomical clock was his introduction of the Ptolemaic ecliptically mounted sighting-tube convenient for studying planetary motions on and near the ecliptic. Wei Shu wrote:

In the 12th year of the Khai-Yuan reign-period [+724] the monk I-Hsing constructed an armillary sphere with an ecliptically mounted sighting-tube in the Library, and when it was finished he presented it [to the emperor]. Earlier, he had received an imperial order to re-organise the calendar, and had said that observations were difficult because there was no apparatus with this ecliptic fitting. Just at that time Liang Ling-Tsan made a small model [of the instrument which was wanted] in wood and presented it. The emperor asked I-Hsing to study it, and he reported that it was highly accurate. Therefore a full-scale [sphere] in bronze and iron was made in the Library grounds, taking two years to complete. When it was offered to the throne the emperor praised it exceedingly and asked [Liang] Ling-Tsan and I-Hsing to study [further] Li Shun-Fêng's book, the *Fa Hsiang Chih* (The Miniature Cosmos),[2] so that later on they drew up complete plans of the armillary sphere. And the emperor wrote an inscription in 'eight-tenths' style characters[3] which was carved on the ecliptic ring and which said:

> The moon in her waxing and waning is never at fault
> Her twenty-eight stewards escort her and never go straying
> Here at last is a trustworthy mirror on earth
> To show us the skies never-hasting and never-delaying.

The scholar Lu Chhü-Thai received an imperial order to write an inscription containing the year and month of construction, and the names of the workers, underneath the plate. The observatory used the apparatus for observations, and it is still employed nowadays...

After this, the emperor ordered the casting of bronze for yet another astronomical instrument. The Chief Secretary of the Left Imperial Guard, Liang Ling-Tsan, and his colleague of the Right, Huan Chih-Kuei, took charge of plans for the separate parts, and a great [demonstrational] armillary sphere (*thien hsiang*) was cast 10 ft in diameter. It showed the lunar mansions (*hsiu*), the equator and all the circumferential degrees. It was made to turn automatically by the force of water acting on a wheel. Discussing it, people said that what Chang Hêng [+2nd century] had described in his *Ling Hsien* (Spiritual Constitution of the Universe) could have been no better.

Now it is kept in the College of All Sages at the eastern capital [Loyang]. In the courtyard there is the observatory (*yang kuan thai*) where I-Hsing used to make his observations.

This brief account was fortunately expanded in the Thang histories. There we read that in +723 I-Hsing and Liang Ling-Tsan 'and other capable

---

[1] *Yü Hai*, ch. 4, p. 24a. Wei Shu's book is available to us today only in extensive quotations in encyclopaedias such as this.

[2] Li Shun-Fêng (fl. +620 to +680) had been the most eminent astronomer and mathematician of the preceding century. Though he died before I-Hsing was born, the latter considered himself a disciple of Li.

[3] This calligraphic style was so named because it was considered intermediate between seal characters and *li shu* characters in that proportion.

technical men' were commissioned to cast and make new bronze astronomical instruments.

One of [these] was made in the image of the round heavens and on it were shown the lunar mansions in their order, the equator and the degrees of the heavenly circumference. Water, flowing [into scoops], turned a wheel automatically, rotating it one complete revolution in one day and night. Besides this, there were two rings [lit. wheels] fitted round the celestial [sphere] outside, having the sun and moon threaded on them, and these were made to move in circling orbit. Each day as the celestial [sphere] turned one revolution westwards, the sun made its way one degree eastwards, and the moon 13 7/19ths degrees [eastwards]. After 29 and a fraction rotations [of the celestial sphere] the sun and moon met. After it made 365 rotations the sun accomplished its complete circuit. And they made a wooden casing the surface of which represented the horizon, since the instrument was half sunk in it. This permitted the exact determination of the times of dawns and dusks, full and new moons, tarrying and hurrying. Moreover there were two wooden jacks standing on the horizon surface, having one a bell and the other a drum in front of it, the bell being struck automatically to indicate the hours, and the drum being beaten automatically to indicate the quarters.

All these motions were brought about [by machinery] within the casing, each depending on wheels and shafts, hooks, pins and interlocking rods, stopping devices and locks checking mutually [i.e. the escapement].

Since [the clock] showed good agreement with the Tao of Heaven, everyone at that time praised its ingenuity. When it was all completed [in +725] it was called the 'Water-Driven Spherical Bird's-Eye View Map of the Heavens' (*Shui Yün Hun Thien Fu Shih Thu*) or 'Celestial Sphere Model Water-Engine', and was set up in front of the Wu Chhêng Hall [of the Palace] to be seen by the multitude of officials. Candidates in the imperial examinations [in +730] were asked to write an essay on the new armillary [clock].

But not very long afterwards the mechanism of bronze and iron began to corrode and rust, so that the instrument could no longer rotate automatically. It was therefore relegated to the [museum of the] College of All Sages and went out of use.

Such are the details of the instrument which, so far as we can see, was the first of all escapement clocks. The reference to the checking link-work is plain, the technical terms used being closely similar to those in the descriptions of the clock of Su Sung. Although the automatic movement of the sun and moon models is not stated with absolute clarity, it is almost certainly implied; the machine had therefore some of the features of an orrery or planetarium. According to the account of Chang Ssu-Hsün's clock, the mechanisation of the movement of the sun and moon models was first effected by him (in +978). But the nuance of the present text indicates automatic motion rather unmistakably. In this case I-Hsing and Liang Ling-Tsan rather than Chang Ssu-Hsün, in the +8th century rather than the +10th, were the first to replace the manually adjusted models of earlier times (pegged into holes or threaded on strings) by an 'orrery' system of mechanical motions. In any case there is no doubt about the mechanical movement built by Wang Fu and his associates in about + 1124. An interesting hypothetical reconstruction (Fig. 75) of the 'orrery' movements from I-Hsing to Wang Fu has been offered by Liu Hsien-Chou (1956). As will be seen, this involves both concentric shafting and a number of gearwheels cut with odd numbers of teeth. We have felt some uncertainty about the feasibility of so complex a design in the time of I-Hsing, though not doubting it for Wang Fu and perhaps also Chang Ssu-Hsün. Wiedemann (1913) and Price (1955, 1956) have drawn attention to the odd-toothed gearing in Arabic and European astrolabes as one of the most important elements in the pre-history of clockwork. The oldest evidence about these gear-trains is textual, a MS of Abū ibn Aḥmad al-Bīrūnī written c. + 1000, but an Arabic geared astrolabe made in + 1221 by Muḥammad ibn Abū Bakr of Ispahan is preserved at Oxford, and a French example of c. + 1300 at London. There is reason to believe that Su Sung in his

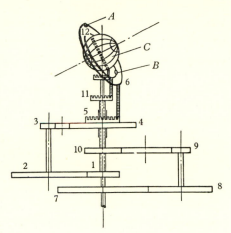

Fig. 75. Reconstruction of the orrery movement in the clocks of I-Hsing (+ 725), Chang Ssu-Hsün (+ 978) and Wang Fu (+ 1124), after Liu Hsien-Chou. A is the sun model, B the moon, and C the celestial sphere or solid globe. The gear-wheels 1–6 which work the solar motion have the following numbers of teeth respectively: 12, 60, 6, 72, 12, 73. The gear-wheels 7–12 which work the lunar motion have 127, 73, 6, 15, 6 and 114 teeth respectively.

clock of + 1090 employed at least one odd-toothed gear-wheel, namely that of 487 teeth. If this is so, no difficulty would have been encountered by Wang Fu and his friends, but whether such wheels could have been made accurately enough in + 725 remains uncertain. Yet Liu Hsien-Chou may well be right in what he attributes to I-Hsing, who may thus have even more claim upon our respect and admiration.

It is not generally known that a replica of the original 'orrery' was carried to China by the embassy of Lord Macartney in + 1793. The first planetarium of this kind, demonstrating the heliocentric system, was made by George Graham (see Lloyd, 1951) about + 1706 for Prince Eugene of Austria (cf. Taylor & Wilson, 1945), in circumstances described not long afterwards by Desaguliers.[1] It was

---

[1] 1st ed. vol. 1, pp. 430 ff.; 2nd ed. vol. 1, pp. 448 ff.

immediately copied for the Earl of Orrery, whence the name of the instrument (cf. Orrery, 1903). A magnificent replica was ordered by the East India Company in +1714, and described by John Harris a few years later (+1719). Whether this was from the beginning destined for China we have not ascertained, but the instrument seems to have been the same as that taken there many years later. Staunton tells us[1] that in +1792 two Chinese students from the College at Naples came to England to accompany the Macartney mission as interpreters, and began by advising as to the presents to be taken. It was decided that it would be unwise to try to compete with the mechanical toys, and clocks with elaborate jack-work, which the 'sing-song' trade had been pouring into China for close on half a century; some objects of intellectual interest would be more suitable. 'Astronomy being a science,' wrote Staunton, 'peculiarly esteemed in China, and deemed worthy of the attention and occupation of the Government, the latest and most improved instruments for assisting its operations, as well as the most perfect imitation that had yet been made of the celestial movements, could scarcely fail of being acceptable.' Accordingly, in the fullness of time, Dr Dinwiddie and M. Petitpierre-Boy were to be seen in Peking unpacking, besides the orrery, a celestial and terrestrial globe, a reflecting telescope, a clock showing the lunar phases, and an air-pump.[2] Thus the high value traditionally placed on astronomical science in China imposed itself on European diplomacy at the end of the +18th century, and Graham's masterpiece of astronomical clockwork found its way as a veritable tribute, even if the term was abjured, and the intention partly unconscious, to the land of I-Hsing, Chang Ssu-Hsün and Su Sung.

The emperor served by Li Shun-Fêng was Thai Tsung, who ruled with much brilliance from +626 onwards for a quarter of a century. Interested in history and technology as well as in the military arts, he knew how to encourage astronomers, and welcomed Nestorian clergy as well as Taoist priests and Buddhist monks. He entertained cordial diplomatic relations as far west as Byzantium, receiving in +643, for example, an embassy from the emperor Theodosius. Such missions may well have brought news of the striking water-clocks at places like Gaza and Antioch. Of course, this can have been no more than a 'stimulus diffusion', for there is no reason for thinking that the Byzantine works employed anything more than the sinking float principle. However, the stimulus would have come just at the right time to encourage Chinese engineers to try to outdo the mechanical toys which formed the striking jack-work of the water-clocks of

---

[1] 1797, vol. 1, p. 42.   [2] Staunton, 1797, vol. 1, p. 492, vol. 2, pp. 165, 287.

the Eastern Roman Empire. And indeed the description of I-Hsing's clock does seem to be the first mention of horological escapement-operated jacks in Chinese history. Here he was much better placed than his Greek colleagues, if we are right in our supposition that the water-wheel, providing so much more power than the float, had already long before been characteristic of Chinese astro-mechanical technique.

We have concluded, then, that what may be called the father and mother of all escapements originated in the first decades of the + 8th century. But the history of clockwork which we are sketching cannot stop at this point on its backward course. For between + 725 and the beginning of the Christian era there were many other examples of astronomical globes or spheres being slowly rotated by water-power. If we define the escapement as the essence of true clockwork, these earlier devices were not clocks, but they may well be considered predecessors of the clock, 'pre-clocks' or 'proto-clocks'. Moreover, they were purely astro-nomical in character and lacked all auditory time-telling components.

Since the meagre power of a sinking float could not have rotated an obliquely mounted sphere or globe, even if made relatively lightly in wood, it is probable that the mechanism consisted of a vertical water-wheel with cups like a noria, doubtless of more simple construction than that of Su Sung. The water-wheel would be attached to a shaft with one trip-lug, quite similar in principle to the water-driven trip-hammer assemblies so common in the Han. Clepsydra drip into the cups would accumulate periodically the torque necessary to turn the lug against the resistance of a leaf-tooth wheel, either itself forming the equatorial ring, or attached to a shaft in the polar axis. Needless to say, the time-keeping properties of such an arrangement would be extremely poor, but the men of those early centuries were probably satisfied with rough approximations.

Examples can be taken from almost every century between the + 8th and the + 2nd. One of the most outstanding technicians of the + 6th century was Kêng Hsün, whose name is often met with on account of his work with clepsydras. A man of matchless technical skill, and witty in argument, he became involved early in life with a rebellion of southern tribal folk, but when eventually captured was pardoned by the general Wang Shih-Chi on account of his great ingenuity.

After a long time Kêng Hsün met his old friend Kao Chih-Pao, whose knowledge of the heavens had brought him to the position of Astronomer-Royal, and from him [Kêng] Hsün received instruction in astronomy and mathematics. [Kêng] Hsün then conceived the idea of making an armillary sphere (*hun thien i*) which should be turned not by human hands but by power of [falling] water. When it had been made he set it up in a closed room and asked [Kao] Chih-Pao to stand outside and observe the time [as shown by the] heavens [i.e. the star transits]. [His instrument] agreed [with the heavens] like the two halves of a tally. [Wang] Shih-Chi, knowing this, reported the matter to the emperor Kao Tsu who made [Kêng] Hsün a government slave and attached him to the Bureau of Astronomy and Calendar.

This account,[1] referring to work which was going on in the neighbourhood of +590, is closely similar to what we are told in all the other cases. Generally no wheel is mentioned, but no float either, and the mechanised instrument is typically set up inside a closed room, with two observers, one inside calling out the indications of the machine, the other outside checking these against the heavenly movements themselves. There may even be no mention of water in the automatic movement, but a water-drive is always to be surmised.

About seventy years earlier the great Taoist physician, alchemist and pharmaceutical naturalist Thao Hung-Ching (+452 to +536) had done something of the same kind. The reports say[2] that he 'made a [demonstrational] armillary sphere more than 3 ft high, with the earth situated in the middle. The "heavens" rotated and the "earth" remained stationary—and it was all moved by a mechanism. Everything agreed exactly with the [actual] heavens.' Not only that, but he wrote a book about it, the *Thien I Shuo Yao* (Essential Details of Astronomical Instruments); this has long been lost. His mechanised sphere may be dated *c.* +520.

We are now within a short distance of our goal, the work of Chang Hêng in the Later Han period. For according to out texts[3] he was the first of this line of men who accomplished the continuous slow rotation of astronomical instruments (globes or demonstrational spheres) with the best approximation which they could make to constancy of speed. To students of Chinese history, Chang Hêng (+78 to +142) is a familiar figure; there was hardly a science with which he was not concerned (mathematics, astronomy, cartography, etc.). Particularly relevant is the seismograph which he set up at the capital in +132, the first instrument of the kind in any civilisation. The ingenuity of this device, with its inverted pendulum, which continued in use for many centuries afterwards, was so striking that there is nothing inherently improbable in his application of water-power to a drive for an astronomical instrument.

It looks as if the invention of Chang Hêng about +120 derived from the growing doubts which had led Hipparchus to the discovery of the equinoctial

---

[1] From *Sui Shu*, ch. 40, pp. 10a ff.; *Pei Shih*, ch. 89, pp. 31a ff.
[2] E.g. *Nan Shih*, ch. 76, p. 11a.
[3] The full discussion of these will be found in the monograph of Needham, Wang & Price (1960).

precession in − 134 and were to lead Yü Hsi to state the same doctrine about + 320. But by + 700 the more accurate measurement of time had become a burning problem, and the social need of a predominantly agrarian culture for an accurate calendar as well as the intrinsic evolution of astronomical science itself led to the answer found by I-Hsing. It is indeed a strange conclusion that an apparatus so deeply enrooted in western mechanical industrial civilisation as the clock should have originated in connection with the calendar required by an eastern agricultural people. But other perspectives no less remarkable must be mentioned. It has often been emphasised that Chinese astronomy was founded on a polar and equatorial system, while Hellenistic astronomy was primarily ecliptic and planetary. Each had its peculiar advantages and its corresponding triumphs. If Hipparchus was able to state the fact of precession four and a half centuries before Yü Hsi, it was because he was measuring and comparing star positions on ecliptic co-ordinates, and therefore it became evident that their distances from the equinoctial points had changed. But if astronomical instruments were rotated mechanically by Chang Hêng fifteen centuries before the conception of the clock-drive arose in Renaissance Europe, and if in this I-Hsing with his first approximation to real mechanised time-keeping had a priority of nine or ten, it was because Chinese astronomers thought always in terms of equatorial co-ordinates and therefore of declination parallels. Along these lines all stellar revolution proceeds, but the tracks of ecliptic latitude and longitude form a geometrical waste-land on which nothing ever moves. In China, therefore, it was an entirely natural thought to arrange the rotation of a celestial globe or a demonstrational armillary sphere if the plan promised to be useful. What was not perhaps so easy was how to do it.

Lastly, a few words to set against their proper background the means which we think Chang Hêng took to solve his problem. To harness the waste dripping of clepsydra water there was one obvious recourse, the art of the millwrights. In + 1st-century China their workmanship was doubtless primitive, but during the previous century the water-powered trip-hammer (cf. Fig. 76) had come into widespread use. Although the most characteristic form of mill-wheel in China later was the horizontally mounted type, the vertical 'Vitruvian' form always persisted, and for the trip-hammer was the more suitable of the two. Moreover, metallurgical blowing-engines powered by water (cf. Figs. 19, 34) had become common in China during the + 1st century. If the Chinese vertical water-wheel really derived from the noria, the transition had been made a good while before

Chang's time, so that his chief originality lay in arranging for a constant drip into scoops rather than a strong flow and fall on to paddles. The trip-lug on his shaft merely corresponded to those which worked the grain-pounding trip-hammers all around him. However, there was originality also in making it push each time

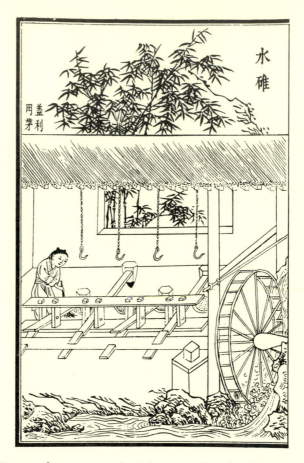

Fig. 76. Battery of water-powered trip-hammers, from the *Thien Kung Khai Wu* (Exploitation of the Works of Nature), of +1637 (Chhing drawing).

the tooth of a ring or gear-wheel, probably bearing leaf-teeth, and controlled by a ratchet. There seems nothing at all in the arrangement which would have been beyond the powers of Han technicians. Indeed the trip-lug, which constituted a pinion of one, has a distinctly Alexandrian air. One may compare it with the peg on the axle of Heron's hodometer or taximeter in the previous century, or with those on the shaft of his organ-blowing windmill. What connections there could have been between the engineers of Alexandria and Han China remains of course a completely unsolved question. It might have been a case of carrying coals to Newcastle, for the Chinese hodometers were contemporary, and their acceptable reconstructions involve small pinions of one, two or three teeth.

Presumably Chang Hêng's simple machine was at rest during each period when water was slowly accumulating in one of the scoops. As soon as enough had collected, its weight overcame the resistance of the toothed wheel and armillary sphere, and the trip-lug turned it round by one tooth, then coming to rest against the next. Although we are told optimistically that 'everything agreed like the two halves of a tally' we are bound to assume that the chronometric properties of the device were extremely poor. So much of it would have depended upon play and resistance, the exact size of each scoop, the nature of the bearings of the polar axis, and similar factors. Very probably it was maintained in regular motion only with some difficulty, and perhaps the successive astronomers who made instruments of the same kind were all searching for the right conditions for doing what no one until I-Hsing was really able to do. Or perhaps they made the machine work better than we are inclined to believe. In any case it is hard to exclude Chang Hêng's apparatus absolutely from the horological definition of von Bertele with which this Lecture began. The problem of securing slow and regular motion by intersecting the progress of a powered machine into intervals of equal duration was not solved in any Alexandrian design as far as we know, and if Chang Hêng did not solve it himself, he opened the gate to the path which led in the end to its solution.

Many inventions, as we have seen (p. 207 above), contributed to the first escapement clock in Europe (c. +1300). The weight-drive descended no doubt from the floats of the Hellenistic anaphoric clocks and mechanical puppet theatres, and was certainly known in its free form in the +13th century, as the Moorish drum water-clock of the Alfonsine corpus demonstrates. The use of gearing for simulating time-measurement descended from remote antiquity, for even if we know little of the nature of the planetarium ascribed to Archimedes (c. −250),[1] the Anti-Kythera object, with its elaborate gear-wheels, remains to show the extraordinary attainment which Hellenistic (−1st-century) technique could reach.[2]

[1] See Wiedemann & Hauser (1918 a).
[2] A monograph on this remarkable object by Dr Derek J. Price is now in preparation. [In the meantime we have his paper in *Sci. Amer.* 1959, **200** (no. 6), 60.]

Then in the Arab realm there was the application of calendrical gearing to the computational astrolabe, as already mentioned (p. 227). The clock dial again may be considered a derivative of the astrolabe's face and therefore ultimately of the revolving dial of the anaphoric clock. For jack-work devices there were plenty of precedents in the Byzantine striking water-clocks and their Arabic successors. Only the 'soul' of the mechanical clock, i.e. the escapement itself, had to be provided by the key inventors, whoever they were, of + 1300. The form which this actually took, namely, the verge-and-foliots, may reasonably be derived from the radial bob flywheel, familiar since the early days of Graeco-Roman screw-presses, though now converted by the pallets from discontinuous rotary to regular oscillating motion. But just how original was the basic idea? The preceding six centuries of Chinese escapements suggest that at least a diffusion stimulus travelled from east to west.

To gain a little light on this transmission, if such it was, we must concentrate attention on the years between about + 1000 and + 1300 and see whether any help can be found from the Islamic culture-area. It is noteworthy that arrangements logically equivalent to those of Chang Hêng and his successors are to be found in later Arabic writings. For example, one of the mechanisms described by Ismā'īl ibn al-Razzāz al-Jazarī in his treatise on striking clepsydras in + 1206, as interpreted by Wiedemann & Hauser (1915), consists of a tipping bucket attached to a hinged ratchet which pushes round a gear-wheel by one tooth each time that the bucket fills with water and comes to the emptying point (Fig. 30). This gear-wheel is connected by a cord with what seems to be the plate of an anaphoric clock. And the arrangment possessed a strange longevity, for it is found in exactly the same form in the book which Isaac de Caus published on mechanical contrivances in + 1644 (Fig. 29, pl.). Indeed the principle was still employed by J. B. Embriaco (using two alternate buckets and a pendulum escapement) for a public clock set up in the Pincio Gardens at Rome in 1872. Now al-Jazarī also had water-wheels in his clocks, a fact which can hardly be without significance in relation to earlier Chinese practice. But here they never seem to be used as the motive-power for trains of gears in continuous rotation; they simply come into action intermittently whenever the bucket tips out its water, and they operate peacocks and other jack figures, either by trip-levers or intermediate gear-wheels (Fig. 77). Where we do find gearing continuously turned by a water-wheel is in certain machines for delivering a constant supply of air to an organ.[1] In these a vertical

[1] See Wiedemann & Hauser (1918b).

shaft carries a semi-lunar cam which raises two valves alternately as it rotates, thus admitting water to two closed spaces in turn so that one can be expelling air while the other is taking it in. The complexity of this method, ascribed, e.g. by al-Jazarī, to the +9th century Banū Mūsā brothers, is in striking contrast with the elegant simplicity of the Chinese double-acting piston-bellows. But it was never a time-keeper. Thus there is nothing in the evidence so far available which suggests any Arabic influence on the Chinese developments. From the beginning of the +8th century onwards the Chinese clocks were undergoing a steady evolution in such a continuous line that external influence seems very improbable. On the other hand, the Arabic material does seem to indicate the passage westwards of certain Chinese elements.

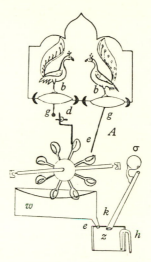

Fig. 77. Al-Jazarī's striking-train water-wheel for peacock jacks, c. +1206 (from Wiedemann & Hauser).

Somewhat humiliating is the fact that one of the darkest patches of our ignorance concerns the nature of the 'horologes' used, mostly in abbeys and cathedrals, in +12th- and +13th-century Europe. Howgrave-Graham (1927) has distinguished a cluster of records from +1284 onwards which suggest the rise of a new invention or the unwonted popularity of some existing device at this time. The monumental clocks concerned were almost surely not driven by falling weights. Of the clock in the cathedral at Wells, for example, the Welsh poet Dafydd ap Gwilym (+1343 to +1400) wrote in curious terms—it had 'orifices' in it, as well as wheels, weights, ropes, hammers, and also 'heads' and 'tongues'; words strangely reminiscent of Su Sung's 'stopping tongue'. To Drover (1954) we owe the study of a remarkable manuscript dating from about +1250, which has a picture apparently showing a water-wheel clock (Fig. 78, pl.). The wheel is distinctly reminiscent of the driving-wheel of Su Sung, water can be seen pouring from an animal head into a sump below, and there are other objects, including a row of five bells, more difficult to make out. Drover has also analysed the accounts of some of the mysterious *horologia*. For instance, the *Chronicle* of Jocelyn of Brakelond tells how in +1198, when a fire broke out in the abbey church of Bury St Edmunds in East Anglia, the monks ran to the clock to get water. Inscriptions in the abbey of Villers in Belgium, dating from +1268, which have been studied by Sheridan, also indicate without doubt a water-clock. On the other hand, Cologne in +1235 had an 'Uhrengasse' which was inhabited by smiths. The issue therefore is whether the clocks of the late +12th and early +13th century in Europe were water-wheel clocks or not. We must hope that further discoveries may throw light on this question. If the existence of water-wheel clocks could be proved, it would suggest a transmission at the time of the Crusades, such as seems to have occurred with the invention of the windmill, rather than

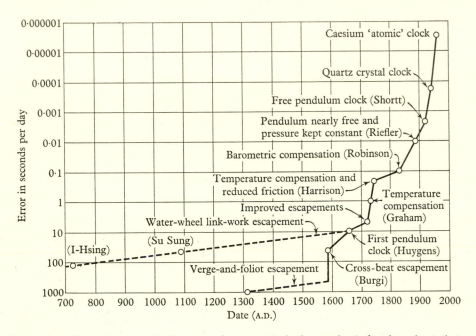

Fig. 79. A graph to show the development of accuracy in hydro-mechanical and mechanical time-keeping through the centuries (amplified by J. Needham, with his approval, from the original graph of F. A. B. Ward, after consultation with J. Combridge and H. von Bertele).

at the later time of Marco Polo and Petrus Peregrinus. In that case the inventors of the verge-and-foliot would have had about a century or so to try out their improvements on the water-wheel link-work escapement. But in some ways it is easier to imagine that the transmission was of the diffusion stimulus type only, and that firm conviction of the prior successful solution of the problem elsewhere led European scholar-artisans to solve it themselves in a different manner. We really do not know.

In any case, whether the Chinese escapement came to Europe in person, or only as a rumour, I-Hsing's great contribution was to introduce a truly chronometric principle into the mechanical as opposed to the clepsydric part of the clock. At first this was not great, for it is clear that in the Chinese clocks the major part of the time-keeping was effected by the constancy of the flow of the water. The mechanism could only intervene in so far as changing the weight on the weigh-bridge would permit the scoops to fall before they were quite full. This is why we must recognise in it a missing link between the clepsydra and the purely mechanical clock. For when the inventions of early + 14th-century Europe had been made, the verge-and-foliot took over the greater part of the time-keeping duty. This was still not completely embodied in the escapement, since any considerable change in the weight hanging from the drum would affect the fastness or slowness of the clock. Not until the introduction of the pendulum in the + 17th century was an approach to a truly isochronous mechanism made [Fig. 79]. That this had taken two millennia is not surprising when one considers the leisurely growth of human technology as a whole before the Renaissance. But the Chinese contribution was vital. Its recognition enables us henceforward to estimate at their true value statements such as the following—so often found. 'The Chinese', writes Lübke, 'never made any discoveries comparable with those of Europeans in the technique of clock-making. The clocks (collected in such great numbers in the Forbidden City) have of course nothing to do with time-measurement in Old China.' And, says Planchon, with superb irony, 'the Chinese have never produced any mechanical clockwork properly so called—in this field they have only been bad imitators'.

Emphasis has rightly been placed on the trade of clock- and watch-maker for the growth of science in Renaissance Europe. These craftsmen became for science what the millwright was for industry—a fruitful source of ingenuity and workmanship. The millwrights had been there all through the Middle Ages, and the clockmakers from the beginning of the + 14th century. Their presence was certainly one of the important roots of Renaissance science, pure and applied, for a supply of artisans was ready to generate makers of machines and instruments as soon as these things were demanded and devised. Yet by now it is abundantly clear that China also had such artisans, in skill and ingenuity at least as eminent. If therefore China had no Renaissance and no development of modern science and technology, the presence of artisans was evidently not in itself enough. And though clock-making in China seems never to have become a mass industry before the time of the Jesuits (as it did in + 15th- and + 16th-century Europe), the building of mill-work and water-raising machinery of all kinds was spread throughout the length and breadth of the empire. The manifold activity of skilled millwrights was therefore not enough either. 'It was the work of the millwright', says Bernal, 'that gave rise to the first genuinely European invention, that of the clock...' Although in the light of the knowledge here set forth, the second part of the sentence can no longer be sustained, the remarkable insight shown in the first pays a debt which we all owe to the engineers of the Middle Ages.

12

THE CHINESE CONTRIBUTION
TO THE DEVELOPMENT OF THE
MARINER'S COMPASS[1]

[1960]

On such a historic occasion as the 500th anniversary of the death of Prince Henry the Navigator, it is surely fitting that we should unite to 'praise famous men' not only of our own Western civilisation of Portugal and Europe, but those also of the Asian cultures who helped to lay the foundations of all civilisation and of all geographical knowledge. My theme, therefore, will be the invention of the first of all pointer-readings, so characteristic of modern science, the discovery of the directive property of the lodestone and the magnet. The belief that this knowledge originated in the Chinese culture-area has become almost proverbial in the West, but it has often been accepted for entirely wrong reasons. The Chinese legendary corpus contains certain stories of a 'south-pointing carriage' which guided men to their destinations in fog or storm. The texts containing these are quite well known, but during the past thirty years only has modern sinology been able to put them in their proper perspective. It is now an assured result of archaeological scholarship that such south-pointing carriages there were, but that they had nothing to do with the magnetic compass—they were assemblies of gear-wheels something like those in modern military combat vehicles which maintained (once set) an indicator pointing in a particular azimuth direction. Moreover, the first to be built belongs to the $+3$rd century, and not to the -23rd, as fable had it. Yet there is good ground for believing, all the same, that the directive property of the lodestone had indeed been discovered by Chinese magicians and scholars by the beginning of our era.

Chronological exactitude is important if we would compare the achievements of different civilisations, titrating, as it were, one against the other, and seeking to trace through all diffusions back to its original home some one or other

[1] A Communication to the Congresso Internacional de Historia dos Descobrimentos, Lisbon, 1960.

particular discovery. Thus in the present question, while we may say broadly that there was nothing to choose between the cultures of the ancient Mediterranean and those of Eastern Asia in statements concerning the *attractive* property of the lodestone, the matter stands quite otherwise for the *directive* property. A century of research has pushed back the first mention of the magnetic compass in Europe to Alexander Neckam about + 1190, followed soon afterwards by Guyot de Provins in + 1205 and Jacques de Vitry in + 1218. Then comes the great, though brief, treatise of Petrus Peregrinus in + 1269. All other European claims have been excluded by detailed study, and the last decade of the + 12th century remains the first fixed point in our survey. What is particularly striking also is that no earlier mentions in either Arabic or Indian literature, reliably demonstrable, have come to light.

But if Alexander Neckam is a fixed point, so is Shen Kua. In his *Mêng Chhi Pi Than* (Dream Pool Essays), written about + 1080, more than a century before Neckam, he gave the first clear description of the magnetic needle in any language. More than that, he clearly stated the phenomenon of magnetic declination. 'Magicians', he said, 'rub the point of a needle with the lodestone, after which it is able to point to the south. But it always inclines slightly to the east, and does not point directly at the south.' I shall never forget the excitement which I experienced when I first read these words in the Chinese text. Shen Kua recommended particularly the suspension of the needle on a freshly spun fibre of pure silk.

Between Shen Kua and Alexander Neckam there are two further Chinese texts of importance. In + 1116 Khou Tsung-Shih in his *Pên Tshao Yen I* (The Meaning of the Pharmacopoeia Elucidated) repeated what Shen Kua had said, but added a description of the needle compass floated on water, and not only gave a precise measure of the declination (15° E) but attempted an explanation for it. Secondly, Chhen Yuan-Chhing, in his *Shih Lin Kuang Chi* (Guide through the Forest of Affairs), an encyclopaedia compiled about + 1135, described two further different forms of magnetic compass. One of these consisted of a small wooden fish containing a piece of magnetite and floated on water; the other was a dry suspension—a little wooden turtle containing magnetite and rotating on a pivot made of sharpened bamboo.

Equally full of interest is a text considerably earlier than that of Shen Kua, the *Wu Ching Tsung Yao* (Essentials of Military Technology), compiled by Tsêng Kung-Liang in + 1040. Here we read of the 'south-pointing fish', a leaf-like piece of iron magnetised not by rubbing on the lodestone but by being heated and cooled while oriented in the earth's magnetic field, and then floated on water to do its office. About the same time there is a reference to the needle compass in a poem by the geomancer Wang Chi (b. *c.* +990). Other geomantic texts, such as the *Shen Pao Ching* (Precious Manual) of Hsieh Ho-Chhing, attest the use of the needle during the Sung dynasty. We are thus fully authorised to accept the practical application of knowledge of the compass in China about two centuries before its first mention in the West.

Earlier than that, however, the Chinese references are more obscure. Let us first go back to the beginning of the story in the + 1st century. In his *Lun Hêng* (Discourses Weighed in the Balance), the 'Lucretius of China', Wang Chhung, says: 'When the south-controlling spoon is thrown upon the ground, it comes to rest pointing at the south.' Here the great sceptic is contrasting fabled phenomena in which he did not believe, with an effect which he had himself observed. It is to the researches of the Chinese archaeologist Wang Chen-To that we owe a convincing elucidation of this obscure statement. 'Thrown upon the ground' means in fact 'placed upon the ground-plate of the diviner's board', and what the spoon was made of was in fact lodestone. The diviner's board (*shih*) was a curious apparatus commonly used for prognostication in the Han period, well known from examples which have been found in tombs, as well as from textual descriptions. It was composed of two boards or plates of lacquered wood or bronze, the lower one being square (to represent the earth), and the upper one being round (to symbolise the heavens). The latter revolved upon a central pivot, and both were marked with compass-points, denary and duodenary cyclical characters, lunar mansions, *kua* and other signs. The diagram of the Great Bear was always marked upon the heaven-plate. At some time, possibly in the − 1st century, this was replaced by an actual model of the Great Bear or Northern Dipper carved into the shape of a dipper or spoon. Still to this day the typical Chinese spoon has a much shorter and thicker handle than the spoons of Europe, and if the bowl of the spoon is well rounded at the bottom it will balance itself nicely on a surface made as smooth as possible. If the whole is carved from a piece of magnetite, the torque on the handle will swing it round until it faces south or north if the plate and spoon be slightly joggled to and fro. This I am able to assert from personal observation, for I saw the experiment carried out by Dr Wang Chen-To himself in his laboratory at Peking in 1952. This interpretation of the *Lun Hêng* text must retain a certain speculative character until an actual spoon made of magnetite has been recovered from a tomb of Han date, but this is an ever-present possibility.

How now can we fill the gap between +80 (the date of Wang Chhung's book) and +980 (the approximate time of the birth of Wang Chi)? Evidence to our purpose is in fact quite abundant, though most of it is indirect. First there are the numerous literary references to a 'south-pointer', not precisely specified. Between +130, when the great astronomer and mathematician Chang Hêng said in an essay 'I shall take you, Sir, as my south-pointer', and +630 when the monk Fa-Lin urged his hearers to take the Buddhist sūtras as their 'south-pointer', a dozen or more of such references may be collected, and a few indeed are rather reliably earlier than Chang Hêng. The clear implication of this literary metaphor is that during these centuries the 'south-pointer' was an actual thing, a compass-like instrument, known as a famous device even by those who had not seen or used one. As these texts never refer to the legend of the carriage, it is more likely that they implied the lodestone spoon on the ground-plate of the diviner's board.

Secondly there are a number of interesting references to the needle in suspicious connections. It is possible to conduct a census of texts which speak of the attraction of the lodestone for iron. In this way it is easy to show that before about +450 the texts always say that the lodestone attracts iron, while after that time they generally say that the lodestone attracts needles. As early as −120 the *Huai Nan Wan Pi Shu* (Ten Thousand Infallible Arts of the Prince of Huai-Nan) talks of the flotation on water of needles (for divination purposes) greased with sweat or the natural oil of the hair. As early as +142 the great alchemical text of Wei Po-Yang (the *Tshan Thung Chhi*) uses the words: 'Bright like the candle burning in the midst of the darkness; like the needle shining in the midst of the bewildering sea.' The *Shu Shu Chi I* (Memoir on some Traditions of Mathematical Art), written either by Hsü Yo about +190 or by its putative commentator Chen Luan about +570, has a method of computation (something like an abacus) in which a rotating needle points in turn at one or other of the markings on a circle or dial. Another alchemical book, the *Thai-Chhing Shih Pi Chi* (Records in the Rock Chamber), dating at least from *c.* +500, has 'needle attraction on a fixed platform' as a cover-name or synonym for magnetite. This suggests again the diviner's board, if not the needle magnetised from its lodestone spoon.

Perhaps the most extraordinary of these finds is the following. Those who know the literature on the mariner's compass in Europe are well aware of the fact that one of its names was 'calamita', perhaps a reed, more probably the small frog or tadpole among the reeds. If they also read Chinese, then, they receive a considerable shock when they find in the *Ku Chin Chu* (Commentary on Things Old and New) the statement: 'Hsia-ma-tzu, the tadpole, is also called the "mysterious needle", or the "mysterious fish", and another name for it is the "spoon-shaped beastie"...' This text is regarded by many as the work of Tshui Pao of the +4th century, but its genuineness matters the less for our argument since an almost identical passage occurs in the *Chung Hua Ku Chin Chu* written by Ma Kao between +923 and +936. On the conventional view this is quite serious enough. One can only suppose that at some time between the +2nd and the +10th century the south-pointing lodestone was superseded by the south-pointing needle magnetised from it, and that by an association of ideas the needle came to be called a frog or tadpole while the real tadpole itself acquired the popular name of the 'mysterious needle'. The significant similarity between the shape of spoon and tadpole will not be missed, and indeed the very character for tadpole (*tou*) includes the spoon-ladle radical. Furthermore, the expression 'mysterious needle' (*hsüan chen*) sounds very like 'suspended needle' in Chinese. So indeed the editors of the *Thai-Phing Yü Lan* encyclopaedia wrote it in +983. And in the *Thai Chi Chen-Jen Tsa Tan Yao Fang* (Tractate of the Supreme-Pole Adept on Miscellaneous Elixir Recipes), very little, if at all, later, we hear of a 'suspended needle aludel', a piece of apparatus probably so called because it was intended to be placed vertically in the furnace. All this makes it clear that the magnetic needle was in use for determining compass directions, no doubt mainly for geomancy, by the middle of the +10th century, and in all probability some centuries earlier.

Still further light can be gained on the dark period before +1000 by a study of statements about magnetic declination. It seems fairly certain that this phenomenon could not have been observed before the needle had taken the place of the lodestone, for the difference concerned was one of less, even much less, than 20°, and only a fine pointer would have detected it. The earlier sinologists believed that the great Tantric monk and astronomer I-Hsing had given a statement of declination about +710, but this it has not been possible to confirm.[1] The geomantic book *Kuan shih Ti Li Chih Mêng* (Mr Kuan's Geomantic Instructor), however, which may be dated as of the late Thang period (*c*. +880), has the first clear account of it. Thus the needle is presupposed; indeed it is specifically mentioned. Then the *Chiu Thien Hsüan Nü Chhing Nang Hai Chio Ching* (Blue-Bag Sea Angle Manual—a significant title), which seems to date from about +900, includes an implicit reference to declination, for it discusses the technical terms *chêng chen* and *fêng chen*. If we look at a modern Chinese geomantic compass we

---

[1] Recent investigations support it, however (Smith & Needham).

shall find on it three concentric circles, of which the two outer ones repeat the 24 points of the innermost one, but staggered, as it were, $7\frac{1}{2}°$ east in the one case and $7\frac{1}{2}°$ west in the other. Astronomical north–south is termed *chêng chen* (orthodox needle), the circle for the easterly declination is termed *fêng chen* (seam needle), and that for the westerly one is called *chung chen* (central needle), probably because this is the midmost of the three circles concentrically. Everything goes to show that these circles are fossilised traces incorporated still on the compass-board and remaining from times when the declination was successively east and west in China. Indeed we have medieval texts which depict vividly the geomancers arguing amongst themselves which was the more correct direction to take for the siting of houses or tombs. Finally, in + 1174, Tsêng San-I, in his *Thung Hua Lu* (Mutual Discussions), embarks upon a theory to account for the declination in detail. It may thus be affirmed without doubt that people in China were worrying about the cause of the declination before people in Europe knew even of the directivity. Moreover, it was still two or three centuries before the declination was appreciated in the West, for not until about + 1440 did the German makers of portable sun-dials embodying compasses by which to set the noon line begin to make special marks on their dials which showed empirical knowledge of the declination.

Let us now turn to the use of the magnetic compass for navigation at sea. Apart from earlier hints, the first circumstantial description in China occurs just about a century before the time of Alexander Neckam. Chu Yü wrote his *Phing-Chou Kho Than* about + 1113, but it referred to events concerning the port of Kuang-chow from + 1090 onwards, for his father had been a high official at Canton from + 1094 and Governor from + 1099. 'The pilots', said Chu Yü, 'steer by the stars at night, and in the day-time by the sun; in dark weather they look at the south-pointing needle...' This passage was long believed by Western sino-logists to refer to foreign (presumably Arab) ships trading to Canton, but it is now known that this idea derived from a mistranslation of the words *chia ling*; this phrase means 'government regulations', and is not, as the first translators thought, the name of some foreign people.

Before the first European mention there are two further Chinese ones. After the fall of the capital of the Sung, Khaifêng, to the Chin Tartars in + 1126, and the move to Hangchow, Mêng Yuan-Lao wrote the *Tung Ching Mêng Hua Lu* (Dreams of the Glories of the Eastern Capital), in which he noted concerning navigation that 'during dark or rainy days, and when the nights are overclouded, sailors rely on the compass (lit. needle plate), which is in charge of the navigator'.

This would refer to some time around + 1125. Three years earlier a diplomatic mission had set out for Korea, and the account of it by Hsü Ching, one of the members of the embassy, contains a mention of the mariner's compass. 'During the night', he wrote, 'it is often not possible to stop (because of wind or current drift), so the pilot has to steer by the stars and the Great Bear. If the night is overcast then he uses the south-pointing floating needle to determine south and north.' Such is the statement in the *Hsüan-Ho Fêng Shih Kao-Li Thu Ching* (Illustrated Account of an Embassy to Korea in the Hsüan-Ho reign-period).

Moreover, after the time of the first European mention Chinese references to the mariner's compass are so frequent as to indicate widespread current use in times preceding. For example, Chao Ju-Kua, in his great work on human geography entitled *Chu Fan Chih* (Record of Foreign Peoples) and finished in + 1225, says: 'To the east (of Hainan) are the "1,000-li sandbanks" and the "myriad-li rocks", and beyond them is the boundless ocean, where the sea and sky blend their colours, and the passing ships sail only by means of the south-pointing compass needle. This has to be watched day and night with attention, for life or death will depend on the slightest fraction of error.' Evidently by this time pilots were striving for the greatest possible accuracy. Half a century later the *Mêng Liang Lu* (Dreams of Old-Time Hangchow while the Rice is Cooking) of Wu Tzu-Mu confirms this, for it says: 'at times of storm and darkness the pilots of the merchant-ships travel trusting to the compass alone. They dare not make the slightest error, since the lives of all in the ship depend upon it. The water of the ocean is shallow near islands and reefs—if a reef is struck the whole ship may well be lost. This depends entirely on the compass-needle, and if a small mistake is made you will be buried in the body of a shark.' By + 1296, when Chou Ta-Kuan wrote his description of a voyage to Cambodia, the *Chen-La Fêng Thu Chi*, not merely mentions of the compass, but actual recorded compass-bearings, have got into the literature, for he lists them on his way. Still more is this the case, of course, from the + 14th century onwards.

Chinese sailors remained faithful to the floating compass for many centuries. Although, as we have seen, the dry pivoted compass had been described in a Chinese text early in the + 12th century, it did not become common on Chinese vessels until it was reintroduced from the West in the + 16th century by the Dutch and Portuguese through Japan. Associated then with it was the compass-card (the wind-rose attached to the magnet instead of surrounding it) which had probably been an Italian invention, doubtless due to the Amalfitani, made at the beginning of the +14th century. By the end of the +18th, visitors to China were much impressed by the way in which Chinese compass-needles were pivoted [Fig. 80, pl.]. Extremely thin, and no longer than an inch, their mounting was highly sensitive, for the needle's centre was attached to a small inverted hemispherical copper bowl resting upon a finely sharpened and polished steel pivot. The bowl being larger than the pivot, the needle tended to retain its position no matter how the compass was moved. Moreover, the fact that the weight was concentrated below the point of suspension meant that it was quite sufficient to overcome the dip, or inclination of the needle downwards. Europeans had avoided this by counterbalancing the opposite ends of the needle, but this was not very satisfactory because of the variation of dip in different parts of the world. So far as we know, the Chinese never measured inclination; in Europe observations on it began only after +1544.

We may now take stock of the position reached. No European, Arabic or Indian reference competes chronologically with those of Tsêng Kung-Liang, Shen Kua and Chu Yü, and the existence of the compass of magnetised iron can be demonstrated in the Chinese culture-area at least two centuries before it appears in any other. At the beginning of the development there we have the spoon-shaped Great-Bear model 'compass' of magnetite, rotating with some slight difficulty on its polished bronze. Between +100 and +1000 there remains much obscurity, but the abundance of curious hints and glimpses builds up a picture which is unmistakable even if still indistinct. First there is the most ancient mention of the needle in the *Huai Nan Wan Pi Shu* (−2nd century). Then there are the obscure descriptions of a compass-like needle method of computing in the +2nd- or +6th-century *Shu Shu Chi I*. Comparative study of the terms employed in descriptions of magnetic attraction gives us a strong indication that magnetised needles were coming in from the +4th to the +6th centuries. Ko Hung, the greatest of Chinese alchemists, makes a strange juxtaposition of the needle and the 'south-pointer' when he says (*c.* +300): 'Those who are prejudiced by affections do not bear criticism. Those who are envious of the beauty of others—their needle is not bright and straight. Ordinary people are bewildered by affections, and have no "south-pointer" to keep them on their course.' As for the 'south-pointer', we have seen that in literary metaphor between the +1st and the +7th centuries it implies the existence of an actual instrument, whether the spoon of magnetite, or the floating fish or needle of iron, or small lodestones cunningly suspended in various ways. Then we have the frog–tadpole–needle nomenclature complex covering the + 4th to the + 10th century and paralleling that of the occidental 'calamita', though so much earlier in China. In the light of this unfolding panorama we may therefore reasonably assume that the transference of the directive property of the lodestone to pieces of iron which it attracted was discovered in China some time between the + 1st and the + 6th century. Some time before the + 11th century it was discovered, too, that magnetisation could be carried out not only by rubbing pieces of iron on the lodestone, but by cooling or quenching them from red heat, through the Curie point, while held in a north–south direction. The needle was replacing the lodestone probably by the Sui and Thang periods (+ 7th and + 8th centuries), and the discovery of the declination, which occurred probably in the + 9th century, was obtained only by its aid. Successsive declinations, first eastern and then western, were embodied in the design of the Chinese geomantic compass as concentric circles which have persisted until our own time. There can be no doubt that the magnetic compass was used in China for geomantic purposes a long time before it was applied at sea, but the mariner's compass was truly a Chinese development which had occurred some time before the + 11th century, perhaps a long time before.

Thus in broad survey we have a long and slow developmental period in China followed by a sudden appearance and faster development in the West. Since the crucial couple of centuries before Alexander Neckam (+ 1190) has so far afforded no trace or clue from intermediary regions such as the Arabic–Persian culture-area or the literatures of the Indian sub-continent, the possibility arises that transmission from China occurred not in the maritime context at all, but by some overland route through the hands of surveyors and astronomers who were primarily interested in establishing the meridian. Certainly Petrus Peregrinus devotes careful descriptions to two azimuth dial instruments with alidades and inserted compasses (floating in one case, dry-suspended on a pivoted spindle in the other). The determination of the meridian was of course important not only for cartography, but also for such operations as the proper adjustment of the only adequate clocks known to Europeans at this time, sun-dials. It is certainly a striking fact that as late as the + 17th century the needles used in the compasses of surveyors and astronomers all pointed to the south, in contradistinction to the north-pointing sailors' needles—exactly as all the Chinese needles had done for perhaps as much as a millennium previously. If this conception were to be adopted, we might have to envisage an overland westward transmission of the 'astronomer's compass', followed by an application of it for use at sea independent of the earlier parallel application made by the sea-captains of China. But what we know of the level of culture of the Russians and their Central Asian neighbours in the two or three centuries preceding the Mongol invasions might at first sight hardly encourage us to believe in the transmission of so scientific a discovery (in contrast, perhaps, with a technological invention—unless indeed we were to consider the magnetic needle in this light). There is evidently room for much further study of the possibilities for transmission from China through the steppe peoples and the Russians to Europe (possibly by way of the Qarā-Khiṭāi or West Liao State in Sinkiang), avoiding the Islamic, Byzantine and Indian culture-areas; in the meantime many may well prefer to believe that in fact the *mariner*'s compass was what was transmitted, and that texts still unknown in Arabic or more easterly sources will yet inform us about how the sailors of the Indian Ocean did it.

These suggestions are speculative enough, but still more so would any attempt be to answer the oldest question of all—how did the *shao* (the lodestone spoon) get on to the *shih* (the diviner's board)? Nevertheless, something may be said, and if it is not entirely astray, we may have to recognise a totally unexpected relation between the magnetic compass and board-games such as chess which originated in divination procedures.

The game of chess, as we know it, has been associated throughout its development with astronomical symbolism, and this was even more overt in related games now long obsolete. Investigation shows that the battle element of chess seems to have developed from a technique of divination in which the Chinese desired to ascertain the balance of ever-contending Yin and Yang forces in the universe. This was in use in + 6th-century China, whence it passed to + 7th-century India to generate the recreational game. This 'image-chess' (*hsiang chhi*) derived in its turn from a number of divination-techniques which involved the throwing of small models, symbolic of the celestial bodies, on to prepared boards. There were also intermediate forms between pure casting, and casting followed by combat moves. All these go back to pre-Han China (the − 3rd century), and many similar techniques have persisted in other cultures. The most significant of these ancient boards was certainly the *shih*, used from the late Warring States period onwards—a double-decked cosmical diagram having a square earth-plate surmounted by a rotatable discoidal heaven-plate, both being marked with cyclical and astronomical signs as well as symbols and technical terms used only in divination. 'Pieces' or symbolic models were employed with this in a variety of different ways, and we may say that the round heaven-plate of the *shih* proved to be the lineal ancestor of all compass-dials. For some time during the Former Han (− 2nd and − 1st centuries), or possibly at the beginning of the Later Han (+ 1st century), the picture of the Great Bear or Northern Dipper always carved on the heaven-plate of the diviner's board was replaced by an actual symbolic model of this constellation, so vitally important in Chinese polar-equatorial astronomy, carved into the shape of a spoon. At first this model spoon was probably made of wood, stone or pottery, but before long the unique properties of magnetite suggested the use of this substance. And so at last what must have been a truly awe-inspiring discovery was made, and the first of all independent pointers began to rotate, and then to come to rest, before man's astonished eyes. Thus the oldest instrument of magnetic-electrical science, the greatest single factor in the voyages of discovery, and the ancestor of all dial-and-pointer readings, may perhaps be said to have begun as a proto-'chess'-man used in a divination-technique.

13

THE CHINESE CONTRIBUTIONS
TO VESSEL CONTROL[1]

[1960]

IN this communication it is proposed to discuss two important features of medieval Chinese nautical technology, the development of steering mechanisms during the first millennium and a half of the Christian era, and secondly the use which was made of these mechanisms in maritime and fluvial navigation.

As every youngster knows, who has paddled his own canoe, the simplest form of rudder is an oar or paddle held steady at a desired angle on either aft quarter (most conveniently the right or starboard quarter) of the boat. In this way the streamline flow of the water is deflected so as to impart a turning moment to the hull. On the other hand, the most highly developed form of rudder is a great vane pivoting upon the stern-post of a ship, and controlled from its bridge by means of chains to which a source of power is applied, chains which substitute more effectively for the leverage exerted in small ships by the bar known as the tiller. In the West, the terminology of the successive stages in direction-control presents little difficulty. First there were steering-oars or quarter-paddles, or occasionally stern-sweeps centrally fixed, then came rudder-shaped paddles permanently attached to one of the stern quarters, and lastly the stern-post rudder itself, hung on pintle and gudgeon. When a Lincoln MS of + 1263 differentiates tolls between *navi cum handerother* and *navi cum helmerother*, we can guess clearly enough what was meant. But the Chinese terminology presents a much more difficult problem, since the things changed while the technical terms did not, as we shall soon see.

A classical monograph was devoted to the invention of the stern-post rudder by Lefebvre des Noëttes, who claimed that because of the weakness of the steering-oar a cardinal limiting factor to nautical development existed before the beginning of the + 13th century. His chief critic, la Roërie, denied this, but the consensus of qualified nautical opinion has crystallised against la Roërie, though

---

[1] A Communication to the Congresso Internacional de Historia dos Descobrimentos and to the Vth International Colloquium of Maritime History, Lisbon, 1960.

adequate credit has often been withheld from des Noëttes, who was admittedly a landsman.

The steering-oar has always remained of value in rapid rivers and landlocked narrow waters, hence its continuing use in contemporary China. To respond to the rudder, a boat must have way on her, must, in other words, be moving relatively to the surrounding water, for otherwise there is no streamline flow to be diverted. But when descending rapids, a boat may be moving at almost the same speed as the water, and in such cases it is highly advantageous to have a long stern-sweep, so long that its effect depends not on streamline flow but on reaction to water resistance, just as in the case of an ordinary oar. The lever, in such a stern-sweep, is much longer on each side of the fulcrum than in any rudder. Imparting to the boat's stern a strong transverse movement, it can equally well be used for turning the vessel about when stationary in a lake or harbour. The limitations of the steering-oar or stern-sweep became particularly severe at sea, or upon great lakes where rough weather was likely to be met with. A ship of any size required a considerable spar for this duty, and all the worse would be the consequences when it broke under the impact of heavy seas. Even if it did not break it was likely to 'take charge' in any kind of bad weather, preventing the adequate use of sail. The attachment of a short but heavy paddle to the aft starboard quarter had other disadvantages; it made an inconvenient projection liable to foul quays and other ships—and that this was felt is shown by the fact that many Roman vessels were built with a kind of streamlined shield to protect the quarter-paddle.

There is no dispute that the ancient and early medieval European world lacked all trace of rudders. Ancient Egyptian ships are generally shown with steering-paddles at the stern, sometimes as many as five a side, or there may be two quarter-paddles connected together by a framework and bar. The stern-sweep was also known in ancient Egypt; it was fixed to the end of the high stern, supported by a post at its forward end, and furnished with a quasi-tiller so that the helmsman standing on the deck could move it. On all known representations of Phoenician, Greek and Roman ships, steering-oars are universal. Hellenistic and Byzantine culture introduced no improvement, but the Viking long-ships, beginning with the steering-oar, went on to attach it to a pivot, and ended by converting the paddle to a rudder-shaped form and hingeing it on the side of the boat. Norman ships continued these methods, but the steering-oars on the Bayeux Tapestry of + 1077 are still very primitive in type.

251

The oldest European MS illustration of a stern-post rudder, with tiller, is in a Latin commentary on the Apocalypse preserved at Breslau, and dates from + 1242, as des Noëttes recognised. But Brindley succeeded in pushing the date of introduction somewhat further back by finding a notable iron-bound rudder on the ship depicted in the seal of the English town of Ipswich, which came into use in the close neighbourhood of + 1200. Other + 13th-century city seals support this epigraphic evidence. But even that is not the first representation, for rudders are clearly seen in certain ships carved on fonts made by a school of artisans from Tournai, and datable at about + 1180. Two fine examples are still to be seen at Zedelghem and at Winchester. We may conclude therefore that the stern-post rudder is attested first for European ships within a very few years, perhaps less than a decade, of the first mention of the mariner's compass. This fact alone might arouse one's suspicion that the stern-post rudder was not an autochthonous European development either, but made its appearance as the result of long travel from somewhere else. In any case, the Chinese evidence had hardly been subjected to analysis before the present study. It has turned out to be of vital importance.

The problem of the history of the rudder in China presents us with a classical case of the difficulty which arises when there is reason to believe that one single word has done duty through the centuries for two or more devices technologically quite distinct. The word *tho* or *to* (with varying orthography involving at least five characters) certainly meant a steering-oar in the −2nd century; equally certainly it meant the axial 'stern-post' rudder in the + 13th. Since, as we have just seen, the first appearance of the latter in Europe antedates + 1200 by very little, any investigation of the possible Chinese contributions cannot rely upon names and words alone. It is necessary to see what everyone who used the word actually said about the thing. This enquiry could clearly be rather tedious, and we shall here do no more than summarise its nature and results. First we examined as many as possible of the miscellaneous early mentions, and then we considered (*a*) what verb was used in connection with the *tho*, (*b*) what its shape and length, (*c*) how it was said to be fixed, and (*d*) of what material it was made.

Some mentions give almost no help, e.g. the *Huai Nan Tzu* book (*c.* − 120) when it says: 'If the will is there, people are capable of destroying a boat to make a *tho*...' The + 3rd-century commentary of the *Chhien Han Shu* (History of the Former Han Dynasty) is a little more technical, for it tells us that 'the *chu* is the after part of the ship, where the *tho* is held'. But it is more interesting to be told by Kuo Pho, in his + 4th-century commentary on the *Fang Yen* (Dictionary of Local Expressions) of − 15, that this *chu* is pronounced in some districts exactly the same as a cognate word which means 'axle', and indeed both characters share the same phonetic. The scent grows warm, for here is something 'turning' like an axle at the stern. Moreover, there are a number of texts in which the verb *chuan* is used for the operation of steering, and this word has the nuance of something swinging round an axis rather than of something pivoting on a single point. One characteristic mention occurs in a story about Sun Chhüan, the emperor of the Wu State in the Three Kingdoms Period (+ 3rd century) given in the *Chiang Piao Chuan* (Biography of Chiang Piao) and quoted in the commentary of the *San Kuo Chih* (History of the Three Kingdoms Period). Other verbs and phrases which can be analysed point in the same direction.

When we come to the consideration of shape and length, we meet with a statement in a late + 5th-century book, the *Hsün-yang Chi* (Memoirs of Hsün-yang), that a certain river was called the Tho-hsia Chhi or 'Downcoming Rudder Stream'. As it is clear from the context that this name derived from a rudder once seen floating down, it implies that by this time it was quite possible to distinguish a rudder from an oar by its shape. This would not have been the case if the steering-oar had still been the only device in use, for it would not readily have been distinguished from any other oar, and one of the ordinary words for oar would have been employed. More important, however, is the statement in the *Kuan shih Ti Li Chih Mêng* (Mr Kuan's Geomantic Instructor), a book which seems not to be later than the + 9th century, and which proved interesting also in connection with the history of the mariner's compass. Here we read: 'If the hairpin is shorter than it ought to be, it cannot be fixed in place. If the key is shorter than it ought to be, the lock of the box cannot be secured. If the *tho* goes deeper than it ought to, then the end of the boat will not carry its cargo...' Here the meaning presumably is that it will go aground or strike a rock. The passage is truly interesting, for we know that Chinese rudders have long been so slung that they hang down well below the level of the ship's bottom and aid in preventing leeward drift. The words are much more applicable to the rudder than to the steering-oar. And by a very little later, about + 940, we reach practical certainty, for, in his *Hua Shu* (Book of Transformations), the Taoist Than Chhiao wrote: 'The control of a ship carrying ten thousand bushels of freight is assured by means of a piece of wood no longer than one fathom.' Such a length is obviously much too short for a steering-oar or stern-sweep (often more than

50 ft on comparatively small river-boats, and attaining 100 ft in many cases). It can only be the axial 'stern-post' rudder, 6–8 ft deep.

The decisive passage preceding the first European evidence comes in +1124, when Hsü Ching accompanied a diplomatic mission to Korea and gave an account of it in his *Hsüan-Ho Fêng Shih Kao-Li Thu Ching*. This text is full of references to rudders, often to mishaps when they broke, or to the operations of changing them, but the main passage is as follows: 'At the stern, there is the rudder (*chêng tho*), of which there are two kinds, the larger and the smaller. According to the differences in the depth of the water, the larger is exchanged for the smaller, or vice versa. Abaft the deck-house, two oars are stuck down into the water from above, and these are called "Third-Assistant Rudders" (*san fu tho*). These are (not) used while the ship sails in the ocean.' This leaves no doubt in the mind that already at the beginning of the +12th century, and on Chinese ships, several sizes of axial rudders were carried and used under different conditions, while at the same time steering-oars were also carried. But with decreasing certainty and precision the earlier texts successively hint also at the presence of the invention. Evidently it had been in use in China long before its first appearance in the West. Exactly how long the textual evidence does not reveal.

Various other lines of argument support this general conclusion. For instance, Thang texts such as the *Thang Yü Lin* (Miscellanea of the Thang Dynasty) habitually speak of the *tho-lou* or 'rudder-tower' at the stern of +8th-century ships. This is the term always later used for the projecting after-castle or poop in or on which the helmsman stood (and still stands) to work the tiller, and which also contains the winches and other arrangements by which the rudder is raised and lowered. Before this period we have not found examples of this expression. Its appearance is quite significant, for with steering-oars there would be no necessity for a *tho-lou*—indeed it would be in the way; and among the many contemporary types of Chinese traditional ships the stern-sweep and *tho-lou* are never found together. When the former is present, it runs forward some distance over the after part of the deck to a kind of light bridge on which the steersman stands.

Mentions and descriptions of rudders on Chinese ships after the turning-point of +1180 are quite easy to find, but it would serve no purpose to discuss them here. We may content ourselves by calling Marco Polo as a witness whose testimony dates from the neighbourhood of +1270. We must pass on to consider the epigraphic evidence for the Chinese developments, the evidence from pictorial representations which have survived, and finally the archaeological evidence. The former, though extremely interesting, cannot take us much further than the textual evidence has done already, but the latter has proved quite decisive, settling the matter in a way more radical than anyone had anticipated. It has shown that the other lines of argument were fully justified, and has proved what they could only surmise.

Before proceeding, however, we must pause for a moment to consider what is known of the way in which rudders have been attached to Chinese ships of traditional construction. As we know from the detailed researches of Worcester and other painstaking specialists, China is the domain of the slung rudder *par excellence*. It is quite doubtful whether any Chinese rudders were ever attached by eyes or gudgeons so as to hinge with pintles on the hull. In Western ships and boats the pintle was always erect, standing parallel with the stern-post and pointing upwards if attached thereto, downwards if attached to the rudder. Such hooks and eyes were foreign to Chinese usage. Throughout the ages their rudders were attached to the hull primarily by tackle, bound to it by cables which could be tightened by suitable means, and suspended from above by other forms of tackle pulling on the shoulder so that they could be raised or lowered in the water. Sometimes the foot of the rudder was even connected by tackle with the bows of the ship. Gudgeon-like fittings (bearings, as it were, for the main rudder post) were however, at least in larger vessels, regularly fixed to the hull, and these could be open, half-open, or altogether closed by outer pieces of shaped timber. Thus they corresponded to the braces (eyes, gudgeons) of occidental ships, though what turned in them was not the pintle but the rudder-post itself. Though cable and wood thus took the place of iron hinges, it should not be thought that iron was not used for rudders in traditional China, for in fact the larger ones, weighing many tons, are heavily bound with iron straps and other strengthenings. The Chinese rudder was by no means necessarily located at the aftermost point of the deck, but sometimes considerably forward of this, and indeed its post frequently descended through a rudder-trunk built into the hull. Such a construction was facilitated by the transom-and-bulkhead anatomy so typical of the Chinese junk, and we shall see in the dénouement how closely the whole conception of the vertical median steering mechanism was connected with this. The essential point is that the rudders of Chinese ships, always remarkably large in relation to the total size of the vessel, were in principle vertical, axial and median. They were in fact 'stern-post rudders' without a stern-post. To this paradox we shall return.

Now for the epigraphic evidence. Let us proceed in the same way as before, starting with the earliest times and coming forwards. We can also work backwards from the most reliable pictures of ships which date from times later than the European appearance of the stern-post rudder. In this way we shall be focusing, as it were in a microscope, both from below and from above. The counterparts of our Han texts are of course the relief carvings in the famous tomb-shrines (Wu Liang, Hsiao-thang Shan, etc.), and the reliefs on the Indo-Chinese bronze drums. In all these we find boats small and large invariably directed with steering-oars. This takes us down to the +3rd century. From then until the Thang period (c. +7th century) we are mainly dependent upon the carvings and paintings of Buddhist iconography, such as the steles of the Liu Chhao and the frescoes of Mo-Kao Khu (the cave-temples near Tunhuang more commonly known as Chhien-Fo Tung). In these again the steering-oar uniformly persists, even when the craft are seemingly quite large. One might be inclined to suspect Indian influence here, since bellying square-sails are shown and never the Chinese mat-and-batten sails, yet perhaps we need not seek to deny them Chinese nationality since the stern-sweep persisted so long and so powerfully at least in river craft until today. In any case no rudders appear in these waters.

What of the other end? Here we are always troubled by the question of authenticity, for Chinese artists did not always faithfully reproduce the technical detail of the paintings which they copied, and the number of authentic examples from the Sung is now quite small. Nevertheless Yuan paintings (+14th century) always show rudders below the high curving sterns of the ships, as for instance a famous one by Wang Chen-Phêng. The Suzuki collection in Japan has a Sung painting dating from before +1250 of a junk with a fine balanced rudder. Scroll-paintings from an original of +1185 showing the battle of Dan-no-Ura, when the Minamoto defeated the Taira, depict axial rudders very clearly. From almost this same year comes another, more important because unimpeachably original, representation, the Chinese junk carved so grandly by Cambodian artists on the walls of the Bayon in the royal city of Angkor Thom. That it is a Chinese merchant-ship is clear from many details, notably the batten-sails with their multiple sheets, the grapnel anchor and other features. Unfortunately some slight doubt persists as to whether the structure shown at the stern is really an axial rudder, but we accept it as such. Then in +1124 we have the celebrated scroll-painting by Chang Tsê-Tuan entitled Chhing-Ming Shang Ho Thu (Coming up the River (to the capital, Khaifêng) after the Spring Festival), with all its wealth of information on the daily life of the people and the details of their techniques. Here many slung and balanced rudders are depicted with the utmost clarity, but if the genuineness of the Peking painting is doubted, we have to depend on copies of the later (Southern) Sung and Ming. These show considerable divergences in the details of the shipping. However, most Chinese scholars accept the authenticity of the Peking copy. One may note at once that the date corresponds exactly with the decisive evidence from the diplomatic mission to Korea, so this evidence is mutually confirmatory.

Between + 1100 and + 100 only one other piece of epigraphic evidence presented itself, a painting of a ship by Ku Khai-Chih, the famous artist of the second half of the + 4th century. It illustrates the *Lo Shen Fu* (Rhapsodic Ode on the Nymph of the Lo River), written by Tshao Chih a hundred years earlier. At the stern of the ship we see a curious trapeziform object which looks remarkably like an axial rudder hoisted up to the highest position out of the water, and beside it there is a backward and downward pointing spar which might be a 'yuloh' or propelling-oar, or perhaps a stern-sweep quite appropriate for a river-ship. While this painting seems to embody certain archaic ideas of perspective which are characteristic of the draughtsmanship of Ku and his times, the earliest extant version of it is a Sung copy of the + 12th century. It was therefore impossible to be sure that the rudder-like object shown was really in the original, though it might well have been in one of the Thang copies through which no doubt the picture was handed down.

Since 1958 all such doubts could be set at rest. For in the previous year excavations undertaken by the Kuangtung Provincial Museum and Academia Sinica in tombs of Hou Han date (+ 1st and + 2nd centuries) in the city of Canton itself, in connection with rebuilding operations, brought to light a magnificent series of pottery ship models which demonstrated the existence of the axial rudder already at that time. Previous to these discoveries, the tomb ship models obtained in modern archaeological research had been of the Warring States or Early Han period (− 4th to − 1st centuries), and all had evidence of steering-oars. Now however the pottery model, nearly two feet long, was found to be equipped in very modern style (Fig. 81*a*, pl.). The masts were not recovered, but a grapnel anchor hangs from the bow, abaft of which there is a curious cowling like that of a lifeboat or a motor-boat. The vessel is not fully decked over, but carries a series of deckhouses or cabins extending for most of the beam and surrounded by a poling gallery. The stern extends aft a considerable distance beyond the last transom-bulkhead in the form of an after-gallery (indeed a *tho-lou*) the floor of which is formed by a criss-cross of timbers through which the rudder-post descends into the water (Fig. 81*b*, pl.). For indeed the true rudder is there, trapeziform as would be expected, and having no resemblance to a steering-oar, but most clearly exemplifying the remark about 'eight foot of timber' which Than Chhiao was to make nearly a thousand years later. Most gratifyingly, its shoulder is pierced by a hole exactly where the suspending tackle should be attached to it. Probably the model, made no doubt for some wealthy merchant-venturer or ship-owner of Han Canton, once showed all the cables by which the rudder was secured, but these have long ago rotted away and we can only guess how it was done. On the main issue, however, guesswork is ended, and we have positive demonstration that by the + 1st century the true median rudder had come fully into being. How strange it is that this was just the time, too, to which one may trace back the first beginnings of the magnetic compass. How strange also that though the latter was much slower in its development than the steering mechanism, it also appeared in Europe at the same time, just a millennium later. The only obvious difference between them is that the compass seems to have appeared first in the Mediterranean, the axial rudder in northern Europe.

About the transmission of the technique (and surely such there was) very little can be said. It would seem overwhelmingly likely *a priori* that such an invention came round by way of mariners' contacts in the South Asian seas, though it is not impossible that a Chinese artisan who had built ships for the Liao dynasty handed on certain ideas to Russian merchant-shipwrights trading to Sinkiang in the realm of the Western Liao (Qarā-Khiṭāi) between + 1120 and + 1160. However, for the rudder there is more light from the Islamic world than there is for the compass. A well-known illustration in a Persian MS of + 1237 shows an axial rudder on a sewn ship; it comes from the *Maqāmāt* (Conversations) of al- Ḥarīrī (+ 1054 to + 1122). Or again, in a description of a difficult passage in the Red Sea, al-Muqaddasī, in his *Aḥsan al-Taqāsīm* (The Best Divisions for the Knowledge of the Climates), written in + 985, speaks of lanyards or ropes which were operated by the helmsman. This would be rather incompatible with steering-oars of any kind, but could well apply to the tackle-controlled axial rudders which have lasted in use in Arabic waters down to the present day. We may therefore very well find that the Chinese invention had already been spread in the Islamic world by the + 10th century. But the transition from the Muslims to northern Europe remains at first sight puzzling. Perhaps some sea-captain from the Channel or the Baltic was more alert and observant during the Second Crusade (+ 1145 to + 1149) than any of his Mediterranean colleagues.

Now comes what I have called the dénouement. The invention of the stern-post rudder involves a remarkable paradox—it was developed by a people whose ships had characteristically no stern-posts. If we look again at pictures of the ships of ancient Egypt, of the Greeks, or of the Norsemen, we see invariably that the stern sloped gradually upwards in a curve from the water-line. It was in fact, to use anatomical terminology, a 'posterior sternum' corresponding to the 'anterior sternum' of the stem, and a direct prolongation, like the latter, of the keel. But the junk had never any keel. Its bottom, relatively flat, was joined to the sides by a series of bulkheads forming a set of watertight compartments, and instead of stem- and stern-posts there were transom ends. Now the bulkhead build provided the Chinese with the essential vertical members to which the post of the true rudder could conveniently be attached, not necessarily the aftermost transom but perhaps one or two bulkheads forward of it. This principle held good from the smallest to the largest sailing-ships. It might be called that of the 'invisible stern-post'. Of course, in later times, rudders were fashioned in curving shapes so as to fit various kinds of curving stern-posts, but our argument suggests that the difficulty of doing this was one of the chief factors which inhibited the earlier development of the invention in the West. The bulkhead-attached rudder posts can be excellently seen in many of the Sung pictures, as well as in drawings of contemporary traditional Chinese craft. The Cantonese ship model of the Han does not show the attachment so well, but this may be partly because we cannot be quite sure how it slung its rudder —in any case the shape of the latter speaks for itself. To sum up the matter, we can not only feel sure that the

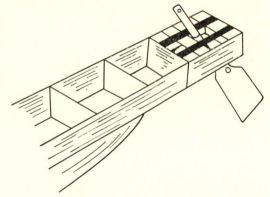

'stern-post rudder' originated in the Chinese culture-area at the beginning of our era, but we can form a pretty good idea of just why it did so.

One word more about rudders. The civilisation (so often miscalled 'static') which initiated them also gave them a far-reaching development. From time to time we have already mentioned balanced rudders. People generally think of the rudder as an object in which the whole of the blade or flat part is abaft the post itself. But large modern ships, on the contrary, have rudders in which there is a flat portion forward of the post as well, and this construction is termed balanced. Such an arrangement not only balances the weight on the bearings but also facilitates the work of the helmsman, since the water exerts pressure in his favour on the forward portion. Europeans were slow in adopting this principle, which did not become general until iron construction afforded means of thoroughly securing the base of the rudder, but it has long been widespread in many forms on Chinese ships. Examples may be seen today on many types of river-junk. Although we have not so far been able to find any literary references to this advanced variety of rudder, it can hardly be doubted that the system is entirely traditional and goes back to the Middle Ages in China. Possibly indeed the balanced rudder was the first to evolve, for the placing of a steering-paddle in an upright position against the aftermost bulkhead would lead directly to it. The rudder of the Han Cantonese ship is distinctly balanced. India also possesses balanced rudders, as in the Ganges *patela* where the two parts are symmetrical, but whether these also derived from China we do not know.

Perhaps the most remarkable of Chinese developments was the fenestrated rudder. When European sailors first frequented those waters they were surprised to see some junk rudders riddled with holes. But this was designedly so, and the holes are of various forms, often diamond-shaped. These fenestrations ease the steering by reducing the pressure against which the tiller has to act, while at the same time leaving the efficiency of the rudder unaffected since the stream lines of the viscous medium flow on past the holes. Drag due to turbulence is also greatly minimised. The principle has been widely adopted in modern iron ships during the present century, and may have led indirectly to the important invention of the slotted wing in aircraft. No doubt the device was originally quite empirical, based upon knotty wood or damaged gear, and one may suppose that some Taoist sailor, finding that his work was eased, was fully content to let well alone and to recommend the arrangement to his mates.

Although out title has promised it, we can hardly here attempt a full discussion of the use which the Chinese made of their vessel control in navigation. From the earliest times their shipmasters sailed by the stars and the sun when out of sight of landmarks. The great astronomer Chang Hêng referred in his *Ling Hsien* about +118 to the 'sum of 2,500 greater stars in all, not including those which the sea people observe'. The first great revolution in navigation occurred in the +11th century, or somewhat earlier, when the compass of the geomancers was taken on board ship. Our first text which shows this also mentions astronomical navigation and soundings, together with the examination of sea-bottom samples. From the +14th century onwards we are well informed concerning the compass-bearings which the Chinese pilots used, and the terminology in which they were recorded. Those who have studied these records have formed a high opinion of the knowledge and precision of their makers. Some idea of the pilots' skill may be gained by the fact that in circumnavigating Malaya they laid their course through the present Singapore Main Strait, which was not discovered (or at least not used) by Europeans until they had been in those waters for more than a century.

Diagrams of pilots' directions in the *Wu Pei Chih*, an early +17th-century compilation of naval and military technology based on +15th-century records, give details of the 'guiding stars' required for particular voyages. The sailors did not use the degrees of the astronomers for measuring polar altitudes, but rather another graduation in finger-breadths (*chih*) and their quarters (*chio*). When the pole fell very low on the horizon, they switched to a circumpolar constellation. All this was closely similar to the practices of the Muslim shipmasters of the medieval Indian Ocean, except that the substitute markpoints for the pole-star were different. Who began these methods and how they developed we hardly know enough as yet to say, but judging from the characteristic features of the astronomy of the two civilisations the Arabs may have taught the use of polar altitudes and the Chinese that of circumpolar stars. Altitude instruments were all modifications of the cross-staff or Jacob's Staff, an instrument which doubtless originated from the crossbow, most probably in Chinese culture. There is much scope for further study of the navigational methods of China's sea-captains, for a considerable literature, both manuscript and printed, survives.

Our subject has remarkable historical implications. 'The Portuguese success in these maritime undertakings and adventures in navigation', wrote J. B. Trend,[1] referring to the late +15th century, 'was due to science; and the science of the day, however rudimentary, had led to a series of technical improvements in ships and how to sail them. Most important were the axled, hinged, rudder; and the mariner's compass...Without it, it is safe to say that the Portuguese discoveries would have been impossible; but their nautical science was up-to-date because they were aware of what was being done in other countries, and willing to invite foreign specialists to come and help them.' Whether they or any others were

[1] *Portugal*, Benn, London, 1957.

aware of the ultimate origin of these discoveries and inventions we may well doubt, but they certainly applied them in the most courageous way. It is a remarkable coincidence that the first half of the same century had been marked by a quite parallel series of explorations carried out by Chinese fleets under the great admiral Chêng Ho, which covered all regions from Zanzibar to Kamchatka and the isles of the Western Pacific with the uncertain exception of Australia. But there were certain contrasts as well as parallels; the Chinese were not seeking to outflank a great but foreign civilisation which lay athwart their trade-routes, they were interested in strange things, rarities and nominal tribute rather than in trade at all, they felt the drive of no proselytising religion, they set up no forts and established no colonies. For less than half a century their presence was felt and established, then suddenly they came no more, and China resumed her inward-looking agrarian personality. But while they roamed and ranged, whether it was to search for exiled princes, or to gather gems and drugs for the imperial cabinets, or to register new tributary realms, their spirit was essentially pacific, bent on peaceful intercourse. In the concluding passage of the chapter on navigation in his *Tung Hsi Yang Khao* (Studies on the Oceans, East and West), Chang Hsieh wrote (+ 1618): 'According to the writer's opinion, those who make carriages build them in workshops, but when they come forth to the open road, they are already adjusted to the ruts. So it is with good sea-captains. The wings of cicadas make no distinction between one place and another, while even the small scale of a beetle will measure the vast empty spaces. If you treat the barbarian kings like harmless seagulls (without any evil intentions), then the trough-princes and the crest-sirens will let you pass everywhere riding on the wings (of the wind). Verily the Atlas-tortoise with mountain-islands for its hat is no different from (an ant) carrying a grain of wheat. Coming into contact with barbarian peoples you have nothing more to fear than touching the left horn of a snail. The only things one should really be anxious about are the means of mastery of the waves of the seas—and, worst of all dangers, the minds of those avid for profit and greedy of gain.'

14

MEDICINE AND CHINESE CULTURE[1]

[1966]

THIS symposium is devoted to the relations between the great medical systems of humanity and the cultures or civilisations in which they arose. It is surely a very hopeful circumstance that Europeans are now giving up their rather self-satisfied parochialism and are eager to look at other systems of medicine, not only in the past before our modern civilisation came into being, but also in other parts of the Old World which have highly continuous and complex civilisations paralleling our own. We assume that we are to speak about Chinese medicine, and one may immediately say that its attachment to its own culture is so strong that it has not yet entirely come out of it. All the sciences of ancient times and the Middle Ages had their very distinct ethnic characteristics, whether European, Arabic, Indian or Chinese, and it is only modern science which has subsumed these ethnic entities into a universal mathematised culture. But while all the physical and some of the simpler biological sciences in China and Europe have long ago fused into one, this has not yet happened with the medical systems of the two civilisations. As we shall later see, there is much in Chinese medicine which cannot yet be explained in modern terms, but that means neither that it is valueless, nor that it lacks profound interest. We hope that the present contribution may lead to greater mutual understanding in the current inter-cultural and inter-civilisational confrontations of our times.

We shall divide what we have to say into three sections. First, the general position of medicine in traditional Chinese society, secondly, the influence of philosophical and religious doctrines upon Chinese medicine, and thirdly, the effects of the transition in contemporary times from traditional society to Marxist socialism. Under the first of these heads there are several questions about which something must be said. There is, for example, the social position of physicians, the nature of the fundamental theories with which they worked, the dates of the Fathers of the doctrine and something on the doctrine itself. Equally important

[1] A Contribution to the Symposium on 'Medicine and Culture' organised by the Wellcome Historical Medical Museum and the Wenner-Gren Foundation, London, 1966.

is the fact that Chinese medicine grew up in a social order widely different from that which was known in the West: namely, feudal bureaucratism, not ancient slave-based city-state imperialism or aristocratic military feudalism. This had profound repercussions in many directions, in due course to be pointed out. Under the second heading we must consider the position of Confucianism, Taoism and Buddhism. But besides this we ought also to say something of the position of mental health in the culture throughout the ages. In the last section we must consider the collaboration of old-style physicians with modern-Western-trained physicians and the principal features of Chinese medicine which are not yet understood in modern-Western terms.[1]

MEDICINE IN TRADITIONAL CHINESE SOCIETY

Pride of place in any sociological investigation of medicine and the medical profession must be taken by the problem of their social position. Greek appreciation of doctors is well known, as witness the quotation which my father was always citing to me:

> A good physician skilled our woes to heal
> Is worth an army to the public weal.

In China there can be little doubt that physicians (*i*) came from the same origin as wizards (*wu*). They were therefore connected with one of the deepest roots of Taoism. Far back at the dawn of Chinese history in the −2nd millennium, probably before the beginning of the Shang kingdom, Chinese society had its 'medicine-men', something like the shamans of the North Asian tribal peoples. During the course of the ages these differentiated into all kinds of specialised professions, not only physicians but also Taoist alchemists, invocators and liturgiologists for the ouranic religion of the Imperial Court, pharmacists, veterinary leeches, priests, religious leaders, mystics and many other sorts of people. By Confucius' time, the end of the −6th century, the differentiation of physicians had already fully occurred. He himself made a celebrated reference to them when he said that 'a man without persistence will never make a good magician (*wu*) or a good physician (*i*)'. We find mention of physicians of these ancient times in the *Tso Chuan*, the greatest of the three commentaries on the *Chhun Chhiu* (the 'Spring and Autumn Annals') of the State of Lu (−722 to −481). More than forty-five consultations or descriptions of diseases occur in these celebrated annals.

[1] Cf. pp. 288, 404.

Among the older ones is the incident when Huan the Physician (I Huan) diagnosed correctly in − 580 the illness of a Prince of Chin. But the most important is the consultation dated − 540 which another Prince of Chin had with an eminent practitioner, Ho the Physician (I Ho), who had been sent to him by the Prince of Chhin. Physician Ho, as part of his bedside discourse, included a short lecture on the fundamental principles of medicine, which enables us now to gain much insight into the earliest beginnings of the science in China. We shall mention this again presently.

The whole history of the social position of doctors in China might be summarised as the passage from the *wu*, a sort of technological servitor, to the *shih*, a particular kind of scholar, clad in the full dignity of the Confucian intellectual, and not readily converted into anyone's instrument. As it is said in the Confucian *Analects*, 'the scholar is not an instrument (*chhi*)'. During the − 2nd and − 1st centuries, in the Former Han period, there were many men of an intermediate sort called *fang shih*; these were magicians and technologists of all kinds, some of them certainly pharmaceutical and medical. Some sinologists have translated this expression as 'gentlemen possessing magical recipes' and this, if somewhat stilted, is certainly not wrong.

Centuries later, for reasons we shall mention in a moment, the ranks of the physicians were joined by scholars of high degree. Indeed there was a general move throughout the Middle Ages to raise the intellectual standing of the physician. As early as + 758, in the Thang dynasty, one can find the beginnings of an important development, the examination of medical students in general literature and the philosophical classics. Presently we shall say more about medical qualifying examinations but here we are concerned with general education. In Hangchow from about + 1140 onwards the candidates were expected to pass tests in the literary and philosophical classics as well as in medical subjects. An imperial decree of + 1188 ordered that unqualified medical practitioners must pass the provincial examinations, and these included the general classical writings as well as sphygmology and other medical techniques. Anyone who did really well could gain an opportunity of rising to the ranks of the Han-Lin Medical Academicians. This gradual change was important, for it ensured the existence of considerable numbers of physicians well educated in general literature and with greater culture than their predecessors had possessed. Such men were called *ju i* (literally Confucian physicians) as opposed to *yung i*, common practitioners, and *chhuan i* or *ling i*, wandering medical pedlars, who went about jingling their special kind of

265

bell on a staff and handing out herbal remedies for the smallest of fees. Of course these latter types never ceased to exist and indeed the grandfather of the greatest pharmaceutical naturalist in all Chinese history, Li Shih-Chen (d. +1593), was one of them.[1] We ourselves have often met with them and recall with particular pleasure a brilliant impersonation of the type in a revolutionary opera which we had the pleasure of witnessing at Thaiyuan in Shansi in 1964. We can thus exclude at the outset any idea that the profession as a whole was a despised one in Chinese civilisation.

Now something about the doctrine, the fundamental philosophy of Chinese medicine. We like the saying of Keele that 'it would seem probable that the first civilised people to free themselves from the purely magico-religious concepts of disease were the ancient Chinese', but we cannot follow him in his belief that this liberation was achieved only briefly until the acceptance of Buddhist thought from India. Nor can we agree with him that the ancient Chinese substituted 'metaphysical' modes of thought for the primitive magico-religious conceptions and practices. Everything depends, of course, on what one means by metaphysical, but if we use the term in its generally accepted sense in modern Western philosophy as meaning ontology, the problem of Being and the dispute between realists and idealists, it is surely not applicable here. Surely what we have to deal with is an ancient philosophy of Nature, a set of hypotheses about the universe and the world of man, which can hardly be called metaphysical because they did not succeed in being scientific in the sense of modern science. One must be quite clear in distinguishing between the mathematised hypotheses of modern science as we know it since the time of Galileo, and the non-quantifiable hypotheses of ancient and medieval times, both in East and West.

Here there is no space to explain at length the natural philosophy which was current among the ancient Chinese. We can only say that, as is generally known, this philosophy was based upon the idea of two fundamental forces, the Yang and the Yin, the former representing the bright, dry, masculine aspect of the universe, the latter the dark, moist, feminine aspect. This conception is probably not much older than the −6th century, but it was certainly dominant in the minds of the early royal physicians whom we mentioned just now. We have already referred to the short lecture given by Ho the Physician in −540 to his patient the Prince of Chin. Here we can see Chinese medical thought *in statu nascendi*. Especially important is his division of all disease into six classes derived

[1] [See the new biography by Lu Gwei-Djen, *Physis*, 1966, **8**, 383.]

from excess of one or other of six fundamental, almost meteorological, *pneumata* (*chhi*). Excess of Yin, he says, causes *han chi*, excess of Yang, *jê chi*, excess of wind *mo chi*, excess of rain *fu chi*, excess of twilight influence causes *huo chi* and excess of the brightness of day causes *hsin chi*. The first four of these are subsumed in the later classifications under *jê ping*, diseases involving fever; the fifth implies psychological disease and the sixth cardiac disease.

This classification into six is of extreme importance because it shows how ancient Chinese medical science grew up to some extent independently of the theories of the Naturalists, which classified all natural phenomena into five groups associated with the Five Elements. These ideas were first systematised in the school headed by Tsou Yen in the − 4th century, and the doctrine of the Five Elements became later universally accepted in all branches of Chinese traditional science and technology. It is well known that these elements differ from those of the Greeks and other peoples in that they comprised not only fire, water and earth but also wood and metal. Chinese medicine, however, never lost entirely its sixfold classification, and in spite of the Five Element theory the Yin and Yang viscera (*tsang, fu*) were always mustered as six of each. In view of the duodecimally based mathematics and world outlook of the Babylonians, one cannot but suspect an influence from ancient Mesopotamia on early China in this respect.

It is not the only example of such an effect either. The twelve double-hours of the Chinese day and night, which go back to the beginning of Chinese culture, have long been thought to be Babylonian in origin, and some evidence has been brought forward also for close parallels in State astrology. As far as medicine is concerned, we can look for them too in another direction, namely, in the very prominent part played by the conception of *chhi*, closely analogous to the Greek *pneuma*. Both words are almost untranslatable but we know that they had significations such as 'life-breath', 'subtle influence', 'gaseous emanation' and the like. Somewhat later on Chinese medical theories also dealt much with another word of very similar meaning, *fêng* (wind). Now Filliozat, in a classical monograph, has shown that the *pneuma* of Greek medicine can be matched word for word and statement by statement in the *prāṇa* of the great Indian medical writers. Thus we see, as in perhaps hardly any other science except astronomy, a widespread community in high antiquity between the peripheral areas of the Old World; from Greece, through India, round to China, there is 'pneumatic medicine'. We are well aware that until now, so far as the cuneiform texts have unravelled it, Babylonian medicine has been largely magico-religious in character, but one cannot help feeling that there must have been some schools of proto-scientific medicine in Mesopotamia which bequeathed their ideas about the subtle breaths, both of normal function and pathological condition, with which the physician must contend. One cannot help feeling that some civilisation older than either Greece, India or China must have originated such conceptions and sent them out in all directions. The Iranian culture-area can hardly qualify on account of its relatively younger age, so that Mesopotamia must have been their home.

Another doctrine prominent in ancient Chinese thought was that of the Macrocosm and the Microcosm. A great interdependence of the State on its people and of the health of the people on the cosmic changes of the Four Seasons was envisaged. The Five Elements were associated together in 'symbolic correlations' with many other natural phenomena in the groups of five, and these conceptions were applied in a remarkably systematic way to the structure and function of the living body of man. As might be expected in a society which was developing the characteristic form of bureaucratic feudalism, great importance was attached to the prevention of trouble, both in the political and personal life of the people, rather than to its control when it arose. And thus in the field of medical thought prevention was considered better than cure. In spite of all influences which may have acted on Chinese medicine from the outside from the beginning onwards, it retained an extremely individual and characteristic quality, and this is still clearly present today. We must, of course, willingly grant to Keele that the practice of using charms, incantations and invocatory prayers to deities persisted through most of Chinese history, particularly among the poorer strata of society and in the exorcistic activities of Taoist adepts and Buddhist monks. In +585, for example, under the Sui dynasty, the Directorate of Medical Administration included, besides two professors of medicine (i), and two professors of physiotherapy (an-mo), two professors of apotropaics (chu-chin); thus there was official sanction for magico-religious techniques.

But Keele is absolutely right in giving the impression that all these phenomena were 'fringe activities' of Chinese traditional medicine. They were quite peripheral to the practice of medicine as such, kept far indeed from the centre of the stage, and it can confidently be asserted that from the beginning Chinese medicine was rational through and through. 'The transmutation from magico-religious to metaphysical pathology was an achievement', writes Keele, 'but it was not enough to provide a basis for progress in medicine, for it was not scientific, either in its method of observation or in its reasoning, in that it entirely failed to make use of the method of induction.' Rewriting this in our own language we should say that the advance from magic and religion to primitive scientific theory was an immense achievement, but that for a wealth of reasons, which we cannot go into here, Europe was the only civilisation in which ancient and medieval science could give way to modern science. We would not say that the old Chinese scientific theories gave no basis for progress in medicine, nor that they were unscientific in observation or reasoning. Undoubtedly they did make use of the method of induction, but they remained pre-Renaissance science and never became modern science.

So much for social position and perennial philosophy, now a word about the Fathers and their history. A comparison between the early classical period of Chinese and Greek medicine is of much interest. In China there is a figure paralleling Hippocrates (-460 to -370), but not quite so much is known of his personality and he is not directly connected with what corresponds to the Hippocratic *corpus*. This was Pien Chhio, for whose life we have an authoritative source in the *Shih Chi* (Historical Memoirs) of Ssuma Chhien, the first of the wonderful series of Chinese dynastic histories. He must have been of the generations preceding Hippocrates, for we have a firm date for a famous consultation of his, -501; this links him with the more ancient physicians already mentioned because he in his turn was in this year called to attend a Prince of Chin. On this occasion the holistic character of traditional Chinese diagnosis was clearly shown, for Pien Chhio was asked by the court chamberlain whether he followed the methods of the legendary physician Yü Fu. Pien Chhio looked up and sighed, saying: 'the methods of which you speak are no better than viewing the sky through a thin tube or considering paintings by looking through a narrow crack. In my way of going about the business I have no need to feel the pulse nor to look at the colour of the patient, nor hear sounds or examine behaviour, in order to say where the disease is located.' And he went on to explain that he judged by the history and condition of the patient as a whole. But the passage is also important because it shows that already at this early time the four important diagnostic observations (*ssu chen*), typical of Chinese medicine, were in use. These comprised first the inspection of the general physical state of the patient, including colour and glossoscopy (*wang*), primitive forms of auscultation and osphristics (*wên*), anamnesis, including the patient's medical history (*wên*, another character) and finally palpation and sphygmology (*chhieh*). The text also shows that here at the time of Confucius himself the physicians were using acupuncture needles, gentle radiant heat (moxa, cf. Fig. 82, pl.), counter-irritants, aqueous and alcoholic decoctions of drugs, massage, gymnastics and medicated plasters. It is striking to find so many therapeutic methods already elaborated before the time of Hippocrates.

Now what corresponded in China to the Hippocratic *corpus*? We know that the books in that great collection were written during a period of time covering much more than the life of Hippocrates himself, i.e. from the beginning of the − 5th century down to the end of the − 2nd. Only a few of them are now considered 'genuine', in the sense of having come from the pen or the dictation of Hippocrates himself. The corresponding collection in China is the *Nei Ching*, and the fact that it is divided into separate chapters in the forms we have now, appearing to be a single book rather than a series of tractates, must not disguise the fact that it is to some extent a parallel compilation. It deals indeed, just as the Hippocratic *corpus* does, with all aspects of the normal and abnormal functioning of the human body, with diagnosis, prognosis, therapy and regimen. The *Nei Ching* was, we think, approximately already in its present shape by the − 1st century, in the Former Han dynasty. No one disputes that it systematised the clinical experience and the physio-pathological theory of the physicians of the preceding five or six centuries. A minor difference from the Hippocratic tractates is that in the *Nei Ching* a great deal of the text is cast in the form of a dialogue between the legendary Emperor Huang Ti and his biological-medical preceptors and advisers (equally semi-legendary) such as Chhi Po.

The full title under which the *corpus* is commonly known is the *Huang Ti Nei Ching* (The Yellow Emperor's Manual of Corporeal [Medicine]). It consists of two parts. The *Su Wên* (The Plain Questions [and Answers]) and the *Ling Shu* (The Vital Axis). This separation was the recension which came from the editorship of Wang Ping in the Thang dynasty, but it is certain that this was not the form which the *corpus* had in the Han period. Another form, known as the *Huang Ti Nei Ching, Thai Su* (The Yellow Emperor's Manual of Corporeal [Medicine]; The Great Clarity), which was edited a hundred years or so earlier than Wang Ping by Yang Shang-Shan in the Sui period, and which has only in very recent times come to light, may be considered nearer the original text of the Han. It contains almost all the material distributed in the two more usual parts, but organised in a different order.

The *Nei Ching* scheme of diagnosis (systematised in the *Shang Han Lun*) classified disease symptoms into six groups in accordance with their relation to the six (N.B. not five) tracts (*ching*) which were pursued by the *pneuma* (*chhi*) as it coursed through and around the body. Three of these tracts were allotted to the Yang (Thai-Yang, Yang-Ming, Shao-Yang) and three to the Yin (Thai-Yin, Shao-Yin, Chüeh-Yin). Each of them was considered to preside over a 'day', one of six 'days', actually stages, following the first appearance of the feverish illness. In this way differential diagnosis was effected and appropriate treatment decided upon. These tracts were essentially similar to the tracts of the acupuncture specialists, though the acupuncture tracts were composed of two sixfold systems, one relating to the hands and the other to the feet, and crossing each other like the cardinal (*ching*) and decumane (*lo*) streets of a city laid out in rectangular grid arrangement. Moreover, by the time of the *Nei Ching* the physicians had achieved full recognition of the fact that diseases could arise from purely internal as well as from purely external causes; the ancient 'meteorological' system explained by Ho the Physician had therefore been developed into a more sophisticated sixfold series, namely, *fêng, shu, shih, han, sao, huo*. As external factors they could be translated as wind, humid heat, damp, cold, aridity and dry heat; but as internal causes we could name them blast (cf. Van Helmont's *blas*), fotive *chhi*, humid *chhi*, algid *chhi*, exsiccant *chhi* and exustive *chhi*. It is interesting to notice the partial parallelism with the Aristotelian–Galenic qualities which were part of a quite different fourfold system.

It will have been noticed that in the previous paragraph we translated the title *Huang Ti Nei Ching* as 'The Yellow Emperor's Manual of Corporeal [Medicine]'. This raises an extremely interesting question. In recent times there has been a tendency among medical sinologists to translate it as 'The Yellow Emperor's Manual of Internal Medicine'. But this is indisputably wrong and should be abandoned as soon as possible. Not only does it introduce a particular modern conception where it has no place to be, but it also entirely mistakes the significance of the word *Nei*. The chapters of many ancient Chinese books are divided into two groups, the 'inner' and the 'outer'; so, for example, we find the two parts *Nei Phien* and *Wai Phien* in the greatest of Chinese alchemical books, the *Pao Phu Tzu* (Book of the Preservation-of-Solidarity Master) written by Ko Hung about +300. One might be tempted to translate 'inner' and 'outer' by esoteric and exoteric respectively; the former being secret doctrine not to be revealed to people in general, the latter being the overt publicly preached system. But this would involve just as serious a mistake as that which we are trying to correct. The key to the real meaning for which we are seeking is to be found in the classical statement of the Taoists that they 'walked *outside* society'. Again, the

Chuang Tzu book says: '*outside* time and space is the realm of the sages, and I am not speaking of it here'. In other words, *nei* or 'inside' means everything this-worldly, rational, practical, concrete, repeatable, verifiable, in a word, scientific. Similarly, *wai* or 'outside' means everything other-worldly, everything to do with gods and spirits, sages and immortals, everything exceptional, miraculous, strange, uncanny, unearthly, extra-mundane and extra-corporeal or incorporeal. Let it be noted in passing that we are not here using the term supernatural, because it is deeply true to say that in classical Chinese thought there is nothing outside Nature, however strange it may happen to be. This is why we propose the translation 'The Yellow Emperor's Manual of Corporeal [Medicine]'. It is fascinating here to notice that the ancient bibliographies also contain a *Huang Ti Wai Ching*, 'The Yellow Emperor's Manual of Incorporeal (or Extra-Corporeal) [Medicine]', but this completely disappeared during the early centuries of our era. The fact that the *Wai Ching* was lost so early emphasises once again precisely the quite secondary character of the magico-religious aspect of medicine in China; for cures effected by charms, cantraps and invocations must certainly have been included in the 'outside' *corpus*.

Before leaving this subject we should like to refer to certain other uses of the terms *nei* and *wai* which might be made to explain the title of the Chinese Hippocratic *corpus* but in fact cannot be. In recent centuries there has been an everyday distinction between *nei kho* and *wai kho*, the former meaning internal and general medicine in the modern sense, and the latter 'external' medicine, formerly usually called *yang kho*. This included such surgery as the Chinese carried out, but it was much wider than surgery in the modern sense for it included dermatology, the treatment of fractures and dislocations, together with that of boils and eruptions and any pathological conditions which could affect the outer surface of the body. This distinction, however, does not go much further back than the Sung period (+ 10th to + 13th centuries) where it began with three specialities (*kho*) and went on to six and nine; eventually in the Ming a classification into thirteen specialities became usual. All this can have nothing to do with the *Nei Ching*, the text of which recognises no such distinctions. Another *nei–wai* differentiation occurs in historical writing associated with the words *shih* or *chuan* referring to events within the Imperial Palace as opposed to others outside it. Again, this does not apply in any way to the *Huang Ti Nei Ching*. In alchemy there is an important distinction between *nei tan* and *wai tan*, the former term referring to 'physiological alchemy', the production of a medicine of immortality within the body itself by respiratory, gymnastic, sexual, meditational, heliotherapeutic and other yogistic exercises. *Wai tan* on the other hand referred to the actual processes of practical manual operations whereby elixirs of longevity or immortality were prepared in the laboratory. Here, as in the case of esoteric and exoteric, the meaning is almost diametrically opposite, for the 'inner' was the psycho-physiological and the 'outer' was the proto-chemical. Lastly, the words could be used in a perfectly straightforward and unsophisticated way; as in the title of an anatomical book written by Chu Hung in the Sung period—the *Nei Wai Erh Ching Thu* (Illustrated Treatise of Visceral and Superficial Anatomy). We think that this sort of philological excursus into the proper nuances of words is abundantly worth while for the prevention of serious misunderstanding.

It is indeed fortunate for the historian of Chinese medicine that we have an extremely important physician's biography dating from just the time when we believe that the *Huang Ti Nei Ching* was being put together. The biography of Shunyü I by Ssuma Chhien is contained in the same chapter of the *Shih Chi* as that of Pien Chhio already referred to. The second part is much more important however, because it contains twenty-five clinical histories related by Shunyü I as well as his replies to eight specific questions all on the occasion of an imperial decree commanding him to reveal the nature of his practice about − 154. To the life and times of Shunyü I has been consecrated by Bridgman the most scholarly monograph yet produced in the field of the history of medicine in China. Born in − 216 in the old territory of the State of Chhi, Shunyü I had practised widely among princes as well as officials and common people, and after having held the post of Granary Intendant from − 177 onwards he was ten years later accused and taken to court upon charges of malpractice, but acquitted after the supplication of his youngest daughter. This was the famous occasion when mutilative punishments were revoked, alas only temporarily. Shunyü I died between − 150 and − 145. It is possible to explain the nature of nearly all the cases attended by Shunyü I in modern terms, and though a few of these interpretations may be subject to revision, the majority are perfectly clear. We have thus a unique record of medical practice and knowledge in the − 2nd century.

On the occasion of his interrogation in − 167 Shunyü I referred to the chief book which had been handed down to him by his teacher Yang Chhing (or Kungchhêng Yang-Chhing). This was the *Mo Shu Shang Hsia Ching* (Treatise on Sphygmology in two Manuals), one of these being associated with the name of Huang Ti and the other with that of Pien Chhio. It seems fairly certain that this was some early form of the *Huang Ti Nei Ching*. Shunyü I further mentioned what appear to be the titles of separate chapters or tractates within this corpus as follows:

(1) *Wu Sê Chen* (Diagnosis by the Five Colours).

(2) *Chhi Kai Shu* (The Art of Determining the Loci of the [Eight] Auxiliary Tracts). This was undoubtedly a treatise on acupuncture.

(3) *Khuei Tu Yin Yang* (The Determination of the Degree of Yin and Yang [involved in different diseases]).

(4) *Pien Yao* (The Drugs that effect Changes [in the Body]).

(5) *Lun Shih* (Discussions on [the Use of] Mineral Drugs).

The fact that no parallel titles to any of these can be found in the bibliography of the *Chhien Han Shu* (Dynastic History of the Former Han Dynasty) suggests that all these titles were those of chapters or tractates within the *Mo Shu Shang Hsia Ching*. And indeed, as Bridgman showed (though we have not been able to follow him entirely in the definition and interpretation of the titles), close parallels to some of them can be found in the chapter headings of the present *Huang Ti Nei Ching*. All this is very significant because it was just at this time that we think the *Huang Ti Nei Ching* was being constituted.

Bridgman ends his monograph by making a weighty comparison with Greek medicine. Far, he says, from being an assembly of magical practices and inapplicable fantasies, it appears that the examination of the sick person, the investigation of the clinical history, the comparison of data from different examinations and the therapeutic deductions, all formed part of a discipline which constituted a valid and valuable precursor of contemporary clinical science. In this light ancient Chinese medicine can fully sustain any confrontation with Greek or Roman medicine of the same period. With this we wholeheartedly concur.

The Later Han, San Kuo and Chin periods then brought a number of outstanding physicians and medical writers roughly corresponding to Aretaeus, Rufus, Soranus and Galen. The life and work of Galen (+131 to +201) is closely paralleled by that of Chang Chi (Chang Chung-Ching) who probably lived from +152 to +219. One could hardly say that the influence of this younger contemporary of Galen during later Chinese history was less than Galen's in the Western world, for his *Shang Han Lun* (Treatise on Febrile Diseases), produced about +200, was one of the most important medical classics after the *Huang Ti Nei Ching* itself. Next came Hua Tho (+190 to +265) active in the San Kuo or Three Kingdoms period, a man about whom many stories subsequently clustered. Little of what he wrote has come down to us, but the great development of medical gymnastics in China, massage and physiotherapy, can be traced back to him. The third century brought two more men of the highest importance. First Huangfu Mi (+215 to +282) whose *Chen Chiu Chia I Ching* (Systematic Manual of Acupuncture) was a most influential work. No less important, however, was the *Mo Ching* (Pulse Lore Manual) compiled by Wang Shu-Ho about +300, and the foundation of all later works on the pulse. As Wang Shu-Ho was born about +265 and died in +317 we have come down to the time of Oribasius, and the classical period of Chinese medicine draws to a close. Its vast developments in later ages we cannot follow further here.

Let us now turn to the social effects exerted upon the medical profession as a result of its developing within a society based on bureaucratic feudalism. It is far too little understood by Westerners that for some two thousand years Chinese society was constructed in an entirely different way from anything known in the West. The principles of aristocratic military feudalism are familiar to all educated Europeans, though historians are aware that the practice of it was far more complex and diversified than is usually imagined. Nevertheless, broadly speaking, traditional China lacked the apparatus of fiefs and feudal ranks, of primogeniture and inherited lordships; instead of all this the culture was governed by a non-hereditary bureaucracy, an immensely elaborate civil service, the members of which were drawn from the ranks of the educated gentry. Instead of earls and barons there were governors and magistrates. Access to this 'mandarinate' was by means of the official examinations, so that the 'carrière ouverte aux talents' was a Chinese invention made a couple of millennia earlier than its successor in France; 18th-century France was, we know, greatly influenced by knowledge of Chinese customs. All this of course is not to say that China never had a feudal or proto-feudal society; on the contrary, it is very probable that the society of the −1st millennium, the time of the Spring and Autumn period and the period of the Warring States, should be so characterised. But it is at any rate certain that with the passage of time all feudal elements persistently declined and were replaced by the non-hereditary bureaucratic society. Why this happened is a problem that we cannot go into here.

As might be expected, the influence of this very different form of society upon medicine was profound. It is demonstrable that examinations of scholarly proficiency were inaugurated by the Han Emperor Wên Ti in −165 (during the lifetime of Shunyü I); while the Imperial University (Thai Hsüeh) was founded in − 124. Although literary, philosophical and administrative culture was always its primary aim, it is not surprising that instruction should have begun early in those sciences which appeared to be of importance to the State, for example astronomy, hydraulic engineering and medicine. So we find that Regius Professorships and Lectureships in Medicine (Thai I Po Shih, Thai I Chu Chiao), implying examinations for qualification to practise, date from as early as + 493. Then between + 620 and + 630 an Imperial Medical College was established, together with medical colleges in all the chief provincial cities, and medical degrees were awarded from then onwards. Although at first the remarkable precocity of these dates may seem surprising, it is less so once the distinctively bureaucratic-feudal character of Chinese society is understood, together with the age-old respect of the Chinese for learning and for a learned, non-hereditary civil service. The year + 931 constitutes a focal point in transmission of the principle of qualification for practice westwards, for that was the date of the first qualifying examinations in the Arabic world, decreed by the Caliph al-Muqtadir at Baghdad under the superintendence of the eminent physician Sinān ibn Thābit ibn Qurrāh. Of Chinese–Arabic contacts during the preceding two centuries much is known, and there is no difficulty in supposing that the Arabs were taking up energetically a much older Chinese idea. Lastly it passes to the West when, in + 1140, Roger of Sicily issues laws concerning State examinations for physicians, and when the school of Salerno begins to graduate men as *Doctor Medicinae* (+ 1224). It looks as if the Arabic and the Western worlds borrowed the idea of examinations in medicine from Chinese culture just as civil service examinations in the 19th century so long afterwards were introduced with full knowledge of the age-old Chinese parallel in mind. It would hardly be possible to imagine a deeper effect of the environing culture on medicine than this 'bureaucratisation' of medical knowledge, which had the extremely happy effect of protecting people at large from the activities of ignorant physicians.

But we can go much beyond this. Besides the traditions of officially recognised scholarship just referred to in connection with medical qualifying examinations, there was another important factor which helped to bring these into existence so early, namely the generation of what can only be described as a national medical service. From the beginnings of this in the Han period (broadly − 200 to + 200) its later form throughout the centuries can be descried, particularly its division into an Imperial Palace service and a Public or National service, the latter responsible also for provincial and medical administration and for the medical staff of the army. In the history of military medical care the Chinese contribution is at least as important as anything which one can find in Greece and Rome, and perhaps a good deal more so, though never yet adequately considered in the world medical-historical literature. For instance, we probably possess in the bamboo tablets of army administration, which have been preserved in the sands of the Gobi along the *limes* of the Great Wall, at least as much information about the military medical service of the Han armies as we have about those attached to the legions of Rome. We even have details of their standard prescriptions dating from the − 3rd century.

In a bureaucratic society it was quite natural that in the development of the conception of hospitals religious and governmental initiatives should, from time to time, contend together. The general picture emerges that the idea of the hospital in China first arose in the Han before the introduction of Buddhism, but that during the Liu Chhao (Six Dynasties) period religious motives led to the foundation of many institutions, not only by Buddhists but also by Taoists. Then, when Confucianism regained strength towards the end of the Thang and especially during the Sung dynasties the national medical service more and more took over the hospitals. Under the Yuan dynasty at the time of the Mongol conquest of Persia and Iraq, medical organisations of Arabic type and tradition were added, just as a Muslim Bureau of Astronomy was set up as an auxiliary to the age-old department of the Astronomer-Royal. Finally, however, under the Ming and Chhing dynasties social organisms of many kinds decayed, and the hospitals shared in this, so that when Westerners first began to visit China in any numbers (early in the 19th century) they gained an altogether wrong idea of the history of medical administration in China. Nevertheless, many interesting hospitals and public charities did continue in these late times.

As in so many other fields, the beginnings of the hospice go back to the troublous but venturesome times of Wang Mang, the only Hsin Emperor between the Former and the Later Han dynasties. On the occasion of a severe drought and locust plague in + 2 an edict ordered that the sick should be accommodated in empty palaces and given medical treatment. In + 38 Chungli I did the same for his people in a time of epidemic. But these were not, it seems, permanent institutions. The first description of a permanent hospice with a dispensary is that of the foundation of Hsiao Tzu-Liang, a Buddhist prince of the Southern Chhi dynasty, set up in + 491. Characteristically, the first government hospital followed very soon afterwards when, in + 510, Thopa Yü, a prince of the Northern Wei dynasty, ordered the Court of Imperial Sacrifices to select suitable buildings and attach a staff of physicians for all kinds of sick people who might be brought there. The significance of this is that the Court of Imperial Sacrifices (Thai Chhang Pu) had from the beginning of the Han been responsible for the Imperial Medical Service (Thai I Chu). This hospital, which was called a Pieh Fang, had a distinctly charitable purpose, being intended primarily for poor or destitute people suffering disabling diseases, and severe epidemics were again the background of the initiative. Later in the same century we have a good example of the pattern of semi-private benefactions by government officials which afterwards became widespread; Hsin Kung-I, one of the generals who had conquered the house of Chhen and helped to unite the empire under that of Sui, encountered a violent epidemic in the province where he had retired to be governor. He turned his own residence and offices into a hospital and provided drugs and medical attendants for thousands of people (c. + 591). The classical example of such a benefaction no doubt was the action of the great poet Su Tung-Pho when Governor of Hangchow in + 1089; he gave rich endowments to a government hospital which he set up there and which formed a model for other provincial cities.

It is in the Thang however that we can best study the conflict between religious and governmental control of hospitals. In + 653 Buddhist and Taoist monks and nuns were forbidden to practise medicine. In + 717 the minister Sung Ching memorialised the throne saying that ever since Chhang-an had been the capital (i.e. since the beginning of the Western Wei in + 534) hospitals there had been supposedly controlled by government officials, but because of neglect the Buddhist religious had taken over these functions more and more. By + 734 action was taken, at least in the capital, to establish government-supported orphanages and infirmaries for the destitute. By + 845, as part of the great dissolution of the monasteries under Wu Tsung, the hospices long called Pei Thien (Compassion Pastures) were transferred to lay control under the name of Ping Fang (Patients' Compounds). At the same time, much temple property in land and buildings was expropriated by the Emperor and allocated to these hospitals. Meanwhile, since the beginning of the dynasty in + 620, there had been a special hospital and clinic (the Huan Fang, or Affliction Compound) within the Imperial Palace, with its own medical stores under the control of a special superintendent. The Chief Medical Directors of the Imperial Medical Service (Palace) (I Chien), Assistant Directors (I Chêng), and Staff Physicians (I Shih) took turns to be on duty at this institution. The regularisation of hospital services carried out in the

Thang bore great fruit in the Sung when we find (*c.* + 1050 to + 1250) a wide variety of State institutions at work both in capital and provinces. There were infirmaries for the care of the aged and the sick poor (Chü Yang Yuan and An Chi Fang from + 1102, and the Fu Thien Yuan), a hospital mainly for foreigners (the Yang Chi Yuan from + 1132), another for sick officials (the Pao Shou Tshui Ho Kuan from + 1114) and even one for Chin Tartar prisoners of war (another An Chi Fang founded by Huang Chün about + 1165). Besides all these there were orphanages (Tzhu Yu Yuan from + 1247, and the Yu Ying Thang), out-patient clinics (Hui Min Yao Chü from + 1151 and Shih Yao Chü from + 1248), and subsidised government apothecaries (Mai Yao So from + 1076).

Comparative data suggest that in hospital organisation Chinese practice was not so far ahead of the rest of the world as it was in the matter of qualifying examinations and government medical services. Hospitals of some kind (more exact studies are needed to elucidate their nature) are attested by the + 1st century both for India (as in the *Caraka-saṃhitā* or at Mihintale in Ceylon) and for the Roman Empire (the *valetudinaria* of legionaries, gladiators, etc.). The Indian facilities of his time were described by the Chinese Buddhist pilgrim Fa-Hsien in the + 5th century, and it was just at this period too that there arose the great hospital of Gundashapur in Persia, heir of the former University of Edessa and precursor of the splendid foundations of Iraq, especially Baghdad, from the + 8th to the + 12th centuries, which correspond with the institutions we have mentioned in Thang and Sung China. It is curious that the + 5th century was so important in this way in all three of the major Asian cultures. After classical times the oldest European references seem to be rather of the + 7th century, after which monastic initiative played a part in the West very similar to that of the *sangha* in China.

For a bureaucratic society there is also interest in examining the beginnings of quarantine regulations. By way of example, as early as + 356, the Chin Emperor, on the occasion of a disastrous epidemic, applied what were called the 'old rules', which prohibited officials whose families had three or more cases from attending court for a hundred days. Another question arising is the isolation of lepers. Though we are as yet uncertain when this started, it is sure that the Indian monk Naren-drayaśas, who died in China in + 589, established leprosaria for men and women at the Sui capital. During the Thang these institutions continued and another monk, a Chinese, Chih-Yen, acquired much fame by his preaching and nursing in a leper colony where eventually he himself died (+ 654).

Whatever may be said against bureaucratic systems of society they do at least go in for rational systematisation. This is certainly relevant to that wonderful series of pharmacopoeias, or rather pharmaceutical natural histories, or, as we are thinking of calling them, pandects of natural history, which followed each other throughout the centuries between the Former Han dynasty and the Chhing (cf. Fig. 83, pl.). The first of these, the *Shen Nung Pên Tshao Ching* (Pharmacopoeia of the Heavenly Husbandman), which must be referred to the − 1st and − 2nd centuries, was not produced under imperial auspices, but several later ones were. All these treatises, some of which were vast in size, go under the generic name of Pên Tshao, and most of them have these characters in their titles; perhaps the best translation of the phrase would be 'the fundamental simples' or 'the botanical basis (of pharmacy)'. The term first appears in the *Chhien Han Shu* for + 5, when the Hsin Emperor Wang Mang called what might be described as the first national Chinese scientific and medical congress. They also appear in the biography of Lou Hu, an eminent physician who was a friend of this Emperor (+ 9 to + 24). In later centuries the *Hsin Hsiu Pên Tshao* was a striking example of an imperially commissioned pharmaceutical natural history (+ 659); it has not survived except in the form of certain chapters which were copied by a Japanese monk and so preserved. In the Sung there followed Su Sung's *Pên Tshao Thu Ching* (Illustrated Pharmaceutical Natural History) of + 1062, and the many successive editions of the great *Ching Shih Chêng Lei Pei Chi Pên Tshao* (Classified and Consolidated Armamentarium of Pharmaceutical Natural History). What was true of plants and animals was also true of the books of standard prescriptions. For example, in + 723 the *Kuang Chi Fang* (General Formulary of Prescriptions) was composed by the Emperor Hsüan Tsung himself with his assistants, and then published and sent out to each of the provincial medical schools. Some of the prescriptions in this work of an Imperial pharmacist were actually written up on notice-boards at cross-roads so that the ordinary people could take full advantage of them. This practice was observed and described by the Arabic traveller Sulaimān al-Tājir who was in China in + 851. By + 739 it was the law that every provincial city with more than one hundred thousand families was to have twenty medical students (I sêng) and those of less than one hundred thousand were to have twelve. Subsequently the numbers were increased. Then in + 796 the Emperor Tê Tsung published throughout the country his *Chen-Yuan Kuang Li Fang* (Valuable Prescriptions of the Chen-Yuan Emperor).

This systematisation, applied to plant and animal drugs on the one hand and

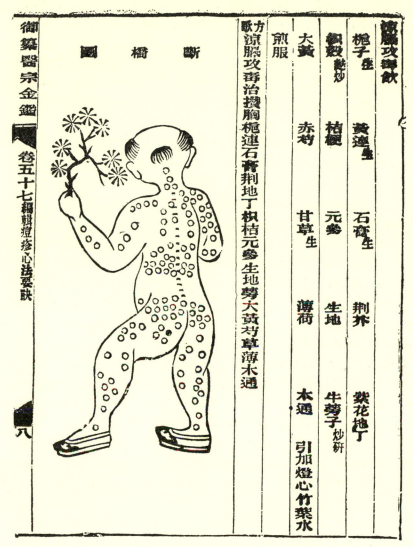

Fig. 84. A form of smallpox in a child; a page from the *I Tsung Chin Chien* (Golden Mirror of Medicine), +1743.

to prescriptions on the other, was extended to diseases by the beginning of the +7th century, for then it was that Chhao Yuan-Fang produced, about +610, his great work, the *Chhao shih Chu Ping Yuan Hou Lun* (Mr Chhao's Systematic Treatise on Diseases and their Aetiology). The interest of this large treatise is that it made a systematic classification of pathological states according to the ideas of the time, without giving any attention to therapeutic methods. It was

281

thus essentially a natural history of disease, and this was a thousand years earlier than the time of Felix Platter, Sydenham and Morgagni. One cannot but feel that the bureaucratic mentality of 'pigeon-holing', and routing things 'through the right channels', had something to do with this early appearance of systematisation in medical science (cf. Fig. 84). Indeed, the classificatory sciences as a whole were strong in traditional China, and the very word for science itself in modern Chinese, *kho hsüeh*, adopted in the 19th century, means nothing other than 'classification knowledge'. Of course the bureaucratic world-outlook affected many other things besides medicine and, as we have shown elsewhere, it is in China that one must look for the beginnings of filing and indexing systems, the differentiation of texts by different coloured inks, and the filling up of forms.

RELIGION AND MEDICINE IN CHINESE CULTURE

This brings us to the influence of the religious systems of China upon medicine. As is generally known, the three great religious systems or doctrines, the San Chiao, were Confucianism, Taoism and Buddhism. Only the first two were autochthonous, for the latter came in after the Han from India. The thought of these religious philosophies affected all aspects of medicine, and they must have influenced entry into the profession. Although a great number of medical men throughout the Middle Ages in China were trained at the government's expense, and often carried out administrative government functions afterwards, even rising to the ranks of Imperial Physicians (Thai I), there must always have been a host of auxiliary practitioners resorted to by the poor, whose knowledge had been obtained by the apprenticeship system. There can be no doubt that there was a tendency for physicians to come from the families of medical men, a process which might extend over several generations; indeed, a famous text contemporary with Confucius himself (early in the − 5th century) has been interpreted to say that one should not take the medicine of a physician whose family had not been physicians for three generations. From what we have already said it is clear that the class-structure in medieval China was quite different from that of Europe, because of the non-hereditary bureaucracy of the scholar-gentry. Social mobility was great, families rose into this estate, and sank out of it, within a few generations. The medical profession, as we have emphasised, was not wholly looked down upon after its early beginnings, for as the centuries went by more and more Confucian scholars tended to enter it. One interesting reason why men of scholarly families tended to take up medicine was because of the duty enjoined by Confucian filial piety of attending upon their parents. This it was for example which made Wang Thao, one of the greatest medical writers of the Thang, embark upon those studies which issued out in the *Wai Thai Pi Yao* (Important Family Practice of a Frontier Official) produced in +752. There are many examples of a like kind. Cases are also known of men who became physicians on account of the challenge of an illness from which they themselves suffered.

We must not forget here the role played by Buddhist compassion. The somewhat forbidding aspect of Buddhism which may be epitomised in the word *śūnya* or emptiness, i.e. utter disillusionment with this world, and the conviction of the necessity of escaping from the wheel of rebirths, was always modified in all varieties of Buddhism by a limitless compassion for all created beings, which may be epitomised in the word *karuṇā*. Thus it came about that no Buddhist abbey was likely to be without its medical specialists, and for many centuries, as we have seen, the Buddhists were active in the foundation and maintenance of hospitals, orphanages, etc. The Taoists also participated in this movement, because as an organised religion Taoism tended more and more to imitate Buddhist practices. But they were not so important in the field of medical organisation.

The profound influence of Taoism on Chinese medicine was exerted in quite a different direction. At an earlier stage we had occasion to speak about the primitive shamans of Chinese society, the *wu*, and there can be no doubt that Taoist philosophy and religion took its origin from a kind of alliance between these ancient magicians and those Chinese philosophers who, in ancient times, believed that the study of Nature was more important for man than the administration of human society, upon which the Confucians so much prided themselves. At the heart of ancient Taoism there was an artisanal element, for both the wizards and the philosophers were convinced that important and useful things could be achieved by using one's hands; they did not participate in the mentality of the Confucian scholar-adminstrator who sat on high in his tribunal issuing orders and never employing his hands except in reading and writing. This is why it came about that wherever in ancient China one finds the sprouts of any of the natural sciences the Taoists are sure to be involved. The *fang shih* or 'gentlemen possessing magical recipes' were certainly Taoist, and they worked in all kinds of directions as star-clerks and weather-forecasters, men of farm-lore and wort-cunning, irrigators and bridge builders, architects and decorators, but above all alchemists.

Indeed the beginning of all alchemy rests with them if we define it, as surely we should, as the combination of macrobiotics and aurifaction.

These words are a little unusual but they are carefully chosen. The ancient Alexandrian proto-chemists in the West were aurifictors, i.e. they believed that they could imitate gold, not that they could make it from other substances, and though they had a mystical or spiritual side to their endeavours, it was not a predominating one.[1] Macrobiotics, on the other hand, is a convenient word for the belief that it is possible to prepare, with the aid of botany, mineralogy, chemistry and aurifaction, actual substances, drugs or elixirs, which will prolong life, giving longevity (*shou*), or material immortality (*pu ssu*). Similarly, aurifaction is the belief that it is possible to make gold from other quite different substances, notably the ignoble metals.[2] These two ideas came together first in the minds of the Chinese alchemists from the time of Tsou Yen in the −4th century onwards; and Europe had no alchemy in the strict sense until this combination had made its way from China through the Arab culture-area to the West. Hence Chinese alchemy (*lien tan shu*) had been, as it were, iatro-chemistry, almost from the first, and many of the most important physicians and medical writers in Chinese history were wholly or partly Taoist. One need only mention Ko Hung about +300 and the great physician Sun Ssu-Mo of the +6th century. There was never any prejudice against the use of mineral drugs in China, such as existed long in Europe, and indeed the Chinese went to the other extreme, preparing all kinds of

[1] [The ancient Fertile Crescent and the Mediterranean basin had plenty of natural metallic gold, and their inhabitants knew how to purify it, identify it, and estimate it by cupellation. This may have been the reason why the Graeco-Egyptians were aurifictors, interested in making counterfeit, or cheaper substitute, gold and gilding. They never concerned themselves with the attainment of longevity or material immortality, though they did analogise the aurifictive processes with advances of the individual soul in spiritual perfection.]

[2] [The Chinese proto-chemists seem to have been from the beginning aurifactors, possibly because China was relatively poorer in natural gold (at least before the time of Wang Mang in the −1st century). They were prominent enough by −144 to justify an imperial edict against illegal coining and the making of 'false yellow gold'; and if some of them had no other interests they too were not alchemists in our sense. But only a few decades later, by −133, when Li Shao-Chün was urging the emperor to support his researches, and −120, when Liu An's group of naturalists was compiling the *Huai Nan Tzu* book, the connection between aurifaction and longevity–immortality (probably originating from Tsou Yen's earlier school) is clearly recognisable. That was the beginning of all 'alchemy' properly so called. In the next century (−60 to −56) Liu Hsiang failed after years of work to attain the alchemical aim, but by +20 the idea of the 'philosopher's stone', an essential ingredient for aurifaction (and argentifaction), makes its appearance for the first time in a story of Huan Than's. By +300 much systematisation has taken place in the writings of Ko Hung. Transmitted to the Arabic world by about +700, Chinese *lien tan shu* set the definitive alchemical style which lasted in European culture from *c.* +1200 till the time of Boyle, Newton and Lavoisier, and gave rise in all three civilisations to a wealth of discoveries in chemistry and chemical technique. Thus Chinese alchemy long antedates Arabic alchemy and even Hellenistic 'alchemy'; a fact which enhances the probability that the very etymological origin of the root 'Chem-' lies as far away as China (cf. p. 77 above).]

dangerous elixirs containing metallic elements which must have caused a great deal of harm.[1] The object of the devout Taoist was to transform himself by all kinds of techniques, not only alchemical and pharmaceutical but also dietetic, respiratory, meditational, sexual and heliotherapeutic, into a *hsien*, in other words an Immortal, purified, ethereal and free, who could spend the rest of eternity wandering as a wraith through the mountains and forests to enjoy the beauty of Nature without end.[2] These are the beings that one can discern, tiny against the immensity of the landscape, flitting across remote ravines in many beautiful Chinese paintings.

As time went by the hope of developing into an Immortal receded somewhat, and from the Sung onwards alchemy shaded imperceptibly into iatro-chemistry. What Chinese iatro-chemistry was capable of can be seen by the extraordinary fact, recently discovered, that the medieval Chinese chemists succeeded in preparing mixtures of androgens and oestrogens in a relatively purified crystalline form and employing them in therapy for many hypo-gonadic conditions. To understand this we have to say a word about the Chinese ideas on sexual endocrinology.[3] This is highly relevant, for sexology and special sexual practices had always been one of the Taoist methods designed to attain material immortality. The importance of the integrity of the sex glands was soon noticed, and the interest of Chinese physicians and Taoist naturalists awakened very early by all the phenomena of hermaphroditism. Already in + 80 the great sceptical naturalist Wang Chhung had an enlightened discussion on the phenomenon of sex-reversal, and it is possible to compile from the dynastic histories records of a great many such cases during the ages. From the + 13th century onwards preparations of testicular tissue taken from various animals were administered for conditions for which androgens would be prescribed today. It is rather striking that the use of human placenta and the placentas of animals was prominent in medieval Chinese opotherapy; first mentioned in the + 8th century but common after the + 14th. The most extraordinary development however in China was the preparation of sex-hormones from urine.

The origins of urinary therapy go far back into ancient Taoism. The Taoists had a philosophical and magico-scientific attitude to sex rather than an ascetic one in the ordinary sense, and in the lives of adepts who lived in the − 3rd century one finds references to the effects of ingestion of urine on sexual health and activity. There can be no doubt that as time went on the theory arose that if urine had

---

[1] [See p. 316 below.]   [2] [Cf. p. 338 below.]   [3] [Cf. p. 294 below.]

valuable properties it owed this to the fact that it was 'of the same category as' (*thung lei*) the blood. More than one medieval writer was clearly of the opinion that even the yellow colour of the urine was related to the red colour of the blood— and they were certainly not wrong. If then the organs were contributing each some valuable constituent to the blood, it was not impossible to hope that some of these useful properties might appear in the urine.

Thus it came about that the iatro-chemists started with large amounts of urine from adults or adolescents of each sex; almost as a pharmaceutical factory might today.[1] The simplest, and probably the oldest, process consisted in evaporating the urine to dryness, including the steroid glucuronides and sulphates with much else. But most of the procedures embodied various precipitations to obtain a colourless and odourless product. One used calcium sulphate, which would have assisted protein precipitation, and therefore that of the conjugates also; another made use of the saponins from the soap-bean tree. The precipitate was afterwards extracted with boiling water so that all the steroids carried down upon the protein were probably released as it denatured, while those combined with the saponins remained insoluble. The striking feature of all the methods is that they ended with a crystalline sublimate; and in fact it is true that the urinary steroid sex-hormones do sublime unchanged in air within a certain temperature range. The end-products of the iatro-chemists were undoubtedly complex mixtures of many different compounds, depending on the fractionation procedures and the original source material. But it is interesting that some of the methods specifically direct that the urine of male and female subjects should be worked up separately, as also from donors of different ages, and in certain cases there are directions about the mixing of the final products in varying proportions. The texts which describe these fascinating procedures date all the way from the +11th to the +16th century. The 'Society of Chymical Physitions' in +17th-century England would indeed have been surprised to know how strong Chinese medicine had been on the iatro-chemical side, and there can be no question that this influence was the fruit of its association with philosophical and religious Taoism.

In connection with the possible influence of religious systems upon medical science we ought perhaps to take up a very different matter, namely, the question of the mental health of the mass of the people in the culture. This opens many wide perspectives. In the absence of adequate statistical analyses we can only give our impression that in traditional and indeed in contemporary Chinese society,

---

[1] [For further information on this, see p. 312 below.]

while the incidence of psychoses is about the same as in the West, that of the neurotic conditions is considerably less. The incidence of suicide may have been about the same in the past, but for different reasons. There is much here that needs further thought and investigation, but it is generally agreed that none of the three Chinese religions gave rise to a sense of sin and guilt as Christianity did in the West. Perhaps China was a 'shame-society' rather than a 'sin-society'. Other facts are interesting in connection with the low incidence of neurosis, e.g. the general acceptance of Nature and natural phenomena inculcated by Taoism, and the extreme permissiveness of Chinese parents in the house-training and home life of young children. If Chinese mentality was on the whole better balanced than that of the West, this was in spite of great uncertainty of life. Since capitalism did not spontaneously develop in China, and there was no bourgeois revolution, bourgeois policed society did not develop either, and even as late as the end of the 19th century public life could be quite dangerously at the mercy of bandits, bullies, loafers, corrupt magistrates and family tyrants. We could not dare to follow the sociological avenues opened up any further here except to say that the universal squeeze, graft and corruption, complained of by the 'Old China Hands' in the last century, was simply the way in which the bureaucratic medieval society had always worked—it only seems strange because Western society, having passed through the stage of 'serving God in the counting-house', had already got away from that level some time before. Of course in making sociological comparisons between Chinese and Western society one must take all periods as well as all aspects into account, and to the credit of the Chinese side must be placed an almost total absence of persecution for the sake of religious opinion. No such phenomenon as the Holy Inquisition can be found in all Chinese history, nor was there anything corresponding to the witchcraft-mania which makes so great a blot on European history between the + 15th and + 17th centuries. Chinese psychology and psycho-therapy remains as yet a closed book to the Western world, but there are many texts available which could be drawn upon to outline it, not least some extremely interesting books of the Middle Ages and later on the interpretation of dreams. A great work remains to be done in this direction.

CHINESE-TRADITIONAL AND MODERN-WESTERN MEDICINE

We now come to the last section of this contribution, in which we have to consider the effect upon traditional Chinese society and medicine of the transition to Marxist socialism which has been accomplished in our own time by the revolution after World War II. The simplest interpretation of this revolution is that it was a recognition by the Chinese people as a whole that while there must be modernisation it need not be Westernisation, and it need not involve going through all the stages of capitalism in the Western world. The old bureaucratic feudalism could give place directly to socialism, with all that that implied, profoundly and on a vast scale, for the bettering of the lot of the people as a whole.

There has necessarily followed a tremendous demand for the improvement of the health of the people and for increased medical facilities. Since there were so few Western-trained physicians in comparison with the size of the population, there has been, since the revolution, a great 'revival' of traditional Chinese medicine. There are now many medical schools which teach it and general encouragement is given to its practice. Refresher courses have been introduced for men trained only in the traditional medicine so that they may play their part in modern health measures, while at the same time the 'modern' physicians have been persuaded to take traditional Chinese medicine seriously. Today the traditional medical men are working side by side with the others in full cooperation. A fusion of their techniques is going to happen more and more, giving rise to a medical science which is truly modern and oecumenical and not qualifiedly 'modern-Western'.

All this is being done under the aegis of a conviction that an integration of the Chinese and the 'Western' or 'modern-Western' systems of medicine must emerge. The mathematics and astronomy of China and the West fused into one science quickly in the +17th century, but the fusion of other sciences, such as chemistry and botany, took much longer.[1] One has to realise that fusion cannot yet be said to have occurred with the highly complicated sciences of the living organism in health and disease, and it is to be expected that such a fusion may take more than a generation. Actually what is at issue is the comprehension in modern terms of the quasi-empirical practices which grew up in China through the centuries. Since the theories of traditional Chinese medicine always remained relatively 'primitive' and 'pre-Renaissance' in type, there cannot be much future

[1] [See further on pp. 396 ff. below.]

for them except in so far as they may be reinterpreted in modern terms. There is danger here for the historian of medicine, who must be careful not to read too much into the ancient theoretical formulations, but at the same time must be careful to avoid making them seem quaint, archaic or senseless. This situation in China contrasts sharply with that in Ceylon and India, as we have also had occasion ourselves to see. Ayurvedic medicine in those countries has some similarity with Chinese traditional medicine, though less original and peculiar in its methods, and laying more emphasis on materia medica. But in South Asia unfortunately there is a positive hostility to any contacts with modern-Western medicine. In Ceylon the Ayurvedic physicians have joined forces with the Buddhist monks in trying to gain an influence over the university in a movement of pure traditionalism. Thus one can see how much Chinese tolerance has to teach the rest of Asia.

The system most fundamentally characteristic of therapy in Chinese medicine is of course acupuncture. It is a system which has been in constant use throughout the Chinese culture-area for some two and a half thousand years; and the labours of hundreds of learned and devoted men through the centuries have turned it into a very highly systematised doctrine and practice. Briefly, this system, as is well known, consists of a large number of points on the surface of the body (we call them *loci*) in which needles of varying length and thickness are inserted by the physician in different specified manners (Figs. 85, 86, pl.). The oldest catalogue of these points occurs in that part of the *Huang Ti Nei Ching* which is called the *Ling Shu*. In the − 2nd century the *loci* (*hsüeh*) so recognised amounted to 360, possibly because of a fancied equivalence with the number of bones in the body, possibly in its turn connected with a rough estimate of days in the year. Each point has a distinctive technical name which has developed through the ages, but there is a good deal of synonymy so that the total number of *loci* which have been identified as distinct names is about 650. At the present day about 450 are recognised and might be said to be in current use, but those most commonly employed are much fewer in number, not exceeding perhaps about 100. Before the Sung period (+ 11th century) we know the titles of some eighty books on the system of acupuncture *loci* but the majority of these were lost. We have already taken note of the *Chia I Ching* by Huangfu Mi written about + 280.

If this were the whole story the system would be indeed empirical but it is far from that. The points became connected with each other in a complicated

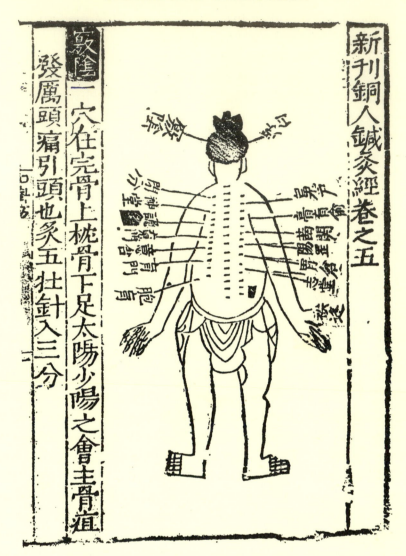

Fig. 85. Acupuncture points on the back, a page from the *Thung Jen Chen Chiu Ching*
(Manual of Acupuncture) by Wang Wei-I, +1026.

reticulate system reminiscent of a map of the London underground railways.
These connections are known as *ching* (cardinal tracts) and *lo* (decumane tracts).
The analogy can be carried somewhat further because the *ching* and *lo* are indeed
invisible, like the principal blood vessels and nerves, running along under the
surface of the city. It is as if one had two transport-system companies with exchange points for the public well defined at the junctions where they meet. We

290

call these junctions (*hui hsüeh*) anastomotic *loci*, and there can be no doubt that the names of the points, at least many of them, ante-dated the system of the *ching* (tracts), for we find named points mentioned in the discussions recorded from Ho the Physician in − 540, and more of them in the case histories of Shunyü I. It is the *Ling Shu* (systematised in the *Chia I Ching*) which connects the *loci* into the tracts and adds correlations with the Yin–Yang forces and the six *chhi* (*pneumata*). There is no doubt that in the *ching–lo* system we have to deal with a very ancient conception of a traffic nexus with a network of trunk and secondary channels and their smaller branches. It seems as if from the beginning these were thought of in hydraulic engineering terms, for there are greater and lesser reservoirs of *chhi*. We are thus in the presence of an important doctrine arising from the idea of the microcosm, the body of man representing the macrocosm in little. And the basic idea of circulation, which originates unmistakably in the Former Han period, may well be derived from a recognition of the meteorological water-cycle—the exhalations of the earth rising into the clouds and falling again as rain.

The question of the origin of the whole system is surely one of extraordinary interest. There must have been close observation of symptoms, especially pain and its relief by various methods. But we suspect that the profound conviction of the organic unity of the body as a whole, which was reflected in the acupuncture system, may have arisen challengingly out of the phenomenon of referred pain. Perhaps some passages in the ancient Chinese texts not yet noted will justify re-course to this as part at least of the explanation. The relation of transitory pain in the extremities or trunk with passing malfunctions of the viscera is so common an observation of every-day physio-pathology that it may well have struck the ancient Chinese physicians with particular force. The sudden relief of cardialgia or flatulence may be preceded by quite severe though very transitory pain in remote parts of the body, and we expect that a close examination of referred pain as it is understood in modern physiology may throw considerable light on the genesis of the Chinese system. One must also remember the zones of the skin in mammals which are related through the sympathetic nervous system with specific viscera; first investigated by Head and named after him. Besides all this there must of course have been a saecular accumulation of clinical experience which convinced the Chinese medical profession indubitably of the efficacy of acupuncture.

Today there are dozens of laboratories in China and Japan actively working with modern methods of a physiological and biochemical character to elucidate what happens in acupuncture. One suggestion now under test has been that the action of the needles may stimulate the production of antibodies by the reticulo-endothelial system. It had always been a matter of surprise for Western biologists that the acupuncturists claim their treatment to be effective, at least in some degree, not only in diseases such as sciatica or rheumatism where no treatment in any part of the world can be considered very successful, but also in cases of infectious disease where an external causative agent is fully recognised. For example, it was difficult to believe that in such an entity as typhoid fever acupuncture could be effective; nevertheless that was the claim of the traditional physicians. However, if the reticulo-endothelial system could be stimulated to produce antibodies in larger quantity, possibly by indirect stimulation through the autonomic nervous system, that could explain the results. Alternatively, there may be a neuro-secretory effect mediated through the autonomic and sympathetic systems upon the suprarenal cortex inducing a rise in cortisone production. Again, there may be a neuro-secretory influence upon the pituitary gland. There is indeed much to be done.

We have already mentioned the importance of the detailed study of the ancient and medieval Chinese system of diagnosis (*liu ching pien chêng*, 'differentiating and diagnosing in accordance with the six *ching*'),[1] never yet properly expounded in a Western language. The term *ching* has been known in the West only as the name for the linear arrays of acupuncture points on the surface of the human body. But it has a much deeper meaning than this, denoting a basic physiological conception in ancient Chinese medicine founded on the theory of the Two Forces (Yin and Yang) and the Five Elements, in which six patterns of physiological function and pathological dysfunction were recognised. During the course of a disease these patterns were affected in diverse succession according to its causative factors. Traditional Chinese medicine classically recognised three fundamental causative factors (*san yin*) in disease: (*a*) external agents [climatic, infective, contagious, i.e. *wai*, a Yang group], (*b*) internal dysfunction[2] or abnormal *krasis* [*nei*, a Yin effect], and (*c*) traumatic and accidental injuries including war wounds [*pu nei wai*, partly Yin and partly Yang]. An important component of the second class was constituted by congenital susceptibility factors (*thai tu*, literally 'morbific force latent in embryonic life'). In so far as these factors were regarded as important in epidemic disease they also participated in the property of being *pu nei wai*, partly Yin and partly Yang. In accordance with this, three great classes

---

[1] [Cf. p. 271 above.]   [2] [Cf. p. 412 below.]

of illness, Yang, Yin, and mixed, were envisaged, and, as we have already emphasised, it was always recognised that treatment must aim at the syndrome as a whole rather than at any particular symptoms, taking very great account of the individual's particular constitution.

This is as far as we can go on the present occasion in our account of Chinese medical ideology and what ought to be done about it. One might feel that if any type case was needed to demonstrate the moulding of medicine by the culture in which it grew up, Chinese medicine would be such a case. But, on second thoughts, is there any reason for regarding it as more 'culture-bound' than Western medicine? To think of the latter as self-evidently universal in application may be an illusion commonly entertained because most of us happen to have been born within that occidental Semitic–Hellenistic culture which was destined by a series of historical accidents to give rise to specifically modern science in the later stages of the Renaissance. Western medicine is only modern because it is based upon the assured results of modern scientific physiology and pathology, in a way which the traditional medicines of the Asian civilisations are not; but it will not be truly and oecumenically modern until it has subsumed all the clinical experience, special techniques, and theoretical insights, achieved in the non-European medical systems. Then will have occurred that fusion of Eastern and Western medicine which we have referred to above. In the last resort, all medical systems have been 'culture-bound'; and modern medicine is only rising above this in so far as it can partake of the universality of modern mathematised natural science. Everything that the Asian civilisations can contribute must and will, in due course, be translated into these absolutely international terms. Only so will medical science be able to free itself from connections with particular cultures and be able to minister universally to a united mankind.

15

PROTO-ENDOCRINOLOGY IN MEDIEVAL CHINA[1]

[1966]

INTRODUCTION

IN this paper we offer a brief account of the pre-history of endocrinology as it can be found in ancient and medieval China. The precise conception of specific chemical substances secreted into the blood-stream by this or that organ is of course an entirely modern one, but many phenomena now known to depend upon the action of hormones were studied and made use of for the benefit of mankind by the physicians and natural philosophers of ancient and medieval China. Their study of Nature was pursued with great single-mindedness over many centuries, and the insight which they showed was often remarkable. It is better perhaps to refer to their discoveries and successes as quasi-empirical rather than empirical in the strictest sense, for they worked with many theories, not of course of modern type like those of science since the Renaissance, but nevertheless of considerable heuristic value.

The first thing to point out is that some of these physiological and pathological theories were remarkably congruent with the basic conceptions of modern endocrinology. We intend elsewhere to give a thorough account of the fundamental theories of the living organism as they developed in China before the period of modern science, and our conviction is that this theory structure was in many ways an enlightened one. First of all, the Chinese conceived of the body, with its numerous organs, as the theatre of complex changes in which each organ had interactions with all the others and specific effects upon them. The theory of the Five Elements (not four as in Ancient Greece)—Metal, Wood, Water, Fire, and Earth (Chin, Mu, Shui, Huo, Thu)—was systematised first in the time of Tsou Yen in the Warring States period (– 4th century). By the Han period (– 2nd century onwards) the Five Viscera (*Wu Tsang*) had already been analogised with the Five Elements (*Wu Hsing*). The idea of five elements in itself would not have taken

[1] Reprinted from *Japanese Studies in the History of Science*, 1966, 5, 150.

294

the ancient natural philosophers very far, but it was only the basis for a set of ideas of a much more sophisticated character. The elements were regarded as affecting one another according to two main cyclical successions. In one of these, the Mutual Production Order, each element was regarded as generating the next one in a particular series. But besides this, there was also the Mutual Conquest Order, a different series, in which each element was believed to conquer or destroy one of the others in a generally accepted scheme. In this way, just as there were continual interactions between the elements in inorganic nature, so also within the living organism, within the body of man, there were continual interactions between the Five Viscera.

Besides this series of five, there was also a series of six; for standing over against the Five Viscera there were also the six other principal organs (*Liu Fu*), mostly connected with the digestive tract. To make up the balance a sixth *tsang* was added, the *hsin pao lo* (pericardium) which corresponded with the sixth *fu*. This latter was one of the most important conceptions in the traditional physiology. The sixth *fu*, the *san chiao* or Three Coctive Domains, is best regarded as three non-visible metabolic regions. Although in late times attempts were made to find a material anatomical basis for these, in principle they did not depend upon any specific anatomical entities; the three consisted of the upper region above the diaphragm, the middle region between the diaphragm and the umbilicus, and the lower region below the umbilicus. This conception was important because it led naturally to the idea of functional, as opposed to organic, disorder, a distinction most clearly appreciated by the Chinese medieval physicians. Here again the regions were regarded as very interdependent, heat and cold, Yang and Yin, Fire and Water were liable to rise or fall abnormally between these levels. The activities of each level were controlled by organs of the level concerned, so abnormalities were seen as due to specific organ defects in each particular region. This again brings one close to the conception of the failure of a particular organ to contribute its influence to the body as a whole.

Besides the Five Elements and the two groups of six organs and regions, there was also the still more fundamental distinction in Chinese natural philosophy between the two basic forces of the universe, Yin and Yang. It is generally known that these very ancient ideas of negative and positive in the universe originated first of all from the antithesis of brightness and darkness, the male and the female, etc., but in the Middle Ages the Yin and Yang were very commonly used by Chinese physicians in a sense closely resembling that which today we would speak of as inhibition and stimulation. The viscera were divided into Yin and Yang, in so far as the *tsang* were all Yin in their capacity of energy-storing organs, while the *fu* were all Yang organs which transmitted intestinal contents or other juices.

Perhaps the way in which medieval Chinese ideas proved to be most clearly adapted to endocrinological conceptions lay in the fact that the Chinese manifested a much greater circulation-mindedness than Western ancient and medieval physiologists. It has long been known that the idea of a continuous steady circulation of *chhi* (*pneuma*) and blood goes back very far in Chinese thought. The *chhi* circulated through special paths in the interstitial tissue known as tracts (*ching*), while the blood circulated in the vessels (*mo*). Interchange between them occurred at special junctions. It is not of course correct to claim that the precise conclusions of William Harvey were in any way anticipated by ancient and medieval Chinese physiology, but the notion of continuous steady circulation is undoubtedly there, and we find nothing resembling the *idée fixe* of 'tides' which was so characteristic of Galenic and other physiology in the West. With the concept of the circulation therefore clearly in mind, it becomes easy to see how the old Chinese physicians could have imagined each organ or viscus contributing its specific influence to the *chhi* and the blood and thus affecting the rest of the body. Indeed, the distinction between arterial and venous blood was to some extent appreciated already in the *Huang Ti Nei Ching, Ling Shu* (The Yellow Emperor's Manual of Corporeal [Medicine]; the Vital Axis), a text much at any rate of which belongs to the − 1st century.

In 1849 A. A. Berthold carried out his classical experiments in which the testes were transplanted in birds from their normal position to the abdominal cavity. He found that there they became vascularised but not innervated, and that they continued to exert their effects on the secondary sexual characters, so that the caponised animals remained perfectly male in all respects. This was perhaps the origin of all modern endocrinology, and it would be interesting to know what were the reasons which impelled Berthold to make the experiment. Probably they arose from the old Greek theory of pangenesis, in which it was thought that every organ in the body contributed some specific particles to form the embryo. As interpreted by Maupertuis, Buffon and de Bordeu in the 18th century, this theory supposed that every organ (even if it had an obviously external secretion also) contributed something to the blood-stream. Berthold's experiment followed from this. Knowing the relation of the testes to the secondary sexual characteristics, he probably thought that it might be feasible to demonstrate their action as being mediated by the blood-stream alone; and so it turned out. From what has already been said, it is evident that the Chinese physiology of the Middle Ages could equally well have encouraged such an experiment. As we shall see later, the Chinese were very familiar with eunuchism, spontaneous sex reversals, inter-sexes, and so on. Perhaps the fact that no such experiment as Berthold's was made in China before the advent of modern science might be due in part to the fact that Chinese medicine was relatively weak on the side of surgery.

One circumstance in particular shows that the Chinese were very conscious of some 'virtues', or precious qualities, circulating in the blood. This was the remarkable use which they made of urine for the preparation of materials having androgenic and oestrogenic activity. According to the traditional doctrine, the nutrient essentials (*ching chhi*) were divided during digestion into the lighter constituents (*chhing chê*) and the grosser constituents (*cho chê*). Then the grosser part of the lighter fraction was excreted as urine, while the lighter part of the grosser fraction went to form the various secretions. The effect of this view was that the urine was considered to be 'of the same category as' (*thung lei*) the blood. Here we come upon another feature of Chinese natural philosophy, the doctrine of categories. Elsewhere we have studied this in its relevance to medieval alchemy.[1] The categories (*lei*) formed a cross-classification additional to the fundamental Yin and Yang division. Reaction would only take place, it was thought (or in some cases, would not take place), if the categories were the same. Since the urine was considered as belonging to the same category as the blood, it was possible to visualise that by the aid of iatro-chemical methods it might be possible to separate from the urine some of the valuable properties which were circulating in the blood. Presumably the handling of blood, with its high protein content, was too difficult an undertaking for the iatro-chemists of the Middle Ages, but the urine was a much more encouraging proposition. It is astonishing to reflect how far this intuition has been justified by the assured findings of modern biochemical science, and the use of urine by the Chinese in their medicine was not at all a superstitious procedure as has no doubt often been thought.

We shall now proceed to discuss three chief subjects of interest in the present context. First, the thyroid and its deficiency, secondly, the problem of diabetes mellitus, and thirdly, the endocrinology of sex.

[1] [Ho Ping-Yü & J. Needham, *Journ. Warburg and Courtauld Institutes*, 1959, **22**, 173.]

THYROID FUNCTION

The first endocrine function which we may take up is that of the thyroid gland. Let us sketch the history of our knowledge of this function in the West. Goitre was noted by Juvenal, Pliny and other Latin authors at least as early as the + 1st century, and the gland itself was described by Galen in the + 2nd. After that there was no further advance until Roger of Palermo, about + 1180, recommended the use of the ashes of seaweed and sponges in cases of goitre—an empirical discovery (whether indigenous to Europe or not we cannot say) which continued in use and mentioned by medical writers long afterwards, as by Russell in + 1755. Giulio Casserio, about + 1600, one of Harvey's teachers, noticed the significant fact that the gland had no duct. From at least the beginning of the 19th century onwards there was a tendency to blame environment for the incidence of goitre, and Chatin in 1860 proved clearly that the disease was correlated with a lack of iodine in the soil and water. In 1884 Schiff implanted thyroid gland into the abdomen of a thyroidectomised animal, and found that it would continue in good health. A dozen years later Baumann (1896) discovered the presence of iodine in the gland, and about the same time Möbius (1891) ascribed exophthalmic goitre to the gland's hyperactivity. This may suffice as a sketch of the progress of knowledge in the West.

The general outlook of the Hippocratic 'Airs, Waters and Places' was very prevalent in ancient China. Elsewhere we have given a translation of a passage about the properties of different kinds of waters which occurs in the *Kuan Tzu* book, and may be placed as early as the − 5th century.[1] In the *Lü Shih Chhun Chhiu* of − 239 we read that

in places where there is too much light (*chhing*) water there is much baldness and goitre; in places where there is too much heavy (*chung*) water there will be found a high incidence of swellings and dropsies which make it hard for people to walk. Where sweet (*kan*) water abounds there will be healthy people, but if there is much acrid (*hsin*) water skin infections will be found. Finally, where there are many bitter (*khu*) waters, there will be many people with ulcers and rickets.

This is not by any means the earliest reference to goitre under its standard name (*ying*), for, in the *Chuang Tzu* book of the − 4th century, we hear of a man with a goitre as big as an earthenware jar who acquired, in spite of it, the favour of one of the feudal lords.

[1] [*SCC*, vol. 2, p. 42.]

The classical doctrines about goitre may be studied in the *Chu Ping Yuan Hou Lun*, that great system of pathology written by Chhao Yuan-Fang in +607, as also in the *Wai Thai Pi Yao* of Wang Thao, written in +752. The opinion then was that there were two causes for the appearance of goitre. On the one hand, emotional upsets which caused the *chhi* (*pneuma*) to congeal, giving rise to hard swellings. But on the other hand, the more usual cause was the drinking of *sha shui*, water from sources where sand existed in quantity. This permitted the *chhi* of the sand to enter the blood and to strike at the epiglottal region. Among various types of neck swelling, goitre was clearly described and differentiated, in that it could be comparatively simply removed by surgical operation. A manual on the preservation of health entitled *Yang Sêng Fang*, or *Yang Sêng Ching*, written by Shangkuan I, probably in the +5th century, says that particularly in mountains where water comes from springs or streams out of black earth, goitre is particularly liable to arise. Thus the association of goitre with certain mountain regions was widespread in Chinese medicine from at least the +5th century onwards.

The use of seaweed, with its high iodine content, as the chief curative agent for goitre, goes back a very long way in Chinese therapy. Out of thirty-six relevant prescriptions in the *Wai Thai Pi Yao*, no less than twenty-seven contain seaweed and other marine products. This in itself would be sufficient to show that the use of the different kinds of seaweeds for the cure of goitre antedates the discovery in Europe. But in fact the Chinese utilisation of the marine products for this purpose goes back much further still. The *Shen Nung Pên Tshao*, earliest of all the Chinese pharmacopoeias, a work which we may place in the −1st century, i.e. the Han period, lists gulf weed (*hai tsao*), identified as *Sargassum siliquastrum*, among its medicinal plants, and it says distinctly that it is used to cure goitre. The *Erh Ya* dictionary, which must be placed rather earlier—about the −3rd century—distinguishes four main kinds of seaweed, but it does not say anything about their curative properties. By the time one reaches the Great Pharmacopoeia (*Pên Tshao Kang Mu*) of +1596, at least a dozen species had been recognised. The great alchemist, Ko Hung, in his *Chou Hou Pei Chi Fang*, written about +340, recommends an alcoholic extract (tincture) of seaweed for goitre; he uses *hai tsao* and he extracts with clear wine the comminuted plant contained in a silk bag. Apparently both the first and second extracts and the residue itself were used. Ko Hung also recommended the use of the sweet tangle (*khun pu*, *Laminaria japonica*), dried and pulverised, then taken in pill form.

After Ko Hung, we hear of an interesting book by a monastic physician of the late + 5th century, the *Shen Shih Liao Ying Fang* (Goitre Prescriptions of the Abbot Shen), who adds nine more items, including the powdered shells of lamellibranch molluscs, no doubt as a source of calcium, as well as iodine.

When we have so much information from the + 4th to the + 8th century, it is needless to give further evidence from later periods of Chinese history. But one might mention the *Yin Shan Chêng Yao*, written by Hu Ssu-Hui about + 1330, the book which shows such a remarkable empirical knowledge of deficiency diseases due to lack of vitamins. Here also he says that seaweeds will effectively cure goitre. All the later books repeat this well-tried therapeutic measure. In the 19th century Chinese naturalists enlarged still further their knowledge of marine algae, and we find, for example, in 1886 a monograph entitled *Hai Tsho I Lu* (Notes on a Hundred and One Marine Products) written by Kuo Po-Tshang in Fukien Province. Here he describes fourteen different kinds of seaweed and recommends a number of them in cases of goitre. One especially, not yet identified scientifically, was called *tho tshai* (ship's rudder herb) because it was found on the rudders of sea-going ships. This plant had already been treated of in Phan Chih-Hêng's monograph on fungi and algae entitled *Kuang Chün Phu*, written in + 1550.

When we survey the date of introduction of iodine-containing marine plants to the occidental pharmacopoeia we encounter a curious and perhaps significant coincidence. If indeed their first use in Europe was towards the end of the + 12th century, transmission of the knowledge from China westwards presents itself as very likely, since we know that the practice had been widespread there in the + 8th, clearly recommended in the + 4th, and already discovered in the + 1st. But not only this, the date of + 1180 for Roger's Salernitan *Practica Chirurgiae* is singularly suspicious, for it was precisely at that moment that Europeans first knew of the magnetic compass, the stern-post rudder, the windmill and paper-making—all techniques with far-reaching consequences, and all demonstrably derived from the Chinese and Iranian culture-areas. Surely the sea-captains of Cathay, Islam and Frankistan had been yarning at length in port somewhere in the Indo-Iranian seas.

The thyroid gland itself was well known to medieval Chinese anatomists. About + 1475, Wang Hsi in his *I Lin Chi Yao* says that the gland lies in front of the larynx and looks like a lump of flesh about the size of a date (a 'jujube', the Chinese date), flattish and of a pink colour. But he says much more than this, for he tells us that for goitre the glands of pigs, sheep, and other animals should be desiccated over a tile and powdered, then taken with wine every night. One of his prescriptions speaks of the air-drying of fifty pigs' thyroid glands and then the consumption of the powder with cold wine by the goitrous patient. Another book of about the same time, the *Hsing Lin Chai Yao* by Wang Ying, gives similar prescriptions; seven pigs' thyroids extracted in wine, evaporated down and taken up in dew water before administration. Thyroid organotherapy was thus in China contemporary with Thomas Linacre.

In the Great Pharmacopoeia of Li Shih-Chen at the end of the + 16th century, Li says that it does not matter what animal the thyroid glands are taken from, because all are of the same function in the various mammals (*i shu lei chih i*). He again distinguishes goitre from other types of neck swelling, and says that only the *ying* type, which moves about, can be cured by these prescriptions. From his text it can be seen that by the end of the + 16th century the glands of pigs, sheep, water buffalo and sika deer were all in use.

The distinction between benign or malignant growths on the one hand, and true goitre on the other, can also be found a thousand years earlier. The physician Tshui Chih-Thi, who flourished about + 650, is quoted by the *Wai Thai Pi Yao* as making a clear distinction between solid neck swellings which could not be cured and movable ones which could.

It seems moreover that thyroid organotherapy in Chinese medicine also goes back that far. One of Tshui's contemporaries, Chen Chhüan, who died in + 643 (early in the Thang dynasty), wrote a book entitled *Ku Chin Lu Yen Fang* (Old and New Tried and Tested Prescriptions), and among these are to be found three which have the thyroid gland in them, in this case taken from gelded rams. For example, he directs that one hundred thyroid glands are to be washed in warm water, the fat removed and the glands dried, then to be chopped up and to be mixed with dates (jujubes) into pills for immediate use. Another prescription takes a single thyroid from the sheep, removes the fat, and directs that it should be held in the mouth raw, and sucked by the goitrous patient to obtain as much juice from it as possible, afterwards eaten. In + 723 the Thang emperor Hsüan Tsung (Li Lung-Chi) issued a book of prescriptions entitled *Kuang Chi Fang*, and here we find the combination of seaweed (*hai tsao* and *khun pu*), the lamellibranch mollusc (*hai ko*) shell powder, and sheep's thyroid. This combined therapy did not however begin with the emperor, for in about + 650 the great physician Sun Ssu-Mo in his *Chhien Chin Yao Fang* had also recommended the combined use of seaweeds, powdered mollusc shells and thyroid gland.

Thus the use of the thyroid for the curing of thyroid insufficiency can be traced back in China, not only to the time of Thomas Linacre, but to the time of Alexander of Tralles. This fact has been appreciated to some extent by Western writers such as Regnault and Brooks, but, so far as we know, details of the kind given here have not hitherto been available in Western literature. There can be no doubt that the success of the Chinese in this matter was due to the *thung lei* principle, that is to say, the appreciation which they had of 'things of the same category'. As we have seen, this classification of natural objects played a very large part in early medieval Chinese alchemy. In the case of the thyroid it proved particularly successful because of the fact that the affection to be cured was localised actually at the site of the organ which normally produces the hormone. A gland such as the pituitary could never have been treated in this way because of the fact that while its activities are so multifarious, it does not itself react macroscopically and obviously in hypo- or hyper-function.

How could the discovery have been made? It may perhaps seem strange that if the medieval Chinese physicians recognised goitre as thyroid enlargement, they should have been prepared to give even more of the gland to bring about the return to normality. To have realised that the hyperplasia was essentially due to an incapacity would seem to have been an impossibly good guess—yet this is just what the old Chinese medical theorists achieved. Accustomed, as always, to thinking in terms of the organism as a whole, and not being dazzled by local phenomena, however impressive, they considered goitre as a *hsü* disease (either Yang or Yin), i.e. one of fundamental insufficiency, as indeed the hypothyroidism of its advanced stages shows it to be today. What the Chinese physicians saw was essentially an abnormality; and what normality meant to them was a correct balance (or *krasis*) between Yang and Yin (excitation and inhibition). The *thung lei* (identity of categories) principle therefore naturally suggested that tissue from the normal necks (of animals) should be employed to cure diseased necks (in man). The balanced state of the healthy gland would, the physicians hoped, exert its beneficent influence on the region of the diseased gland. When in 1890 Murray and others began to administer the thyroid or its extracts for myxoedema they acted following a course of reasoning based on physiological experiments of modern type on animals. Ko Hung and Chen Chhüan did not have this basis for their ideas, but that does not mean that they were destitute of theory; on the contrary, they reasoned out the matter with extreme acuity. Thyroid organotherapy must clearly be counted a signal success of medieval Chinese medicine.

DIABETES MELLITUS

In the nature of the case, diabetes could not be mastered by medieval physicians anywhere in the world as hypothyroidism had been. The double function of the pancreas, and the fact that insulin is a protein destroyed by the external secretion of the same gland which produces it internally, made it impossible to understand the mechanism of pancreatic dysfunction until modern times. But if the Chinese physicians of the Middle Ages had no similar success in diabetic cases, their theoretical discussions of diabetes mellitus were, as we shall see, remarkably good.

The sweetness of the urine in diabetes mellitus was known to the author of the relevant part of the *Suśruta-saṃhitā* in India, but it is not quite certain whether the references to 'honeyed urine' (*madhu-meha*) go back to the earliest stratum of the corpus (*c.* + 200); they may be as late as the + 6th century. Nor is it quite clear how far this symptom was correlated with the syndrome of diabetes mellitus in all its detail. Subsequently the Arabs may have derived this knowledge. It is often said that Avicenna, early in the + 11th century, was acquainted with the sweetness of the diabetic urine, but we have not been able to confirm this statement. In the West the recognition of the sweet taste of the urine in diabetes was firmly established by Thomas Willis about + 1660 (his publication was of + 1679). The identification of the sweet substance as sugar both in urine and in blood was the work of Matthew Dobson in + 1776, and the full identification of it as glucose was due to Chevreul in 1815. The islets were first described by the histologist Langerhans in 1869, and twenty years later Mering & Minkowski carried out the classical experiment which showed that rapidly fatal diabetes was produced after extirpation of the pancreas in animals. Then, as is generally known, insulin was first isolated by Banting & Best in 1921 in usable form and crystallised five years afterwards. So much for the course of knowledge of the disease in the Western and the modern world; its conquest is an exceedingly recent achievement. We shall now show that knowledge of the sweetness of the urine was in China, though perhaps a little later than the Indian discovery, very much earlier than the knowledge of it in Europe.

The modern term for diabetes mellitus in Chinese medicine is *thang niao ping*, but the ancient terms were quite different, *hsiao kho* (dissolutive thirst) and *hsiao chung* (central coctive dissolution). There can be no possible doubt about the identification of this syndrome in the ancient medical texts. A good description of the disease was given in the *Huang Ti Nei Ching*, *Su Wên*, the first of the great

303

Chinese medical classics, dating from the −2nd or −1st century. In chapter 47 of this work, after describing the disease in some detail, the text goes on to say:

a patient suffering from this disease must have been in the habit of eating many sweet delicacies and fatty foods. Fatty foods make it difficult for the internal heat to disperse, while very sweet things give rise to obesity. Therefore the *chhi* (*pneuma*) tends to overflow and thus causes diabetes (*hsiao kho*).

Ssuma Hsiang-Ju, the eminent poet, official and road-builder, died from this disease in −117. Towards the end of the Later Han dynasty, about +190, Chang Chi, the greatest physician of his time, prescribed *Pa Wei Wan* pills for this disease.

It is curious and interesting that although the Chinese could not identify the endocrine function of the pancreas, they nevertheless regarded it as a very important organ. Li Shih-Chen in the *Pên Tshao Kang Mu* (+1596) has a long description of it. He says that the organ, named *i*, is located between the two kidneys, and although appearing like fat or flesh it is neither fat nor flesh (i.e. glandular). The old physicians gave it the name of the 'gate of life' (*ming mên*) and it was considered to be the fundamental organ (*yuan*) or starting-point of the flow in the three coctive regions (*san chiao*). The doctrine of the *san chiao* is too complicated to go into here, but its content is essentially metabolic, and it was therefore a strangely deep insight to view the pancreas as having a close connection with such processes. Li Shih-Chen goes on to say that the pancreas is larger in obese people than it is in thin or normal subjects. Its importance, he adds, is manifested in the way that the character *i* is written, for the orthography is similar to that of another word *i*, meaning the chin and jaws, but also to nourish or to rear. This indicates that the pancreas is essential for the nourishment and maintenance of life. And for longevity too, for in the classics a centenarian is one who has lived for the span of an *i* (*chü i*). This is reminiscent of the English proverb about 'digging one's grave with one's teeth...'. There is no doubt that the medieval Chinese physicians regarded the pancreas as absolutely essential for the maintenance of health and life, though they did not know exactly how it worked.

The first of all the Chinese physicians to mention the sweetness of the urine in diabetes mellitus, so far as we have been able to find out, was Chen Chhüan, who died in +643. He was the author of the *Ku Chin Lu Yen Fang* (Old and New Tried and Tested Prescriptions). This work is quoted by Wang Thao in his *Wai Thai Pi Yao* (+752) as follows:

The *Ku Chin Lu Yen Fang* says that there are three forms of diabetic affection (*hsiao kho*). In the first of these the patient suffers intense thirst, drinks copiously, and excretes large amounts of urine which contains no fat but flakes looking like rolled wheat bran (*fu phien*), and is sweet to the taste. This is diabetes (*hsiao kho ping*). In the second form the patient eats a great deal but has little thirst; the urine though frequently passed is less in quantity, and has a fatty appearance. This is *hsiao chung ping*. In the third form there is thirst but the patient cannot drink much; the lower extremities are oedematous but there is wasting of the feet, impotence and frequent urination. This is called *shen hsiao ping*. (Here) sexual activity is to be avoided. For the first form the dose of *huang lien* should be doubled, for the second the dose of *kua lou* should be doubled, for the third add 0·6 oz of *mang hsiao* (saltpetre or sodium sulphate, perhaps also nitrate) and take the *Chhien Tan Wan* (pills) until the urine has become salty as is normal. To prevent recurrence the patient should take *Hua Tshung Jung Wan* (pills).

Here the first form is quite clearly diabetes mellitus, and we are inclined to recognise in the second the polyphagic form of that disease, with associated lipuria. The third one cannot be diabetes insipidus, and we believe that Chen Chhüan was more probably describing chronic nephritis or Bright's disease.

Before the end of the + 7th century a monograph was consecrated entirely to diabetes. It was entitled *Hsiao Kho Lun*, and its author was Li Hsüan. We are strongly inclined to identify Li Hsüan with Director Li of the Bureau of Imperial Sacrifices (Li Tzhu-Pu Lang-Chung), who wrote another work about the same time dealing with the same subject. This was entitled *Chin Hsiao Chi Yao Lun* (The most important Prescriptions of a Recently Serving Official). Director Li followed Chen Chhüan closely in his account of diabetes, and was perhaps the first to discuss at length the reason for the sweetness of the urine in that disease. It might seem somewhat strange that a physician should be the Director of the Bureau of Imperial Sacrifices, but when one looks at the description of the bureaucracy in the Thang period, one finds that the Director of this Bureau was the official charged with the chief responsibility for the medical qualification examinations. Clearly therefore the Director and staff of this department of government were very likely to be physicians of eminence. Director Li (Li Lang-Chung) is quoted in the *Wai Thai Pi Yao* at length, and we give here a translation of the elegant résumé of this quotation contained in the *I Shuo* (Medical Discourses) written in + 1189 by Chang Kao. It runs as follows:

The *Chhien Chin* (*Yao Fang*) (of Sun Ssu-Mo, *c.* +655) says that in *hsiao kho* illnesses (diabetes) three things must be renounced, wine, sex and eating salted starchy cereal products; if this regimen can be observed, cure may follow without drugs. Whether or not such patients are cured, one must be on the watch for the development of large boils and carbuncles (*yung chü*); should these develop near the joints the prognosis is very bad. I myself witnessed my friend Shao Jen-Tao suffering from this disease for several years, and he died of the ulcers.

In the Thang period, Director Li of the Bureau of Sacrifices, discussing diabetes (*hsiao kho*), said: 'This disease is due to eremosis (weakness) of the renal and urinogenital system (*shen hsü*). In such cases the urine is always sweet. Many physicians do not recognise this symptom, and neither old nor new books mention it. As the *Hung Fan* (chapter of the *Shu Ching*, Historical Classic) says, the cereal foods of the farmers are precursors of sweetness; so arguing from the nature of things (*wu li thui chih*), the methods of making cakes and sweetmeats with malt, vinegar and wine mean that they all very soon turn to sweetness. Thus it is clear that after a person has eaten, (the digestion) turns all the kinds of nourishing food (lit. tastes, *tzu wei*) to sweetness (just as the fermentation makes the sugar for the cakes and the spirit for the wine). (In diabetic affections) this flows into the bladder; if the *chhi* of the renal and urinogenital system at the reins (*yao shen*) had been strong enough, the sweetness would have turned into the flame of life (*chen huo*) and would have distilled upwards to the spleen and stomach, and undergone further change.

'Food and drink (after digestion) separate into watery and solid parts, which are evacuated by the two inferior orifices. The nutrient essentials go to the bone-marrow, and also combine with what is circulating in the vascular and interstitial systems (*jung wei*), to nourish the entire body. Thus in due course fatty parts, blood and flesh are produced. What is in excess overflows as urine; therefore the urine is yellow in colour because it is a residual part of the blood, and its rankness is due to the *chhi* of the five viscera, for it is the nature of the saline quality to be excreted. But since the renal and urinogenital system at the reins is eremotic (i.e. weak and cold, *hsü lêng*) it cannot distil the nutrient essentials, so that all is excreted as urine. Therefore the sweetness in the urine (comes forth), and the latter does not acquire its normal colour.

'(In diabetic affections) there is wasting and dryness of the skin; just as in nursing mothers the nutrient essentials soak upwards and are changed into milk, so in these patients they all go downwards and depart in the urine; the nutrient essentials are not used for the benefit of the body. This is why there is such abundance of urine. Moreover the pulmonary system is the baldachin (*hua kai*) of the five viscera; if below there is a heating *chhi* (in the renal and urinogenital system), there is distillation into the lungs, but if the lower parts (i.e. the *shen chhi*) are excessively weak and cold, then the *Yang chhi* cannot ascend—thus the lungs are dry and thirst results.

'The *I Ching* (Book of Changes) under the *Kua* "*Fou*" says: *Chhien* is above and *Khun* below; when the Yang is without Yin it cannot descend, if Yin lacks Yang it cannot ascend; when above and below do not integrate, that is *Kua* "*Fou*". Or again if you take water in a cauldron and heat it with fire below, the wooden cover will become saturated with steam, but if no heat is underneath then the water-vapour will not rise and the board will never be saturated. Here the force of the fire is analogous to the strength of the renal and urinogenital system, which normally requires warm *chhi* to assist its function. When the food and drink feels the force of the fire it permeates upwards and is easily converted (for use), thus thirst is prevented. Chang Chung-Ching (Chang Chi) says that such patients should take (*fu*) *Shen Chhi Pa Wei Wan* (eight-constituent pills for strengthening the renal and urinogenital system).

'Although (the diabetic affections) are similar to beri-beri (*chiao chhi*) in that there is weakness

306

of the renal and urinogenital system, the latter tends to start in the 2nd and 3rd months, becomes most prevalent in the 5th and 6th, and declines in the 7th and 8th; but the former tends to start in the 7th and 8th, becomes prevalent in the 11th and 12th, and declines in the 2nd and 3rd. Why is this? Beri-beri is a *yung* disease (i.e. Yin, elleiptic, deficient or inhibitory), diabetes is a *hsüan* disease (i.e. Yang, hyperochic, excessive or excitatory). In spring and summer the Yang *chhi* is in the ascendant, so (it is natural that) *yung* (Yin) diseases (like beri-beri) should begin, while *hsüan* (Yang) diseases (like diabetes) fall off; in autumn and winter the Yang *chhi* is declining, so (it is natural that) *hsüan* (Yang) diseases (like diabetes) should have their onset, while *yung* (Yin) diseases (like beri-beri) should have surcease. In examining these two things one can see that diseases have their pattern-principles. It is like the art of government; broad-minded lenience can (sometimes) benefit desperate characters, and severity can (sometimes) benefit the law-abiding; the ruler must act according to the concrete situation.

'(Chang) Chung-Ching (Chang Chi) says that the pedotelic vesical Thai-Yang tract is related to the bladder, and the bladder is the receptacle of the kidney; in cases of super-abundant urine there is overactivity of the *chhi* (of that tract), hence greater food consumption and constipation, but where this *chhi* is weak and sluggish then diabetic affections result. A man suffering from diabetes who drinks a gallon of water will excrete as much as a gallon of urine. Let him take *Shen Chhi Pa Wei Wan* pills.'

This passage is quoted at greater length by Wang Thao in +752 and in part by many later medical writers. The knowledge of diabetic glycosuria then became a permanent acquisition to Chinese medicine, and is mentioned by all writers. From the +8th century onwards there is mention not only of tasting the urine in diagnosis, but also of watching whether bees would visit a sample of it and whether ants would be attracted to it when a sample was placed near a nest.

It is highly instructive to compare this long statement of Li Lang-Chung in the +7th century with the similar discussion of Thomas Willis in the latter part of the +17th. They are really on exactly the same scientific level, and if anything, one is inclined to give the preference in perspicacity to Li Lang-Chung rather than to Thomas Willis. Both of them were of course struggling to understand something which could not be analysed until the methods of modern physiological science had been developed, but the thousand years priority of Li Lang-Chung is a fact which deserves careful consideration, for his attempted analysis of the situation is eminently reasonable and highly suggestive. If comparative investigations show that the recognition of the sweetness of diabetic urine occurred indubitably first in India, then the historical relations of those times would suggest that the Buddhist monk-physicians were the means of its passing into Chinese culture. It remains however to be shown that the Indian statement is earlier. As for possible influences on Europe, nothing concrete can be said, but by the time that Thomas Willis was writing there had been close contacts between China and the West for nearly two hundred years. First the Portuguese travellers had visited many Chinese cities, and from the end of the + 16th century onwards there had been the Jesuit mission, the members of which laid themselves out to study as much as possible of Chinese scientific and technological culture and to report upon it to the West. One may therefore plausibly conclude that the knowledge travelled westwards, though we have no information as yet by what channels it came. In view of the state of Sanskrit study at that time, it seems to us more likely that the stimulus for Thomas Willis was derived from Chinese rather than Indian antecedents.

We may now take up the question of the reduced resistance of diabetic patients to attacks of pathogenic organisms. We have noted this already in the quotation from Li Lang-Chung. Both Sun Ssu-Mo in the + 7th century and Wang Thao in the + 8th emphasise and reiterate the tendency of diabetic patients not only to obesity but also to infective skin lesions, pyogenic boils, furuncles, carbuncles, rodent ulcers and oedema, as well as troubles of the eyesight. Indeed for this reason Sun Ssu-Mo says that acupuncture was contra-indicated after the first three months of the malady. The tendency to retinitis, night-blindness and cataract was early appreciated; it was much emphasised for instance by Liu Wan-Su, in his late + 12th-century *Hsüan Ming Lun Fang*. Today it is realised that the frequency of pyogenic lesions in diabetic patients is so great as to make examination for glycosuria mandatory if such lesions are common on the body. The medieval Chinese physicians fought them as well as they could in the days before antibiotics and sulpha-drugs with herbal applications—of course it is always possible that some of the plants used possessed antibiotic constituents.

Diabetes may of course also be produced by exogenous means. Substances may have specific effects upon the endocrine cells of the pancreas, inhibiting the endocrine secretion. Alloxan is a famous example, but metallic poisons may have similar effects. Hence it is remarkably interesting that the physicians of the Sui and Thang were strongly inclined to associate diabetic affections with elixir poisoning. One finds this in the *Chu Ping Yuan Hou Lun*, the great pathology of Chhao Yuan-Fang (+ 607), as also in the *Chhien Chin (Yao) Fang* of Sun Ssu-Mo (+ 655), as well as the works of the following century. Chhao Yuan-Fang and Wang Thao use the expression *shih fa*, i.e. the metallic or mineral substances manifesting their effects, and they list an abundance of such symptoms. Elsewhere

308

we have studied in detail the evil effects of the elixirs taken by many people in those times at the instigation of Taoist alchemists.[1] All the lamentable consequences of lead, arsenic and mercury poisoning are described in the medieval Chinese medical books. Unfortunately, although diabetes seems so often to have resulted from these interventions, we do not know whether it was injury to the islets of Langerhans in the pancreas producing diabetes mellitus, or to the cells of the brain and pituitary glands affecting the anti-diuretic hormone and producing diabetes insipidus. So far we have not come across in the Chinese literature any radical diagnostic differentiation of insipidus and mellitus in medieval texts. We know that these two diseases were not separated until +1794, when they were first defined by J. P. Frank. Perhaps something very interesting lies behind the close association which the Sui and Thang physicians made between diabetes and elixir poisoning.

There are certain individual cases which may be taken as examples of this. For instance, the great rebel general, An Lu-Shan, who died in +757, was intensely obese and suffered very severely from boils and rodent ulcers; his eyesight also deteriorated to the point of complete blindness. There is good reason for regarding his case as diabetes mellitus, either spontaneous or induced by metallic elixir poisoning. Perhaps if insulin had been available the Thang dynasty would not have gone down to ruin as it did, after the civil wars unleashed by the cruel and irritable Sogdian leader. But we will refrain from encouraging physiology to usurp the role of economics in historical explanation.

SEXUAL ENDOCRINOLOGY

We propose lastly to examine medieval Chinese knowledge of the phenomena due to the action of the sex hormones. Some rather surprising discoveries were awaiting us when we began the study of this subject. There can be little doubt that the first beginnings of this knowledge arose, as in other civilisations no doubt, from the practical procedure of castration. This had been undertaken since very ancient times in Chinese society on human beings for social reasons but also on animals, both for medicinal purposes and for gastronomy, since gelded animals would put on fat and gave a more tender meat. In +1378 Yeh Tzu-Chhi said in his *Tshao Mu Tzu* book that 'the outer glory of the seminal essence is manifested by the beard, that of the *chhi* is manifested by the eyebrows and that of the blood

[1] [P. 316 below.]

by the hair'. This was expanded by Wang Shih-Chên in his *Lei Yuan*, written in + 1575. He spoke as follows:

Hair pertains to the heart, which is born from the *chhi* of (the element) Fire, therefore it grows on the top (of the head). The beard pertains to the kidneys and testes, which are born from (the element) Water, therefore it grows at the lower part (of the head). The eyebrows pertain to the liver, which is born from (the element) Wood, therefore they grow on each side (of the head). Thus it is that the *chhi* of the testes (*wai shen*) produces the beard as its outward manifestation; and hence it is also that women and eunuchs (whether congenital or induced) do not grow beards though they do have hair and eyebrows.

It is certain that by this time theories of this kind were already very old.

The interest of Chinese physicians and naturalists was also awakened very early by the phenomenon of hermaphroditism. Li Shih-Chen in the *Pên Tshao Kang Mu* of + 1596 has an elaborate discussion of the ten principal forms of hermaphroditism. He had long been preceded by the first founder of forensic medicine in any civilisation, Sung Tzhu, in his famous *Hsi Yuan Lu* (The Washing Away of Wrongs, i.e. False Charges) of + 1247. We also find in Chinese literature numerous accounts of sex reversals both in man and in animals. An early example of this was the account of a man in − 6 who turned, as was thought, into a woman, and another striking case was reported in + 202. Wang Chhung, the great sceptical naturalist, who wrote his *Lun Hêng* (Discourses Weighed in the Balance) about + 80, has an enlightened discussion of the phenomena of sex reversal. Naturally such prodigies were taken much note of by prognosticators and diviners engaged in predicting the future concerning affairs of State in those ancient times.

It was quite natural, once the importance of the testes had been recognised, that they should be used in pharmacy. In the *Pên Tshao Kang Mu* we find numerous preparations of testicular tissue, desiccated or raw, taken from the pig, the dog or the sheep, and administered for male sexual debility, spermatorrhoea, hypogonadism, impotence, and other conditions for which androgens would be prescribed today. The practice of using testicular tissues seems to have developed at least as early as the Southern Sung period, for the first references that we have found to this occur in a book entitled *Lei Chêng Phu Chi Pên Shih Fang* (Classified Fundamental Prescriptions of Universal Benefit) printed in + 1253 and ascribed to an eminent physician, Hsü Shu-Wei, whose *floruit* was in the neighbourhood of + 1132. Today the oral route of administration is not employed because we know that testosterone is inactivated in the liver. But in medieval times, before the isolation of the sex hormones, and *a fortiori* before the development of their derivatives not naturally occurring, the administration of testis tissue by the mouth, if in sufficient quantity, could have had reasonably significant effects. All this was happening a very long time before 1849, when Berthold demonstrated the effective replacement of testes by testicular tissue implanted elsewhere in the body, and 1889, when Brown-Séquard made those classical experiments in which he injected testis extracts into himself. Hsü Shu-Wei should be recognised as a predecessor of these 19th-century scientists and should be remembered along with them.

Testicular tissue, to be sure, had figured also in the materia medica of other ancient civilisations (Dioscorides, Suśruta), though by no means always administered for the appropriate conditions. The richest source of oestrogens, placental tissue, used with any precise theory in mind, and for conditions such as amenorrhoea, for which oestrogens are used today, occurs more rarely in comparative therapeutics, but was prominent in China. The use of human placenta was mentioned first by Chhen Tshang-Chhi in his *Pên Tshao Shih I* pharmacopoeia of +725, but its employment apparently remained rare until the Yuan period, when Chu Chen-Hêng encouraged the use of preparations containing it. Early in the +14th century he was prescribing it for debility of all kinds, including sexual weakness, and recommended the incorporation of various other specific and tonic drugs with the placental tissue. The tissue was carefully washed and drained, boiled with wine, reduced to a small volume and combined with a number of vegetable drugs. At the end of the +15th century, for example, Wu Chhiu in his *Ta Tsao Wan* (pill) added, among others, *tang kuei* (*Angelica polymorpha*), a well-known uterine stimulant, and *tu chung* (*Eucommia ulmoides*), a drug known to affect blood pressure.

The use of placenta in medicine may well go back before the +8th century, because Li Shih-Chen quotes the *Tan Shu* (a general term for books on iatrochemical medicine) as discussing the best colour of placentas to be used in medicine. Since the corpus of alchemical literature began in the Han rather than the Thang, it may well be that the practice originated a good time before Chhen Tshang-Chi. What Wu Chhiu says in his *Chu Chêng Pien I* about the matter is interesting. He wrote:

Although the placenta takes its form from the nourishment of the (mother's) food, yet it also contains (or transmits) the congenital endowment (of the foetus). Thus it is far superior to any other drug, mineral or herbal. I have often used it and obtained perfect results, especially in female patients. This is because the organism from which the placental material itself originates

311

(is the very organism which it will benefit) since all things have a tendency to follow their categories. Barren women, those who only produce girls, those with dysmenorrhoea, those with miscarriages, difficult births, etc., if they take *Ta Tsao Wan* (pills) get sons, and even when critically ill one or two doses will keep them going for a couple of days. The merit of the placenta is chiefly to increase the efficiency of the Yin force in the body, including sexual function. It always seems to give good results. If one takes it for a long time it improves the hearing, brightens the eyes, keeps the hair and beard black, increases longevity, and indeed has such merit that it can overcome the natural process (*tsao hua*) of ageing. Hence these pills are known as *Ta Tsao Wan*.

As in the case of the thyroid gland, use was made of the placentas of animals, notably the horse and the cat.

But the most remarkable development of medieval Chinese medicine concerning the sex hormones was the veritable fractionations applied to urine. The use of urine in the pharmacopoeia of many ancient nations has generally been written off in modern times as a typical example of 'Dreckapotheke', the employment of utterly useless and disgusting substances. Historians of medicine, however, might have been more circumspect in such condemnations after the classical discovery of Aschheim & Zondek in 1927 of the presence of large amounts of sex hormones in pregnancy urine, and the subsequent realisation that all urine, but especially that of certain animals such as the mare, contains these active substances. The external and internal application of urine in disease occurs in many ancient civilisations, but in medieval China alone, so far as we know, did pharmacists engage in the preparation from urine almost on a manufacturing scale of active products by means of precipitation, re-solution, evaporation to dryness, sublimation and crystallisation. This forms a really rather extraordinary story.

The origins of urinary therapy go far back into ancient Taoism. The Taoists had a philosophical and magico-scientific attitude to sex rather than an ascetic one in the ordinary sense, and we find a curious passage in the *Hou Han Shu* (History of the Later Han Dynasty) about three adepts who lived *c.* +200. The text says:

Kan Shih, Tungkuo Yen-Nien and Fêng Chün-Ta, these three were all (Taoist) adepts (*fang shih*). They were all expert at following (the techniques of) Jung Chhêng in commerce with women. They could also drink urine, and sometimes used to hang upside down. They were very careful and sparing of their seminal essence (*ching*) and (inherited) *chhi*, and they did not boast with great words of their own powers. What Kan Shih, (Tso) Yuan-Fang and (Tungkuo) Yen-Nien could do was recorded by (Tshao) Tshao (founder of the Wei (San Kuo) dynasty), who asked them about their art and tried to practise it. Fêng Chün-Ta was called the Blue-Ox Master (Chhing Niu Shih). All these people lived to between 100 and 200 years of age.

Reference to the effect of urine on sexual health and activity can be found all through the ages. Early in the +14th century Chu Chen-Hêng tells us that he once attended an old woman over 80 years of age who gave the appearance of being only 40. In reply to his questioning she explained why she had had such good health and had suffered no illnesses. Once when she had been ill she had been instructed to take human urine and this she had done for several decades. Who could maintain therefore, says Chu Chen-Hêng, the old belief that the property of urine is algorific and that it could not be taken for a long time? All cases, he goes on, of *Yin hsü* (impotence, sexual debility, eremosis, excess Yang of burning feverish type), which no medicine can benefit, take a better turn if urine is administered. Urine was also recommended for other reasons, as by Chhu Chhêng (fl. +479 to +501) in the *Chhu Chhêng I Shu*, who praised its styptic properties in laryngeal bleeding.

There can be no doubt that as time went on the theory developed that urine owed its properties to the fact that it was 'of the same category as' (*thung lei*) the blood. As we saw, Li Shih-Chen in the *Pên Tshao Kang Mu* (+1596) says that of the nutrient essentials (*ching chhi*), the lighter part (*chhing chê*) goes to form blood while the grosser part (*cho chê*) goes to form the *chhi*. Then the grosser part of the lighter fraction forms the urine, while the lighter part of the grosser fraction forms all the secretions. Consequently by the Thang period a great deal of interest was being taken in the natural precipitates and sediments of urine (*ni pai hsin* or *jen chung pai*). At the beginning of the +14th century Chu Chen-Hêng said that the urinary precipitate had the ability to drive out the undue Fire element affecting the liver, the *san chiao*, and the urinary bladder by way of the urine. 'This is because the *jen chung pai* was itself originally excreted from the bladder.' This was the principle of *yin tao*, a secretion or excretion leading out something else by the same way that it had itself come. The precipitate, said Li Shih-Chen in the +16th century, 'travels along with the blood' and has the power of bringing out from the body other things by the route which it previously travelled itself. He also says that although it helps sexual debility, princes and patricians disliked the idea of taking it because of its unclean origin. So the iatro-chemists (*fang shih*) began to purify the sediments into derivatives known as *chhiu shih* and *chhiu ping* ('autumn mineral' and 'autumn ice'), i.e. special preparations. The term *chhiu shih* had first been used by the *Huai Nan Tzu* book in the −2nd century as the name of some kind of elixir. We do not know whether it had anything to do with the *chhiu shih* of the later iatro-chemists.

The oldest among the detailed fractionation methods, which we now have accounts of, dates from the beginning of the + 11th century, when Chang Shêng-Tao described the simplest of them in his *Ching Yen Fang* (Tried and Tested Prescriptions). This book is now lost, but the essential passages are quoted in the *Chêng Lei Pên Tshao* of + 1249. Then from the early years of the + 12th century we have two further descriptions in the *Shui Yün Lu* (Water and Clouds Record), written by a famous scholar, Yeh Mêng-Tê (+ 1077 to + 1148). Another book of the same period, the *So Sui Lu*, by an unknown author, theorises about these preparations. The rest are in three books of the Ming period, one by Chhen Chia-Mo, the *Pên Tshao Mêng Chhüan* (Ignorance about the Pharmacopoeia Dissipated), datable at + 1567, and the others possibly of the preceding century.

What now were the outstanding features of these fractionations? In the first place remarkably large amounts of urine were used, either from adult subjects or from boys or girls. Quantities might be up to 200 gallons. The simplest method was to take the whole mass obtained by evaporating the urine to dryness, including urates, phosphates, steroid glucuronides and sulphates, and everything else that was there. Other methods, however, involved preliminary precipitation. One used calcium sulphate, which would probably take down any protein that was present and therefore certainly any steroid conjugates also. The most remarkable precipitation, in view of the classical discovery of Windaus in 1909 that digitonin precipitates certain steroids quantitatively, was the use of saponin from the soap-bean tree (*Gleditschia sinensis*), the bean juice being added to the urine at the rate of one bowlful for every tub. This method, when followed further, involved the boiling of the precipitate with water at a later stage, and suggests therefore that all those steroids which were carried down on the protein would be released as it de-natured, while those which had been combined with the saponin would remain behind. Criticisms of this method contained in the *Pên Tshao Kang Mu* suggest strongly that it was very widely used between the + 10th and the + 16th century. Besides the vegetable protein added in the soap-bean juice, it may be considered rather probable that in all the samples of urine there was some admixture of urine from people suffering from renal lesions, so that there would be a small amount of protein in all these massed raw materials. This would have been important because the conjugated steroids would have precipitated with it.

In nearly all the methods the process ends by a sublimation. At first this seemed very difficult to understand, but in fact the urinary steroid sex-hormones do sublime unchanged in air between 140 °C and 280 °C, there being considerable differences in sublimation temperature between them. Indeed this method is now used for their identification. The description of the technique of the tightly luted sublimation apparatus (*ku chi fa*) is interesting and fairly clear. Small earthenware pots were used with 'hermetically' sealed covers, and great attention was paid to the degree of heating, so that they should be neither too cool nor get too hot.

The Chinese alchemists had been perfecting the technique of sublimation since as far back as the − 3rd century; it had always been particularly important in connection with the chemistry of mercury, a central feature of the spagyrical art. That the Chinese iatro-chemists, knowing nothing of steroid chemistry, were in fact preparing rather purified mixtures is indicated by the fact that the texts so frequently speak of white, crystalline, glittering, lustrous material like translucent jade or pearls, which is strongly suggestive of the pearly character of cholesterol and other substances of the same class when fairly pure. In a number of cases the final process involved emulsification with milk fat, again a procedure which seems quite appropriate in view of the steroid character of the end-products which the iatro-chemists were aiming at. It is not of course to be suggested that the end-products were pure in the modern sense; they must have been mixtures of many different compounds, including no doubt, in certain cases, some harmless ones such as cyanuric acid, which would have been produced by the action of heat on any remaining urea. Yet the iatro-chemists had some control over what they obtained, not only by the precipitations and fractionations used, but even more by working up separately, as they did, the urine of donor groups of different sexes and ages.

Thus there can be little doubt that between the + 11th and the + 18th century the Chinese iatro-chemists were producing preparations of androgens and oestrogens which were probably quite effective in the quasi-empirical therapy of the time. This must surely be considered an extraordinary achievement for any type of scientific medicine before the age of modern science.

16

ELIXIR POISONING IN MEDIEVAL CHINA[1]

[1959]

AMPLE evidence shows that the belief in the existence of certain substances in Nature, or elixirs prepared from them, which would prolong human life for an indefinite period, flourished in the minds of Chinese alchemists and pharmaceutical adepts as early as the Warring States period (− 480 to − 221). But it was most ironical that in their pursuit of longevity many of them reaped just the opposite result from their preparations, since some of their elixirs (especially in the Chin and Thang periods, i.e. from the + 3rd to the + 10th centuries) contained substances such as mercury, lead and even arsenic. It is our purpose here to discuss elixir poisoning, its repercussions and the counter-measures for its prevention taken by the Chinese alchemists themselves.

The dynastic histories inform us about many emperors and high officials who sought for the elixir of life and some who took the preparations offered to them by their alchemists. One of the oldest examples of such an elixir is that mentioned in the *Chan Kuo Tshê* (Records of the Warring States), a book compiled not later than the Chhin period (− 255 to − 206). The event referred to would have occurred in Chhu State during the reign of King Ching Hsiang between − 294 and − 261. This dating is assured by the existence of an almost identical version of the story in the *Han Fei Tzu* book which belongs without doubt to the early part of the − 3rd century. It says:[2]

Once upon a time someone presented an elixir of life (*pu ssu chih yao*) to the Prince of Ching. As the chamberlain was taking it into the palace the guard at the gate asked if it was edible and when he answered yes, the guard took it from him and ate it. The Prince was [extremely][3] annoyed and condemned the guard to death. But the latter sent a friend to persuade the Prince, saying, 'After all the guard did ask the chamberlain whether it could be eaten before he ate it. Hence the blame attaches to the chamberlain and not to him. Besides what the guest presented was an elixir of life, but if you now execute your servant after eating it, it will be an elixir of death [and the guest will be a liar]. Now rather than killing an innocent officer in order to demonstrate a guest's false claim, it would be better to release the guard.' So the Prince let him off.

---

[1] A contribution to *Janus*, 1959, **48**, 221.   [2] *Chan Kuo Tshê*, ch. 5, p. 33 *b*.
[3] The parts within square brackets are found only in the *Han Fei Tzu* (ch. 22, p. 5 *b*). See also Liao Wên-Kuei's (1939) translation, p. 235.

This passage was no doubt preserved as an exercise in sophistic argument, but it takes its place among a whole series of texts of an alchemical or quasi-alchemical character which establish the origins of Chinese alchemy in the Warring States period from at least the time of Tsou Yen onwards.[1]

One of the earliest examples of the use of mineral preparations leading to illness is contained in the biography of Shunyü I (−216 to c. −150).[2] It describes a certain Sui, personal physician to the Prince of Chhi, who himself prepared medicines from the five minerals and took them, but died of ulcers (*chü*) in the chest after 100 days just as Shunyü I predicted.

A clear case of elixir poisoning is that of the Chin emperor Ai Ti,[3] who, as the result of his efforts to avoid growing old, died in his very prime aged only 25. According to the *Chin Shu* (History of the Chin Dynasty):[4]

He had a liking for the art of the alchemists. He abstained from cereal grains, but consumed elixirs. As the result of an overdose he was poisoned and no longer knew what was going on around him (*pu shih wan chi*).

In another example we have a case of an emperor who lost his life indirectly due to elixir poisoning. The *Hsü Thung Chih* (Supplement of the Historical Collections)[5] says:

Deluded by the sayings of the alchemists, Thang Hsien Tsung[6] ingested gold elixirs and his behaviour became very abnormal. He was easily offended by those officials whom he daily met, and thus the prisons were left with little vacant space. (Finally) at midnight on a *kêng-tzu* day in the first month of the 15th year (of the Yuan-Ho reign-period) (14th Feb., +820) Wang Shou-Chhêng and a Palace Attendant (*nei chhang shih*) Chhen Hung-Chih assassinated the emperor in the Chung-Ho Palace Hall.

Li Shun had been a good and capable ruler who had introduced many reforms, and successfully preserved the unity of the empire in civil wars from +814 onwards. But a little later he came under the influence of an alchemist named Liu Pi, who had been recommended by a high official Li Tao-Ku. In +819, after the emperor had for some time been consuming Liu Pi's preparations, an official

[1] Cf. *SCC*, vol. 2, p. 232. We shall deal fully with this evidence, which is of outstanding interest for comparison with Bolus of Mendes, Pseudo-Democritus and the Graeco-Egyptian mystical proto-chemists, in *SCC*, vol. 5. [See Figs. 87, 88, 89, 90, 91, pls, 92.]
[2] Cf. *Shih Chi*, ch. 105, pp. 21*b*, 22*a*, and Hübotter (1927), p. 21 (case 22).
[3] Reigned +361 to +366; personal name Ssuma Phei.
[4] Ch. 8, p. 8*a*.   [5] Ch. 575, p. 1*a*.
[6] Reigned +805 to +820; personal name Li Shun.

of the postal service, Phei Lin, presented a memorial which afterwards became a classic in its genre. Here we give it in abridged translation:[1]

May it please your Majesty: I have heard that he who eradicates evil, himself reaps advantage in proportion to his work; and that he who adds to the pleasure of others, himself enjoys happiness. Such was ever the guiding principle of our ancient kings...Of late years, however, (the capital) has been overrun by a host of pharmacists and alchemists, such as Wei Shan-Fu, Liu Pi and others, recommending one another right and left with ever wilder and more extravagant claims. Now if there really were immortals, and scholars possessing the Tao, would they not conceal their names and hide themselves in mountain recesses far from the ken of man? Their only fear would be to come in contact with humankind. How is it then that we have all these characters hanging about the vestibules of the rich and the great, and boasting of their wonderful pharmaceutical arts—surely they cannot be men who possess the Tao. They have come for nothing but profit. By saying that they can transform themselves into spirits they tempt the powerful and those in high positions, and for a while their astounding claims convince. But when their lies are found out, they think it no shame to decamp without cere-mony. This is the plain truth. How can you credit their words and consume their elixirs?...

The medicines of the sages of old were meant to cure bodily illnesses, and were not meant to be taken constantly like food. How much less so these metallic and mineral substances which are full of burning poison!...

Of old, as the *Li Chi* says, when the prince took physic, his minister tasted it first, and when a parent was sick, his son did likewise. Ministers and sons are in the same position. I humbly pray that all those persons who have elixirs made from transformed metals and minerals, and also those who recommend them, may be compelled to consume (their own elixirs) first for the space of one year. Such an investigation will distinguish truth from falsehood, and automatically clarify the matter by experiment (*khao chhi chen wei, tsê tzu jan ming yen*).

The petition was not accepted and Phei Lin demoted to be Chiang-ling Ling, a provincial official at Chiang-ling.

The same biography of Phei Lin concludes with details about Chang Kao, a scholar who made a similar memorial to Mu Tsung[2] c. +823. After Mu Tsung came to the throne Liu Pi and others were executed but then he himself gradually came under the influence of the alchemists. Chang Kao wrote:

If mental calm is maintained, the blood and the *chhi* (*pneuma*) will be harmonised. If lust and desire gain the upper hand illness will follow. Medicines are for use against illnesses, and should not be taken as food. Formerly Sun Ssu-Mo said that the actions of medicines give assistance by their specificity, but they disquiet the *chhi* of the viscera. Even when one is ill medicines must be used with great circumspection; how much more so when one is not ill. If this is

---

[1] *Chiu Thang Shu*, ch. 171, p. 6b. See also Giles (1923), p. 152.
[2] Reigned +820 to +824; personal name Li Hêng.

true for the common people how much more so will it be for the emperor! Your imperial predecessor believed the nonsense of the alchemists and thus became ill; this your majesty already knows only too well. How could your majesty still repeat the same mistake?[1]

The emperor appreciated his words. But soon afterwards he fell ill and died.

The next Thang emperor who fell victim to elixirs was Wu Tsung.[2] The *Chiu Thang Shu* says:[3]

The emperor (Wu Tsung) favoured alchemists, took some of their elixirs, cultivated the arts of longevity and personally accepted (Taoist) talismans. The medicines made him very irritable, losing all normal self-control in joy or anger; finally when his illness took a turn for the worse he could not speak for ten days at a time. The prime minister Li Tê-Yü and others asked for audience but were refused, and nobody inside or outside the palace knew his real state, so that people were alarmed and sensed danger. On the 23rd day of that month (22nd April, +846) the imperial will was read, and the emperor's uncle, Prince Kuang (Li Shen), ascended the throne (as Hsüan Tsung) in front of the coffin.

After this it will hardly be believed that Li Shen himself was interested in elixirs towards the end of his reign. According to Giles (1898)[4] he died of them, but this is not mentioned in the dynastic history;[5] indeed on the contrary the Taoist hermit and alchemist Hsienyuan Chi of Lo-fou Shan declined to venture elixirs in the cure of the emperor's illness in +858 and restricted himself to moralising speeches. Other sources however support the view of Giles, for instance the *Tung Kuan Tsou Chi* (Record of Memorials from the Eastern Library) written about +890 by Phei Thing-Yü, who names other alchemists less scrupulous. He says:[6]

A medical official, Li Hsüan-Po, presented to the emperor cinnabar which had been heated and subdued by fire, in order to gain favour from him. Thus the ulcerous disease of the emperor (Hsüan Tsung) was all attributable to his crime. When I Tsung came to the throne (in +860) (Li) Hsüan-Po, the hermit Wang Yo and the Taoist Yü Tzu-Chih were all executed in the market-place.

There is plenty of evidence that the taking of metallic elixirs was not confined to the imperial court, but extended over wide circles of society and might affect anyone who could afford to pay. Thus a famous writer Yeh Mêng-Tê gave in +1156 an account of friends personally known to him who had suffered the

---

[1] Abridging *Chiu Thang Shu*, ch. 171, p. 8*a*. The *Pien Huo Pien* (ch. 4, p. 7*b*) has an abridged parallel passage.
[2] Reigned +840 to +846; personal name Li Yen. He was strongly opposed to Buddhism; cf. an edict of his tr. Giles (1923), pp. 153 ff.
[3] Ch. 18A, p. 17*b*.   [4] No. 119.
[5] *Chiu Thang Shu*, ch. 18B.   [6] Ch. 3, p. 7*a*.

consequences of what had by then become a veritable superstition. In his *Pi Shu Lu Hua* he mentions that[1]

in not a few cases ordinary people and scholars in official positions have died from eating cinnabar. With my own eyes I saw what happened to Lin Yen-Chen who was normally very robust and could eat and drink more than a normal person. He lived at Wu and often boasted about his health and muscular strength. But a physician, Chou Kung-Fu, claimed to have obtained a secret formula from Sung Tao-Fang for the preparation of a cinnabar elixir which could prolong life and was at the same time harmless. (Now) (Sung) Tao-Fang had been a good doctor in Hungchow so (Lin) Yen-Chen believed this and took the elixir for three years. Whereupon ulcers developed in his chest, first near the hairs as large as rice-grains, then after a couple of days his neck swelled up so that chin and chest seemed continuous. After ten months he died. When he was very bad his doctors caused the pus and blood to be wiped away with a cloth, and after standing in water cinnabar powder was found at the bottom. Thus it had accumulated in the body and came out with the poison. There was also Hsieh Jen-Po. Whenever he heard of anyone who had some cinnabar subdued by fire he went after it, caring nothing about the distance, and his only fear was that he would not have enough. Last year he also developed ulcers in the chest. At the onset of his illness his friends noticed a change in his appearance and behaviour, but still he did not recognise the influence of the poison, till suddenly it came upon him like a storm of wind and rain, and he died in a single night.

Thus in the past ten years I have personally seen these two cases, which ought to be a warning to others.

We also read about those who were more cautious in such matters. For example, the emperor Liang Wu Ti[2] declined to take the elixirs offered by the alchemist Têng Yü-Chih,[3] but did not hesitate to try those prepared for him by his friend Thao Hung-Ching, because he had confidence in the work of that great herbal pharmacist.[4] The emperor Wên Hsüan[5] of the Northern Chhi, on the other hand, was only prepared to try the life-giving drug at an opportune moment—on his deathbed. The *Pei Shih* (History of the Northern Dynasties) says:[6]

He ordered Chang Yuan-Yu and other alchemists to make the *chiu chuan chin tan* (ninefold cyclically transformed elixir). When it was accomplished the emperor kept it in a box made of jade and said, 'I am still too fond of the pleasures of the world to take flight to the heavens immediately—I intend to consume the elixir only when I am about to die.'

Of scholars we read about the great poet and civil servant Su Tung-Pho (+ 1036 to + 1101) who was acquainted with the art of alchemy, but nevertheless showed

---

[1] Ch. 1, p. 76 b. The same passage was quoted some 30 years later by another Chang Kao in his *I Shuo* (ch. 9, p. 15 a, b) which contains quite an anthology of warning pieces about the taking of elixirs and also a story about Sung Tao-Fang (ch. 9, p. 23 a, b).

[2] Reigned + 502 to + 549; personal name Hsiao Yen.

[3] Cf. *Nan Shih*, ch. 76, p. 8 b.

[4] *Ibid.* p. 9 a.

[5] Reigned + 550 to + 559; personal name Kao Yang.

[6] Ch. 88, p. 12 a.

great discretion when he was offered an elixir. In a letter to a friend, Wang Ting-Kuo, he wrote, 'I have recently received some cinnabar (elixir) which shows a most remarkable colour, but I cannot summon up enough courage to try it...'[1]

To what extent did caution generate procedures which we could call clinical experiment and test? As we have seen, alchemy found great favour with many emperors. Besides those already mentioned we may single out Tao Wu Ti[2] of the Northern Wei, and Hsüan Tsung[3] of the Thang. The events at the beginning of the +5th century were particularly interesting. Thopa Kuei instituted a professorship of Taoism and alchemy with facilities for the study and preparation of elixirs. According to the *Wei Shu* (History of the Northern Wei Dynasty),[4]

In the 3rd year of the Thien-Hsing reign-period (+400) (the emperor Tao Wu Ti) instituted a Hsien Jen Po Shih Kuan (Office of the Alchemist-Royal) to take charge of the preparation of drugs and elixirs (*chu lien pai yao*).

Elsewhere it says:[5]

(The emperor) was fond of the teachings of Lao Tzu...During the Thien-Hsing reign-period he appointed an Alchemist-Royal and built an imperial elaboratory (*hsien fang*) for the concocting of drugs and elixirs. He also reserved the Western mountains (Hsi-shan) for the supply of fire-wood (for the furnaces). Furthermore, he ordered criminals who had been sentenced to death to test (*shih fu*) (the products) against their will. Many of them died and (the experiments gave) no decisive result (*wu yen*).

This raises the whole question of systematic biological experimentation on man and animals, the history of which has been sketched by Green (1954) and Bull (1951). That is something distinct from biological observation and dissection, on which of course there are famous Hellenistic traditions concerning Herophilus and Erasistratus[6] and Cleopatra.[7] Thopa Kuei's +5th-century use of condemned criminals for what were in principle clinical therapeutic trials was followed many times in subsequent ages, as e.g. by the great surgeon Ambroise Paré in the +16th century for testing the supposed beneficial effects of bezoar and unicorn's horn in corrosive sublimate poisoning, and by the physicians of Queen Caroline in +18th-century London when they inoculated some of the Newgate prisoners. The first real clinical trial with carefully matched controls seems to have been that of James Lind who in +1747 investigated the powers of various citrous fruits

---

[1] See *Lu Huo Chien Chieh Lu* (Precautions in the Work of the Stove) by Yü Yen, p. 9a.
[2] Reigned +386 to +408; personal name Thopa Kuei.
[3] Reigned +712 to +762; personal name Li Lung-Chi.
[4] Ch. 113, p. 3a.   [5] Ch. 114, pp. 32b, 33a.
[6] Discussed by Dobson (1925, 1927).   [7] Cf. Needham (1959), p. 65.

as preventives against scurvy. But Avicenna's *Canon Medicinae* (*Qānūn-fī al-Ṭibb*) of about + 1030 is said to give rules for the testing of drugs in paired cases of uncomplicated disease.

Pharmaceutical experiments on animals must also go back very far in China as in other civilisations. The fowl was traditional as a test-animal in forensic medicine, as we know from the + 13th-century *Hsi Yuan Lu* (Washing Away of Wrongs);[1] and this theme was common in later Chinese opera, which appreciated the skill of popular magistrates in the detection of crime. But the most famous example is that of the alchemist Wei Po-Yang and his pupil Yü Sêng in the + 2nd century. It will be remembered that Wei Po-Yang was the author of the celebrated treatise *Tshan Thung Chhi* (The Kinship of the Three), which is regarded as the oldest complete alchemical book extant in any culture. It is said that[2]

Wei Po-Yang entered the mountains to prepare the magical elixir. With him were three disciples, two of whom were lacking in complete faith. When the elixir was made he warned them saying, 'Although the gold elixir is now accomplished we ought first to test it by feeding it to a white dog. If the dog can fly after taking it then it is edible for man; if the dog dies then it is not.'

The story continues paradoxically; the dog dies but Wei and his disciple Yü take the medicine all the same and fall down dead, after which the two cautious disciples depart from the mountains. After they have gone Wei and Yü revive, rejoice in their faith, take more of the elixir and go the way of the immortals. The substance was probably considered to be an amalgam of mercury with gold or lead. Although the story is not in the original *Lieh Hsien Chuan*, some of which goes back to the − 1st century, its presence in the *Yün Chi Chhi Chhien* secures it a date prior to the + 11th. Besides its interest for the pre-history of animal experimentation, it must also have allegorical significance in connection with the death-and-resurrection philosophy of classical Taoism.[3]

Notices of the poisonous effects of elixirs occur from an early time in Chinese literature. We have seen the cautious attitude adopted by the emperors Wu of the Liang and Wên Hsüan of the Northern Chhi when they received elixirs from the alchemists. This was the orthodox position of the pharmaceutical naturalists, whose series of *Pên Tshao* compendia form so prominent a feature of Chinese

---

[1] Ch. 3, p. 34 *b*. See also Giles (1898).
[2] *Yün Chi Chhi Chhien*, ch. 109, pp. 5 *a* ff.; also Wu & Davis (1932).
[3] Cf. *SCC*, vol. 2, pp. 139 ff.

medieval chemical and biological literature. Thus Khou Tsung-Shih (fl. + 1110 to + 1119), the author of the *Pên Tshao Yen I* (The Meaning of the Pharmacopoeia Elucidated), pointed out very plainly the poisonous nature of mercury [cf. Figs. 91 (pl.), 92]. He says:[1]

Although there are medical recipes that include mercury, great care should be exercised because of its toxic effects. For a woman an overdose causes sterility. Nowadays people often use it for treating convulsion, fever, and increasing the flow of saliva (*ching jê hsien chhao*) in children. Not a word about these uses can be found in the (*Shen Nung Pên Tshao*) *Ching*.[2] Hence the matter requires further elucidation. (Mercury when) mixed with lead forms an amalgam (*ning*; lit. coagulates). It has an affinity for (*chieh*; lit. ties with) sulphur (forming mercuric sulphide). It breaks up (*san*; lit. disperses) when ground together with the fleshy part of a date. In one of its applications it can be heated to form face-powder (*ni fên*).[3] Calomel (*fên shuang*) ground with saliva can be used for killing fleas. Copper brightens when (mercury) is applied to it. If poured into a corpse mercury will delay putrefaction. Gold, silver, copper and iron will float when placed on mercury. With lead (*tzu ho chhê*) mercury is subdued (*fu*).

Han Yü of the Thang (period) said:[4] 'Li Chhien,[5] a Professor of the Imperial Academy (Thai Hsüeh Po Shih), met the alchemist Liu Pi,[6] a man from Hsin-an, who could heat mercury and convert it into an elixir. The method consisted of filling a reaction-vessel (*ting*)[7] with lead, leaving some space at the centre for the introduction of mercury, and when the cover was replaced and sealed all round, the contents was heated to form cinnabar.[8] After taking (the elixir) (Li Chhien) passed blood in urine and faeces, and after four years he got worse and died. I do not know exactly when the teaching about elixirs first began. (Although) it has resulted in countless deaths among the people, yet the art is still admired and sought after even today. (Many instances of) such delusion have already been recorded in the literature. Without quoting from hearsay evidence I shall now refer to six or seven persons known to me personally who came to grief after taking elixirs. I witnessed these things myself and set them forth here to serve as a public warning.

Kuei Têng, once Minister of Works (Kung Pu Shang Shu),[9] told me that he fell ill after taking mercury. He felt as if he was being pierced by a red-hot iron rod, from head to toe, so much so that it seemed that flames were coming out of his orifices and joints. He screamed aloud with the pain, and mercury was found in his mattress. After vomiting blood for over ten years he

---

[1] *Pên Tshao Yen I*, ch. 5, pp. 2a ff.; also in *Pên Tshao Kang Mu*, ch. 9, pp. 12a ff.

[2] The first of the pharmacopoeias, written *c*. − 1st century.

[3] It has a high content of calomel; cf. Read & Pak (1936), no. 45.

[4] In *Han Chhang-Li Wên Chi* (Collected Literary Works of Han Yü), ch. 34, pp. 11b ff., and quoted in abridged form by Chang Kao in *I Shuo*, ch. 9, pp. 18a ff. From here to the end a parallel passage, almost a quotation, is given in the *Pien Huo Pien* (ch. 4, pp. 6b ff.) written by Hsieh Ying-Fang in + 1348.

[5] Also known as Li Yü, but corrupted as Li Kan in the *I Shuo*, the *Pien Huo Pien* and the *Pên Tshao Kang Mu*. He died in + 823.

[6] He prepared elixirs for Hsien Tsung. See p. 317 above and also *Chiu Thang Shu*, ch. 16, p. 1b.

[7] See Ho & Needham (1959a).

[8] This should either be a lead amalgam, or else sulphur was included but not mentioned.

[9] See for example Kracke (1957).

Fig. 92. Retort still for mercury, from the *Thien Kung Khai Wu* of +1637 (Chhing drawing).

died. Li Hsü-Chung, a Palace Censor (Tien Chung Yü Shih), died (from mercury poisoning) with boils on his back. Li Sun, once Minister of Justice (Hsing Pu Shang Shu),[1] informed me that he had made a blunder by taking the elixir, and he died subsequently. Li Chien, an Executive Official of the Board of Punishments (Hsing Pu Shih Lang), died instantly (from elixir poisoning) while in excellent health. Mêng Chien, another Minister of Works (Kung Pu Shang Shu), invited me to see him at Mêng-chou and said to me in private, "I have acquired this secret drug, and as I do not wish to become an immortal all by myself, I am offering you some in a container. It should be made into pills with the fleshy part of the jujube date and eaten." One year after we parted from each other he was taken ill. Later someone went to enquire after him and found him saying, "What I ate before was the wrong drug, but I shall recover after getting rid of it (by taking an antidote)." But he died after an illness of two years. Again, (after taking an elixir) Lu Than, the Regional Commandant (Chieh Tu) of Tung-chhuan and Censor-in-Chief (Yü Shih Ta Fu), passed blood and suffered such unbearable pain in his body that he begged to have his life taken away. Li Tao-Ku, Commander of the Imperial Guard (Chin-Wu Chiang-Chün), punished the guilty Liu Pi by obliging him to take the drug himself;[2] so that he died at sea at the age of 50.

These (examples) should (be sufficient) to serve as warnings. Is it wise to try to avoid death in such a way as to hasten one's death instead? The five cereals, the three kinds of animals used for offerings (*san shêng*, i.e. fowls, fish and pigs), salt, vinegar, fruits and vegetables are the things that men normally eat; they constitute the essential foods which give us strength. Now misguided people clamour that the five cereals shorten human life and should be eaten as little as possible. They only feel sorry for themselves when death is approaching. How regrettable this is!'[3]

Nowadays there are people who roast mercury into cinnabar, but without being aware of (the poison) some physicians use (the cinnabar) for coating medicines or as ingredients of prescriptions. Is not this a great mistake and can we ever be off our guard?

Perhaps the strongest and most open condemnation of the use of poisonous substances by the alchemists came later from Li Shih-Chen (+1518 to +1593), the great pharmaceutical naturalist and physician who composed the *Pên Tshao Kang Mu* (Great Pharmacopoeia). It may suffice to quote just a few of the many remarks he made in his celebrated work. About the use of gold as an elixir ingredient he says:[4]

...Gold was rarely mentioned in ancient medical recipes, but has been the talk of the alchemists. In the *San-shih-liu Shui Fa* (Thirty-Six Methods for Bringing Solids into Aqueous Solu-

---

[1] It is strange that Li Hsü-Chung is supposed to have written one of the most important books on the divination of individual fates (see further *SCC*, vol. 2, p. 358).

[2] The *Chiu Thang Shu* (ch. 16, p. 1b) says that Li Tao-Ku and Huangfu Tsun had recommended him to the emperor. Cf. pp. 317, 318, 323.

[3] But the *Thiao Chhi Yü Yin Tshung Hua* by Hu Tzu of the Sung period says that Han Yü himself took sulphur at the end of his life and died. Here ends the quotation from Han.

[4] Ch. 8, pp. 6b ff.

tions)[1] of Huai-nan (Wang) it is converted into a liquid potion. In Ko Hung's *Pao Phu Tzu* the ingestion of (solid) gold is stated to be as effective as (the drinking of) potable gold.[2] According to this method when (gold) is treated one hundred times with the skin and lard of pigs, and with vinegar, it will soften. Alternatively it can be mixed with the bark of the 'stinking cedar' (*shu* or *chhu*)[3] or solubilised with wine made from the *mu ching* shrub (*Vitex negundo*)[4] and magnetite, or ingested with realgar and orpiment. The state of terrestrial immortal (*ti hsien*) can thus be attained. It is also said that cinnabar can be made into 'gold of the sages' (*shêng chin*), which brings about immortality when eaten. The (*Ming I*) *Pieh Lu* (Informal Records of Famous Physicians) (by Thao Hung-Ching) and (the *Pên Tshao Shih I* of) Chhen Tshang-Chhi also mention that taking (gold) over a long period will bring about the state of immortality. Such sayings must have been handed down by the alchemists since the time of Chhin (Shih) Huang (Ti) (reigned −221 to −210) and Han Wu (Ti) (reigned −140 to −87). However, (the alchemists) will never realise that the human body, which thrives on water and the cereals, is unable to sustain such heavy substances as gold and other minerals within the stomach and intestines for any length of time. How blind it is, in the pursuit of longevity, to lose one's life instead!...

Earlier he mentions[5] that the Empress Chia of Hui Ti of the Chin Dynasty was assassinated in +300 by being forced to drink a wine containing gold fragments (*chin hsiao chiu*).[6]

Li Shih-Chen criticised not only the alchemists but also some of the pharmacopoeias for the use of poisonous substances. In another place he says:[7]

...The *Ta Ming* (*Jih Hua Pên Tshao*) (The Pharmacopoeia of Ta Ming, Jih Hua Tzu)[8] alleges that (mercury) is not poisonous; the (*Shen Nung*) *Pên* (*Tshao*) *Ching* states that eating (it) over a long period will make a man immortal; Chen Chhüan maintains[9] that (mercury) is the 'mother' (i.e. the basic ingredient) of the 'cyclically-transformed' elixir (*huan tan*); and Pao Phu Tzu (i.e. Ko Hung) regards (it) as a prime medicine of longevity. I am not able to tell the number of people who since the Six Dynasties period (+3rd to +6th centuries) so coveted life that they took (mercury), but all that happened was that they impaired their health permanently or lost their lives. I need not bother to mention the alchemists, but I cannot bear to see these false statements made in pharmacopoeias. However, while mercury is not to be taken orally, its use as a medicine must not be ignored.

Already in the Sung a prophetic passage had been written by Shen Kua in his *Mêng Chhi Pi Than* (+1086) in which he had given expression to the view that

---

[1] See Tshao, Ho & Needham (1960).
[2] *Nei Phien*, ch. 4, pp. 18*a*, 20*b*, 21*a* and ch. 11, p. 19*a*. See also Feifel (1944, 1946).
[3] See Tshao, Ho & Needham (1960). [4] A medicinal plant; see Read (1936), no. 148.
[5] *Pên Tshao Kang Mu*, ch. 8, p. 6*a*. [6] See *Chin Shu*, ch. 31, p. 11*b*.
[7] *Pên Tshao Kang Mu*, ch. 9, pp. 12*b*, 13*a*.
[8] A pharmacopoeia composed *c.* +972 by Jih Hua Tzu, whose name was probably Ta Ming according to Li Shih-Chen.
[9] Author of the early +7th-century pharmacopoeia *Yao Hsing Pên Tshao*.

mercury compounds were valuable in certain ways and might become even more so if we only knew more about them. He says:[1]

My cousin Li Shan-Shêng and several friends of the same generation once transformed cinnabar into an elixir. More than a year afterwards the product was again put into a reaction-vessel to be purified, but then by mistake one piece was left behind. One of his students made it into a pill and ate it, after which he became delirious and died after one night. Now cinnabar is an extremely good drug and can be taken even by a newborn baby, but once it has been changed by heat it can kill an (adult) person. If we consider the change and transformation of opposites into one another, since (cinnabar) can be changed into a deadly poison why should it not also be changed into something of extreme benefit? Since it can change into something which kills, there is good reason to believe that it may have the pattern-principle (*li*) of saving life; it is simply that we have not yet found out the art (of doing this). Thus we cannot deny the possibility of the existence of methods for transforming people into feathered immortals, but we have to be very careful about what we do.

These words are quite remarkable when one considers the great use of organo-metallic compounds in modern medicine, e.g. mercury in salvarsan or antimony for kala-azar. One is also reminded of the profound conviction of Paracelsus centuries later that 'poisonous action and remedial virtue are intimately bound up with each other', as in the case of arsenic and especially mercury.[2] So also Paracelsus said: 'Die Dosis macht dass ein Ding kein Gift sei.'[3] Today it is fully recognised that a powerful remedy is almost of necessity also in certain circumstances a powerful poison.[4]

All the descriptions which we have of the symptoms exhibited by those suffering from elixir poisoning agree broadly with classical descriptions of intoxication with mercury, lead and arsenic. In mercury poisoning there is excessive salivation and ulceration of the gums, fever and bloody vomiting or purgation, together with severe 'hot' pains. Muscles may be affected by tremor or paralysis and there may be severe impairment of the special senses. The effects on the mind are very marked, leading through abnormal irritability to idiotic, melancholic or manic conditions. It will be remembered that of some of the elixir-taking emperors it was said *pu shih wan chi*—'he no longer knew what was going on around him'. The fact that mercury is excreted in the sweat and urine would account for the statement in another description that the metal was recovered from the mat on which the patient had lain.

[1] *Mêng Chhi Pi Than*, ch. 24, p. 7*b*. The story was told again by Khou Tsung-Shih in his *Pên Tshao Yen I* (+1116), ch. 4, p. 2*a*.   [2] Pagel (1958), p. 145.
[3] Lieben (1935), pp. 13 ff.   [4] Green (1954), p. 19.

Symptoms of chronic lead poisoning may also be detected in the descriptions of the effects of elixirs. Though no observation seems to have been made of the famous blue line on the gums, loss of appetite, headache, tearing pains, colic, cramps, paralysis and 'wrist-drop' all appear. As in the 'Poitiers colic' described by Citois in +1616, there are accounts of formication or needling pains all over the skin, and of loss of control of elbows, hands and leg joints. Colic, cramps, vomiting and purgation occur also in arsenic poisoning. The increased appetite in the early stages of the process may have been something of a trap for the alchemists.

It would be well worth while to follow up in the medical literature discussion of poisoning by metals and minerals; here we shall only remark that Sung Tzhu in his *Hsi Yuan Lu* (The Washing Away of Wrongs)[1] (+1247) was well acquainted with their effects in the context of forensic medicine. Thus he gives as a test for mercury poisoning the formation of a superficial amalgam on a piece of gold plunged into the intestine or tissues.[2] He also describes[3] the colic and cramps of arsenic poisoning, as well as the discharge of blood, and gives[4] several antidotes including emetics which may be used. He also treats[5] of calomel poisoning. Similar material is found in later works such as the *Wu Yuan Lu* (Cancelling of Wrongs) of Wang Yü (+1308).

Mercury compounds were of course habitually used in Chinese medicine,[6] but in general one gets the impression that the dosages prescribed by the physicians were much smaller than those used by the alchemists. For example, Sun Ssu-Mo's chief medical work, the *Chhien Chin I Fang* (Supplement to the Thousand Golden Remedies), about +665, gives a series of medicaments containing 1 part of cinna-

---

[1] This has been translated by Giles (1924).
[2] Ch. 3, p. 39a. That this goes back to the Han or at least the Chin is shown by a statement in the *Chin Kuei Yao Lüeh* (ch. 25, p. 99) by Chang Chung-Ching of the +2nd century, edited by Wang Shu-Ho of the +3rd.
[3] Ch. 3, p. 38a.   [4] Ch. 4, p. 15a.   [5] Ch. 4, p. 17a, b.
[6] It is worth noting that in the history of Chinese medicine there is nothing corresponding to the European +17th-century quarrel between the Galenists and the Chymists. The Chinese had never confined their pharmacopoeia to plant drugs—mineral and animal material had always been admitted. In using mineral substances the medieval Chinese physicians were seeking, like their colleagues in medieval and Renaissance Europe, for emetics, cathartics, diuretics, diaphoretics, tonics, anodynes, and drugs promoting suppuration. The armamentarium attained by the European chymists of the early +17th century, as shown by Croll's *Basilica Chemica*, has been analysed by Multhauf (1954), and would be well worth comparing with the pharmacopoeias of Sung and Yuan. One interesting point is that antimony preparations are almost absent from medieval Chinese medical chemistry, though in that of the Renaissance as prominent as mercury and arsenic, and this is particularly strange, for China has in Southern Hunan and Kuangsi the largest deposits of antimony in the world. We suspect that there were local uses which did not get into the pharmacopoeias, as in the outstanding case of *Rauwolfia*.

bar in from 14 to 40 parts of other mineral and plant drugs (2½–7 per cent).[1] Yet in the beginning of the book, in his entry for cinnabar among the minerals, he repeats the time-honoured doctrines that it is not poisonous, and can transform people into immortals. And in the alchemical literature formulae are attributed to him containing only minerals and including from 35 to 90 per cent of cinnabar and calomel. This we may see for instance in the *Thai-Chhing Chen-Jen Ta Tan* (The Great Elixirs of the Adepts; a Thai-Chhing Scripture).[2] Unfortunately the dosages of the alchemists are very seldom given, and the recipes just mentioned use 'the lard of an old sow' as the vehicle of administration so that one can hardly tell how much the aspirants to immortality took. As time went on the physicians' dosages seem to have become more and more conservative, for in the *Wai Thai Pi Yao* (Important Medical Family Practice of a Frontier Official) of Wang Thao (+752) one finds prescriptions in which cinnabar occurs to the extent of only 1 part in 356 of other mineral and plant material (about 0·28 per cent).[3] These formulae are described among the general tonics or panaceas.

Perhaps the medieval Chinese alchemists should not be too strongly condemned for their heroic medication with mercury. After its introduction as a cure for venereal diseases in +16th-century Europe it was applied indiscriminately and excessively down to the end of the +19th century, for malaria, dysentery and other febrile conditions, even to the extent of doses as lethal as those of the Taoists. For syphilis indeed mercury justly remained down to the time of introduction of salvarsan (1910) the principal, and almost the only, treatment.[4]

Actually many a Chinese alchemist must himself have perished as the result of his experiments. It is, however, difficult to find such instances recorded in Taoist writings, since they would have militated against the characteristic doctrine that physical death was avoidable. We can only infer such instances from certain descriptions of *shih chieh* ('release from the mortal part'),[5] *yü hua* ('taking flight to attain the state of immortality'), *kao hua* ('announcing the change to the state of immortality'), etc.,[6] although we can seldom tell whether these were the direct

[1] Ch. 20, p. 232.2.
[2] Reproduced in *Yün Chi Chhi Chhien*, ch. 71, pp. 1a ff. (esp. 2b and 3b).
[3] Ch. 31, p. 849.
[4] Calomel too was so greatly valued for its diuretic properties in the treatment of dropsy that over-prescription led to abandonment, and its reintroduction by Jendrassik in the late 19th century has been described by Sollmann as amounting virtually to a re-discovery.
[5] See *SCC*, vol. 2, p. 141.
[6] No less than 19 such terms are listed in the *Li Shih Chen Hsien Thi Tao Thung Chien* (Comprehensive Mirror of the Adepts who attained the Tao throughout History), no. 293 in Wieger (1911), a vast tractate probably of the Yuan dynasty, by Chao Tao-I.

consequences of elixir poisoning. In the *Yün Chi Chhi Chhien* (The Seven Tablets of the Cloudy Satchel), a great compendium compiled *c.* +1022 by Chang Chün-Fang, it is mentioned[1] that no visible sign of putrefaction was noticed in the body of Sun Ssu-Mo more than a month after his death. Possibly this great +7th-century alchemist and pharmacist had taken one of the many elixirs containing mercury or arsenic described in his own work *Thai-Chhing Chen-Jen Ta Tan* already mentioned. His secret recipe for the making of the gold elixir (*chin tan*), for example, consists of 8 oz of gold, 8 oz of mercury, 1 lb of realgar (arsenic disulphide) and 1 lb of orpiment (arsenic trisulphide).[2]

The alchemists themselves were quite aware of the phenomena of elixir poisoning. Accounts of about thirty-five common mistakes made in the preparation of elixirs are mentioned in the *Chen Yuan Miao Tao Yao Lüeh* (Classified Essentials of the Mysterious Tao of the True Origin of Things)[3] a text probably of the +8th or +9th century, but parts of which may go back to the +4th century, the time of its putative author Chêng Ssu-Yuan. Among the warnings against mistakes we find three clear types of elixir poisoning. The author records cases where people died from eating elixirs made from cinnabar, mercury, lead and silver; cases where people suffered from boils on the head and sores on the back by ingesting cinnabar obtained from roasting together mercury and sulphur; and cases where people became seriously ill through drinking 'liquid lead' made by heating 'black lead'. The Thang alchemist Chhen Shao-Wei asserts in his work *Ta-Tung Lien Chen Pao Ching, Chiu Huan Chin Tan Miao Chüeh* (Mysterious Teachings on the Ninefold Cyclically-transformed Gold Elixir according to the Precious Manual of the Re-casting of the Primary (Vitalities); a Ta-Tung Scripture)[4] that the metals iron, copper, silver, lead, tin and gold are all poisonous.

According to the attitude which they adopted towards elixir poisoning the Chinese alchemists may be divided, broadly speaking, into two different schools. The first ignored the poison danger altogether and considered the symptoms supervening after taking elixirs to be quite normal and even essential. The second recognised the poisonous nature of some of the constituents and tried to neutralise it in one way or another, or else to use only substances which were harmless.

The *Thai-Chhing Shih Pi Chi* (Records in the Rock Chamber: a Thai-Chhing

[1] Ch. 113, p. 20*a*.　　　　　　　　[2] *Yün Chi Chhi Chhien*, ch. 71, p. 9*b*.

[3] No. 917 in Wieger (1911). This is the book which contains the oldest formula in any civilisation for a proto-gunpowder; it included arsenic besides saltpetre and sulphur, with dried honey as the source of carbon. See further *SCC*, vol. 5. Cf. p. 80 above.

[4] No. 884 in Wieger (1911).

Scripture),[1] a text written probably in the + 6th century, but containing materials of the + 3rd, describes after-effects which it regards as quite normal, and recommends methods of bringing relief when they occur. It says:[2]

After taking the elixir one feels an itch all over the body and the face, rather like having the sensation of insects crawling over one. The body, the face, the hands and the legs may become swollen. One may experience a feeling of repulsion at the sight of food, and vomiting usually follows after a meal. One feels rather weak in the four limbs. Other symptoms include frequent defaecation, vomiting, headache and pains in the abdomen. No alarm should be caused by these effects, because they are due to the work of the elixir in dispelling all the inherent disorders (in the human body).

To help the working of the elixir the book recommends that after eating it one should avoid tasting mutton or carp or any (cooked) blood, or inhaling anything with a 'false' odour. The treatment it suggests for relieving the symptoms is as follows:

When the elixir takes effect one should immediately bathe oneself with hot and cold water and take a mixture of scallion, soya-bean sauce and wine. The same cure is to be recommended (for relieving the after-effects of) *hsün huang* (a dark variety of realgar).[3] If relief does not come, then a hornets' nest (*lu fêng fang*),[4] some *Euphorbia sieboldiana* (*kan sui*),[5] some Solomon's Seal (*Polygonatum officinale*) (*wei jui*)[6] and some *Ephedra sinica* (*ma huang*)[7] may be separately extracted with boiling water and combined for use as a medicine. One dose is sufficient to bring relief.

The second school is typified by Chang Yin-Chü, an alchemist of the + 8th century, who in his book *Chang Chen-Jen Chin Shih Ling Sha Lun* (A Discourse on Metals, Minerals and Cinnabar, by the Adept Chang)[8] emphasised the poisonous nature of gold, silver, lead, mercury and especially arsenious acid, and said that he had witnessed many cases of premature death brought about by consuming cinnabar. He believed however that the poison could be rendered harmless by a proper choice and combination of ingredients; for example gold should be used together with mercury, while silver can only be used when combined with gold, copper carbonate and realgar for the preparation of the gold elixir (*chin tan*). Wu Wu, a + 12th-century alchemist and author of the famous text on laboratory procedures, the *Tan Fang Hsü Chih* (Chemical Elaboratory Practice),[9] says in

---

[1] No. 874 in Wieger (1911). [A draft translation of this book has been made by Ho Ping-Yü.]
[2] Ch. 2, p. 7*a*.   [3] See Read & Pak (1936), no. 49*a*.
[4] See Read (1941), no. 6.   [5] See Stuart (1911), no. 169.
[6] *Ibid.* no. 340, and Read (1936), no. 688; used in Chinese medicine as a tonic and sedative.
[7] An antipyretic containing the important alkaloid ephedrine; see Stuart (1911), no. 161.
[8] No. 880 in Wieger (1911).   [9] No. 893 in Wieger (1911); see Ho & Needham (1959*a*).

another work, the *Chih Kuei Chi* (Guide to the Way of Return),[1] that lead and mercury are the two essential elixir ingredients and that the 'four yellow substances' (sulphur, orpiment, realgar and arsenious acid)[2] and the 'eight minerals' (sulphur, realgar, orpiment, gold, silver, copper, iron and lead)[3] are poisonous. Wu also says:[4]

Moreover, alum is potent enough to kill a tiger and sal ammoniac contains sufficient poison to make copper deteriorate. If such ingredients are added to mercury and then taken into the human stomach ten thousand deaths will follow in every ten thousand instances.

The *Tan Lun Chüeh Chih Hsin Chien* (Handbook of the Secret Teaching concerning Elixirs),[5] a text of uncertain date, but written by Chang Yuan-Tê not later than +1020,[6] had similarly pointed out earlier that the 'four yellow substances' and the 'eight minerals' were poisonous, and advocated the use of only lead and mercury as elixir ingredients.

How poisons could be removed from elixir ingredients or rendered harmless was described in great detail in the large tractate *Huang Ti Chiu Ting Shen Tan Ching Chüeh* (Explanation of the Yellow Emperor's Manual of the Nine-Vessel Magical Elixir),[7] a text compiled during Thang or early Sung, but incorporating some material as old as the +2nd century. It says:[8]

The five metals and the three *hung* (mercuries) together with the nine *chhien* (leads) and the eight minerals are all poisonous. Without procuring the original formulae of the ancient (masters) any attempt to use newly acquired recipes described verbally in a few words is doomed to failure. Hence Hu (Kang) Tzu[9] says, 'the five metals have to be purified from all poisons caused by heating. If they are not properly treated the poison will turn to a powder form, and if (the ingredients) are used for making elixirs without having their poison removed, and ingested for any length of time, death will be caused when the rules are not followed.'

Elsewhere this text says:[10]

The ancient masters (lit. sages) all attained longevity and preserved their lives (lit. bones) by consuming elixirs. But later disciples (lit. scholars) have suffered loss of life and decay of their bones as the result of taking them.

---

[1] Cf. p. 1*a*; no. 914 in Wieger (1911).   [2] See Ho & Needham (1959*b*).
[3] *Ibid.*   [4] P. 4*a*.
[5] No. 928 in Wieger (1911).
[6] Because quoted in the *Yün Chi Chhi Chhien* (ch. 66, p. 1*a*), which gives the author's name as Chang Hsüan-Tê.
[7] No. 878 in Wieger (1911).   [8] Ch. 3, p. 6*b*.
[9] Cf. *SCC*, vol. 4, pt. 1, p. 308, on magnetism and the geomantic compass.   [10] Ch. 10, p. 1*a*.

The general method known to the ancients of rendering the ingredients harmless, according to this tractate,[1] was to treat them with wine made from the *mu ching* shrub (*Vitex negundo*)[2] or with saltpetre (*hsiao shih*) and vinegar. Another method of removing the poison from mercury was to '(put it) in wine three years old, add sal ammoniac and boil it for 100 days'.[3] Yet another consisted of 'boiling (mercury) in vinegar containing fragments of gold and silver'.[4] The poison in realgar was to be removed by 'warming it in vinegar contained in a copper vessel over a gentle fire'[5] and that of lead by 'heating it in vinegar together with red salt and cinnabar'.[6]

The preceding school of thought can be understood better by reading the *Yin Chen-Chün Chin Shih Wu Hsiang Lei* (The Similarities and Categories of the Five (Substances) among Metals and Minerals, by the Adept Yin),[7] a text purporting to be of the +2nd century, which points out that elixir poisoning is due to the lack of understanding of alchemical theory. This, as we shall soon see, means precisely the theory of categories which we have elsewhere described.[8] Of sal ammoniac (*nao sha*) we read:[9]

If too much of it is used the (final) product will certainly cause death. It rapidly attacks the five viscera and also the five metals. These effects never fail. But if the finer principles are known the real use of sal ammoniac can be understood, indeed, the adepts use this substance to attain the great Tao.

Of saltpetre we read:[10] 'It is the essence of the Yin minerals... The adepts employ it to control (*chih*) the poison (of substances belonging to) the Greater

---

[1] Cf. ch. 8, p. 9*b*.

[2] No. 456 in Stuart (1911); used as astringent and sedative.

[3] Ch. 11, p. 5*b*.

[4] Ch. 11, p. 6*b*. If all these instructions meant that the aspirant should drink only such 'extracts' this may well have been a means of taking practically no mercury at all.

[5] Ch. 14, p. 3*b*.

[6] This +10th-century use of vinegar is interesting in connection with the preparation of acetates, sulphides and chlorides of metals by John of Rupescissa and Paracelsus in later Europe (cf. Multhauf (1954*a*), p. 365). So also the cathartic action of wine heated with mercury (*purgatio mercurii optimi*) will have been as well known to the physicians of the Sung as to the chymists of the European +17th century (cf. Multhauf (1954*b*), p. 112). Some of the Chinese instructions give one the impression that insoluble salts of the poisonous metals may sometimes have been obtained, in which case the ingestion of the filtrates or distillates would again have been fairly harmless. There is a close Paracelsian parallel here, for when from the +14th century onwards salts of metals were distilled to form 'healing quintessences', nothing came over except water and the volatile mineral acids, so that the preparations contained nothing of the metals (cf. Sherlock (1948), pp. 52, 56; Multhauf (1956), pp. 339 ff.; (1954*b*), p. 117; and Pagel (1958), p. 274).

[7] No. 899 in Wieger (1911).   [8] See Ho & Needham (1959*b*).

[9] P. 20*a*.   [10] P. 20*b*.

Yang.' The same paragraph adds that 'sal ammoniac is the essence of the Yang minerals', which implies that it could be used to remove or to neutralise the poison of the Greater Yin, for example mercury. This is in effect the method for rendering mercury non-poisonous as described in the *Huang Ti Chiu Ting Shen Tan Ching Chüeh*.[1]

In other words, the ingredients selected had to be governed by the theory of categories. The importance of understanding the theory is emphasised in the following sentence:[2]

Sulphur (*liu huang*) is also employed in certain procedures for attaining a state of middle-grade immortality. (Substances) of similar categories (*hsiang lei*) must be used as its walls and chamber, (or to serve) as its officials and assistants, (or to supply it with) energy and strength. Those who know (these principles) will become middle-grade immortals (*chung hsien*)—those who use (sulphur) without understanding (them) will instantly turn into corpses.

The *Hsüan Chieh Lu* (Mysterious Antidotarium),[3] a work prefaced by an anonymous writer in +855, recognises the poison in the elixir ingredients, but recommends a potent herbal composition which would serve both as an elixir and as an antidote for ordinary elixir poisoning. The text takes the form of a dialogue between an adept named Chiu Hsiao Chün and one Liu Hung about the year +122. It deals with poisons in the ingredients and says that they cause death, or, in less severe cases, malignant boils. It then describes the preparation of such an antidote, *Shou Hsien Wu Tzu Wan* ('the five-herbs immortality-safeguarding pills') as follows:

Procedure for the making of the *Shou Hsien Wu Tzu Wan* pills: *yü kan tzu* (*Phyllanthus emblica* or Indian gooseberry),[4] *fu phên tzu* (*Rubus coreanus* or wild raspberry),[5] *thu ssu tzu* (dodder or *Cuscuta japonica*),[6] *wu wei tzu* (*Schizandra chinensis* and *Kadsura japonica*),[7] *chhê chhien tzu* (plantain or *Plantago major*).[8] Take 5 oz of each of the above 5 ingredients and pound them separately to a powder like flour. Take the young stems and leaves of the *kou chhi* (*Lycium chinense*),[9] during the second or the third months, pound them until 2 *shêng* (about 2 pints) of juice is obtained. Mix the juice with the powder and leave them to dry well. Take the *lien tzu tshao* (ink plant or *Eclipta alba*)[10] during the seventh or the eighth month and extract its juice. Mix one *shêng* (about 1 pint) of the juice with the ingredients and leave to dry.

---

[1] See p. 332.
[2] *Yin Chen-Chün...*, p. 25b.
[3] No. 921 in Wieger (1911).
[4] Read (1936), no. 330.
[5] *Ibid.* no. 457, also Stuart (1911), no. 383; a tonic.
[6] No. 156 in Read (1936) and no. 140 in Stuart (1911); a diaphoretic, tonic and purgative.
[7] Nos. 512 and 507 in Read (1936) and no. 398 in Stuart (1911); a tonic and lenitive.
[8] No. 90 in Read (1936) and no. 335 in Stuart (1911); a diuretic, antirheumatic, astringent and tonic.
[9] No. 250 in Stuart (1911); a tonic.
[10] No. 160 in Stuart (1911); an astringent said to tighten the teeth.

Take 1 *shêng* (1 pint)[1] of almonds (*hsing jen*), put it in 5 *shêng* (pints) of good wine contained in a silver vessel and heat until the almond loses its bitter taste. Heat half a *shêng* (about half a pint) of the juice extracted from *sêng ti huang* (*Rehmannia glutinosa*),[2] 5 oz of soya bean curd (*chen su*) and 5 oz of deer glue (*lu chio chiao*)[3] together, grind the mixture to powder and put it into the previous liquid.

Warm the liquid and the mixture gently and then introduce the five ingredients. Stir vigorously with a comb made from the wood of a willow-tree (*liu pi*). At the appropriate degree of dryness make the product into pills by hand. The pills should be about the size of the *wu thung tzu* (nuts from *Sterculia platanifolia*). The dosage is 30 pills a day taken with wine, but this can be varied to suit the circumstances. Avoid eating pork, garlic, mustard and turnips (at the same time).

We cannot find anything in the above formulary other than mild tonic agents.

The *Hsüan Chieh Lu* is the oldest printed book in any civilisation on a scientific subject.[4] It was first issued by Hokan Chi (possibly its author) between +847 and +850.

Let us first recapitulate the points which have been touched upon in this study. The belief that it was possible to prepare an elixir of immortality from metallic and mineral substances, as vital as quicksilver and as permanent as the rocks and hills, was ancient and classical in China. Partly because the operations required for its preparation were expensive, and partly because emperors and high officials throughout the ages considered themselves eminently suitable for survival, the alchemists tended naturally enough to congregate in the neighbourhood of courts and to throng the vestibules of the great. In the Thang period the prolongation of life by elixirs became a veritable *idée fixe*, and each imperial reign tended to follow a cyclical order of events, the incoming emperor fortified by Confucian advisers executing the alchemists to whom was attributed the demise of his predecessor, but gradually himself falling a prey to the pretensions of new adepts and expiring in his turn from the effects of their labours. If we are right in believing that the elixir poisoning was mainly due to the toxic properties of mercury, arsenic and lead, the last days of many Chinese emperors must have been exceedingly disagreeable.

Examples have been given in this paper of what might almost qualify as a particular literary genre, the memorials presented by Confucian scholars and

---

[1] Probably 1 lb. The two words *shêng* (pint) and *chin* (lb) if not written carefully will look very much alike.
[2] No. 371 in Stuart (1911); an alterative and tonic. Cf. *Pên Tshao Kang Mu*, ch. 16, p. 3*a* to p. 7*a*.
[3] No. 364 in Stuart (1911).
[4] Another text in the Taoist Patrology, the *Yen Mên Kung Miao Chieh Lu* (The Venerable Yen Mên's Record of Marvellous Antidotes), no. 937 in Wieger (1911), is identical with the *Hsüan Chieh Lu*, but without the prescription for the potent anti-elixir which was itself an elixir.

ministers against the taking of elixirs. As we have seen, moreover, many emperors and high officials adopted a wisely cautious attitude towards the preparations which were offered to them. In certain cases this led to procedures which take their place in the history of regular therapeutic trials and experiments involving the use both of men and animals, a development of biological methodology in which the Chinese seem to have been at least as advanced as people in other cultures. There was also a good deal of speculation about the deleterious effects of elixirs, sometimes following the lines of the characteristic category theories of Chinese alchemy, but sometimes recognising (as early as the + 11th century) that intimate association between high toxicity and powerful beneficial action which was to impress Paracelsus so greatly five hundred years later. As the physicians in China were never affected by the Galenic orthodoxy which permitted the use of plant drugs alone, they continued from early times to prescribe metallic, mineral and animal substances, but it seems fairly clear from a comparison of medical with alchemical tractates and treatises that the posology of the doctors was generally much more conservative and cautious than that of the alchemists. At the same time many of the latter were very aware of the dangers of elixir poisoning, and sought, either by means of category theory or by the compounding of what they believed to be effective antidotes, to minimise or to avoid it. By the Ming period (+ 16th century) alchemy in China had gone into a profound decline, leaving the victory to the pharmaceutical naturalists, who fulminated against elixirs of any kind whatsoever.[1]

It is evident that this general picture does not correspond at all with the course of events in Europe. While in the Western Middle Ages and early Renaissance poisoning by alchemical elixirs was by no means unknown, it never played the same prominent part as in China.

If one turns over the pages of a work such as Lynn Thorndike's *History of Magic and Experimental Science* one is hard put to it to find much to the purpose. For example, about + 1360 Thomas of Bologna made an elixir from gold and mercury which 'was reputed something sinister', and he was suspected of trying to poison kings and dukes.[2] A hundred years or so later anonymous alchemical letters assert that a single drop of the elixir of projection has wonderful medicinal

---

[1] This was not the converse of the victory of the Chymists over the Galenists in Europe, for the pharmaceutical naturalists throughout Chinese history had dealt with mineral and animal drugs equally with those of plant origin—what they finally succeeded in stopping was the massive alchemical dosages of poisonous metals and minerals.

[2] Cf. Thorndike (1923–58), vol. 3, pp. 611 ff.

properties, but if it is made from mercury and sulphur instead of 'water of gold' it will be corrosive and poisonous. At Florence one Alexander Tarentinus and his servant Arnelius died instantly from the effects of such a poisonous elixir.[1] But so little did these toxic phenomena enter into medieval European alchemy that the very word 'poison' will scarcely be found in the indexes of works devoted to the history of the subject.

Perhaps the great reason for this difference is that Europe did not have the same conception of *material* immortality as China. In the West there was a rather clear idea of human survival after death which derived from origins both Hebrew and Christian; heaven, hell and even purgatory were real for both Latins and Greeks in Christendom. The elixir of life, though acculturated to some extent from the Islamic world, where it had been hardly more at home and certainly derived from further east, was always far less important than the philosopher's stone which would transmute the ignoble metals into gold. From the time of the mystical proto-chemists[2] of the Hellenistic world onwards, projection by the Stone was primarily for the purpose of acquiring material wealth (even though this might be idealised as eleemosynary in aim). At the same time we need not deny that the operations could be, and doubtless often were, undertaken with the parallel purpose in mind of purifying the soul of the operator from spiritual dross just as the lead or iron was freed from its base elements and raised to the level of gold.[3] Prolongation of life, rejuvenation, longevity and well-being in old age some tincture derived from the Stone might give, but where could an alchemical immortality, properly speaking, be spent? This world, so often castigated by the preachers as a justly uncomfortable antechamber to heaven, was hardly inviting enough to recommend itself, and the next was already departmentalised to the full.

Far different were the Chinese conceptions. Of an individual 'soul' there was no clear conviction in any of the great Three Doctrines. Confucianism had always declined to discuss personal survival, in the explicit interest of high social morality in the here and now,[4] while for Buddhist philosophy the belief in an individual

---

[1] *Ibid.* vol. 4, p. 347.

[2] [On these 'aurifictors', cf. pp. 76, 284 above, and p. 416 below.]

[3] As is well known, the psychological interpretation of alchemy has received an epoch-making treatment in the hands of C. G. Jung. Here we shall only refer however to the brilliant book of Eliade, which sets this very strikingly in the context of all alchemy and practical chemical technology from ancient times onwards. [In China also there was a long-standing distinction between the chemical art (*wai tan*) and the physiological training (*nei tan*), but, as we shall show elsewhere, the latter was equally practical, not allegorical-mystical as in the West.] [4] Cf. *SCC*, vol. 2, p. 13.

persisting soul was a positive heresy.[1] The Taoists, most relevant in the context of alchemy, recognised a considerable number of spiritual essences, even godlings, in the human body–soul complex, almost as many indeed as the limbs and viscera of the human organism itself, but there was no place other than earth for them to inhabit as a coherent entity, and after death they simply dispersed, some rising to join the *pneuma* (*chhi*) of the heavens, some sinking to mingle with that of the earth, and others disappearing altogether.[2] But there had also been from the beginning of Taoism in the Warring States period, as far back perhaps as the − 5th century, a firm belief in immortals (*hsien*), ethereal purified beings, originally feathered like birds, possessed of magical powers and wandering for ever without material needs among the mountains and forests, there eternally to enjoy the contemplation of the beauty of Nature, the outward and visible form of the Tao. This state of bliss, in which his spiritual components were purified but not dispersed so that a man remained recognisably himself, could be attained by him through the practice of specific Taoist techniques. By no means all of these were forms of asceticism as the West understood it, for they comprised not only dietary regimen and abstentions, but also gymnastic and sexual techniques, forms of heliotherapy, liturgical rite and sacrifice, and last but not least pharmaceutical procedures in which the making of elixirs was paramount.[3] Now one begins to understand the importance of the 'water of life', *soma–hraoma*,[4] Kuan-Yin's *amṛta*,[5] the drug of deathlessness (*pu ssu chih tshao*), the gold elixir (*chin tan*), in Chinese thought. Where else could the individual in medieval China turn? Heaven or paradise in any seriously credible sense did not exist, but the visible world was eternal and uncreated, nor would it ever cease, and he who could make himself worthy might continue to enjoy it with sense-perceptions perpetuated but purified. This was the inner meaning of the proverbial salutation *wan shou wu chiang* ('Life world without end!'). Thus the temptation to believe the claims of the alchemists was in China particularly strong, and one can see that an almost heroic Confucian austerity must sometimes have been needed to prevent men of high poetic sensibility not only from taking an elixir themselves but from inducing those near and dear to them to take it too.

As for the alchemists themselves, we are not disposed to over-estimate the

[1] Cf. *SCC*, vol. 2, p. 401.
[2] Cf. *SCC*, vol. 2, p. 153 and further Maspero (1950).
[3] See further Maspero (1937).
[4] See Renou & Filliozat (1947), vol. 1, pp. 347 and 355; cf. also Dubs (1947).
[5] Cf. Masson-Oursel *et al.* (1933), p. 159.

number of conscious charlatans among them. The urge to investigate the chemical behaviour of bodies was probably at least as great in medieval China as in other times and places, and so strong was the faith in the feasibility of elixirs that many an alchemist must himself have fallen a martyr to his own beliefs, or even to mistakes in following the obscure and contorted instructions of his predecessors. Indeed it would probably not be going too far to suggest that the elixir mania may have acted as a real inhibiting factor for the progress of chemical knowledge during the medieval Chinese centuries, for often no doubt the most experienced or industrious experimenters were the most enthusiastic believers and in the end the surest victims. The story of the death and resurrection of Wei Po-Yang is the epitome and type of this.

But if such were the miseries of Chinese alchemy, its grandeur lay in the fact that unlike that of Europe it was from the beginning iatro-chemistry. Ko Hung and Tuku Thao needed no Paracelsian Luther[1] to persuade them that the true business of alchemy was not to make gold but to make medicines. The Chinese alchemists, not only in outstanding examples like Thao Hung-Ching and Sun Ssu-Mo, but normally and universally, had always been pharmacists at the same time. And if the elixir of material immortality proved in the end to be as much of a will-o'-the-wisp as perpetual motion (or has it yet done so?), Chinese iatro-chemistry was destined to be subsumed in due time into the universal science of modern chemo-therapy.

---

[1] Though Paracelsus in the +16th century will always remain the standard-bearer of this movement, it is now realised that certain alchemists of the +14th really initiated it, notably John of Rupescissa (fl. +1325 to +1350); cf. Multhauf (1954a), Jacob (1956), Pagel (1958), pp. 263 ff.

HYGIENE AND PREVENTIVE MEDICINE
IN ANCIENT CHINA[1]

[1962]

INTRODUCTION

MUCH has been written in Western languages (not all of it very well informed) on the history of medicine and the allied sciences in the Chinese culture-area. In the course of work now in progress on a general history of science, scientific thought, and technology in that civilisation, we are making a study of the most important contributions of Chinese thought and practice to medicine and the biological sciences throughout the historical period. This subject not only has great comparative significance for the history of biology as such, but it is also of some topical interest in that the Chinese today are making a great effort to recover and validate all that was of value in their own cultural heritage in this field. They are systematically investigating the contributions of the past in the hope of integrating them into a modern medical science.[2]

Here we should like to offer to the readers of this volume some observations on the historical concepts of hygiene and preventive medicine in ancient China. By this we have in mind mainly the period down to the end of the Han (c. +200). After that time Indian influence, entering China with Buddhism, brought considerable foreign importations in medical philosophy; these we shall not here consider. Moreover, in the early times the relations between medical philosophy and the thoughts of the writers on philosophy, ethics, and logic were especially close; for example, it is interesting that throughout the ages the duty of the

---

[1] Reprinted from *Journal of the History of Medicine and Allied Sciences*, 1962, **17**, 429.

[2] This situation contrasts rather sharply with that in India and Ceylon, as one of us (J.N.) had good opportunity of seeing during the spring of 1958. Ayurvedic medicine in the Indian culture-area deserves just as careful study as that which Chinese traditional medicine is receiving, but unfortunately most of the Indian and Ceylonese traditional practitioners are unwilling to seek for any integration with modern Western medicine. Thus the investigation by modern methods of the traditional pharmacopoeia and all the persisting therapeutic procedures is rendered very difficult. It is easy to think of sociological reasons accounting for this difference between the Chinese and Indian culture-areas, and it is to be hoped that the inhibitions in the latter will soon be overcome.

Confucian literati to care for their parents according to the precepts of filial piety was a powerful influence in assisting the growth of medical science. It must, of course, be understood that we have to deal with ancient and early medieval conceptions which in practice had only a limited application to the life of the mass of the people as a whole. Nevertheless, it is reasonable to believe that the success of Chinese culture in maintaining its characteristic ethos throughout so large a population over the centuries was in part due to the contributions of those who concerned themselves with hygiene and preventive medicine. With these considerations in mind and notwithstanding our chosen limitation of period, we shall not hesitate to quote from sources as late as the end of the Thang dynasty (i.e substantially from the whole of the + 1st millennium).

Here we cannot enlarge upon the philosophy of Nature which was current among the ancient Chinese. It must suffice to say that this philosophy was based upon the idea of two fundamental forces, the Yang and the Yin, the former representing the bright, dry masculine aspect of the universe, the latter, the dark, moist feminine aspect. This conception is probably not older than the − 6th century.[1] Besides this, there had arisen from the school of Tsou Yen in the − 4th century the universally accepted doctrine of the Five Elements.[2] As is well known, these differ from the elements of the Greeks and other peoples in that they comprised not only fire, water and earth, but also wood and metal. Another doctrine prominent in ancient Chinese thought was that of the Macrocosm and the Microcosm.[3] A great interdependence of the State on its people, and of the health of the people on the cosmic changes of the Four Seasons, was envisaged. The elements were associated together in 'symbolic correlations' with many other natural phenomena in groups of five,[4] and these conceptions were applied in a remarkably systematic way to the structure and function of the living body of man. As might be expected in a society which was developing the characteristic form of bureaucratic feudalism, great importance was attached to the *prevention* of trouble, both in the political and personal life of the people, rather than to its control when it arose, and thus in the field of medical thought prevention was considered better than cure.

One further aspect of ancient Chinese nature-philosophy demands allusion here, namely, the doctrine of *pneuma* (*chhi* and *fêng*). As Filliozat has shown in

---

[1] Cf. *SCC*, vol. 2, pp. 273 ff. Ancient Chinese medicine was full of other antitheses, for example, symptoms manifested externally (*piao*) or internally (*li*), and fever (*jê*) as against chill (*han*).
[2] *SCC*, vol. 2, pp. 243 ff. [3] *Ibid.* pp. 294 ff.
[4] *Ibid.* pp. 261 ff.

a classical monograph, the conceptions of Greek pneumatic medicine can be paralleled with much precision in the Indian ideas of *prāṇa* found in the great Suśruta and Caraka corpuses. *Chhi* and *fêng* in China are nothing other, but the former at least cannot have reached China from India since it appears there too early; one wonders therefore whether a Babylonian influence did not (as in so many other cases) spread both east and west. While in later times (e.g. in the thought of the + 13th-century Neo-Confucians) *chhi* came to mean all forms of matter, from the most condensed to the most tenuous; in ancient China it referred rather to subtle matter (comprising what we should now think of as gases and vapours, radio-active emanations, radiant energy, etc.) and invisible biological influences (including nerve-impulses, hormonal actions, infection and contagion). In medical thought *chhi* was something like a vital force in living mind–body organisms, acted upon favourably or unfavourably by other *chhi* from the environment, but also itself sometimes capable of spontaneous malfunction. We shall come across the word *chhi* very frequently in what follows, and like Tao it is better left untranslated.

Ancient Chinese medicine was closely associated with the beliefs of the philosophers who may broadly be termed Taoist. In contradistinction to the Confucians who were interested primarily in human society alone, the Taoists devoted themselves to the study of Nature, believing that man's life should be lived in conformity with her, and they developed a system of religious mysticism which has been termed the only one ever known in the world which was not essentially anti-scientific. The Taoists believed in the possibility of attaining a material immortality so that they could continue to exist as etherealised beings on the earth, enjoying the beauties of nature.[1] For this purpose, they engaged in the study of alchemy, sought drugs which would confer longevity or immortality, and practised all kinds of techniques (some ascetic, some not) which they thought might contribute to this end. Their relation with preventive medicine was therefore particularly intimate. They spoke of the art of nourishing the life (*yang sêng*), concentrating special attention on the inner causes of illness, which they pictured as the result of an improper balance between the Yin and the Yang. It was necessary to harmonise these two in order to remove the causes of disease. What

---

[1] *SCC*, vol. 2, pp. 139 ff. It should be remembered that while later Taoism (after the + 3rd century) generated a veritable church, and lost itself in a maze of polytheistic superstition, the early Taoists, deeply engaged as proto-scientists with their 'natural magic', were very practical men. And all their operations were conducted within the framework of a world-view which regarded the Tao as essentially immanent—the Order of Nature itself.

the *Tao Tê Ching* (Canon of the Virtue of the Tao)[1] calls 'having a true hold on life' (*shê sêng*) was their aim. To attain longevity and perhaps immortality, the 'way of life' (*sêng tao*) was to be pursued. It was necessary to practise 'the nourishing of the mind and the emotions' (*ching shen ti hsiu yang*)[2] and to train the body to live in accordance with the cyclical changes of the four seasons (*ying ssu shih ti pien hua*). One must have a normal and regular way of life (*chêng chhang kuei lü*). It was thus believed that the life span of a normal person should be at least a hundred years. All these methods and techniques were grouped together under the term *wei sêng* which may be translated without hesitation as what we mean by 'hygiene', literally, the 'protection of life'. Besides the search for the elixir of immortality, and for drugs (often mineral) which would promote longevity, the other Taoist techniques (*shu*) thought likely to contribute to that desirable end included various forms of gymnastics, special sexual practices, fasting and abstinence, and even exposure of the body to sunlight, moonlight, and wind. Many of the techniques, both of the Taoists and the ancient physicians, were necessarily mixed with beliefs in what we should today call magic. Nevertheless the practical benefits to health of much of what they recommended are undeniable.

EARLY CONCEPTS OF PREVENTION

We shall now present a series of texts illustrating these concepts from the Chou period down to the beginning of the Thang. It will be seen that the philosophers and the medical writers are at one on the subject. First of all, in the *I Ching* (Book of Changes), the basic text of which goes back to the −7th or −8th century, we have the fundamental statement[3] that 'the *chün-tzu* (great-souled man)[4] always meditates on trouble in advance and takes steps to prevent it'. Then in the *Tao Tê Ching*, the great poem attributed to Lao Tzu which we have already

---

[1] Ch. 50.

[2] [Although this translation is not wrong, these terms designate two out of the three components of the organism highly important in later Taoism; the 'three primary vitalities' (*san yuan*), *ching* (seminal essence), *shen* (mental essence), and *chhi* (pneumatic essence).]

[3] Baynes tr., vol. 1, p. 261, the *hsiang chuan* of the *kua* Chi-Chi. This statement occurs in what is essentially a commentary of the −4th or −3rd century.

[4] Like Tao and *chhi*, we prefer to leave *chün-tzu* untranslated. Though it originally meant the ruler or the lord in bronze-age proto-feudalism, it came to have all the aura of Aristotle's 'great-souled man', or what is implied by the term 'gentleman', even 'knight', of aristocratic birth but not necessarily, a scholar but not necessarily, an officer of state but not necessarily. For Westerners, Sir Thomas More might be cited as the type of all *chün-tzu*.

mentioned and which may date from about the −4th century, we have the following verse:[1]

> To know when one does not know is best;
> To (think) one knows when one does not know is a disease.
> Only he who recognises this disease as a disease
> Can prevent himself from having such diseases.
> The sage does not suffer from these affections;
> He recognises his (incipient) diseases as diseases
> And therefore does not have them.

Here we have the recognition of abnormal functioning at its small beginnings and of causes bodily as well as mental. It is interesting that the conception should be applied to what modern science also recognises as the worst of all intellectual diseases.

Coming now to the early Han period we have the following statement in the *Huai Nan Tzu* book, compiled about −120 by a group of scholars and proto-scientific magicians under the patronage of the Prince of Huai-Nan, Liu An. There we read:[2] 'A skilful doctor cures illness where there is no sign of disease, and thus the disease never comes' (*liang i chhang chih wu ping chih ping, ku wu ping*).

To this period belongs the greatest of all the Chinese medical classics, the *Huang Ti Nei Ching, Su Wên* (The Yellow Emperor's Manual of Corporeal (Medicine); the Plain Questions and Answers). The date of this work has given rise to a great deal of discussion but we may accept it as being mainly of the early Han period (−1st and −2nd centuries), though of course with some later interpolations. In one of the earliest chapters of this book, we find the statement,[3] 'It is more important to prevent illness than to cure the illness when it has arisen' (*fang ping chung yü chih ping*); and in an adjacent place we read:

The sage does not cure the sick only when they are sick but he prevents the illness from arising. In the same way he does not put right upheavals in the body politic, he prevents them from ever taking place. Surely it is too late to administer drugs after the illness has declared itself or to try to suppress a revolt after it has come about. Is it not like beginning to dig a well when one feels thirsty, or starting to manufacture weapons of war after the battles have begun?

To this we may add a statement from another book of great fame, the *Chin Kuei Yao Lüeh* (Systematic Treasury of Medicine)[4] written by the 'Chinese Galen', Chang Chi (Chang Chung-Ching), about +200: 'Someone asked: "What is

---

[1] Ch. 71. We diverge from the translation of Waley. The extremely laconic and epigrammatic quality of this famous text may be appreciated from the last three lines: *Shêng jen pu ping, i chhi ping ping, shih i pu ping.*   [2] Ch. 16, p. 4*b.*

[3] Ch. 2, p. 12*b.*   [4] Ch. 1, p. 1*a.*

meant by experts when they speak of 'curing a disease before it appears'?" The teacher replied: "When one observes a disease of the liver and knows it will spread to the spleen, it is wise treatment to strengthen the latter first."'

Aphorisms of this kind may be found also in the *Pao Phu Tzu* book written by the great alchemist and physician, Ko Hung, in the close neighbourhood of +300. He says:[1]

Thus the adept disperses troubles before they begin, and cures diseases before they show any symptoms at all. The prescriptions of physicians are best given before any serious signs have appeared; then it will not be necessary to run after the disease when the patient is already dying.

And in the works of the great physician Sun Ssu-Mo, of the +7th century, we find the following words:[2]

After enjoying excellent health for ten days, it is advisable to employ moxibustion at the *san shu* point[3] in order to expel noxious humours. Day by day it is necessary to harmonise the *chhi* and nourish the body. But massage and gymnastics is best of all. In time of health (lit. peace) do not forget danger. Always try to prevent the coming of disease beforehand.

And finally, there is an interesting passage in the *Ho Kuan Tzu* (Book of the Pheasant-Cap Master). This is a very mixed compilation of a philosophical character, many parts of which date from as early as the −3rd century but which was added to from time to time and did not reach the form in which we have it now until the +3rd or the +4th century. One of the discussions there recorded[4] (we need not insist too much upon its historical certainty) is between Chao, the Prince of Cho-Hsiang, the son of Duke Hsiao of the State of Chao, and his general, Phang Nuan.

Phang Nuan said to the prince of Cho-Hsiang, 'Have you not heard that Duke Wên of Wei State asked the great physician, Pien Chhio, saying, "Of your three brothers, which is the best physician?" Pien Chhio answered, "The eldest is the best, then the second, and I am the least worthy of the three." Duke Wên said, "Might I hear about this?" Pien Chhio replied, "My eldest brother treats diseases in a most godlike way. Before the disease has shown any symptoms, he has already got rid of it. Thus his fame never reached beyond our own clan. My next brother also treats disease very exactly and precisely, so his name has not become known beyond our own district. As for myself, I practise with needles, I examine the pulse and I prescribe drugs, in addition to which I have insight into what is happening between the skin and the flesh. Thus my name has become known among all the Princes of the Kingdoms and all the Feudal Lords."'

---

[1] *Nei Phien*, ch. 18, p. 5b.   [2] *Pei Chi Chhien Chin Yao Fang*, ch. 27 (pp. 481–2).

[3] The 'point' named is one of the numerous points on the surface of the body where needles were inserted in acupuncture or mild moxa cautery employed.

[4] Ch. 16, pp. 10b ff.

In this connection it is interesting to return to Sun Ssu-Mo in the +7th century and to learn that he suggested the universal provision of what we should now call first-aid kits. In his *Pei Chi Chhien Chin Yao Fang* (The Thousand Golden Remedies for Emergency Use) he recommended[1] that people, whether at home or travelling, should always have medicines ready. They should have moxa (*shu ai*) for moxibustion, various kinds of emergency pills, rhubarb, mercury, *kan-tshao*, *kuei-hsin*, etc., ginger and other drugs handy for making up prescriptions. In addition to this, every large family should possess one or two treatises on emergency medicaments.

ANCIENT LITERATURE

Before proceeding further, it will be desirable to have a look at certain interesting specimens of ancient literature in the Chinese tradition; first, the *Chou Li* and, secondly, the *Shan Hai Ching*. The *Chou Li* (Record of Institutions of the Chou Dynasty) is a large and interesting work compiled by scholars of the early Han Dynasty and to be dated about the −2nd century. It purports to be a register of the officials and their duties under the High King of the Chou Dynasty, namely, the period in the middle of the −1st millennium. It cannot, however, be a record of actual facts of that date but constitutes rather a system showing what the Han Confucian scholars considered the ideal form for a unified Chinese imperial bureaucracy. Some parts of the book, especially the *Khao Kung Chi* (Artificers' Record), are probably genuine documents going back to the −4th century or a little before, in this case from the State of Chhi, but the greater part of the work is essentially Han.

Now the interest of this compilation in the present context is the detailed account of the medical and health officials attached to the imperial court. It is immediately interesting from the hygienic point of view to find not only an Imperial Physician (I Shih) who superintends the imperial medical staff,[2] but also an Imperial Dietician[3] (Shih I). His duties and those of his assistants were to take charge of the diet of the emperor and the court, bearing in mind the nature of the various foods to be combined into a balanced diet and adapted at the same time to the cycle of the seasons. Besides these (and it is interesting that the Imperial

[1] Ch. 27 (p. 411.2).
[2] Ch. 1, p. 4*a*; ch. 2, p. 1*a*. All the passages referred to in this paragraph will be found in the Biot tr., vol. 1, pp. 8 ff., 92 ff. The reason why each official has two references is that he and his staff are mustered in one chapter, while the description of their duties is given in another.
[3] Ch. 1, p. 4*a*; ch. 2, p. 1*b*.

Dietician comes next in order to the Imperial Physician), there was the Physician-in-Ordinary for Internal Medicine (Chi I) who dealt with the infections and epidemics characteristic of the four seasons, certifying and recording causes of death as well as applying remedies to the sick;[1] and the Physicians-in-Ordinary for External Medicine (Yang I). These officials cannot quite be called surgeons because they rarely operated, but their domain concerned war wounds and fractures as well as ulcers and oedemas.[2] Lastly came the Imperial Veterinarian (Shou I) and his staff.[3]

In addition to all these groups, however, there are quite a number of sanitary officials whom we might call 'fumigators' or 'vermin exterminators'. For example, the Chu Shih is charged with combating poisonous parasites harmful to man (*tu ku*) using drugs in the form of smokes, but also conjurations.[4] Similarly, the Chien Shih is charged with combating insect pests (e.g. silver fish, moths, boring beetles, ants, termites, and so on).[5] Again conjuration plays a part as well as fumigation and there is mention of the plant *mang* (*Illicium religiosum*, the bastard anise).[6] This was a well-known poison for insects and fish in antiquity. We find, for example, in the *Huai Nan Wan Pi Shu*, an interesting collection of recipes attributed to the group of experts surrounding Liu An, the Prince of Huai Nan, and very reasonably attributable to that date (− 120), the statement that *mang tshao* is a powerful fish poison (*mang tshao fou yü*).[7] Returning to the *Chou Li*, the Chhih Pa Shih is charged with combating vermin in walls and houses;[8] for this purpose he uses lime prepared from mollusc shells and 'scattered ashes', words which may conceal the use of caustic alkali prepared by the burning of wood and the 'sharpening' of its lye. Then the Kuo Shih has the duty of combating insects which make disagreeable noises such as frogs or cicadas.[9] Fumigation again occurs here with the very significant use of *mu chü* (*Pyrethrum seticuspe*, the winter aster)[10] both for sprinkling as powder and also as fumigatory smoke. Although the use of *Pyrethrum* has come down to us in the present text in connection with frogs and cicadas, this is probably only a chance and we cannot but take notice of the

[1] Ch. 1, p. 4*b*; ch. 2, p. 2*a*.
[2] Ch. 1, p. 4*b*; ch. 2, p. 3*a*.   [3] Ch. 1, p. 4*b*; ch. 2, p. 4*a*.
[4] Ch. 9, p. 5*b*; ch. 10, p. 7*a*. All the passages referred to in this paragraph will be found in the Biot tr., vol. 2, pp. 386 ff.
[5] Ch. 9, p. 6*b*; ch. 10, p. 8*b*.   [6] Read, no. 505; Stuart, p. 489.
[7] Entry no. 53. Cit. in *Thai-Phing Yü Lan*, ch. 993, p. 2*b*.
[8] Ch. 9, p. 6*b*; ch. 10, p. 9*a*.   [9] Ch. 9, p. 6*b*; ch. 10, p. 9*a*.
[10] Read, no. 26; Stuart, p. 260; Bretschneider, vol. 2, nos. 130, 404. *Pyrethrum* is mentioned again, and in the same context, in the −2nd-century *Huai Nan Wan Pi Shu*, (entry no. 107); cit. in *Thai-Phing Yü Lan*, ch. 996, p. 2*b*.

interest of the fact that such a valuable plant insecticide is mentioned among the paraphernalia of the sanitary officials in the *Chou Li*. Lastly, the Hu Cho Shih had to get rid of aquatic pests (*chhung*).[1] He beats pottery drums and throws burning stones (or perhaps burnt stones, lime?) in order to frighten them away, and he is concerned with the magic of elm-wood branches and elephant teeth. We are obviously here in the presence of a mixture of ancient magic and effective insecticides. At the least it remains remarkable that this ideal if somewhat archaised reconstruction of what was imagined to be the perfect governmental staff includes so many sanitary officials.

Indeed, this was not the end of them. Elsewhere in the same book we meet with the waterways police[2] (Phing Shih). Their duty was to warn of dangerous places as well as protecting fisheries out of season. They might thus be considered to be concerned with public safety in more senses than one since there are not a few species of fish in Chinese waters which are poisonous at certain seasons of the year. More directly important for our present subject, perhaps, are the sanitary police[3] (Chhü Shih, the word *chhü* here meaning a rotting corpse, or maggots therein). This official and his staff are charged with the removal of all rotting corpses whether of men or animals. When someone dies on the road, they have to bury the body and report to the magistrate on the day and month of the event, bringing the clothes and possessions of the deceased, with a view to the reclaim of the latter by his or her family. Last of all come the Yeh Lu Shih or traffic police[4] whose duty was to inspect communications, direct traffic both on streets and waterways, keep rest-houses in good condition, conduct visitors to and from the frontiers and—again for us important—to organise the sweeping of the streets in the capital and other cities. Thus there is a great deal of interest from the point of view of preventive medicine in this ancient survey of the ideal imperial bureaucracy.

It is always difficult to be sure how far the organisation described in the *Chou Li* refers only to the imperial capital and the imperial court itself and how far it was regarded as extending outward in a provincial network throughout the whole of the country which owed allegiance to the High King or Emperor. The question is somewhat academic for the Chou period itself because this elaborate organisation never existed during that period. But during the early Han period, when the

---

[1] Ch. 9, p. 7*a*; ch. 10, p. 9*b*.
[2] Ch. 9, p. 5*a*; ch. 10, p. 4*a*. All the passages referred to in this paragraph will be found in the Biot tr., vol. 2, pp. 380 ff.
[3] Ch. 9, p. 4*b*; ch. 10, p. 3*a*.   [4] Ch. 9, p. 4*b*; ch. 10, p 2*a*.

Table 1. *Preventive medicine in the 'Shan Hai Ching'*

Agent or measure	*fu chih* to use in one way or another on the body	*shih chih* to use by ingesting
Prevention of infection by parasites (worms or insects) (*ku* poison) (*fang ku*)	8	5
Prevention of epidemic or infectious diseases (*fang i*)	4	3
Promotion of health (*chhiang chuang*) by preventing hunger or avoiding emotions such as fear and jealousy	25	8
Protection against external diseases of the sense organs (*fang wu kuan ping*)	8	5
Protection against external diseases of the skin and limbs (*fang phi fu wai kho chu ping*)	8	7
Protection against internal affections, viscera, etc. (*fang tsang chhi chu ping*)	4	2
Prevention of conception (*pi ying*)	2	1
Protection of animals from disease (*fang shou ping*)	1	0
Total	60	31

compilation was actually being made, there was growing up a highly bureaucratic organisation of prefectures and commanderies all over the Empire subject to the one universal monarch, the imperial head of the whole Chinese *oikoumene*. Thus we may interpret the meaning of the scholarly writers of the *Chou Li* as having regard to the bureaucracy throughout the provinces as well as in the capital itself. Of course, some of the officials such as the imperial medical staff were primarily those attached to the court.

Another ancient literary work of equal interest is the *Shan Hai Ching* (Classic of the Mountains and Rivers), a work which has sometimes been described as the oldest geographical treatise of China. It is a very archaic text going back at least as far as the − 4th century, but it contains some material far older than that, perhaps even in origin as old as the Shang. This book is ostensibly a geographical account of all the regions of the Chinese culture-area. It contains indeed a good deal of mythological material about strange beings, gods and local spirits who

Table 2. *Materia medica in the 'Shan Hai Ching'*

Mammals	5	Herbs	10	Minerals	1
Fish	13	Shrubs	12		
Birds	9				
Reptiles	2				
	—		—		—
Total of vertebrate animals	29	Total of plants	22	Total of minerals	1

were worshipped in different places, yet the tone is surprisingly matter-of-fact and the material includes a large quantity of very rational descriptions, for example, the minerals found in different places are recognised, the kinds of trees and animals found there, and the difficulties of communications. Now when we look at what is said of herbs and minerals in this treatise, we find, rather surprisingly perhaps, that the idea of prevention rather than cure is outstandingly present. The *Shan Hai Ching* usually recommends particular drugs, not for curing diseases but for preventing their onset. No less than sixty items of this kind are stated to promote health and to prevent illness. The word *fang*, which we might translate 'will ward off', is extremely prominent here. The agents and measures referred to are shown in Table 1.

Apart from the terms mentioned at the head of the columns in this table, the text also speaks of *phei chih*, meaning to wear the object; and very occasionally, to pay homage to it or worship it. The interest of this analysis is that one can see the kinds of diseases which were feared in the Warring States period especially by travellers. No doubt Fan Hsing-Chun is right in believing that '*ku* poison' was schistosomiasis and that one of the chief diseases listed under affections of the sense organs, *mi*, was probably trachoma. One can also visualise the great extent of forest country and uncultivated waste land in those days by the fact that so many animal substances rather than plants are mentioned. We may make a table according to this breakdown as above (Table 2).

We thus see how thirty-one items were actually eaten for protection and only a slightly less number used in other ways such as being worn on the body or being smeared over the body. It is also noteworthy that the idea of the promotion of general health in accordance with the Taoist principle of 'conserving the life' (*yang sêng chih tao*) appears in the third group of Table 1 above, namely, the general promotion of health, both mental and physical. The promotion of mental health is indicated by the fact that in certain cases the advantageous effect of the medicament is said to lie in removing fear of storms, thunder, wild beasts, etc. It is very likely that maintenance and improvement of good nutritional status were important functions of the items in group 3, for no better means would have existed for keeping in good heart those ancient travellers, official, military or private, through the wild mountains and forests between the isolated and far scattered urban foci of ancient Chinese culture. Thus in general one can see how the *Shan Hai Ching* represents a more archaic stage in the development of materia medica in China than the *Shen Nung Pên Tshao Ching*, the first of the long line of pharmacopoeias (*c.* −1st century).

Whether there were any written manuals of public health and the prevention of disease in the Warring States period, we do not know. Certainly none has come down to us. There is, however, the possibility of seeing a reference to such manuals in one of the stories of the *Chuang Tzu* book written by Chuang Chou about −300. There he is speaking[1] of an interview between one of the disciples of Lao Tzu, Kêngsang Chhu, and an inquirer, Nanjung Chu: 'Nanjung Chu said, "What I hear about the Great Tao makes me feel like a patient taking medicine and his illness being intensified thereby. What I would like to hear about are the (teachings of the) manuals on the protection and preservation of life (*wei sêng*)."' Here the expression actually used is *wei sêng chih ching* and although this can be translated as 'the standard methods of guarding and preserving life', it is not at all impossible that actual written manuals are referred to.

In later times, the expression *wei sêng* was employed in the titles of many medical books but generally with a rather wider implication. One might mention, for example, the Sung book, *Wei Sêng Shih Chhüan Fang* (Perfect Prescriptions for Preserving Life) by Hsia Tê-Mou, or another book, by a medical writer of the Yuan dynasty, the *Wei Sêng Pao Chien* (Precious Mirror of the Preservation of Health) by Lo Thien-I. Where we find a very clear concentration on prevention rather than cure is in the book *I Hsien* (Pre-therapeutic Therapy) by Wang Wên-Lu of the Ming; this was written about +1550.

THE YELLOW EMPEROR'S TREATISE

We now come to the ideas on preventive medicine in the greatest of all the Chinese medical classics, the *Huang Ti Nei Ching*, *Su Wên* (The Yellow Emperor's Manual of Corporeal (Medicine); the Plain Questions and Answers). To this work

[1] Ch. 23; *Chuang Tzu Pu Chu*, ch. 8 A, p. 9 a.

we have already referred. Its nucleus must date from the − 2nd century but it cannot be considered a work compiled only by one individual, nor is it from one period. It certainly summarises the experience of the Chou and Warring States physicians,[1] but it also contains interpolations from the Later Han and even subsequent times. The attribution of the work to the mythical Yellow Emperor (a favourite Taoist figure) is of relatively little significance since the work forms a compendium of all the practical knowledge of the ancients and an elaboration of the *philosophia perennis* of Chinese medicine. All later writings in this field derive and develop from the *Nei Ching*. It is quite natural that such a compendious treatise should have been made in the Chhin and Han periods, for the institution of the first unified Empire in the Chhin brought about not only a centralisation of government but also a standardisation of weights and measures, even down to the gauge of carriage-wheels; in sum, a general orderly systematisation of all Chinese practices.[2]

The *Nei Ching* undoubtedly contains the fundamental principles of traditional Chinese medicine. In the *Su Wên* it recognises and describes many specific disease entities, noting the regular association of different symptoms which permit their diagnosis; it traces their aetiology in terms of the classical physiological theories which it itself enunciates, having due regard to external influences; and as for therapeutics it concerns itself chiefly with acupuncture. Unfortunately, although the basic principles of the *Nei Ching* are not difficult, the language of it is archaic and hard to understand. Nor were the ancient commentaries very easy to understand either. Hence during the centuries, only scholars of high quality could master it and become truly learned physicians. Such physicians were termed *tai fu* or *liang i*, in contrast to which one finds many other terms for itinerant practitioners such as *yung i* or *ling i* (meaning those who went about with a little bell seeking for patients) or even *chhuan i* (a term referring, no doubt, to the herbs which they carried strung together on strings as part of their paraphernalia). The difficulty of the *Nei Ching* is that the technical terms are often ordinary words with special meanings, and sometimes occur along with the same word used in its ordinary sense in the same passage. Much misunderstanding of Chinese medicine has probably arisen owing to misunderstanding of the *Nei Ching*.

[1] Some gleanings from the Chou texts themselves on medicine and hygiene have been assembled by Shu Shih-Chhêng.
[2] On the *Nei Ching* and its history, we will only mention here out of a large literature the papers by Huang Man and Vogralik. [Cf. pp. 270 ff. above.]

Now throughout the work emphasis is laid on the responsibility of the physician not only to attend to the sick but to maintain the health of the healthy person. At the very opening of the book, we find the statement:[1] 'Regimen being adopted and the mind and emotions being guarded, how could diseases arise at all? (*Ching shen nei shou, ping an tshung lai*).' The philosophy of the *Nei Ching* is, of course, 'pneumatic' and the term *chen chhi* (literally, true *pneuma*) means primarily the innate constitution or endowment of the mind–body organism at birth. Synonyms for this often found are *yuan chhi* (original *pneuma*) or *ching chhi* (essential *pneuma*). The expression *chen chhi*, however, has a wider connotation because it may be regarded also as a health *chhi* or salutary *chhi* if the individual meets with optimum nutritional and other conditions of life, in other words, a *chêng chhi* (rightly adjusted *pneuma*). In this case, the four seasons (*ssu chi*) act upon the body at the right time and to the right extent; their influence is thus rightly adjusted (*chêng fêng*). On the other hand, if influences adverse to health are met with, unseasonable or harmful 'winds' (*hsü fêng, hsü hsieh, tsei fêng*),[2] then the *chen chhi* will not be able to reach its fullest expansion and completion. All these unwholesome or unseasonable influences are characterised by the term *hsieh chhi*, 'bane breath' or 'crooked *pneuma*'. We may find a good statement of this conception in another part of the *Nei Ching*, the *Huang Ti Nei Ching, Ling Shu* (The Yellow Emperor's Manual of Corporeal (Medicine); the Vital Axis) a text about which there has been great difference of opinion as to dating, but which we regard as Han or possibly Later Han. Here we find[3] the statement that 'the *chen chhi* is an inborn nature received from Heaven, and the body in which it exists is fortified by the *pneuma* of food (literally, the cereal *pneuma, ku chhi*)'. Thus we see that the *chen chhi* is a composite term made up of the inborn constitution and the resistance built up by the nutritional regimen and regularity of life which the individual follows or encounters.

These two conceptions were sometimes referred to by the expressions *hsien thien* (prior to Heaven) and *hou thien* (posterior to Heaven). These two technical terms are connected with the *I Ching* (Book of Changes), to which we have already referred, where they have to do with two celebrated arrangements of the hexagrams. The *hsien thien* and *hou thien* arrangements may be pre-Han but they are certainly associated primarily with two famous mutationist diviners of

[1] Ch. 1, p. 3*a*.
[2] Or of course the poisons of animate or inanimate things.
[3] Ch. 75, p. 13*b*.

that dynasty, Chiao Kan and Ching Fang in the − 1st century. These terms occur in many passages in the *Nei Ching*.[1] Thus 'prior to Heaven' factors are those of innate constitution plus the effects of adequate nutrition and living conditions, while the 'posterior to Heaven' factors are those of specific external influences which act harmfully or beneficially upon the organism.

The second chapter of the *Huang Ti Nei Ching, Su Wên* is entitled 'The Importance of Adjusting the Mind (and Body) to the Climatic Conditions of the Four (Seasons) (*ssu chhi thiao shen ta lun*, i.e. modernised, *ssu-chi chhi-hou thiao-ho ching-shen ta lun*)'. It gives accounts of the results of failure to observe this vital adjustment and says, for instance, that diseases of the *nio* (or *yao*) class will result in the autumn if it has not been achieved in the summer. The leading idea is that the right regimen in the appropriate season prevents the occurrence of diseases in the following season. The term *nio* is that which later on became standardised for fevers of the malarial type, but it is unlikely that we may take it in such a specialised sense at this early date. Reference is made to chronic diseases (*chiu nio*), daily fevers (*jih nio*), intestinal diseases (*sun hsieh*), and weakness or bad circulation in the extremities (*wei chüeh*).

HYGIENE, MENTAL AND PHYSICAL

It is remarkable to see how far what we should now call the psychosomatic causation of disease was empirically recognised in ancient China. The philosophers of all the schools agreed without hesitation on the necessity of the cultivation of the mind and the control of the emotions. A typical statement may be found in the book *Shen Chien* (Precepts Presented to the Emperor) written about + 190 by Hsün Yüeh. There he says:[2] 'In order to nourish the spirit, pleasure, anger, pity, happiness, cares and anxieties must at all costs be moderated.' He goes on to say that those who are good at controlling their *shen chhi* (their mental *pneuma*) can direct it as Yü the Great (the legendary hydraulic engineer and emperor) did with water, the importance of regulation being to avoid excess and deficiency. The same attitude appears in the *Lü Shih Chhun Chhiu* (Master Lü's Spring and Autumn Annals), that wonderful compendium of natural philosophy compiled by the proto-scientists and magicians gathered under the auspices

---

[1] For example, ch. 69, p. 2*a*; ch. 70, p. 34*a*. Further details of the mutationist use of them will be found in connection with the ancient history of the magnetic compass and the compass points in *SCC*, vol. 4, pt. 1, p. 296.
[2] Ch. 3, p. 2*b*.

of Lü Pu-Wei around −239 in the State of Chhin. In this book we find the statement:[1]

Thus for the nourishment of health, nothing exceeds in importance the recognition of the fundamental nature of the human being (*mo jo chih pên*). If this is once understood, then there is no room left for diseases to enter in.

The chapter ends with a splendid passage:

Nowadays people are always appealing to divination and offering up prayers for health and recovery, yet diseases rampage all the more. It is like an archer at a shooting-match who failing to hit the target calls for another one to replace it—how will this help him to score a bull's-eye? If one wants a kettle of soup to stop boiling, one must take away the fire from underneath. As for all these leeches and physicians, with their drugs and potions supposed to expel diseases, to slight them and master them, the wise men of old esteemed such men but little, considering that they dealt with the branch and not with the root.

Here we may need to ask what was meant by the 'fundamental nature of the human being'. The immediate answer of the philosophers, which was common to most of the schools but especially characteristic of the Confucians, was that the fundamental nature of man is goodness. In this case one immediately sees that the interpretation suggested by the *Lü Shih Chhun Chhiu* writer has the implication of trusting people and not being suspicious of them, in other words, a healthy well-balanced mental state of social optimism. On the other hand, the medical writers such as those who compiled the *Nei Ching*, for example, would undoubtedly have meant by the fundamental nature of the human being the correct balance (like a *krasis*, κρᾶσις) between the Yin and the Yang. Failure to achieve this adjustment arises from going against (*ni chih*) the seasonally correct mode of life. Here we meet with the philosophically important word *ni*, which throughout ancient Chinese thought refers to all actions or influences going against the true natural patterns into which the whole universe of man and non-human nature spontaneously tends to organise itself. Thus we can see by these examples that, so far as the philosophers were concerned, it was essential to foster a serenity and calmness of mind not merely as an ataraxia (ἀταραξία), an impassibility untouched by the misfortune of others, but a calm assurance based on belief in the fundamental goodness of human beings and the perfectibility of human society.

[1] Ch. 12 (vol. 1, p. 25). The writer has just been saying that harm to health arises from excess in any one of the five flavoured foods, or in any one of the five emotions, or in any one of the seven climatic conditions. Cf. the passage in Wilhelm, p. 30.

Turning now to the medical writers, let us see what the *Huang Ti Nei Ching, Su Wên* says. Right at the beginning of the book we come across the following passage:[1] 'When one feels naturally happy and free from self-seeking and upsetting personal desires or greedy ambitions, then the salutary *chhi* of necessity responds and follows (*thien tan wu hsü, chen chhi tshung chih*). Vitality thus guarding from within, how can diseases originate?' Another passage puts very explicitly the effect of mental health upon the actual resistance of the body to external attack. The text says:[2]

Pneuma (literally, wind, *fêng*) is the beginning of all diseases. If there is purity and calm in the mind, then the flesh and the interstitial tissues bar the doors and resist entry (*chhing ching, tsê jou tshou pi chü*). Then though powerful winds and virulent poisons may be at work, they are quite unable to do any harm.

The balance of the mind being thus safeguarded, the Taoists and the physicians advocated a number of different kinds of comparatively mild gymnastics. These exercises were generally called *tao yin*, that is to say, extending and contracting the body. In later times, the terms *kung fu* and *nei kung*, implying work, or inwardly directed work, were used for it. Possibly the exercises derived from the dances of the ancient rain-bringing shamans, but in any case they were certainly associated with the idea, as old in Chinese as in Greek medicine, that the pores of the body were liable to become obstructed, thus causing stasis (*yü*) and disease.

Already in the *Lü Shih Chhun Chhiu*[3] we can find the aphorism that 'Running water does not become stale nor does a door-pivot ever get worm-eaten' (*liu shui pu fu, hou pu lou*). This was the basic idea of the gymnastic exercises, the origin of which we see quite clearly in the biography of Hua Tho in the *San Kuo Chih* (History of the Three Kingdoms). Hua Tho was an outstanding physician and surgeon (fl. +190 to +265) about whose name many legends subsequently gathered. In his biography in the dynastic history, we read the following interesting passage:[4]

Wu Phu of Kuang-ling and Fan A of Phêng-chhêng were both pupils of Hua Tho. Wu Phu followed exactly the arts of Hua so that his patients generally got well. Hua Tho taught him that the body should be exercised in every part but that this should not be over-done in any way. 'Exercise', he said, 'brings about good digestion (literally, causes the dispersion of the cereal *pneuma, ku chhi tê hsiao*) and a free flow of the blood (*hsüeh mo liu thung*). It is like a door-pivot never rotting. Therefore the ancient sages engaged themselves in *tao yin* exercises (for example) by moving the head in the manner of a bear and looking back without turning

[1] Ch. 1, p. 3a.
[2] Ch. 3, p. 16b.
[3] Ch. 12 (vol. 1, p. 25).
[4] *Wei Chih*, ch. 29, pp. 6a ff.

356

the neck. By stretching at the waist and moving the different joints to left and right one can make it difficult for people (to grow) old. I have a method,' said Hua Tho, 'known as the "play of the five animals (*wu chhin chih hsi*)", the tiger, the deer, the bear, the ape and the bird. It can be used to get rid of diseases and is beneficial for all stiffness of the joints or ankles. When the body feels ill, one should do one of these exercises. After perspiring, one will sense the body grow light and the stomach will manifest hunger.' Wu Phu followed this advice himself and attained an age greater than 90 yet with excellent hearing, vision and teeth.

These techniques generated a rather large literature. One may mention the *Thai-Chhing Tao Yin Yang Sêng Ching* (Manual of Nourishing the Life by Gymnastics), the date of which is uncertain but perhaps Sung,[1] and secondly, the *Tsun Sêng Pa Chien* (Eight Explanations on Putting Oneself in Accord with the Life Force) by Kao Lien of +1591. This latter book was analysed in great detail by Dudgeon who also describes some of the minor works on this subject. Probably the art of Chinese boxing, *chhuan po*, a technique with quite different rules from that of the West and embodying a considerable element of ritual dance, originated from Taoist medical gymnastics. Although we cannot here pursue the matter further,[2] there is considerable reason to believe that the growth of medical gymnastics in modern Europe was influenced by information about the Chinese practices. This influence was marked from the eighteenth century onwards, but of course there were earlier works in Europe, deriving presumably from Greek sources. The whole story of the origins of modern hygienic and remedial exercises remains to be written.

PRINCIPLES OF NUTRITIONAL REGIMEN

The importance of a complete and balanced diet was fully recognised in the medical classics of the Han. For example, the *Huang Ti Nei Ching, Su Wên* says:[3]

Taking the five cereals as nutriment, the five fruits as assistants, the five meats as chief benefactors, and the five vegetables as supplements, and combining together the *chhi* and the tastes (*wei*) in the diet; this blending is what benefits the mind and body.

Elsewhere the same work says:[4] 'When eating and drinking is doubled the stomach and intestines are seriously harmed.' And again we read of a certain disease:[5] 'The origin of this is the patient's abnormal way of food intake and mode of life.' On nutrition in general the text says:[6] 'Eat moderately, live with

---

[1] No. 811 in Wieger's *Tao Tsang* catalogue. [2] Cf. *SCC*, vol. 2, pp. 145 ff.
[3] Ch. 22, pp. 10*a, b*. [4] Ch. 43, p. 10*b*.
[5] Ch. 43, p. 8*a*. [6] Ch. 1, p. 2*a*.

regularity, never get too tired either physically or mentally. In this way good health is guarded from within. How could diseases arise?'

In later times the tradition of dietary medicine developed and expanded very fully. For example, in the Sung period (*c.* + 1080), Chhen Chih said that old people are generally averse to taking medicine but are fond of food.[1] It is therefore far better to treat their complaint with proper food than with drugs. Nutritional therapy should be resorted to first and drugs prescribed only after proper feeding has failed. This he considers the great rule in caring for the aged. 'Experts at curing diseases (*chih chi*)', he said, 'are inferior to specialists who warn against diseases (*shen chi*). Experts in the use of medicines are inferior to those who recommend proper diet.' In the same work, Tsou Hsüan (about + 1307) wrote that

physicians must first recognise the causes of an illness and know what transgression of the normal regular (balance of the Yin and Yang) has taken place. To correct this imbalance adequate diet is the first necessity. Only when this has failed should drugs be prescribed.[2]

In the Thang, also, the works of physicians such as Sun Ssu-Mo contain many forms of nutritional treatment.

How large the nutritional literature was in the Chinese Middle Ages may be gauged from a glance at the titles of some of the more important of the lost books. *Shih Ching* (Nutrition Manuals), or what Shih Shêng-Han calls 'Catering Guides', appear as early as the San Kuo period (+ 3rd century) with the *Shih Liu Chhi Ching* (Manual of the Ingestion of the Six Chhi), a very Taoist book. The context in the bibliography[3] shows that this work probably had to do partly with quasi-magical heliotherapy and similar techniques of union with Nature, but it would also have included principles of dietary regimen, e.g. how to deal with the cereal *pneumata* (*ku chhi*). Of regular practical treatises on diet there was no lack, however. The *Sui Shu* (History of the Sui Dynasty) bibliography, compiled by + 635, lists no less than nine *Shih Ching*, one of the most famous being by Tshui Shih, the mother of Tshui Hao, one of the ministers of the Northern Wei Dynasty, who died in + 450. The Thang bibliographies list several more, and by the Sung we hear of a *Shih Chhin Ching* (Manual of Forbidden (or Dangerous) Foods) by Kao Shen, and a *Shih I Hsin Chien* (Essential Mirror of Nutritional Medicine) by Tsan Yin. Among the 'pharmacopoeias', several concerned themselves specifically with foods, notably the *Shih Liao Pên Tshao* by Mêng Shen

[1] *Shou Chhin Yang Lao Hsin Shu* (On the Treatment of the Aged), ch. 1, p. 22*a* (sect. 13).
[2] Ch. 2, p. 21*a*.
[3] *San Kuo I Wên Chih*, p. 108; in *Erh-Shih-wu Shih Pu Chu*, vol. 3, p. 3296.3.

in the Thang (+670) and the *Shih Hsing Pên Tshao* by Chhen Shih-Liang in the Sung (+10th century). Some of these are still extant.

Here space is lacking to elucidate the empirical discovery of vitamins through the study of deficiency diseases in medieval China. This has already been the subject of a paper,[1] and we hope to return to it in much greater detail elsewhere.

Fig. 93. Frontispiece of the book of Hu Ssu-Hui, *Yin Shan Chêng Yao* (Principles of Correct Diet), c. +1330. The caption says: 'Many diseases can be cured by diet alone', and the picture shows a nutritional specialist giving a consultation.

The greatest name in this field is no doubt that of Hu Ssu-Hui, Imperial Dietician between +1315 and +1330. His book, the *Yin Shan Chêng Yao* (Principles of Correct Diet), finished in the latter year, describes in much detail the wet and the dry forms of beri-beri and advocates foods now known to be rich in vitamins for the treatment of those suffering from deficiency diseases [Fig. 93]. The empirical recognition of these is summarised in Hu's aphorism, 'many diseases can be cured by diet alone (*shih liao chu ping*)'.

[1] Cf. Lu & Needham.

WATER AND TEA

The importance of supplies of pure drinking water was appreciated very early amongst the Chinese. Already in one of the most ancient texts, the *I Ching* (Book of Changes) which we might date somewhere about the − 7th century, we find the commonplace, 'Men do not drink water from foul wells.'[1] The *Shih Ming* dictionary (+ 100) says,[2] punning, 'A well (*ching*) means essentially clear and clean (*chhing*), the clear produce of a spring.' Regular custom in ancient China demanded the periodic cleaning of wells. For example, in the *Kuan Tzu* book dating from about the − 4th century, we have two passages on this; the first says:[3]

In the third month (of the year) they *chhiu* dwellings and dry out (*than*) new houses. [The commentary says:] *Than* means to dry with fire. In the third month the Yang *chhi* is expanding, so that pestilences easily arise; (therefore it is necessary to) fumigate dwellings with *chhiu* wood[4] which takes away bad smells and removes poisonous vapours (*tu chhi*). The same wood is burned to dry out new houses, and this forms part of the purification ceremony. [The text continues:] At this time, new fire is obtained by the use of the wood drill and the burning mirror, and new water is got by thoroughly cleaning out (lit. shuttling, *chu*) the wells. All this brings about the expulsion of poisonous influences harboured from the past.

The second passage says that among the regular work of spring and summer is the clearing out of wells and the removal of all extraneous matter from them, so that the water becomes more pure to drink. These customs continued right through the Han period. In the *Hou Han Shu* (History of the Later Han Dynasty) we find[5] that the summer solstice (June) was then considered the right time for 'changing the water', i.e. digging new wells and cleaning out the old ones. At this time of the year there was a prohibition against making big fires and melting or casting metals, which lasted until August. This cleaning out of wells in the summer was again analogised with the obtaining of new fire by boring of wood or the use of a burning-mirror at other times of the year, especially the winter solstice.

In the + 11th century, Shen Kua, the famous scientific scholar, writing in his *Wang Huai Lu* (Occasional Observations of a Carefree and Disinterested Mind),[6] describes to us the care taken even at that time for the preservation of purity of water sources. He says: 'Covers are made for wells, and people lock

---

[1] Baynes tr., vol. 1, p. 199, the *hsi tzhu* of the *Kua* Ching.
[2] Ch. 17 (Palaces and dwellings).   [3] Ch. 53, p. 11 *b*.
[4] *Catalpa kaempferi*, Read, no. 99.   [5] Ch. 15, p. 5 *a*.
[6] In *Shuo Fu*, ch. 19, p. 14 *b*.

them up to prevent insects (worms, reptiles, etc.) and rats from falling into them, not to speak of people's children.' In connection with wells, Shen Kua says in the same book that people were fond of 'medicated wells' (*yao ching*). These were something like what we should call today 'sand filters'. He says that in the mountains there is quartz, magnetite, and mica as well as stalactites in caves, and these materials were pounded into chips not too fine and filled into the wells for several feet. Some, following lines of thought derived from alchemy, put down cinnabar and sulphur and chipped jade. In the Thang period, the family of Li Wên-Shêng had been famous for having one of these alchemical 'medicated' wells. Li Fang's contemporary, the monk Lu Tsan-Ning, mentions the purification of well water by filtration through sand on the opening page of his *Wu Lei Hsiang Kan Chih* (On the Mutual Responses of Things according to their Categories), *c.* +980. A mention of clearing water or fermented beverages by glue or isinglass occurs[1] in the −2nd-century *Huai Nan Wan Pi Shu*.

The extent of piped water supplies in ancient China has been very much under-estimated by most later writers, both Chinese and Western. It is only now with the great expansion of archaeological investigation and excavation in China that evidences of this are coming to light. Throughout the Chhin and Han periods, palaces and cities were supplied with piped water not only in bamboo tubing, very easily installed and replaced, but also by conduits of earthenware. In Chinese museums at the present time abundant examples of these water supply pipes can be found. There are large earthenware rings about three feet across for lining the walls of wells, and a great quantity of piped lengths usually about two feet long, sometimes with right angles or two-way divisions, and usually with flanges for fitting the lengths of piping together. Fuller particulars of these we expect to give shortly elsewhere.[2] Such facts lend particular interest to one of the oldest accounts of water-raising machinery which we have. It concerns the Han Dynasty engineer Pi Lan who in +186 was entrusted with the water supplies of the capital city of Loyang. The following passage comes from the biography of Chang Jang in the *Hou Han Shu*:[3]

Pi Lan was commissioned to make (lit. cast) norias (*thien lu hsia ma*) (which would) spout forth water. These were set up to the east of the bridge outside the Phing Mên (Peace Gate) where they revolved (continually, sending) water up to the palaces. He also made square-pallet chain-pumps (*fan chhê*) and 'siphons' (*kho wu*, possibly lift-pumps) which were set up to the

---

[1] Entry no. 84. Cit. in *Thai-Phing Yü Lan*, ch. 736, p. 8*a*.
[2] See *SCC*, vol. 4, pt. 2, p. 130.　　　　　[3] Ch. 108, p. 24*b*.

west of the bridge (outside the same gate) to spray (*sa*) water along the north–south roads of the city, thus saving the expense incurred by the common people (in sprinkling water on these roads, or carrying water to the people living along them).

This passage is one of the first in Chinese literature which clearly describes the norias and the chain-pumps so characteristic of later Chinese technology. The identification of the latter is certain and that of the norias (wheels driven by the force of the current and having bamboo containers attached to their periphery thus lifting the water to a flume high above the river level) is almost so.

A rather astonishing statement is found in the writings of Chuang Chho (fl. *c.* +1126), for he says: 'Even when the common people are travelling they take care only to drink boiled water.' We have not yet been able to locate the exact source of this statement, which is quoted by Fan Hsing-Chun but seems not to be in any of the books of Chuang Chho available to us; yet there is nothing in the least improbable about it since the practice of drinking tea, and consequently boiling the water in which it was infused, had already a thousand years of life before the time of Chuang Chho. In fact, the practice of tea-drinking originated during the Later Han period[1] and references to it begin to become numerous towards the end of the Three Kingdoms period and the Chin, around +270. Originally it was introduced, in all probability, as a drug, and its effect of inducing transient insomnia in certain subjects was very early known. But by the +4th century it had universally conquered Chinese society and was everywhere used as the refreshing drink we know. An interesting story comes from the *Nan Pu Hsin Shu* of Chhien Hsi-Pai, a Sung writer. He tells us that

in the third year of the Ta-Chung reign-period (+849), Tung Tu presented to the Emperor an old monk more than 120 years of age. Hsüan Tsung, the emperor, asked him, 'What medicines did you take in order to attain this longevity?' He replied, 'Your servant was born in a poor family and never attained to any understanding of the nature of medicines. But I have always been extremely fond of drinking tea. Wherever I go I look for it and when I visit another temple I often drink more than 100 cups in one day. Even normally I drink between 40 and 50.' Whereupon the Emperor bestowed upon him 50 catties of the best tea and prepared a lodging for him to live in the Pao-Shou Temple.

Such was the virtue attributed to the 'cup that cheers but not inebriates'.

[1] See the recent studies of Bodde and of Goodrich & Wilbur.

COOKING AND NUTRITIONAL HYGIENE

One of the most important early developments in hygiene is the recognition of the spoilage of food. Food prohibitions played a prominent part in ancient Chinese life, if not perhaps so extensively developed as in ancient Israel. About − 500, we meet with a relevant passage in the *Lun Yü* (Confucian Analects), that celebrated book which gives the sayings of the greatest of all Chinese philosophers, Khung Chhiu (Confucius) and reminiscences of him by disciples. Here we are told that

rice which has been injured by heat or damp and turned sour (*i erh ai*) one must not eat, nor fish or flesh which has spoiled. That which is discoloured or smells badly one must not eat, nor any food which has been insufficiently cooked (*shih jen*) or which has been kept too long.[1]

Safety regulations about food are found in very greatly enlarged form in the late + 2nd-century work[2] of the eminent physician, Chang Chung-Ching. He tells us that eating and drinking is for the enjoyment of taste and the nourishment of life. Yet if it is not correctly ordered, what is eaten can be very harmful. There are certain things which must be abstained from. Only if the selection of diet is wise will the body benefit by it, and if the selection is unwise, great dangers, illnesses very difficult to cure, may arise. Then Chang Chung-Ching gives long lists of what should not be eaten. Among numerous taboos, the sense of which is not obvious to us today, we find many important prohibitions; for example, all meat and fish which dogs or birds will not eat, or again meat which has on it red or coloured spots; dirty rice, or meat or fish which has a disagreeable smell. Chang also warns against all animals which have died spontaneously, and says that raw meat or milk will turn to white or red parasites (*pai chhung, hsüeh chhung*). Elsewhere he gives a number of antidotes for poisoning by mushrooms, and cautions against certain toxic wild yams. All this is contained in the book which he entitled *Chin Kuei Yao Lüeh* (Systematic Treasury of Medicine) and his relevant chapters deserve much more study than they have so far received.

It should be recognised that no people have paid more attention to the importance of cooking than the Chinese. We do not refer here to the gastronomic triumphs of Chinese cuisine which are well known to all the world, but rather to the principles of Chinese cooking whereby so much oil at high temperature

[1] x, viii, 1–4, cf. also 8. Legge tr., p. 96, Waley tr., p. 148. This chapter used to be regarded as referring only to the behaviour of Confucius himself, but it is now considered to be a ritual text concerned with the rules for *chün-tzu* (educated men) in general, and here particularly with the consumption of sacrificial food. It was probably incorporated in the *Analects* in the Warring States or Chhin and Han periods.

[2] See especially *Chin Kuei Yao Lüeh*, chs. 24, 25 (pp. 89 ff., 95, 96, etc.).

was used, the frying being carried on typically in large thin-walled cast-iron vessels (*kua*). It seems likely that the disinclination for cold food which seems to have characterised Chinese cooking for twenty centuries was a powerful hygienic factor in preventing the spread of infections. Even under the relatively unhygienic conditions of the old China, those Westerners who knew the country realised that the method of preparing food was extremely hygienic, so that as long as one always ate hot dishes the risk of infection was slight. The recognition of the importance of cooking may be illustrated by a statement from one of the apocryphal treatises of the early Han period. In the *Li Wei Han Wên Chia* (Apocryphal Treatise on the (Record of) Rites; Excellences of Cherished Literature), we read, 'It was Sui Jen who first drilled wood to obtain fire and taught the people to cook food from raw materials in order that they might suffer no diseases of the stomach, and to raise them above the level of the beasts.' Here the reference is to a purely legendary character, a culture-hero, but the recognition of the hygienic function of cooking is quite clear. It was from these roots that there came forth the proverbial saying, 'Anything thoroughly boiled or cooked cannot be poisonous (*pai fei wu tu*).'

Closely allied to the recognition of the role of heat went the mistrust of the presence of insects or other small animals. For example, about +83, Wang Chhung says in his *Lun Hêng* (Discourses Weighed in the Balance),[1] 'If insects (or worms) fall into the vessel when wine is being made, reject the lot; if rats have run over rice baskets, throw it all away and don't eat it.' This is much elaborated in the *Chou Hou Pei Chi Fang* (Pocket-book of Medicines for Emergencies) written by the great physician and alchemist, Ko Hung, about +300. He is talking about the *ho-luan* group of diseases which include cholera (for which the term is specialised today) and also the dysenteries and diarrhoea. Ko Hung says:[2]

Ho-luan diseases arise from food and drink, often because various raw or cold things have been eaten, or greasy food or uncooked fish soaked in wine; or else because of exposure to draughts and dampness, or by sitting in the wet fields with insufficient clothing, or by having inadequate bedding at night.

Here we see the suspicions aroused by uncooked food and possible traces left on it by animals. In the work of the Thang scholar, Wang Thao, who completed in +752 his *Wai Thai Pi Yao* (Important Family Practice of a Frontier Official),

[1] Ch. 5, p. 1*b*.
[2] Ch. 12, p. 1*a* (ch. 2, p. 27). Chhao Yuan-Fang quotes Ko Hung on this, emphasising again the dangers of raw flesh and cold uncooked food, *Chu Ping Yuan Hou Lun*, ch. 22, pp. 1*a* ff.

we find this very explicit. He is talking about *lou* diseases; the term *lou* alone meant originally swollen glands, and it covered a considerable number of affections which it is a little hard to be precise about now. But he says:[1]

Lou diseases come from eating and drinking foods which have been contaminated by the essential *chhi* (*ching chhi*) of rats, snakes, bees (or flies?), toads, worms and all such kind of vermin; their poisons entering the body cause changes which lead to these diseases. When the local lesions break down, the poisons creep into the blood system and may kill people. Rats and ants are the most noxious for they come more into contact with human beings.

Wang Thao quotes the *Chou Hou Pei Chi Fang* of his predecessor Ko Hung, but in that book's discussion of *lou* diseases only cures are given[2] and there is no discussion of the infections produced by the action of these animals. In the century before Wang Thao, Chhao Yuan-Fang had written his magnificent systematic pathology, the *Chu Ping Yuan Hou Lun* (Systematic Treatise on Diseases and their Aetiology). Under the heading *lou* he again mentions[3] a great many insects and says that one *lou* comes from eating fruit or melons contaminated by them. In this connection, he quotes a *Yang Sêng Fang* (Prescriptions for the Preservation of Life) as saying that in the sixth month one should never eat fruit which has dropped on the ground and been left overnight. Furthermore, in winter time one should never eat flesh which has been fouled by dogs or rats. Now in the bibliographies of the dynastic histories, no *Yang Sêng Fang* is to be found but there was a *Yang Sêng Shu*, a book on this subject, current in the Three Kingdoms period (+ 3rd century). Perhaps the most probable work to which Chhao Yuan-Fang was referring was the *Yang Sêng Ching* (Manual of the Preservation of Life) written somewhere about the + 5th century by Shangkuan I.

PERSONAL HYGIENE AND SANITATION

The desirability of frequent and adequate ablutions is a commonplace in the old Chinese medical texts as no doubt among most ancient nations. From the general literature, too, it is possible to collect abundant material for a history of bathing customs in the Chinese culture-area.[4] Such a history would consider

---

[1] Ch. 23, p. 11 *b* (p. 641). One of the most ancient prescriptions for *lou* disease caused by rats must be that in the *Huai Nan Wan Pi Shu* of the −2nd century (entry no. 67). From the *Wai Thai Pi Yao* and other texts it is clear that the forms of swellings produced were often supposed to be similar to the shapes of the animals which were thought to have given rise to the diseases.

[2] Ch. 41 (ch. 5, pp. 168 ff.).   [3] Ch. 34, p. 3 *a* (p. 179).

[4] This has been done recently in the learned and interesting paper of Schafer. Some of the same ground is covered in Fan Hsing-Chun, pp. 58 ff.

baths and washing facilities first in relation to the family, then in monastic establishments and colleges, continuing with the development of public bath-houses and ending by an account of the elaborate and celebrated bathing-pools of the imperial palaces in successive dynasties. A special chapter would also have to be devoted to the hot springs, numerous in China. Here we need only say that the terminology of washing and bathing was already stabilised in the late Chou period, indeed in Confucian times (− 6th century).[1] We get a striking impression of the cleanliness which traditional etiquette and ceremonial required of the *chün-tzu* (patrician, gentleman, or scholar)[2] of the Warring States period and the Han, from the *Li Chi* (Record of Rites),[3] a compilation completed by − 50 though dating in some parts as far back as *c.* − 450. Washing the hands five times a day, with a hot bath every fifth day and a hair-washing every third day, was obliga-tory. This latter operation was considered especially important since the hair was worn long, and many a story in the classical books depends upon the high social value placed upon this duty. The *Lun Hêng* (Discourses Weighed in the Balance) of + 83 shows us that the scholars of the Han took frequent baths,[4] for Wang Chhung directs a sceptical attack on the superstition, enshrined in a manual called the *Mu Shu*, that there were lucky and unlucky days for doing so.[5]

With the growth of Buddhism after the + 2nd century an emphasis on personal cleanliness originally Indian gave powerful reinforcement to the indigenous pre-scriptions. Buddhist abbeys (as in India and Ceylon) normally contained a bath-house or a bathing-pool, and its use was by no means restricted to the monks and nuns themselves, though at least in the earlier centuries the age-old

[1] For a discussion of the terms, see Schafer.

[2] Cf. p. 363 above. We do not know so much about the customs of the *plebs* in ancient times, but there are many indications in the texts (including poetry) that they took full advantage of natural bathing-places in lakes and rivers.

[3] See especially chs. 12 (Nei Tsê), 13 (Yü Tsao) and 41 (Ju Hsing); Legge tr., vol. 1, pp. 449 ff.; vol. 2, pp. 1 ff., 402 ff.

[4] Ch. 70. In the Han and later there was a *lustrum* of five or ten days; at these intervals officials took a day off for 'relaxation and hair-washing' (see *Shih Wu Chi Yuan*, ch. 1, pp. 36a, b).

[5] Chinese literature contained a number of books devoted to bathing and personal hygiene. The bibliography of the *Sui Shu* (ch. 34, p. 26b) lists a *Mu Yü Shu* (Treatise on the Bath), which would thus have dated from before the + 7th century. Furthermore, an emperor himself did not disdain to write a Manual of Bal-neology, the *Mu Yü Ching*; this was the work of Hsiao Kang, who reigned for one year (+ 550) as Chien Wên Ti of the Liang dynasty. He was a scholar, poet and philosopher who also wrote commentaries on the Taoist classics, and many other works. The subject of balneotherapy in China has been touched upon by Schafer, but deserves more detailed study. There were certainly parallels for those treatments which in Ceylon have left such remarkable traces behind them as the stone medicinal baths like crusaders' coffins still found on the sites of medieval hospitals at Anurādhapura and Mihintale.

Confucian prohibition of the mingling of the sexes in bathing was strictly observed.[1]

The Taoists also had bathing customs which figured prominently in connection with their fasts and purification ceremonies before important festivals. This practice had since ancient times been a matter of course for the emperor and high officials before great State liturgies and sacrifices. Bath-houses and bathing-pools were also attached to colleges, as we know from a number of stories concerning (for example) poems written on their walls.[2] Possibly because of a decline in the prosperity of Buddhist foundations after the Thang, the Sung saw the rise and rapid development of public bath-houses. In the Wu Tai period (about +914), the monk Chih-Hui became famous for the water-raising machinery constructed by him for the public baths at Loyang which accommodated thousands of people.[3] Perhaps these were still Buddhist baths, public but not yet commercial. However, from the +11th century onwards, references to bath-houses which one could enter on payment of a fee become quite frequent. By the time when the Sung capital had been transferred to Hangchow the common name for such public baths was *hsiang shui hang*, 'perfumed water establishments'.[4] They advertised their presence by hanging up a water-pot or a kettle as their shop sign.[5] These were the baths so much admired by Marco Polo. In later centuries the public *thermae* became almost as notable a feature of Chinese urban life as of that of ancient Rome, but unfortunately they often no longer contributed to hygiene, for the custom was to use the same water over and over again, heating it in boilers and continuously circulating it by chain-pump to the pool. Possibly for this reason the baths came to be known as *hun thang*, 'chaos halls', from the revolving motion of primitive chaos, but it is more likely that the word *hun* here referred simply to the atmosphere of social promiscuity. Since working

---

[1] See e.g. the *Lo-yang Chhieh Lan Chi* (Description of the Buddhist temples and monasteries at Loyang), *c.* +530, ch. 4 (p. 78), for a story of the jubilation occasioned by the discovery at Pao-Kuang Ssu of a well and bathing-pool which had been built in Chin times. This was in the north. Broadly speaking, in ancient times and in the early Middle Ages bathing customs tended to be rather more emphasised in the south, with its hotter climate, than in the north.

[2] See *Kuei Hsin Tsa Chih* (Pieh Chi), ch. 1, p. 15a; a work finished by +1308.

[3] See Chu Chhi-Chhien & Liu Tun-Chen, p. 163.

[4] As in the descriptions of the city given in *Tu Chhêng Chi Shêng* (+1235) and *Mêng Liang Lu*, ch. 13, p. 3a (+1275); on which see the article by Moule which includes the reference to the bath-houses.

[5] So the mid +12th-century *Nêng Kai Chai Man Lu*, ch. 1, p. 3a. Wu Tshêng explains that this custom originated of old from the Chhieh Hu Shih official ('Hoister of the Water-pot') mentioned in the *Chou Li* (ch. 7, p. 27a; ch. 30, Biot tr., vol. 2, p. 201). His duty was to signalise in this way the position of the well or other water-source in an army encampment, and besides this of course he had to look after another sort of water-pot, i.e. the time-keeping clepsydra. Cf. Needham, Wang & Price, p. 159.

people engaged in the dirtiest of occupations, together with pedlars, caravan-drivers, and butchers now all used the baths, access to which was not forbidden to those suffering from contagious or infectious diseases, what had begun as a laudable measure of hygiene probably sometimes did more harm than good.[1] A feature of the better public baths, especially those built round hot or medicinal springs, was the presence of masseurs (*tsha pei jen* or *khai pei jen*), as we know for the +11th century already from the poems[2] of Su Tung-Pho.

Lastly, we shall pass over the heady history of the bathing-pools of the imperial palaces, about which much is known from the Han onwards.[3] The emperor and his entourage of beauties could certainly be trusted to make use of the best facilities which the age had to offer, and whether it was a matter of casting into the pools red-hot bronze statues to heat the water, or causing fountains and cooling streams to play in pavilions during summer heats, architects, engineers, and other servitors were never lacking to organise whatever was possible. It is more germane to our purpose to ask a question, perhaps more prosaic but certainly of greater scientific interest: what soap was used by the emperor and all the other bathers we have been discussing?

Here the point of greatest interest is the rather surprising fact that although the modern world is now in the 'detergent age', China was always dependent on detergents rather than on true soaps. We can find very little reference to the making of soaps by saponification of fatty acids with caustic alkali in any period of Chinese history. As is well known, the history of soap-making in the West is also strangely obscure.[4] There is some evidence that oil and alkali were boiled together to make soap in ancient Mesopotamia, especially by the Sumerians of Ur, and the same has been claimed for the ancient Egyptians but apparently with less justification.[5] The Greeks and Romans used mainly oil, together with

---

[1] A graphic account of the *hun thang* was given in the Ming by Lang Ying in his *Chhi Hsiu Lei Kao*, ch. 16, pp. 12*a*, *b*; *c.* +1530. He tells us that in spite of the decline in hygienic level, poor scholars at any rate still used the public baths, attracted by the warmth and facilities which they could not afford at home. Of course, where flowing water was used, as in the hot springs, whether at Fuchow in Fukien, Anning in Yunnan, or Pei-Wên-Chhüan in Szechuan, conditions always remained good, and one of us (J.N.) has the pleasantest memories of visits to them with friends.   [2] Cf. his *Ju Mêng Ling*.

[3] Schafer has dealt well with this subject, and the reader is referred to him. He deals in particular detail with the baths at Lintung near Sian, associated so romantically with the Thang emperor Hsüan Tsung and Yang Kuei Fei. We had the pleasure of seeing this still delightful place in the summer of 1958. It would not be without interest to extend the study by comparison with the imperial bathing-pools and gardens of other cultures, e.g. those of Anurādhapura in Ceylon, or the remarkable +5th-century royal pleasaunce still conserving its original geometrical layout at Sigiriya, also in that country, places which one of us (J.N.) had the good fortune to visit in the spring of 1958.

[4] Cf. Sherwood Taylor, pp. 129 ff.   [5] Cf. Forbes, in Singer, vol. 1, p. 261.

mechanical detergents, and lacked true soap. They employed, however, to some extent, the soap-wort (*Saponaria officinalis*) and the soap-root (*Gypsophylla struthium*), resembling therein, as we shall see, the Chinese. References in later times to the re-invention of soap are quite obscure;[1] some indications point to Gaul and some to the Scythians or Tartars of Western Asia. All one can say is that by the time of Galen (+150), true soap was normally in use and continued to be made thereafter. In medieval Europe, it was known (if little used) from at least +800 onwards.

Now in China the dependence on the saponins of plants seems to have been complete from early times onward. There were three chief sources of saponins for use as detergents in personal washing and the laundering of clothes. The oldest was the soap-bean tree (*Gleditschia sinensis*),[2] called *tsao chia*. Although we have not succeeded in finding any references to this in the classical literature of Chhin and Han, it nevertheless clearly appears in the *Shen Nung Pên Tshao Ching*, the first of the pharmacopoeias,[3] and we may thus securely date its use in the Han.[4] A special variety of this tree, the *kuei tsao chia*, was recommended as yielding a saponin especially good for the bathing of the body and the washing of hair. This was introduced in the *Pên Tshao Shih I*, a pharmacopoeia of the +8th century. The second main source of saponins was the tree *Sapindus mukorossi* which produced the beads used in the rosaries of Buddhist monks (*bodhi* seeds).[5] Its Chinese name was *wu huan tzu* or, on account of its unpleasant smell, *kuei chien chhou*. For this reason it was not allowed in public bath-houses. This source seems not to have been much used before the +10th century when it is first mentioned in the *Khai-Pao Pên Tshao* of +970. The third and latest saponin source was the soap-pod tree (*Gymnocladus sinensis*),[6] the *fei tsao chia*, which

---

[1] Cf. Sherwood Taylor & Singer, in Singer, vol. 2, p. 355 [and now Schmauderer]. For the post-Renaissance soap industry, see Gibbs, in Singer, vol. 3, pp. 703 ff.

[2] Read, no. 387; Stuart, p. 188. See *Chêng Lei Pên Tshao* (+1249), ch. 14 (p. 341). Related species which were not effective as sources of detergents were recognised early—thus the *chu ya tsao chia* (pig's tooth soap-bean) or *Gleditschia japonica* (Stuart, p. 188) was noted as useless by Su Kung in the *Thang Pên Tshao*, *c.* +660. See the whole entry in *Pên Tshao Kang Mu*, ch. 35B, pp. 4a ff.

[3] Mori ed., ch. 3, p. 91.

[4] This then must have been the saponin detergent of which we hear in the +5th century when a military governor no doubt more familiar with camps than courts drank the water presented to him together with 'washing beans' (*tsao tou*) by servant-girls in a rich household after returning from the lavatory (*Shih Shuo Hsin Yü*, ch. 34, p. 44a). His name was Wang Tun (d. +324). An almost identical story, which must also have involved the same detergent, is told about Lu Chhang in the Thang before +863 (*Yu-Yang Tsa Tsu*, Suppl. ch. 4, p. 6b). The servant-girls always laugh on these occasions.

[5] Read, no. 304; Stuart, p. 395. See *Chêng Lei Pên Tshao*, ch. 14, p. 350; *Pên Tshao Kang Mu*, ch. 35B, pp. 14a ff.

[6] Read, no. 393; Stuart, p. 198. See *Pên Tshao Kang Mu*, ch. 35B, pp. 13a, b.

first appears in the *Pên Tshao Kang Mu* (The Great Pharmacopoeia) completed by Li Shih-Chen in +1593. Naturally these vegetable products had other uses, for example as emetics, but their principal employment was as detergents. They were made up into balls with different kinds of flour, mineral powders, and perfumes analogous to cakes of soap. The poisonous action of saponins upon fish was well known in the +11th century, for Su Tung-Pho, the famous poet, mentions it in his small book of natural philosophy[1] the *Wu Lei Hsiang Kan Chih*.

Plant biochemists have long recognised that the saponins were a valuable product, especially interesting because of the fact that they have no deleterious effects upon the most delicate textile fabrics.[2] One of us (G. D. L.) remembers how her grandmother objected to the introduction of fat soap in modern China partly because it gave a precipitate in hard water and partly because the old detergents had been so safe in action on fabrics; for example, with the saponins, the whiteness of silk was long conserved. Families used to buy the black beans and make their own aqueous extracts from them, laundering silk first, then cotton things; for toilet purposes, people used the dark, moist, scented balls prepared by the druggists, which gave a very beautiful and comfortable soft lather. Other preparations included honey and various kinds of natural fats, notably the *chu i tzu*, fat from the omentum of pigs.

A fascinating chapter of the *Wai Thai Pi Yao* is entirely devoted to washing and cosmetic preparations for personal hygiene under the rubric of *tsao tou* (bath beans). Wang Thao begins by quoting from the *Chhien Chin I Fang* by Sun Ssu-Mo (about +665). Sun Ssu-Mo had said:

Creams used for face and hands, and preparations for making clothes fragrant, and beans used for washing (and bathing), are things which scholars greatly value; therefore they are important articles. In recent times, some pharmacists have kept the recipes as dear family secrets, children being forbidden to reveal them, and (sometimes) not even handed down from father to son. Yet when the sages discovered anything, they always wanted it to be universally known. How could one hope to conceal such things from all the people and prevent the sagely Tao from spreading? That would be strangely contrary to the ideas of the great men of old.

After this, Wang Thao proceeds to record about 220 recipes including many kinds of saponin detergent. There are two especially for face washing (*hsi mien fang*), five for hair washing (*mu thou chhü fêng fang*); and eight for bath soap (*tsao tou fang*). Not all of these and the other medicated detergents, which have various

---

[1] P. 25. On this work and the literary tradition to which it belonged, see Ho & Needham.
[2] Cf. Haas & Hill, vol. 1, pp. 261 ff. Their detergent power may have been weak in comparison with modern detergents, and comparative estimates would be of some interest.

drugs and perfumes added, contain *Gleditschia* (*tsao chia*), but those which do not have this have other vegetable products which probably contained saponins in sufficiently active form. Among these may be mentioned *pai chu* (*Aristolochia recurvilabra* or *Atractylis sinensis*, in both cases the roots),[1] and a powder, *tou mo*, prepared from soya bean (*Glycine soja*)[2] in which saponins must certainly also have been contained. Among the vehicles for the saponins, besides the omentum fat just mentioned, one finds talc, steatite, root powders, beeswax, and other kinds of fatty materials.[3] Wang Thao never mentions caustic alkali. However, there is plenty of evidence from all these texts that use was made of sodium carbonate (*chien*) just as *natron* was used in occidental cultures, especially for laundering clothes.

As is well known, the saponins of plants are mainly of two kinds, those of steroid structure and those of triterpene nature. Both are of course complex condensed-ring hydrocarbons and in the native state are generally bound as glucosides, but all contain that double endowment of hydrophilic and hydrophobic groups without which they could not manifest detergent properties. Monomolecular films being formed between the surface to be washed and the adhering particles of dirt or oil, the latter are lifted off as the surface tension is reduced and so the detergent effect is brought about.[4] A typical steroid saponin is the tigogenin from the plant *Digitalis* and a typical triterpene one is hederagenin.

The modern synthetic detergents are, as is well known, quite different in structure, being very largely sulphonic acid derivatives of long-chain non-cyclic hydrocarbons such as lauric acid or the hydrocarbons of castor oil. Some of them, however, are purely organic soaps, ammonia replacing the usual alkali metal. Standing in a way intermediate between the sulphonated chains or the true soaps and the classical Chinese saponins of highly complex condensed-ring character is the detergent abietic acid, a very surface-active compound obtained from the depolymerisation of pine-resin. This has a hydrophenanthrene system of three rings only and is used today in yellow laundry soaps as well as in paper size, varnishes, and plastics. It thus forms a link between the new detergents and the old saponins.

---

[1] Read, nos. 585 and 14; Stuart, pp. 49 and 58.   [2] Read, no. 388.

[3] Among the materials used as vehicles for preparing saponin balls and cosmetic creams, the flour of the common pea (*Pisum sativum*, *wan tou* or *pi tou*: Read, no. 402; Stuart, p. 335) was used, at any rate by the time of Li Shih-Chen (cf. *Pên Tshao Kang Mu*, ch. 24, p. 20*b*). He recommended them for removing the traces of smallpox. But pea flour was certainly not, as has been supposed (Schafer, p. 64), the essential ingredient of 'washing-legume', better termed 'bath-bean', detergent.

[4] Cf. the introductory account of Adam & Stevenson.

In one specific department of cleanliness, the care of the teeth, a good deal could be said about Chinese practices. Chhao Yuan-Fang, in his early + 7th-century treatise, the *Chu Ping Yuan Hou Lun*, refers to this many times. Quoting the (perhaps + 5th-century) *Yang Sêng Fang*, he says: 'Early in the morning, gnash (or rub) the teeth and then they will not decay. After eating, always wash out the mouth several times; if this is not done the teeth will go rotten and give much trouble.'[1]

The use of the soap-bean *Gleditschia* saponin (*tsao chia*) as a tooth-powder or tooth-paste is referred to in the *Wei Sêng Chia Pao Fang*, a Sung book[2] probably of the + 11th century, and also[3] in the *Phu Chi Fang* (Simple Prescriptions for Everyman) of + 1410. This almost certainly implies the use of some instrument, a roughened stick if not an actual toothbrush. The history of the toothbrush in China has recently been investigated, and evidence brought forward that toothbrushes made with bristles originated in the Liao dynasty (+ 937 to + 1125).[4] Indian influence, entering China with Buddhism, would undoubtedly have brought the use of a simple stick at a much earlier date, and indeed it is possible to see amongst the fresco paintings at the Thousand-Buddha cave-temples at Mo-kao-khu[5] near Tunhuang in Kansu Province on the border of Central Asia, in a cave which may be dated in the near neighbourhood of + 775, a very clear picture of a monk cleaning his teeth. Others have shown[6] that in the early part of the + 13th century, both tooth powder (*sanūn*) and toothbrushes made from a plant called *siwak* (*Salvadora persica*) were in use in the Arabic culture-area, especially Egypt. As a first approximation, it would seem likely that the use of a roughened stick spread outwards from India and that the toothbrush, in the strict sense, may have developed both in China and in the Arabic lands. The study of the subject, however, has only just begun.

Another question which has received too little attention is that of the sanitation of lavatories (*tshê*), both public and private, in China at different periods.[7] In

---

[1] Ch. 29 (p. 156.1). The modern term *sou khou*, to brush the teeth and rinse out the mouth, is already used in this ancient text.

[2] Cited in *Pên Tshao Kang Mu*, ch. 35 B, p. 13 b.   [3] *Ibid.* p. 8 b.

[4] See the papers of Chou Tsung-Chhi.

[5] Usually known in the West as Chhien-fo-tung. We had the pleasure of studying this painting personally during the summer of 1958. It is in cave no. 159. The monk is attended by a servant holding a towel and while brushing his teeth with his right hand holds in his left something at first sight remarkably like a tooth-paste tube, in fact perhaps a syringe. Another similar scene is to be found in cave no. 196 where the frescoes are of slightly later date but still Thang.

[6] See the papers of Wiedemann.

[7] See on this the account of Fan Hsing-Chun, pp. 74 ff.

very ancient times, instruments of bamboo, possibly spatulae (*tshê chhou*, *tshê pi*, or *tshê chien*), may have been used with the assistance of water in cleaning the body after defaecation. At other times and places, it seems that pieces of earthenware or pottery were so used. Undoubtedly one material which found employment in this respect was waste silk rag, but after the invention of paper about + 100, with which the name of Tshai Lun is always associated, the spreading of more hygienic habits among the whole body of the people was assisted by the increasing availability of this expendable vegetal tissue. No doubt only the upper class had been able to use silk. The rougher kinds of paper for this purpose were called *tshao chih*. By the time of the Sui dynasty the use of paper in lavatories was probably universal, as we can see from a passage in the *Yen shih Chia Hsün* (Family Oeconomica)[1] written by Yen Chih-Thui about + 590. From early times, there must have been some sort of systematic arrangements for the removal of night-soil from urban agglomerations each morning; and in the + 13th century we hear the name they bore, *chhing chiao thou* (the 'turners-over of unwanted refuse'). This occurs in the *Mêng Liang Lu* (Dreaming of the Past),[2] a valuable work about Hangchow under the Sung, written in + 1275 by Wu Tzu-Mu. The lavatories of some especially luxurious people came down in song and story; mention may particularly be made of those in the home of Shih Chhung, a famous merchant of great wealth in the Chin period (c. + 300).[3] Details were reported in the *Yü Lin* (Forest of Anecdotes) by Phei Chi. Shih Chhung's lavatories were very elaborate with wash-water provided by women attendants, who assisted the toilet and the ablutions of the guests. One interesting detail is the provision of large boxes of Chinese 'dates' (the fruit of the jujube tree)[4] which were perhaps intended to absorb evil odours and keep the place sweet. There are stories of rude army commanders, more used to the field than to the niceties of cultured life, eating these dates under a misapprehension as to their purpose, and causing mirth among the servant-girls thereby.[5]

[1] Ch. 5, p. 13 b. The passage involves a prohibition of using paper on which characters had been written. It will be remembered that the written or printed word in all its forms carried in traditional China some of the numinous associations of the Logos, and to this day highly ornamental stoves or grates of glazed ceramics exist in temple precincts for destroying written or printed paper in an honourable manner, though they are no longer used.

[2] Ch. 13, p. 13 a.

[3] See the account in Yang Yuan-Chi, p. 170; and *Shih Shuo Hsin Yü*, ch. 30, p. 33 a.

[4] *Zizyphus sinensis* or *jujuba*, Read, nos. 292, 293; Stuart, p. 466. If the fruits absorbed ammonia, they would have been of service in earth latrines, but it seems they were often used as nose-stoppers.

[5] E.g. in the + 5th century, *Shih Shuo Hsin Yü*, ch. 34, p. 44 a.

SPECIFIC DISEASES; RABIES AND SMALLPOX

Although in ancient and medieval times it was of course impossible to organise efficiently the defence of the population against specific diseases such as hydrophobia and smallpox, it is nevertheless interesting to see what measures were taken in this respect. The recognition of hydrophobia as a specific disease goes back a long way in Chinese history. Indeed, we find what is perhaps the earliest reference to it in a passage referring to − 556. This comes from the *Tso Chuan* (Commentary on the Spring and Autumn Annals), an annotated record of events in the feudal States. The passage reads as follows:[1]

In the eleventh month of the seventeenth year of Duke Hsiang, people (in the capital of Sung State) were chasing after rabid dogs (*chi kou*). Some entered into the (house of) Hua Chhen. People followed them into Hua Chhen's compound and he was scared so he moved away and took refuge in the capital of Chhen State.

We are not necessarily to conclude that Hua Chhen's fright was due to fear of the disease, for the whole context of the story indicates that he had good cause to be afraid of many other things and he probably took this as an omen. Nevertheless, it is a very early reference to the disease, especially interesting because it indicates that an organised effort was being made to get rid of the dangerous animals, so much so that they could not rely on sanctuary even in the private houses and gardens of high officials. By the + 3rd century, we find it very prominent in Ko Hung's *Chou Hou Pei Chi Fang* and other medical works.[2] The treatment was usually the expression of blood from the bitten area and subsequent moxibustion.

In the absence of adequate social organisation, the sequestration of rabid dogs was not a practical possibility, so the population was warned at certain times of the year. In the Thang, Sun Ssu-Mo tells us in his *Chhien Chin Yao Fang* that

dogs generally go mad during the late spring and early summer. At this time children, weak women or aged people must be warned to carry sticks for their protection. For those who still cannot avoid being bitten in spite of such preventive measures, nothing wards off the danger better than moxibustion daily for 100 days. If at an early stage the wound begins to heal and it is thought that all is over, nevertheless death may come within a very short time. [And elsewhere:] When one is bitten by a rabid dog, the symptoms of the sickness appear on the seventh day. If nothing has happened within three weeks one may escape, but only after 100 days have passed without signs of the disease can one be sure that one is safe.[3]

[1] Ch. 9, cf. Couvreur tr., vol. 2, p. 330.
[2] *Chou Hou Pei Chi Fang*, ch. 7 (p. 212).   [3] Ch. 25 (p. 453.2).

The story of smallpox is rather different from this because China participated prominently in the pre-history of inoculation, in fact the development of variolation procedures. As is well known, brief descriptions of what was probably smallpox occur in the West in the + 6th century.[1] But the definitive description is usually considered to be that of the great al-Razī early in the + 10th.[2] Knowledge of the disease, however, is certainly earlier in China than any of these periods. In his *Wai Thai Pi Yao* of + 725, Wang Thao gave a very clear description of a smallpox epidemic[3] quoting from a book entitled *Chou Hou Fang*. The circumstantial information provided enables this epidemic to be dated, in probability, at + 497; so the book from which Wang Thao was quoting is not so much the *Chou Hou Pei Chi Fang* of Ko Hung as the enlargement of it written in the + 6th century by Thao Hung-Ching. In any case, the description is an exceptionally clear one.[4] We intend to give the translation of it in another paper.

In later ages, the practice of inoculating the contents of smallpox pustules into the nose (*chung hua* or *chung tou* or *chung miao*) was traditionally ascribed to a Taoist or Buddhist adept from O-mei Shan who successfully treated the family of the Prime Minister, Wang Tan, in the time of the Sung emperor Chen Tsung (+ 997 to + 1022). This tradition is reported, for example, in the imperial medical compendium[5] of + 1743, the *I Tsung Chin Chien*. However, if we accept only evidence from contemporary literary sources, we must conclude that the technique originated, or in any case began to be prevalent, in the Ming, perhaps not much before the + 16th century. In his *Tou Kho Chin Chhing Fu Chi Chieh*, Yü Thien-Chhih says that the Lung-Chhing reign-period (+ 1567 to + 1572) was the time when it became generally known and used. A particularly clear account exists from + 1643 when Yü Chhang in his *Yü I Tshao* (Miscellaneous Ideas in Medicine)[6] described his consultations over the inoculation of the sons of Ku Shih-Ming.

It seems likely that then, at some time during the + 17th century, the technique spread from China to the Turkish regions; and it is a well-known story[7] how Lady Mary Wortley Montagu (+ 1689 to + 1762), the wife of the British ambassador at Constantinople, allowed the technique to be used on her own family in + 1718. Details of it had already been published in Western memoirs[8] by +1714, and the

---

[1] Garrison, p. 125. [2] Major, p. 196. [3] Ch. 3 (p. 119.2).

[4] The problem of the dating, to which we shall return elsewhere, is discussed at length by Fan Hsing-Chun, pp. 106 ff., and others.

[5] Ch. 60, p. 1 *a*. On the tradition, see further Chhen Pang-Hsien, p. 371. Cf. Fig. 84 on p. 281 above.

[6] P. 52 *b*. [7] Cf. Halsband. [8] See those of Timoni and Pilarini.

general introduction of variolation or 'engrafting', as it was often called, in the West occurred between that date[1] and about +1721. It is interesting that after the Treaty of Nerchinsk in +1689, a special Russian medical mission was sent to China in order to study the method of inoculation used there, and indeed an English medical man in the Russian service, Thomas Harwin, participated in it.[2] The work of Edward Jenner using cowpox occurred of course much later, towards the end of the 18th century (+1796). The time is now ripe for a full investigation of the development of the method in China and we hope to devote another publication to this subject. At present, it certainly seems as if the technique originated in China in the +16th century if not in the +15th, or earlier still. If this is so, it formed a remarkable climax to the many centuries of Chinese effort in preventive medicine which the foregoing pages have attempted to outline.

COMPARISONS AND CONCLUSIONS

Looking back over what has been said, one is moved to compare the achievements of ancient and medieval Chinese thought and practice in the field of hygiene with the highlights of other civilisations in the same field. Most of these are, to be sure, common knowledge, for example, in ancient Israel the development of sanitary legislation took the form especially of food prohibitions; and both in Israel and in Islam, great emphasis was laid upon ablutions before prayer, and before and after meals. The Hebrews in ancient times were notable also for their circumcision operation, for their laws on leprosy and for the health-giving Sabbath rest. Ancient India also participated, doubtless moved by the heat of the climate, stressing the need for cleansing the body, and there perhaps came the earliest realisation of the value of good care of the teeth. To India also much credit is due for the insistence on the cremation of the dead. Then in the West everyone knows of the sanitary latrines and water-conduits of Minoan Crete, while later the love of the Greeks and Romans for gymnastics in the *palaestra* is part of our own heritage. The drainage systems of Roman cities, for example the Cloaca Maxima at Rome itself, and the public baths there supplied by fourteen noble aqueducts built between the −4th and the +2nd century are also known to all the world. In Rome, the *quatuor viri viis purgandis*, the officials charged with street-cleaning and the care of the public latrines, recall the similar officials

---

[1] Cf. the papers of Blake and of Stearns & Pasti.
[2] Cf. Cordier, vol. 3, pp. 273 ff., and the interesting paper of Wu Yün-Jui.

and workers mentioned in the *Chou Li* and everywhere found during the Sung period.

Viewing the general attitude of the ancient and medieval Chinese physicians and scholars to hygiene and preventive medicine, one gains the general impression that it compares most favourably with what was done in Greek and Roman civilisations. It was certainly much better than anything the medieval West could show until the Renaissance.[1] Here we meet with an interesting circumstance which has been noted in a number of other connections. There are many branches of science and technology in which the Chinese of ancient times are found to equal the contribution of the classical Western world, after which Europe falls behind for some dozen centuries until the coming of the Renaissance and the renewal of cultural life and scientific thought in the West. A good parallel may be found in the field of quantitative cartography. While in the West the triumphs of the Greek astronomers and geographers, culminating in the map-making of Ptolemy at Alexandria (+ 2nd century), were forgotten during the Middle Ages, and this inheritance only revived again at the Renaissance or after the introduction of the magnetic compass to Europe and the rise of the portolan charts; in China matters took a very different course. There also quantitative map-making began in the + 2nd century with Chang Hêng but it was continued especially by the great geographer, Phei Hsiu, in the + 3rd, and the use of a rectangular grid for quantitative cartography persisted throughout the Middle Ages until the coming of the Jesuits in the + 17th century and the unifying of geography, East and West. In other words, while in Hellenistic times the Mediterranean world was well advanced, this advantage was all subsequently lost, and in the + 10th or + 11th century, for instance, the Chinese were far ahead of the Europeans in scientific geography. It seems as if something very similar may turn out to be true of the medical sciences as a whole, and certainly, as we see here, with hygiene and preventive medicine in particular.

[1] All such generalisations are of course very difficult to make. Roman bathing customs persisting in the *hammām* of Islam strongly influenced the crusaders, so that bath-houses (in which the sexes were not separated) became very popular in late medieval Europe. After this there was a marked backsliding in the + 16th and + 17th centuries, quite paradoxical in view of the spread of classical humanistic culture, and progress was not resumed until the romantic age in the + 18th. In judging the comparative achievement of European and Chinese preventive medicine through the centuries, personal hygiene and bathing customs should probably be treated as an independent variable. Assuredly life in a Buddhist or Taoist abbey of the Thang was far more cleanly than in a Christian one of the same period, and Marco Polo noted a similar superiority at the end of the + 13th century, yet in the + 14th and + 15th there may have been less difference. Not until the last half of the 19th century did Europe draw decisively ahead, and now this brief monopoly of efficient plumbing is rapidly ending.

Nevertheless, there were certain aphorisms in the preservation of health which find echoes in all the civilisations. When we read the *Regimen Sanitatis Salernitanum*, for example, those precepts on health which were offered to a king of England, seemingly about + 1100, by the doctors of the medical school of Salerno in Italy, we seem to be in presence of fundamental ideas which the Chinese physicians would have endorsed as well as those of many other cultures. In Harington's translation:

> The Salerne School doth by these lines impart
> All health to England's King, and doth advise
> From care his head to keep, from wrath his heart.
> Drink not much wine, sup light, and soon arise,
> When meat is gone, long sitting breedeth smart;
> And after-noon still waking keep your eyes.
> When mov'd you find yourself to Nature's needs
> Forbear them not, for that much danger breeds.
> Use three Physicians still, first Doctor Quiet,
> Next Doctor Merry-man, and Doctor Diet.
> Rise early in the morn, and straight remember,
> With water cold to wash your hands and eyes
> In gentle fashion reaching every member,
> And to refresh your brain whenas you rise
> In heat, in cold, in July and December
> Both comb your head, and rub your teeth likewise...

We must not quote more, but words of this kind remind us that men of intelligence in all the cultures throughout the ancient and medieval times would have been at one with Hippocrates and Galen, with Pien Chhio and Chang Chung-Ching, in their advice on health to Everyman.

18

CHINA AND THE ORIGIN OF QUALIFYING EXAMINATIONS IN MEDICINE[1]

[1962]

THE origin of examinations in medical and surgical proficiency for the protection of the public from unskilled practitioners is a remarkably interesting subject. The germ of the idea goes back a very long way, for the beginnings of punishment for malpractice can be found in the famous law-code of Hammurabi (King of Babylon, − 2003 to − 1961). This continued in Achæmenid Persia where we find about the − 5th century the Avestan surgeon being bound to practise first on three non-Mazdæans; then only, if successful, could he perform operations on Zoroastrians. But the science and art of medicine had to advance much further before we find any elaborate system of medical examinations. The evidence available (which we shall mention further in due course) indicates clearly that our European system of examinations derived from Arabic culture through the School of Salerno. The question arises, however,[2] whether the physicians of + 10th-century Baghdad could have been influenced by still earlier practices in regions farther East.

Chinese civilisation constitutes a milieu of choice for seeking the origins of medical examinations in the modern sense. It is natural to suspect that the bureaucratic feudalism of ancient and medieval China, so different from any kind of society known in the Western regions of the Old World, originated examinations for medical qualification. There are striking parallels. When Bentley introduced written examinations in Cambridge in + 1702, for the first time in Europe, he was certainly not unaware of the age-old Chinese civil service examinations, which had been described in detail by many + 17th-century Jesuit writers in all the chief European languages. When Civil Service examinations were introduced in the nineteenth century in the West, the inspiration came again from

[1] A lecture at the History of Medicine Section, Royal Society of Medicine, London, 1962 (*Proceedings of the Royal Society of Medicine*, 1963, **56**, 63).

[2] The present paper originated as the response to a query of this kind raised in stimulating form by Mr Walter Pyke-Lees of the General Medical Council, to whom our best thanks are therefore owing.

the mandarinate examinations conducted for two thousand years previously in China. And indeed, as soon as we begin to investigate the matter by a study of the original texts transmitted in great quantity from classical Chinese sources, we find that two things were inextricably combined there. First, there was a very early development of the idea of a State Medical Service, and secondly, the development, also very early, of the conception of an Imperial University for educating the Chinese equivalent of 'persons well qualified to serve God in Church and State'. Of course, as Chinese cosmic religion was theocratically atheist, so to say, and the State was not separate from the Church, this phrase needs to be taken with a grain of salt; nevertheless it expresses the situation well enough. In what follows we shall see that these two currents combined at a certain point to generate specifically medical colleges, and the introduction of qualifying examinations for medical students naturally followed. The problem is, exactly when?

Let us begin with the situation at the end of the Chou period, the time when the Warring States were engaged in their internecine strife before the foundation of the First Empire, the Chhin. During the succeeding Han period, about the $-$2nd century, an important work was produced which stands as a monument of what people then thought was the ideal bureaucratic organisation of society. This is the *Chou Li* (Record of the Institutions (lit. Rites) of the Chou Dynasty). A very interesting chapter in this classical text is concerned with what can only be described as a State Medical Service. Besides the Imperial Physicians (I Shih) and the Imperial Dieticians (Shih I), who were evidently connected intimately with the service of the emperor, there were also State Physicians (Chi I) charged with attending to the diseases of the mass of the people, and State Practitioners of External Medicine[1] (Yang I). There has always been some uncertainty as to how far the ideal organisation sketched in the *Chou Li* was applicable to the country as a whole beyond the palace and the capital, but from the many other types of officials mentioned, such as Fumigators (Chu Shih) and Sanitary Police (Chhü Shih), it is clear that the beginnings of a State Medical and Public Health Service were envisaged.

Elsewhere the *Chou Li* in its lists of complements of official departments gives 2 Senior and 2 Junior Imperial Physicians with 2 Bursars (Fu), 2 Registrars (Shih)

---

[1] This is a term (*wai kho*) special to Chinese medicine. Besides surgery (which was never as highly developed as in Indian culture) it comprised ophthalmic and orthopaedic surgery, dermatology (including diseases with prominent peripheral symptoms such as smallpox), and the treatment of all kinds of tumours and ulcers, war wounds and other traumatic lesions.

and 20 Apprentices or Disciples (Thu). This we should not overlook because it shows that at the very outset of the Chinese State Medical Administration, young aspirants to medical skill are found directly attached to the imperial organisation.

After the first unification of the Chhin, the Han Empire lasted no less than four centuries, the Chhien Han from −202 to +9, and then the Hou Han from +25 to +220, paralleling Imperial Rome. Here we begin to see more clearly the entry of the distinction between two independent medical services, those of the palace and those of the nation.[1] For in the Former Han (−2nd century) we have the Thai I Chhêng-Ling, i.e. the Deputy Director of Medical Administration under the Chhang Fêng (Court of Imperial Sacrifices), and another Thai I Chhêng-Ling, Deputy Director of Medical Administration (Palace) under the Shao Fu (Directorate of Imperial Workshops). These are the first appearances of the expression Thai I Ling, which parallels the title of Astronomer-Royal (Thai Shih Ling) and runs on down the centuries thenceforward. Special titles for the physicians in total attendance upon the emperor now appear, such as Shih I, equivalent to the later Yü I. In the Later Han period we find a number of officials listed in the Department of the Thai I Ling, for example, 2 Chief Pharmacists (Yao Chhêng), 2 Chief Prescribing Physicians also Deputy Directors (Fang Chhêng), 293 Junior Physicians (Yuan I) and 19 Administrative Personnel (Yuan Li). At this time also we begin to see the initiation of provincial medical organisations. Besides the officials at the capital there were in each of the provinces a Superintendent of Medical Services (I Kung Chang). Many entries in the chapters on official organisation for this period show that all the government offices at the capital and throughout the provinces had fixed complements of medical officers distributed among them. If we hear less of the position of the Apprentices or Disciples there can be no doubt that this system, together with at least the skeleton of the National Medical Administration, continued into the three States of the San Kuo period, a time of division which followed the united empire of the Han.

When we come to the second unification, that of the Chin dynasty (+265 to +420), we must pause to look at a parallel development, of the utmost importance for our story, which had long been going on. This was the constitution of the Imperial University. On the one hand the title of Po-Shih (doctor or professor)

[1] A parallel may be found in the two independent astronomical observatories maintained under certain dynasties (*SCC*, vol. 3, p. 191).

had appeared as early as the Chhin dynasty (− 3rd century), and on the other hand the principle of examinations for scholarly competence had been brilliantly inaugurated by the emperor Han Wên Ti, who himself set the questions, in − 165, probably the earliest occasion of the kind in any culture. Then in − 124 Han Wu Ti, at the suggestion of Kungsun Hung and Tung Chung-Shu, had provided a body of these 'Erudite Gentlemen' (Po-Shih) with disciples (Ti-tzu) supported at government expense, thus establishing for the first time the Imperial University (Thai-Hsüeh). Starting with 50 students, it had grown steadily to no less than 3,000 by about − 10; and in + 4 Wang Mang convoked for it a grand congress of scholars from all over the empire to place the sciences and humanities upon a more definitive basis than before. The first emperor of the Later Han, Kuang Wu Ti, re-established the university in + 29 after troublous times, and by the San Kuo period, when Wei Wên Ti moved it to Loyang in + 224, Po-Shih had become an honoured rank in the bureaucratic hierarchy. Now in the newly unified empire of the Chin, between + 275 and + 278, Wu Ti reformed the university and divided it into two parts, the Kuo-Tzu Hsüeh for the sons of noble families and high officials, the Thai-Hsüeh for promising students of less distinguished stock.

In 1958 we had the pleasure of studying a magnificent stele which is still to be seen at the Kuan Kung temple at Loyang, dated + 278. This inscription gives an account of the Imperial University of that time, with its 3,000 students, many from abroad, some from 'East of the Sea' (Liaotung) and some from 'beyond the shifting sands' (*liu sha*), i.e. Sinkiang. There is a long list of students' and professors' names, and a full translation of the inscription would be of much interest. The student body was divided into freshmen (Mên-jen), those who had passed in one classic (Ti-tzu), those who had passed in two (Pu-wên-hsüeh), and those who had passed in three (Thai-tzu shê-jen). The final degree was that of Lang-chung, taken after about seven years, and eventually the candidate was ready to be gazetted to a government post. There was never any subsequent age when the Imperial University went altogether out of existence, though in times of disunion there were several institutions under competing dynasties, and it was only in periods of prosperity that the major institution could carry out its tasks to the full.

During the Chin the Thai-Hsüeh was strengthened by successive emperors, as for instance by Yuan Ti in + 317 and later, when further chairs for the study of particular classics (including the *Chou Li*) were established; and then in the following century, under the Northern Wei, provincial colleges were set up in + 466 in every commandery (*chün*).[1] The larger of these provincial colleges contained 2 professors (Po-Shih), 4 lecturers (Chu-Chiao) and 100 students (Hsüeh-Sêng). Towards the end of the Northern Wei, about + 490, the emperor Hsiao Wên Ti changed the name of the Imperial University to Kuo-Tzu (lit. Sons of the Nation), and then, under the dynasty of the Sui, about + 610, the name was finally changed to the Kuo-Tzu Chien. This designation lasted a very long time, down in fact to the end of the Empire in our own period. It is not too much to say that the universities of modern type existing all over contemporary China may trace their descent to the Kuo-Tzu Chien of the Middle Ages, enlarged, modified and duplicated as it was under Western influence in the 19th century. But that is another story into which we cannot go here. The important point is that throughout the + 1st millennium the conception of an institution of higher education within the framework of the national bureaucracy was thoroughly rooted in Chinese culture.

We must now retrace our steps to follow the developments in the national medical administration in the Chin period and subsequently. There was not much alteration. When we look at the Northern Wei records, we find little difference of importance until we suddenly come upon two striking new posts, Thai I Po-Shih, which we can only translate as Regius Professor of Medicine, and Thai I Chu-Chiao, again Regius Lecturer in Medicine. These appear as part of a great reorganisation and settlement of the official hierarchy carried out by Kao Tsu (the emperor Hsiao Wên Ti) and most probably completed in + 493. The posts were of the lower 7th rank among the officials and lower 9th rank respectively, and they parallel a number of other didactic posts at the Imperial University not only in general education but also in astronomical and calendrical science, even including geographical communications. The picture is of such interest that it is worth tabulating:

Rank	Post
5 ii A	Regius Professor of Classics, College of Princes
	Libationer, Imp. Univ.
6 i B	Senior Regius Professor of Astronomy, Imp. Univ.
	Regius Professor of Classics, Imp. Univ.
6 i C	Regius Professor of Music, Imp. Coll. Music
7 i B	Student, College of Princes

[1] Provincial academies and seminaries had first been established (or at any rate first recorded) by a decree of +3 issued under the emperor Hsiao Phing Ti.

Rank *Post*

7 ii C   Manager, Imp. Univ.
      Junior Regius Professor of Astronomy, Imp. Univ.
      Regius Professor of Divination, Imp. Univ.
      Manager, Imp. Coll. Music
      Regius Professor of Medicine, Imp. Coll. Med.
8 i B   Regius Lecturer in Classics, Imp. Univ.
9 i B   Regius Lecturer in Astronomy, Imp. Univ.
      Regius Lecturer in Medicine, Imp. Coll. Med.
9 ii B   Regius Professor of Geographical (Post-Station Service) Communications, Imp.
      Univ.

Thus the medical men were in dignified, if not the most exalted, company. How many of the high medical teaching posts there were in each category we do not know, but the idea spread rapidly.[1] The implications here are important. As we have seen, the Thai-Hsüeh of the Han was enlarged by the Kuo-Tzu Hsüeh of the Chin, and already by the + 3rd century there had been classical examinations for nearly five hundred years. In the + 4th, many new professorships had been established and the University enlarged in the + 5th. Thus when the official dynastic history now speaks of high medical teaching officers, the implication is quite clear that there were not only lectures but also examinations for competence in medical knowledge. At this time we still do not have documentary evidence. But it comes soon.

The third great unification of China occurred in + 581 when a single House, that of the Sui, took over again the dominance of the entire country. Apart from the usual administrative staff of the Thai I Ling we read that there were 2 Chief Pharmacists (Chu Yao) and 200 regular Physicians (I Shih), but also 2 Curators of Physick Gardens (Yao-Yuan Shih) where medicinal plants were systematically cultivated. On the educational side around + 585 we find an expansion, in that 2 Professors of Medicine (I Po-Shih) are mentioned with 2 Lecturers (Chu-Chiao), also 2 Professors of Physiotherapy (An-Mo Po-Shih) and 2 Professors of Apotropaics (Chu-Chin Po-Shih). We know that the former must have taught medical gymnastics and massage among other things, while the latter represented a bureaucratic recognition of all the various incantatory and talismanic methods of driving away diseases which had existed among the Chinese people,

---

[1] In + 553 the Korean kingdom of Paekche sent an I Po-Shih named Wangyu Rungtha to Japan to re-organise medical education there; his mission included two Masters of Drug Production, Pan Kyŏngnye and Chŏng Yutha. By + 702 it bore its full fruit, for in that year the emperor Mommu established an Imperial Medical College in Japan with five departments, and regular monthly and annual examinations.

as among all ancient nations, from high antiquity downwards. Veterinary science now also comes in because the Court of Imperial Equipages (Thai Phu) had among its staff 120 Teachers of Veterinary Science (Shou-I Po-Shih Yuan) [Fig. 94].

For the great Thang dynasty (+618 to +906) we have truly abundant information, and it is in the middle of the +8th century that we shall see the full development of medical examinations. We might first take a look at the Imperial University under the Thang. The Kuo-Tzu Chien, as it was now called, seems to have consisted of two parts, one more socially exalted than the other (as indeed had been the case in the Northern Wei), perhaps something like 'Collegians' and 'Oppidans'. The 'Noblemen's' group consisted partly of relatives of the Imperial House itself, and partly of sons of officials of the 3rd rank and higher. These were Kuo-Tzu students, numbering 300. But besides them there were the ordinary students of the Imperial University, the Thai-Hsüeh students, numbering 500, sons or nephews of officials of the 5th rank and above; and in addition to these as many as 1,300 from all over the country, known as Ssu-Mên students. Five hundred of these came from the families of officials of the 7th rank and above, but 800 were essentially 'Scholars' or 'Sizars' drawn from the most intelligent and best families of the common people. Fifty places were reserved for students specialising in music and acoustics, and 30 for students of mathematics, all from commoners' families 'who understood these subjects'. Lastly there were 80 places specially reserved for metropolitan students, something rather like the scholarships tied to localities which existed till recently in Cambridge and Oxford colleges. The full complement of students was thus 2,100.

Turning to the National Medical Administration (Thai I Shu), we find that it was one of the eight administrations under the Thai Chhang Ssu (Court of Imperial Sacrifices). Below the two Directors, Senior and Junior (Thai I Ling) with their Deputy Directors (Thai I Chhêng), there were 4 Chief Medical Directors (I Chien) and 8 Assistant Medical Directors (I Chêng). Among the lesser staff of the Administration we hear of 2 Managers (Fu), 4 Secretaries (Shih), 8 Pharmacists (Chu Yao), 24 Apprentices in Materia Medica (Yao Thung), 2 Curators of Physick Gardens (Yao-Yuan Shih), 8 Apprentices in Physick Gardens (Yao-Yuan Sêng), and lastly 4 Clerks (Chang Ku). It is rather remarkable to find that the rank and file of State physicians were still assessed bureaucratically according to the success of their medical practice, a procedure which, as we know, goes back to the end of the Chou period in China.

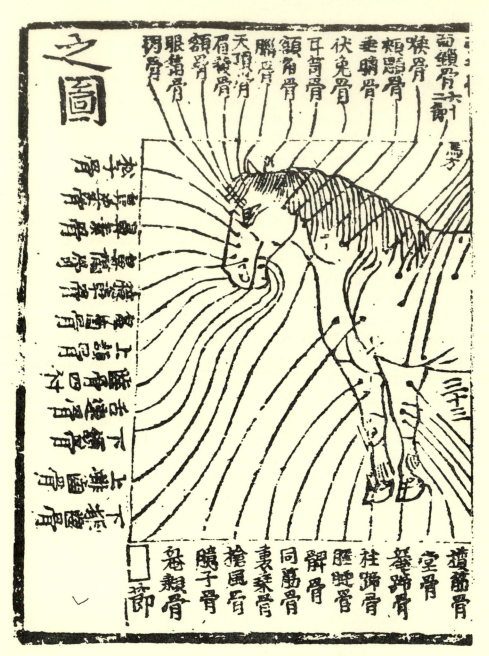

Fig. 94. Veterinary equine surface anatomy; a page from the
Ma Niu I Fang of +1399.

386

The *Hsin Thang Shu* says:

For the Physicians (I Shih), the Assistant Medical Directors (I Chêng) and the Practitioners (I Kung) their duties were clinical practice and the results were carefully recorded [by secretaries]. The data were analysed each year to examine their merits [and their salaries adjusted by the proportion of successful cures effected during the year]. Drugs and medicines were provided [by the Government] for the protection of the people's health, and stored in the temples of the imperial ancestral tombs. [Within the palace] the Imperial Infirmary (Huan Fang) had its own medical storehouses, and the distribution of medicines was under the responsibility of the Superintendent (Chien-Mên). One of the I Shih, I Chêng or I Chien took turns to be on duty in this hospital. All the provincial regions which paid tax on medical produce had one Master of Drug Production (Tshai-Yao Shih). In and near the capital city the best fields were set apart for [experimental] gardens for the cultivation of medicinal herbs. Any ordinary commoner over 16 years of age could become an apprentice student in a Physick Garden (Yao-Yuan Sêng) for the cultivation of medicinal plants, and when their training was accomplished, they could become Curators of Physick Gardens (Yao-Yuan Shih). The duty of these men was to be able to recognise every item in Materia Medica and to be able to say where it came from and what part of the plant should be used. They were also charged with selecting the best material from what was sent in as payment of taxes, to be the portion of the government.

What interests us here still more is the situation in the teaching field. We read again:

The Thai I Ling were in charge of all methods used in therapeutic treatment, assisted by their Deputies. Under them too were the four [teaching] departments, namely those for the Physicians (I Shih), the Acupuncturists (Chen Shih), the Physiotherapists (An-Mo Shih) and the Apotropaists (Chou-Chin Shih). All these were instructed by Professors (Po-Shih) and obtained their official posts by passing examinations in the same manner as the students of the Imperial University (Kuo-Tzu Chien).

Here we have a quite specific statement about examinations which must refer to the middle decades of the +7th century. Each of the teaching divisions included 1 Professor of Medicine (I Po-Shih) and 1 Medical Lecturer (Chu-Chiao). The Thang dynastic histories give us a great deal more information about the four teaching departments of this Imperial College of Medicine and include data on the number of staff and students in each.

Thus the Department of General Medicine comprised, besides the Professor (I Po-Shih) and his Lecturer, 20 physicians (I Shih), 100 practitioners (I Kung), 40 students (I Sêng) and 2 dispensers (Tien Yao). Only the Professor and Lecturer had rank in the official hierarchy. They taught the young students and practitioners from the already great medical literature, including the books of the *Pên Tshao* or pharmacopoeia type, the sphygmological classics such as the *Mo Ching*, and, of course, the fundamental classics such as the *Nei Ching* (*Su Wên* and *Ling Shu*). The courses into which the lectures were divided have also come down to us. They were five in number: General Medicine (*thi liao*); Oncology (*chhuang chung*); Pædiatrics (*shao hsiao*) [cf. Fig. 95, pl.]; Otology, Ophthalmology, Stomatology and Dentistry (*erh mu khou chhih*) and Cupping (*chio fa*). As phlebotomy was never a characteristic procedure in Chinese medicine, it may be that this last term is a mistaken reading for *chiu fa*, moxibustion. Next, the Department of Acupuncture, besides the Professor (Chen Po-Shih) and his Lecturer (Chu-Chiao), contained 10 Acupuncturists (Chen Shih), 20 Acupuncture Practitioners (Chen Kung) and 20 Students (Chen Sêng). Here the students and young practitioners specialised on sphygmology and acupuncture, learning the system of points on the surface of the body where 'needling' should take place in accordance with the signs indicated by the pulse and other diagnostic aids. All the mysteries of the different kinds of instruments, too, were opened to them. The other two Departments were somewhat less important. The Professor of Physiotherapy (An-Mo Po-Shih) had no Lecturer to second him, but his staff included 4 Physiotherapists (An-Mo Shih); these were in charge not only of medical gymnastics and massage but also dealt with traumatic injuries and fractures, having under them 16 Physiotherapic Practitioners (An-Mo Kung) and 15 Students (An-Mo Sêng). Lastly there was the Department of Apotropaics. The Professor (Chou-Chin Po-Shih) was seconded by 2 Apotropaists (Chou-Chin Shih) under whom were 8 Exorcists (Chou-Chin Kung) and 10 Students (Chou-Chin Sêng). This gives a grand total of 271 established posts and studentships in the Imperial College of Medicine, 162 in General Medicine, 52 in Acupuncture, 36 in Massage and Gymnastics and 21 in Apotropaics.

We get a vivid glimpse of the position regarding examinations in the year +758. In the *Thang Hui Yao* we read:

On the 5th day of the second month of the 1st year of the Chhien-Yuan reign-period it was decreed that thenceforth those who qualified as medical officers, undergoing tests for their medical knowledge and skill, should receive the same treatment as those graduating with the Ming-ching degree. Early in the following year [+759], the Chief Administrator of the Left Imperial Guard, Wang Shu, memorialised to the throne, requesting that the selection of personnel in the field of medicine be done by examination in accordance with [the rules already in use for] the Ming-fa degree. From that time until now there have been State examinations in the Medical Classics, Therapeutics and various techniques, altogether ten papers (*tao*). These are as follows: *Pên Tshao* (Materia Medica) 2 papers; *Mo Ching* (Sphygmology) 2 papers; *Su Wên* (Basic Medical Theory) 10 papers; *Shang Han* (Febrile Diseases) 2 papers;

Miscellaneous branches of Medicine, 2 papers. Those who succeed in seven or more out of the ten compulsory papers are qualified, and the remaining candidates fail and are sent home. He (Wang Shu) also asked that [the examinations for the personnel of] the Bureau of Nutrition and the Bureau of Medical Services within the Palace should be like those of the Bureau of the Catering Service of the Crown Prince; and that those under the National Medical Administration (Thai I Shu) should be similar to those under the Imperial Bureau of Music (Thai Yo Shu).

This very interesting passage requires a little commentary. The Thang dynasty established six special degrees in the selection of talented scholars for the administration. The first of these (Hsiu-tshai) was the basic degree, but besides that there were five others: Ming-ching for classical studies involving textual interpretations of the rituals and institutions; Chin-shih, primarily literary; Ming-fa to do with statutes, regulations, administrative technology and jurisprudence; Shu on calligraphy and history; and Suan for mathematics. It is quite interesting that the type of degree to be awarded for medicine should change from the Ming-ching to the Ming-fa. Could this indicate, one wonders, that the types of learning with which physicians were concerned were regarded as more relevant to the interpretation of natural regularities (like the fixed system of human law) than to philological criticism?[1] For the Ming-fa degree, 80 per cent correct answers were required for 1st Class. We find also that there was a total of 18 papers in the medical examination, of which 10 were compulsory, and passing in at least 7 necessary for qualification. The final remark about the Imperial Bureau of Music explains itself when one looks at the regulations for that organisation. There it appears that the professors themselves were periodically examined, as well as the executive musicians and acoustics experts, so that Wang Shu was presumably asking for a periodical test of the learning and efficiency of the teachers themselves. It does not look as though the trials of the embryo medical officers ended with the passing of their written examinations, for elsewhere in the *Hsin Thang Shu* we read:

As for the sons and advanced students of famous physicians, they get clinical experience and test their therapeutic skill under the supervision of high-ranking [medical] officers (Chhang Kuan) for three years. The names of those who succeed in this probationary practice then go forward [to the Government].

We gain further insight into the Thang system of medical education by following the institution of provincial colleges of medicine. As early as +629, an imperial edict decreed that in every important provincial city (*chou*) of the country, a

---

[1] On the relation between human law and the laws of Nature much will be found in *SCC*, vol. 2, pp. 518 ff.

Medical College (I-Hsüeh) with a Professor of Medicine (I Po-Shih) was to be established. In +723 a further imperial edict added Lecturers (Chu-Chiao) to the Professors in the provinces. At the same time it was ordered that pharmacopoeias (*Pên Tshao*) and certain medical books, especially one entitled *Pai I Chi Yen Fang*,[1] should be kept in all these colleges alongside the classics and histories. In the same year the *Kuang Chi Fang* (General Formulary of Prescriptions), composed by the emperor Hsüan Tsung himself, was published and sent out to each of the provincial medical schools. Some of the prescriptions in this work of an imperial pharmacist were actually written on notice-boards at cross-roads so that the masses could take advantage of them.[2] By +739 it was the law that every provincial city (*chou*) with more than 100,000 families was to have 20 medical students (I Sêng) and those of less than 100,000 were to have 12. In subsequent years the number of professors and students was increased. In +796 the emperor Tê Tsung published throughout the country his *Chen-Yuan Kuang Li Fang* (Valuable Prescriptions of the Chen-Yuan Emperor). In the same year an imperial edict ran as follows:

Early during the Chen-Kuan reign-period (+629) professorships of medicine (I Po-Shih) were established in all parts of the country. By the Khai-Yuan reign-period (+723) additional posts of lecturers (Chu-Chiao) were added. They held examinations regularly for scholars on medical subjects and expounded the proper organisation of rounds of medical visits within the area.

The emperor now observing that the system had not developed as it should have done

ordered that henceforth every large city (*chou*) must re-establish the medical professorship [for a properly qualified man], and all chief administrators (Chhang Shih) are required to give personal attention to the search for candidates to sit for examinations, and then select the best qualified for the posts [of medical officer]. Those already examined and qualified need not be examined again.

All this has a considerable bearing on our view of the date of foundation of the first Imperial Medical College. Since the provincial medical colleges were initiated as early as +629, it is clear that the central medical college must have been functioning within the first decade of the foundation of the dynasty, i.e. from +618. As we have seen, the institution was duplicated in Japan from +702 onwards.

For reasons which will be clear in a moment we do not propose to follow

---

[1] Probably the same as the *Chou Hou Pai I Fang* (Handbook of Medicine) by the famous alchemist-physician Ko Hung, c. +300.

[2] This was remarked upon by the Arabic traveller Sulaimān al-Tājir, who was in China in +851.

much further here the subsequent developments of the medical administration and its examinations during the dynasty of the Sung (+960 to +1280), the Yuan (Mongol) dynasty (+1280 to +1368), the Ming (+1368 to +1644) or the Chhing (+1644 to +1911). We have already said enough to establish the point we wish to make in this paper. However, something may usefully be said of the situation in the Sung.

The main educational work was carried on by the Imperial Medical College (Thai I Chü) which had reached its definitive form in +1076 during the premiership of the enlightened Wang An-Shih. Under the Director (Thai I Ling) and Deputy Director (Thai I Chhêng) there were a number of Professors and Lecturers, teaching no less than 300 students (I Sêng) in a number of specialities.

In this connection it is most interesting to see the further growth of an important development, the examination of medical students in general literature and the philosophical classics. We noted the beginnings of this already in the Thang, as early as +758 (p. 388). In Hangchow from about +1140 onwards the candidates were expected to pass examinations in the literary and philosophical classics as well as medical subjects. The imperial decree of +1188 ordered that unqualified medical practitioners must pass provincial examinations which included the general classical writings as well as sphygmology and other medical techniques. Anyone who did really well could gain an opportunity of rising to the ranks of the Han-lin Medical Academicians. This change was of great importance, for it ensured the existence of considerable numbers of physicians well educated in the classical literature and with greater general culture than their predecessors had possessed. Such men were called *ju i* (lit. Confucian physicians) as opposed to *yung i*, common practitioners, and *chhuan i*, wandering medical pedlars. Whether this was entirely a gain may be questioned, for a new race of men appeared who were very well educated but lacked perhaps practical experience which some of the rougher leeches of the past had had. Such was the situation at the end of the +12th century.

A well-known physician, Ho Ta-Jen, who published in +1216 an important treatise on pædiatrics, preserved for posterity at least two important collections bearing on our subject. One was entitled *Thai I Chü Chu Kho Chhêng Wên Ko* (Miscellaneous Records of the Imperial Medical College) and the other *Thai I Chü I-Sêng Fu Shih Wên Ta* (Questions and Answers in the Examination Papers of the Imperial Medical College). These texts, though still extant, are rare, and we have not seen them.

Although the *Sung Shih* history gives prominence to three main divisions in medical education, other texts show that there were more elaborate classifications into six and nine. The latter, which became particularly well known in later times, is given in the *Yuan-Fêng Pei Tui*, together with the number of students following each course, as follows:

(1) Internal and general medicine (*ta fang mo kho*) with 120 students.
(2) Convulsive and paralytic diseases (*fêng kho*) with 80 students.
(3) Pædiatrics (*hsiao fang mo kho*) with 20 students.
(4) Ophthalmology (*yen kho*) with 20 students.
(5) External medicine (*chhuang chung chien chê yang kho*) with 20 students.
(6) Gynaecology and obstetrics (*chhan kho*) with 10 students.
(7) Stomatology, dentistry and laryngology (*khou chhih yen hou kho*) with 10 students.
(8) Acupuncture (*chen chiu kho*) with 10 students.
(9) The treatment of war wounds and the use of apotropaics (*chin tsu chien shu chin kho*) with 10 students.

It will be seen that the total number of students amounts to three hundred. It is interesting to note that in our own time the examinations in traditional medicine currently in use in China follow much the same division into categories, with the addition of physiotherapy and history of medicine.

During the Sung period the Imperial Medical College was housed in spacious buildings. This can be seen from the *Sung Shih* history which, however, just falls short of telling us whether the College of the Northern Sung was fully residential, though the use of the term 'Hall' for the student bodies of the various years suggests that it was. After the fall of Khaifêng to the Chin Tartars in + 1126, and the setting up of the new Southern Sung capital at Hangchow, the residential aspect of the organisation comes into greater prominence, for the *Mêng Liang Lu* has much to say of the elaborate buildings provided in that city for the Imperial Colleges. The Medical College had lecture-halls for the teaching of the 4 Professors, and a temple for the worship of the tutelary deities of medicine; while the 250 students, who got excellent food in the refectory, were accommodated in 8 'Study-Houses' (*chai shê*), the very names of which have come down to us. They were all equipped with a special cap and belt which distinguished them from ordinary citizens, but they had to face examinations every month and every quarter. Such was the state of the Imperial Medical College about + 1275, when Marco Polo was on his way to China.

It will have been noticed that we have run over the centuries which corresponded to the rise and fall of the School of Salerno. This famous seed-bed of

Western European medicine began in the + 9th century when the Thang dynasty was drawing to its close. It reached its apogee in the + 12th century during the highly cultured period of the Southern Sung after the fall of the capital, and it continued until the end of the + 14th, which would take us to the close of the Yuan (Mongol) period in China. It appears that Arabic influence in Salernitan medicine is not to be detected much before + 1050, but after that it becomes extremely strong, with the *Antidotarium* of about + 1080. That was the time of Constantine the African, after which a massive transmission of Arabic knowledge and practice occurred, culminating in the *De Aegritudinem Curatione* of the + 12th century.

We now approach the dénouement of this paper. The pattern of general education without vocational trend, followed by a course of theoretical medical study, and a year or two of practice under supervision, was foreshadowed in Europe as far back as + 1224 in an edict of the Emperor Frederick II, which held good for Sicily, South Italy and Germany. Medical students at the School of Salerno were required first to study the logical treatises of Aristotle for three years and then to learn medicine from the books of Hippocrates, Galen and Avicenna for five years, finally carrying on clinical practice for one year under an experienced physician. The candidate was eventually subjected to a searching examination on the works of the Greeks and the Arabs, thereby obtaining a licence and graduating as *Doctor medicinae*, a term which originated in Salerno. The edict of Frederick II was apparently not quite the first of this kind, for in + 1140 Roger of Sicily had decreed laws concerning State examinations for physicians. Moreover, examinations for surgeons appeared first in Paris at the Collège de St Côme in + 1210. After the + 13th century the monopoly of Salerno declined with the rise of the Schools of Montpellier, Paris and Bologna. We hear also of medical examinations among the Arabs in Cairo in + 1283 and it is almost certain that the examinations and the licensing of Western Europe were borrowed from Arabic practice.

The year + 931 constitutes a focal point in this matter. As a result of a death occurring through the mistake of a medical practitioner, the Caliph al-Muqtadir placed an eminent physician, Sinān ibn Thābit ibn Qurrāh (*c.* + 880 to *c.* + 942), in charge of medical examinations for all existing practitioners and students. Sinān was the son of the great astronomer and mathematician, Thābit ibn Qurrāh (+ 825 to + 901) and the decree which he received was personally written out by the Caliph himself. Sinān proceeded therefore to examine all those who came before him, passing 860 practitioners, both old and new, in the first year. The examinations were continued by his son Abū Ishāq Ibrāhīm ibn Qurrāh (+908 to +947), and there is reason to think that similar tests continued regularly in Egypt, if not in Persia, until the end of the Caliphate and afterwards. By +980 there was a large new hospital in Baghdad[1] founded by the Buwayhid Emir, 'Adud al-Dawlah, where twenty-five physicians taught, examining pupils and attesting to their proficiency. In view of all that we have found concerning the long history of Chinese medical examinations, we cannot but ask ourselves whether some influence from further East could have initiated the important development in Baghdad in +931.

There is indeed evidence that contacts existed between the Arabs of Iraq and the Chinese at least from the beginning of the +8th century. After the Battle of the Talas River in +751, which marked the farthest limit of Islamic expansion eastwards, many Chinese artisans settled in Baghdad, including paper-makers and metal-workers, and it would even be surprising if there were not physicians among them. Some prisoners returned home in Chinese ships from the Gulf in +762 but many remained and founded families in Baghdad. Then, to say nothing of a mass of other material which exists on Chinese–Arab contacts, we have, just at the moment we need it, evidence of medical contacts between the two civilisations. In the *Fihrist al-'Ulūm* (Index of the Sciences), by the famous author Abū'l-Faraj al-Nadīm, there is a story concerning the great Rhazes, perhaps the greatest physician and alchemist of his time (Muḥammad ibn Zakarīyā al-Razī, +850 to +925, or even +932). This story concerns his friendship with a Chinese physician, who asked him to read Galen aloud as fast as he could while the Chinese translated rapidly and took down notes or whole passages verbatim in the shorthand script known as 'grass-writing'.[2] This glimpse of Arabic–Chinese contact provides exactly the bridge that we need, and it is very easy to imagine that one or other of al-Razī's Chinese medical friends suggested to him that periodical examinations should be held, as they were in China, by means of which the fitness of young physicians could be tested.[3] It would seem that the

---

[1] The background of all the Islamic hospitals and medical education was the great Sassanid medical school of Gundī-shapur (Jundī-Shāpūr) founded in the +5th century. It enjoyed the continuance of the Greek tradition because it absorbed the medical school of the University of Edessa (founded by Seleucus Nicator in −304) when that was closed by the emperor Zeno in +489; but it also drew abundantly from Indian, and later Chinese, sources, because Nestorian Christianity was its dominant faith.   [2] [P. 18 above.]

[3] Very recently, Iskandar has discovered at Alexandria a remarkable Arabic MS of a text 'On Examining Physicians' attributed to Galen, and quoted as such by al-Razī himself in a tractate with the same title. It remains to be seen how reliable this attribution is, but at least the text must be contemporary with al-Razī or somewhat earlier. In any case it does not deal exactly with qualifying examinations, but rather with the selection of physicians for important responsibilities from medical men already practising.

Arabs took up this suggestion with energy and enthusiasm, handing on the torch of public safety and medical self-respect to the Western world.

Thus it is clear, to summarise, that examinations of proficiency were held in China from − 165 onwards; the Imperial University was founded in − 124; Regius professorships and lectureships in medicine, implying examinations for qualification, date from + 493; an Imperial Medical College and provincial medical colleges were established between + 620 and + 630; and medical degrees were awarded from then onwards. In the light of all that we now know, we are able to estimate at its true worth the opinion of John Barrow, who wrote in 1804: 'The Chinese...pay little respect to the therapeutick art. They have established no public schools for the study of medicine, nor does the pursuit of it lead to honours, rank or fortune.'

THE ROLES OF EUROPE AND CHINA IN THE EVOLUTION OF OECUMENICAL SCIENCE[1]

[1966]

MANY historians of ideas and culture still blandly assume that the Asian civilisations 'had nothing that we should call science'. If slightly better informed, they are apt to say that China had humanistic but not natural sciences, or technology but not theoretical science, or even correctly that China did not generate *modern* science (as opposed to the ancient and medieval sciences). This is not the place to set such ideas to rights in any detail, but my own experience has shown that it is comparatively easy to produce a whole series of bulky volumes about the scientific and technological achievements which the Chinese are supposed not to have had. If, as is demonstrably the case, they were recording sun-spot cycles a millennium and a half before Europeans noted the existence of such blemishes upon the solar orb,[2] if every component of the parhelic system received a technical name a thousand years before Europeans began to study them,[3] and if that key instrument of the scientific revolution, the mechanical clock, began its career in early + 8th-century China rather than (as usually supposed) in + 14th-century Europe,[4] there must be something wrong with conventional ideas about the uniquely scientific genius of Western civilisation. Nevertheless it remains true that modern science, i.e. the testing by systematic experiment of mathematical hypotheses about natural phenomena, originated only in the West. Yet it cannot even be maintained that China contributed nothing to this great break-through of modern science when it occurred in the later stages of the European Renaissance, for while Euclidean geometry and Ptolemaic planetary

---

[1] An address at the opening of the Permanent Exhibition of Chinese Medicine at the Wellcome Historical Medical Museum, London (see Poynter, Barber-Lomax & Crellin), afterwards enlarged for the Presidential Address to Section X of the British Association, Leeds, 1967 (*Advancement of Science*, 1967, **24**, 83); also printed in *Journal of Asian History*, 1967, **1**, 1.

[2] *SCC*, vol. 3, pp. 434 ff.

[3] *SCC*, vol. 3, pp. 474 ff. and Ho Ping-Yü & Needham.

[4] *SCC*, vol. 4, pt. 2, pp. 435 ff. and Needham, Wang & Price. [Also p. 203 above.]

astronomy were undeniably Greek in origin, there was a third very vital component, the knowledge of magnetic phenomena, and the foundations of this had all been laid in China.[1] There people had been worrying about the nature of magnetic declination and induction before Westerners even knew of the existence of magnetic polarity.

But from the time of Galileo (+ 1600) onwards, the 'new, or experimental, philosophy' of the West ineluctably overtook the levels reached by the natural philosophy of China, leading in due course to the exponential rise of modern science in the nineteenth and twentieth centuries. What metaphor then can we use to describe the way in which the medieval sciences of both West and East were subsumed in modern science? The sort of image which occurs most naturally to those who work in this field is that of the rivers and the sea. There is an old Chinese expression about 'the Rivers going to pay court to the Sea',[2] and indeed one can well consider the older streams of science in the different civilisations like rivers flowing into the ocean of modern science. Modern science is indeed composed of contributions from all the peoples of the Old World, and each contribution has flowed continuously into it, whether from Greek and Roman antiquity, or from the Arabic world or from the cultures of China and of India.

Here I shall confine myself to the Chinese case. In considering the situation before us there are two quite distinct questions to ask, first, when in history did a particular science in its Western form fuse with its Chinese form so that all ethnic characteristics melted into the universality of modern science; and secondly, at what point in history did the Western form decisively overtake the Chinese form? We may thus try to define the date of what may be called the 'fusion point' on the one hand, and that of the 'transcurrent point' on the other. Since by a historical coincidence the rise of modern science in Europe was closely accompanied by the activities of the Jesuit mission in China (Matteo Ricci S. J. (Li Ma-Tou) died in Peking in + 1610), there was relatively little delay in the juxtaposition of the two great traditions. Since the break-through occurred in the West, the transcurrent point for each of the sciences naturally preceded the fusion point, but, as we shall see, the interest lies largely in the lag or delay between the two.

First let us consider the fusion points, those estuaries in time when the rivers flowed into the sea, and when full mixture took place. Here at once we find a remarkable difference between what happened in the physical sciences and in the

---

[1] *SCC*, vol. 4, pt. 1, pp. 239 ff., 334 and Needham (*Legacy*), p. 255. [Also p. 239 above.]
[2] Cf. *SCC*, vol. 3, p. 484.

biological sciences. On the physical side, the mathematics, astronomy and physics of West and East united very quickly after they first came together. By + 1644, the end of the Ming dynasty, there was no longer any perceptible difference between the mathematics, astronomy and physics of China and Europe; they had completely fused, they had coalesced.

If at first it seemed that Western mathematics had been at a higher level than Chinese mathematics this was found to be due, as the decades went by, to a loss of the expertise which the Sung and Yuan algebraists had had, and the restoration of their techniques redressed the balance—though the lack of deductive geometry remained a debit on the Chinese side. Chinese mathematics had always been by preference algebraic rather than geometrical.[1] Astronomy differed between the civilisations in an equally fundamental way, for while Greek astronomy had always been ecliptic, planetary, angular, true and annual; Chinese astronomy had always been polar, equatorial, horary, mean and diurnal.[2] The two systems were not in any way opposed or incompatible; it was just as in the mathematics, the attention of Chinese and Europeans had been concentrated upon different aspects of Nature. If the Chinese had never had the passion for geometrical models which produced the Ptolemaic epicycles and ultimately the Copernican solar system, their medieval cosmology had been far more modern than that of Europe, for instead of crystalline celestial spheres they thought in terms of infinite empty space and an almost infinitude of time.[3] By + 1673 when Ferdinand Verbiest S.J. (Nan Huai-Jen) was reconstructing the Peking Observatory and equipping it with new instruments even more accurate than the splendid + 13th-century ones of Kuo Shou-Ching,[4] this was perfectly understood; and the seal was set upon the matter when early in the eighteenth Antoine Gaubil S.J. (Sung Chün-Jung) published his great works on the history and theory of Chinese astronomy.[5]

Since the break-through occurred in Europe first, Europe contributed rather more, giving up its crystalline celestial spheres but introducing more refined calendrical computations, giving up its Greek ecliptic co-ordinates[6] but opening the way into those undreamt-of worlds which the telescope would shortly reveal,[7] and above all introducing the new celestial mechanics and dynamics of the age

---

[1] A full account of the history of Chinese mathematics is given in *SCC*, vol. 3, pp. 1–168.
[2] This epigrammatic formulation was due to Leopold de Saussure; see *SCC*, vol. 3, p. 229.
[3] Cf. *SCC*, vol. 3, pp. 408, 438 ff.
[4] *SCC*, vol. 3, pp. 350 ff., 367 ff., 451 ff.
[5] *Histoire de l'Astronomie Chinoise, Traité de l'Astronomie Chinoise*, etc., see *SCC*, vol. 3, pp. 760 ff.
[6] *SCC*, vol. 3, pp. 266 ff.
[7] Cf. Pasquale d'Elia; Needham & Lu Gwei-Djen (*Optick Artists*).

of Galileo. Unified astronomy of course profited greatly by the records of celestial phenomena (eclipses, novae and supernovae, comets, etc.) which Chinese astronomers had kept, as accurately as they could, since the − 5th century, and in greater abundance than any other culture.[1] Lastly one must take account of the fact that if oecumenical astronomy today uses exclusively the Greek constellation patterns this is not in the slightest degree due to any inherent superiority of these over the entirely different Chinese ones; it is simply a side-effect which followed upon the meteoric rise of modern science in the West as a whole. Men like Flamsteed or Herschel would have thought it bizarre to speak of *Wei hsiu* or *Thien chi* instead of Scorpio,[2] but there is no intrinsic reason for this; it was an incidental result of the rise of modern science in Western civilisation first. One must always be on the look-out for such side-effects. At all events by the mid + 17th century the union of the two astronomies had occurred.

One can see how complete this fusion was by taking an example such as the large + 18th-century Korean astronomical screen from the Yi Dynasty's royal palace in Seoul, lately deposited in the Whipple Museum of the History of Science in Cambridge.[3] To the right it reproduces the classical planisphere of + 1395, prepared for Yi Thaejo by Kwǒn Kǔn and his colleagues, but though equatorial in projection this bears the Chinese names of the Western zodiacal houses round its periphery. In the centre there are two Jesuit planispheres, ecliptically projected and using the Western 360° graduation instead of the Chinese $365\frac{1}{4}°$ one, yet conserving not only the entire pattern of Chinese constellations (quite different from the Western, as we have noted) but even the age-old division of the stars into three colours based on the ancient star lists drawn up by the − 4th-century astronomers, Shih Shen, Kan Tê and Wu Hsien.[4] To the left there are diagrams of the several planets, with text describing the new discovery of their moons and phases by Galileo and Cassini; and sun-spots depicted on the sun, as Chinese astronomers had noted them from the − 1st century onwards. Further text describes the resolution of nebulae, star clusters and the Milky Way, with the aid of the telescope. The Jesuit work preserved on the screen centres round one of the Directors of the Chinese Bureau of Astronomy, Ignatius Kögler S.J. (Tai

[1] *SCC*, vol. 3, pp. 409 ff. These records are in constant use by astronomers at the present day; cf. e.g. the current discussion on a nova of + 1006 by Goldstein, Goldstein & Ho Ping-Yü, Minkowski, Marsden, Gardner & Milne; and another on the saecular deceleration of the earth involving ancient eclipse records by Curott.

[2] Schlegel, pp. 153 ff.

[3] See Needham & Lu Gwei-Djen (*Korean Screen*).

[4] *SCC*, vol. 3, p. 263.

Chin-Hsien), and it must have been painted about +1757, not long after his death.[1]

Another striking example can be seen in two Chinese optical virtuosi, Po Yü and Sun Yün-Chhiu, who lived and worked in Suchow between +1620 and +1650, making apparatus such as telescopes, compound microscopes, magnifying glasses, magic lanterns, etc.[2] Po Yü, indeed, may be numbered with Leonard Digges and J.-B. della Porta, Lippershey, James Metius and Cornelius Drebbel, and all those other figures involved in the invention of the telescope and the microscope in the West. It is quite astonishing to find that within a couple of decades seekers in China were hot on the same trail. We do not yet know, and perhaps we shall never find out, exactly how independent they were, possible Jesuit intermediation being in this case obscure, but one can certainly say that at a surprisingly early date, +1635, the telescope was being applied to artillery in battle in China; and there is a strong possibility that Po Yü himself independently invented the telescope by juggling about with combinations of biconvex lenses just as a number of the inventors named above did in the West. Sun Yün-Chhiu even wrote a treatise entitled *Ching Shih* (History of Optick Glasses). Such was the extreme rapidity with which the mathematical and physical sciences of the Western and Chinese cultures fused after they first came in contact.

When you consider an intermediate science like botany, you find a totally different picture. It has been most interesting to observe, in work which my collaborators and I have been doing recently on the history of botany,[3] that there was a long delay after the first contacts, and one might say that the fusion point in botany did not occur until about 1880. Down to that time Chinese botany continued on its classical way. The naming, classifying and describing of plants went on along traditional lines. Even as late as 1848 the indigenous style persisted in the important work of Wu Chhi-Chün called the *Chih Wu Ming Shih Thu Khao* (Illustrated Investigation of the Names and Natures of Plants). Though written at such a recent date, this splendid and well-illustrated treatise was entirely traditional in character, and did not take any account of the advances in botany which had been made by Camerarius and Linnaeus. It is important to notice here that the Jesuit mission of the +17th century did relatively little for botanical

---

[1] The circumstances of the matter have a special interest because in +1741 a Korean astronomical official on a mission to China, An Kukpin, had formed a friendship with one of Kögler's aides, Andrew Jackson S.J. (Hsü Mou-Tê), the only Englishman who is numbered among the roll of the Jesuit Mission in China.

[2] See Needham & Lu Gwei-Djen (*Optick Artists*).

[3] *SCC*, vol. 6, pt. 1, in preparation.

contacts; indeed what it transmitted was westwards rather than eastwards, as witness e.g. the *Flora Sinensis* of Michael Boym S.J. (Pu Mi-Ko) printed in + 1656. Moreover, it could not have transmitted modern botany because its activity was both pre-Camerarian and pre-Linnaean in time. But when one comes to 1880, when Emil Bretschneider, the great medical officer of the Russian Ecclesiastical Exarchate in Peking, was doing his work on Chinese botany, then there began to be Chinese botanists who could speak the same language, could talk about Linnaean families and natural families, men who understood like European naturalists the function of the flower, and what the microscope could reveal of plant morphology. This was the time too at which centred the great effort of many investigators, indispensable for further development, to establish correlation as complete as possible between the Chinese traditional plant names and the Linnaean binomials.[1] Thus one might say that it was not before 1880 that the fusion point took place in botany, and a decade or so later might be a better guess.

Beyond this, when one passes on to medicine, one finds a situation in which the fusion of the sciences, pure and applied, in East and West, has not taken place even yet. I dare say that this is the case because, although physicists don't quite like you to say so, and astronomers equally may demur, nevertheless the phenomena of these sciences are surely much simpler than those with which biologists have to deal, and *a fortiori* physiologists, pathologists and medical men. Wherever the living cell is concerned, and *a fortiori* the living cell in its metazoan forms of high organisation, the puzzles are profounder, the tools both practical and conceptual more inadequate, the room for doubt greater. However optimistic one may feel as a young biologist or biochemist, the secret of life is still not yet just round the next bend. I speak from experience. Thus the coming together of the two cultural traditions, the fusion of them into a unitary modern medical science, has not even now been effected.

Many people, of course, when they think of Chinese medicine today, imagine it as some kind of 'folk-medicine', something bizarre and quite outdated, some sort of meaningless curiosity, but in truth these are all entirely wrong ways of reacting to it. It is, one must say, the product of a very great culture, a civilisation equal in complexity and subtlety to that of Europe.[2] While conserving a medieval body of theory, it contains a wealth of empirical experience which has got to

---

[1] Cf. the works of Bretschneider, still quite indispensable.
[2] Pálos rightly emphasises this.

be taken account of. Just as in other sciences we can find many Chinese priorities; for example, the compilation of a great classificatory description of disease entities without therapeutic material, the *Chu Ping Yuan Hou Lun* (Systematic Treatise on Diseases and their Aetiology), by Chhao Yuan-Fang in +610,[1] a whole millennium before Felix Platter[2] and Thomas Sydenham.[3] Or again the first handbook of forensic medicine in any civilisation, the *Hsi Yuan Lu* (Washing Away of False Charges) by Sung Tzhu (+1247),[4] appeared a considerable time before the European foundation-stones of the subject, the books of Fortunato Fedele[5] and Paolo Zacchia.[6] However, the rationale of some of the most important Chinese therapeutic practices, such as acupuncture, to which I will return in a moment, is not yet clearly understood; and obviously not all the drugs of the very rich traditional Chinese pharmacopoeia have yet been thoroughly examined from the biochemical and pharmacological point of view.

Equally important, there has until now been little unification of concepts. The original stimulus for this study arose indeed from problems of translation and technical terminology. In all the inorganic sciences Dr Wang Ching-Ning and I found long ago that once you know exactly what the ancient or medieval Chinese writer is talking about you can find the occidental equivalent for his words without too much difficulty, and everything makes good sense. So for example one can converse with him across the centuries about solstices (*chih*) or equinoxes (*fên*), square roots (*khai fang*) or comets (*hui hsing*), Lowitz arcs (*thi*), rock-salt and brachiopods (*shih yen*, written in two different ways); equally in the technological world about norias (*thung chhê*), water-powered reciprocating engines (*shui phai*) or chain-and-link work (*thieh ho hsi*). This is true only to a slightly lesser extent in fields such as botany and zoology, where *sui* is a spike or raceme and *thai* a capitulum or flower-head, while the corolla (*pa*) is clearly distinguished from the calyx (*o*). Similarly *wei* cannot mean anything but the stomach of an animal nor *fan chhu wei* anything but a rumen. Alchemy and early chemistry

---

[1] There is no adequate translation as yet of any part of this remarkable work, but in the meantime a paper by Rall may be referred to with due reserve.

[2] +1536 to +1614 (Garrison, pp. 271 ff., Castiglioni, pp. 429, 441, 452); his *Praxis Medicae* (+1608) has been described as the first attempt at a systematic classification of diseases.

[3] +1624 to +1689 (Garrison, pp. 269 ff., Castiglioni, pp. 546 ff.); his *Observationes Medicae* (+1675), *De Podagra et Hydrope* (+1683) and *Opera Universa* (+1685) were all outstanding for their pathognostic descriptions.

[4] Partial translation by Giles.

[5] +1550 to +1630 (Castiglioni, p. 557); his *De Relationes Medicorum* (+1602) defined the subject.

[6] +1584 to +1659 (Castiglioni, p. 557); his *Quaestiones Medico-Legales* (+1635) is the great landmark in the European history of forensic medicine.

have their own special problems exactly as they do in the West and for the same reasons, of purposive concealment and the like, but even amidst the spagyrical flights of poetic fancy there is far greater regularity than might be supposed, so that the 'river chariot' (*ho chhê*) always means metallic lead,[1] and the 'food left behind by Yü the Great' (*Yü yü liang*) always means brown nodular masses of haematite (ferric oxide).[2] China, moreover, had her Martin Ruhland, but her *Lexicon Alchemiae* was nearly a thousand years older than his, the *Shih Yao Erh Ya* (Synonymic Dictionary of Minerals and Drugs) by Mei Piao, *c.* +806, and it is extremely useful still today. Similarly, on the technological side *fan* is always alum, *shih tan* always copper sulphate, and *huo yao*, the 'fire chemical', never means anything else than a gunpowder composition.[3] On the whole the terminology of medieval Chinese chemistry, though far from unravelled as yet, presents no fundamental difficulties. Dr Ho Ping-Yü and I have found that it is quite possible to make medieval Chinese alchemical writing intelligible, though many years must pass before all its secrets become known.

But it is when one comes to the medical sciences that the translator finds himself in a really embarrassing position. Medical texts bristle with technical terms for which no equivalents in Western languages exist, some being ordinary words like *han* (cold) used in a highly technical sense, others specially constructed ideographs (often using the 'disease' radical), e.g. *i* (infectious epidemic illness) or *nio*, *yao* (malaria-like fevers) or *li* (dysenteries of various origin). The key words of the highly systematised medical philosophy are the most difficult, for even Chinese lexicographic works do not dare to define them, since it was always expected that physicians would acquire their correct usage during long apprenticeship. Nowadays, to be sure, there are many works produced by the schools of traditional medicine in China which help to expound the terms, though not of course to translate them. Since there can be no exact equivalents in the Western world, where the evolution of physiological and medical thought followed very different courses, my collaborators, especially Dr Lu Gwei-Djen, and I, are adopting a new technique in translation, i.e. constructing an entirely fresh series of 'words of art' from Greek and Latin roots designed to express the innermost senses of the Chinese medical technical terms, and then using them systematically. All other possible procedures are open to the gravest objections; to leave the

---

[1] But *caveat emptor*, of course, for in medico-physiological terminology the same words mean the placenta.
[2] *Caveat emptor* again, for in botany the same words refer to certain plants, forming an alternative name for the liliaceous medicinal herb *Liriope spicata* (sometimes called the black leek), and for the sedge *Carex macrocephala* (Cyperaceae).　　　[3] [Subject to the reservation on p. 90 above.]

terms untranslated makes the result unreadable; to translate them by mechanical use of the dictionary makes it archaic, quaint and ridiculous as well as incorrect; while to seek to replace them by modern technical terms in current use is liable to distort the traditional ideas in a very dangerous way. The effect of our method is to make the medieval Chinese physicians talk like + 16th- and + 17th-century European medical writers, Ambrose Paré or Thomas Willis, in the same genre as it were, but obviously in a completely different tradition, and this is just the effect desired [cf. the example on pp. 305 ff. above].

Now what all this essentially means is that the medical philosophies and theories of China and the West are by no means as yet mutually expressible, so that we have a situation quite different to what pertains in the inorganic and the simpler organic sciences. Indeed there are important technical terms in Chinese medicine, such as *hsü* and *shih*, *piao* and *li*, which almost defy any rendering in a Western language—almost but, we believe, not quite.[1] Such considerations came very prominently to mind during the discussions which Dr Lu and I had with Professor Chhen Pang-Hsien, the eminent historian of medicine, and other colleagues, in Peking in 1958; and it was out of them that the ideas expressed in the present paper first arose.

One finds oneself obliged to speak (as herein I do) of 'modern-Western' medicine. It is not fair to call it just 'Western' as if it was wholly on a par with 'Chinese' or 'Indian' medicine, because it is palpably based on modern science in a way which the medicine of the non-European civilisations is not; but it is equally unfair to call it blandly 'modern' medicine, because that implies that no non-European civilisation has anything to contribute to it. On the contrary, truly modern and oecumenical medicine will not come into being until all these contributions are gathered in. Therefore I contrast 'modern-Western' with 'Chinese-traditional' medicine.

Let us now examine more closely the transcurrent points rather than the fusion points. One can hope, I believe, to define a certain number of moments in history when modern science, as we know it in the West since the time of Galileo, took

---

[1] Thus we propose for their equivalents respectively 'eremotic, plerotic, patefact and subdite'. Needless to say, the systematic use of such language in the translation of medical texts will be preceded by a deep analysis of the content of the Chinese terms and a semantic justification of the linguistic components chosen in the construction of their equivalents. Obviously the whole task is one of extreme difficulty, not only because of the conceptual discordance between the cultures, but also because naturally Chinese medical terminology changed slowly through the centuries. Nevertheless the ideas can, we believe, be harmonised if sufficiently understood; and there is a sufficient consensus in the language of Chinese medicine to allow of first-approximation Graeco-Latin equivalents very widely valid.

clearly and decisively the lead over against Chinese science. One has to remember of course the earlier situation, pertaining in the Middle Ages, when nearly every science and every technique, from cartography[1] to chemical explosives,[2] was much more developed in China than in the West. From the beginning of our era down almost to the time of Columbus, Chinese science and technology had very often been far ahead of anything that Europeans knew. Just to take only one or two examples, seismology was cultivated in China generations before the West, Chang Hêng in the + 2nd century devising apparatus for locating the azimuth direction of the epicentre and the force of the shock.[3] So also while the Roman agriculturists despaired of classifying and describing soils, their Chinese equivalents before the + 2nd century had brought into use more than fifty definable pedological terms, at the same time laying the foundations of all oecology and plant geography.[4] Or again, nobody in Europe (that Europe of the later boasted 'iron horse' and irresistible 'ironclads') could reliably obtain a single pig of cast iron until about + 1380, while the Chinese had been great masters of the art of iron-casting ever since the − 1st century.[5] The standard method of interconversion of rotary and longitudinal motion, the eccentric, connecting-rod and piston-rod assembly, was not known in Europe before about + 1450 but it had been fully at work in China since + 970 and the combination of the first two components of it goes back there to + 600 at least.[6] This is why I have elsewhere said that Chinese physical science attained a Vincian but not a Galilean level.[7] When then can we find these turning-points of transition, these transcurrent points, as I call them, when modern science and technology originating in the West decisively took over from the Chinese level?

In the case of mathematics, astronomy and physics, I think one can say that it happened almost at the same time as the fusion point, or only a very short time before. What the Jesuits brought to China included of course Euclidean geometry and Ptolemaic planetary astronomy, both of which were very ancient, and

[1] See *SCC*, vol. 3, pp. 525 ff.

[2] *SCC*, vol. 5, in preparation; meanwhile Wang Ling, and a summary account in Needham (*Legacy*).

[3] *SCC*, vol. 3, pp. 626 ff.

[4] There are good studies in Chinese of this birth time of pedology, oecology and plant geography, but nothing of any moment in a Western language pending *SCC*, vol. 6, the relevant sections of which are already written.   [5] See Needham (*Iron and Steel*).

[6] See *SCC*, vol. 4, pt. 2, pp. 369 ff., 380 ff.; Needham (*Steam Engine*); and Chêng Wei. In *SCC* about two centuries of priority could be substantiated, but the discovery by Chêng Wei of a scroll-painting of *c.* + 970 by Wei Hsien, depicting a large water-mill which includes a reciprocating bolter, gives about five centuries. As for the connecting-rod and eccentric only, see *SCC*, vol. 4, pt. 2, p. 759.

[7] *SCC*, vol. 3, p. 160.

certainly not part of modern science. But they also brought the algebraic notation of François Viète, which had only just been developed in the middle of the sixteenth century,[1] and later the logarithms of John Napier; and above all they brought the new dynamics, mechanics and optics of Kepler and Galileo. It is interesting to reflect that Tycho Brahe, the observational father of modern astronomy, filling his ledgers of data on the island of Hveen, was a patently Chinese figure, not in technique or conception much more advanced than Shen Kua or Su Sung; it was only the next generations that began to overpass the Chinese levels. Although the Jesuits played down the Copernican theory itself they gave full publicity to the results which Galileo obtained with the telescope from + 1610 onwards.[2] It will be remembered that he got the idea of the instrument from Holland and then made his own, after which things began to happen very quickly. In mathematics, astronomy and physics, therefore, it would seem that the transcurrent point came only a few decades before the fusion point.

In botany, on the other hand, there was, as we have seen, a great time lag, because the fusion point did not occur until after 1880. The transcurrent point we should have to put as occurring some time between + 1695, when Camerarius first demonstrated the nature of the flower, Linnaeus' prime in + 1735, and the restorative work of the great Adanson in + 1780. Were it not for the fact that the Linnaean sexual system of classification was a sort of branch-line or siding, and not in the main line of advance, one might be tempted to say that Chinese botany attained a Magnolian or Tournefortian, but not a Linnaean, level. Yet it would perhaps be fairer to take Adanson as the turning-point about + 1780 and to say that then it was that botany in the West began to be decisively ahead of Chinese botany. But after that there was a lag of some hundred years, between + 1780 and 1880, hence the feelings of superiority of the early nineteenth-century plant collectors, admirers though they were of the Chinese horticulturists whose gardens they delightedly pillaged.[3]

Next then we come to consider the question: when did Western medicine decisively draw ahead of Chinese medicine? I confess that the more we think about it the later we are inclined to put this moment. I am beginning to doubt whether the transcurrent point was really much earlier than about 1900, perhaps 1850 or 1870. There are many things to be considered.[4] One has to weigh, for

[1] *SCC*, vol. 3, p. 438.
[2] Cf. Needham & Lu Gwei-Djen (*Korean Screen*).   [3] See the monograph by Cox.
[4] Here a valuable help is the monograph of Keele on the evolution of clinical diagnostic methods.

example, the clinical discoveries (Morgagni and Auenbrugger, + 1761, Corvisart, 1808, Laënnec, 1819);[1] the rise of pharmaceutical chemistry (Pelletier & Caventou, 1820) with the study of alkaloids as its centre;[2] the new understanding of neuro-physiology (Bell, 1811, Magendie, 1822); the development of bacteriology after Pasteur (1857);[3] the growth of immunology from Jenner, *c.* + 1798 (itself originating from a Chinese technique, variolation); the development of antiseptic surgery (Lister, 1865) and anaesthesia (1846); radiology (Röntgen, 1896), radio-therapy (the Curies, 1901) and radio-isotopes (Joliot-Curie, 1931); then para-sitology with the discovery of the malaria plasmodium and its life-cycle (Laveran, 1880, Ross, 1898); eventually the coming of vitamins (Hopkins, 1912),[4] sulpha-drugs (Domagk, 1932), antibiotics (1940) and so on.[5] All this requires more thought, but if therapeutic success rather than diagnostic understanding is taken as the criterion, I suspect that it was not much before 1900 that medicine in the West drew decisively ahead of medicine in China.[6] Naturally, terms need careful definition. The work of Vesalius was not done in vain, and surgery and morbid anatomy were therefore correspondingly far ahead of China already by 1800. It may well be that all the sciences basic to medicine were much more advanced throughout the nineteenth century than what was known in China, and this must certainly be true of physiology as well as anatomy; yet from the point of view of the patient these branches of knowledge were rather slow in application, so that if we judge strictly clinically, the patient may not have been much better off in Europe than in China before the beginning of the twentieth century.[7] In one single day in 1890 my father, himself a physician, lost both his first wife and beloved teenage daughter of diphtheria, no antitoxin being then

[1] Pathological anatomy, percussion, auscultation, the invention of the stethoscope. The thermometer came somewhat earlier, the sphygmograph much later.

[2] Isolation of strychnine and quinine; there is a fine statue commemorating these two chemists on the Boulevard St Michel in Paris.

[3] Necessarily involving microscopy. A landmark in this was Beale's *The Microscope in Medicine* (1854).

[4] Biochemistry was to contribute fundamentally of course to clinical diagnosis as well as treatment but not much before the twentieth century. Qualitative and quantitative urine analysis was how it began. This was first systematised by Neubauer & Vogel in 1860; a book to which I remember referring myself in my young days.

[5] The other side of the medal is how long medieval practices continued in Europe. Galenic pulse-lore was still being systematised in 1828, and the Manchester Royal Infirmary ceased its bulk purchasing of leeches only in 1882.

[6] To take two last examples, the electrocardiogram dates only from 1903 and the electroencephalogram only from 1929.

[7] In 1826, the year of Laënnec's death and the appearance of D. M. P. Martinet's significant *Manuel de Patho-logie*, the basic sciences, says Keele, received more lip-service than application in the practice of medicine, even in France where there was most awareness of their value.

available, though in later years he could constantly use it. I mention this family tragedy in order to emphasise that it may be an entire illusion to suppose, as so many do, that European medicine enjoyed a serene superiority over Chinese medicine throughout the eighteenth and nineteenth centuries. The 'enteric fever' of the Boer War could be another striking example. So when one finds a traveller like Dr Dinwiddie, who accompanied the Macartney Embassy to China in 1793, putting on airs of great superiority about Chinese science and medicine, one realises today that he had very little reason for being so pleased with himself.[1] By 1900, of course, perhaps even by 1870 or 1885, there were very good and solid reasons. But if the change-over or transcurrent point came so late, what is certain is that the fusion point has not even yet been attained. Indeed many decades must doubtless pass before this is achieved.

Today the traditional medical men in China are working side by side with what we may call the 'modern-Western' physicians in full co-operation.[2] This is a very remarkable fact, which my collaborators and I have ourselves seen, in 1952, 1958 and 1964. It has been brought about in China by national renaissance, social conditions, and the paucity of medical men trained in modern style, during the past fifteen years. The two types of physicians and surgeons have joint observations, joint clinical examinations, and there is the possibility for patients to choose whether they will have their treatment in the traditional or the modern way; in other cases the physicians themselves decide which is best and proceed to apply it. And if one reads the *Chinese Medical Journal*, for example, carefully, one will find certain fields, as for instance the treatment of fractures,[3] where prolonged consideration has decided that in fact there were many valuable features in the traditional methods, and what is in use now is a combination of the two, the Chinese and the Western. Such fusion is going to happen more and more, giving rise to a medical science which is truly modern and oecumenical and not qualifiedly modern-Western. Here is only one example of it.

Now I shall say briefly something more on the question of acupuncture. As is generally known, this is a method of therapy, developed some two thousand years ago, which involves the implantation of very thin needles into the body in

---

[1] *A fortiori* his even more self-satisfied colleague the surgeon Dr Gillan; see the papers of the Macartney Embassy recently published by Cranmer-Byng.

[2] This raises questions about traditional Chinese conceptions of clinical diagnostic method, regimen and therapy, as well as the nature of the medical philosophy itself. We cannot go into these here but must refer the reader to *SCC*, vol. 6. In the meantime we may cite as the least misleading among Western books about Chinese medicine, the following publications: Hume; Morse; Beau; Pálos; Chamfrault & Ung Kang-Sam.　　　　　　　[3] See the series of papers by Fang Hsien-Chih *et al.*

different places according to a scheme or chart based on traditional physiological ideas and thoroughly systematised at an early date, certainly by the Thang and Sung periods.[1] We ourselves have seen the way in which this implantation of needles is done, attending acupuncture clinics in several Chinese cities. The method is still used very widely indeed in China at the present day. The problem arises of how its action can come about, and one may say without fear of contradiction that there are dozens of laboratories in China and Japan at the present day which are actively working with modern methods of a physiological and biochemical character to elucidate what happens. There are many possibilities; for example, that the stimulation of the autonomic nervous system by this method may increase the antibody titre in the blood, or increase the cortisone production by the supra-renal cortex, or it may exert a neuro-secretory influence upon the pituitary gland. A wealth of experimental approaches lies open.[2] Moreover, it must be recognised that the acupuncture system connects in many ways with assured facts in neuro-physiology, notably the Head Zones of the skin in mammals, which are related with specific viscera, and the remarkable phenomena of referred pain.

No one will ever really know the effectiveness of acupuncture, or the other special Chinese treatments, until accurate clinical statistics have been kept for several decades. The Chinese are not getting around to this at the present time because the practical job of looking after the health of 700,000,000 people does not readily permit it, but I have no doubt that within a century accurate clinical

---

[1] Though there are many more recent books in Western languages, those of Soulié de Morant have not been superseded. From 1901 he studied directly under two eminent physicians named Yang and Chang respectively at Peking and Shanghai; and thirty years later, on returning to France, he set forth at length the classical system of acupuncture. Among the writings derivative from this tradition are those of Baratoux and the Lavergnes. Since then several different strains of transmission have led to Europe. From Formosa the influence of Wu Hui-Phing has generated the books of Lavier, Moss and the Lawson-Woods. From Vietnam that of Nguyen van Nha has affected those of Mann. Japanese studies have also exerted influence in Europe (Nakayama and Sakurazawa). In approaching acupuncture through the works of representatives of the present-day European practitioners some reserve should be exercised, for (a) very few of them have had linguistic access to the voluminous Chinese sources of many different periods, (b) it is often not quite clear how far their training has given them direct continuity with the living Chinese clinical traditions, (c) the history in their works is generally quite unscholarly, and accounts of theory very inadequate, (d) their works are naturally much influenced by Western concepts of disease aetiology and semeio-graphy so that they seem not to practise the classical Chinese methods of holistic classification and diagnosis. Nevertheless, pending the historical and theoretical account which we ourselves hope to give in SCC, vol. 6, this literature has its value. A brief and anonymous but authoritative statement issued by the National Academy and Research Institute of Chinese Traditional Medicine at Peking a few years ago is an important document.

[2] It is hard to give any adequate references here since so much of the literature is published in Chinese and Japanese, and in journals both difficult of access and not sought after by Western medical libraries. Much time must yet, I fear, elapse before all this is digested into a form available to the world scientific and medical public.

statistics will be kept, and this will be a fundamental contribution to our knowledge of traditional Chinese medicine.

A view commonly expressed (mostly by Westerners) is that acupuncture acts purely by suggestion, like many other things in what they often call 'fringe' medicine. This is, I believe, a question of what one might call relative credibility (or perhaps a calculus of credulity), a choice of what is the most difficult thing to believe. It may well be more difficult to believe that a treatment which has been engaged in and accepted by so many millions of people for something like twenty centuries has no basis in physiology and pathology, than to believe that it has been of purely psychological value. Of course it is true that the practices of bleeding (phlebotomy) and urinoscopy in the West had exceedingly little physiological and pathological basis on which to sustain their extraordinary and long-enduring popularity, but none of these had the subtlety of the acupuncture system. Possibly blood-letting had some slight value in hypertension, and extremely abnormal urines could tell their story,[1] but neither contributed much to modern practice.[2] I can only say that for my own part I find the purely psychological explanation of acupuncture much more hard to credit than an explanation couched in terms of physiology and pathology. Experiments on animals, where the psychological factor is ruled out, are now being actively pursued in Chinese and Japanese laboratories, and results so far support this opinion. Surely in due course the scientific rationale of the method will be found. But until it is, Chinese and modern-Western medicine will not have fused.

Something more remains to be said about the theoretical setting of acupuncture and other traditional methods, such as the medical gymnastics for example which originated very early in China.[3] I have in mind the relative value placed in Chinese and Western medicine on aid to the healing power of the body on the one side and the repulsion of attack by invading organisms on the other. Now both in Western medicine and in Chinese medicine these conceptions are both to be found.[4] On the one hand, in the West, besides the seemingly dominant idea of

[1] On these see the books of Keele and Brockbank. Keele rightly congratulates the ancient Chinese on their freedom from 'magico-religious concepts of disease', though he calls their traditional medical philosophy 'metaphysical', but he might have added that they were also always free from that individual genethliacal astrology which played so painfully prominent a part in medieval European medicine, as he himself shows.

[2] [Venesection is, however, still used in polycythaemia, haemochromatosis, and haemosiderosis; and by some in congestive heart failure, acute pulmonary oedema, etc. But this is far from the universal application which it once had.]

[3] Cf. Dudgeon.

[4] The principal works in Western languages on the history of Chinese medicine are those of Hübotter; Wang Chi-Min & Wu Lien-Tê; Huard & Huang Kuang-Ming. But only certain monographs on special

direct assault on the pathogen, we have also the conception of the *vis medicatrix naturae*, which my father was always telling me about when I was a boy; for resistance and the strengthening of resistance to disease is an idea strongly embedded in Western medicine from Hippocrates and Galen onwards. On the other hand, one can also affirm that in China, where the holistic approach might be thought to have dominated, there was the idea of combating external disease agents,

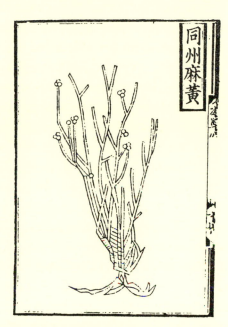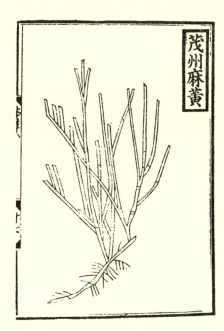

Fig. 96. *Ephedra sinica* (*ma huang*), the plant from which the drug ephedrine gets its name, two varieties; from the *Ta-Kuan Pên Tshao* (Pharmacopoeia of the Ta-Kuan reign-period), +1108.

whether these were sinister *pneumata*, the *hsieh chhi* from outside, of unknown nature, or whether they were distinct venoms or toxins left behind for example when insects had been crawling over food—this is a very old conception in China—so that the combating of external agents was certainly present in Chinese medical thought too.[1] This may be called the *i liao* aspect (or, in the ordinary parlance, *chih ping*); and the other one, the *vis medicatrix naturae*, was what was meant in China by *yang shêng*, the strengthening of resistance. Nevertheless I think it is clear that whatever the acupuncture procedure does, it must be along the lines of strengthening the patient's resistance (e.g. by increasing the antibody or

topics and periods, e.g. that of Bridgman on Han medicine, attain the standard of modern scholarship requisite today, the standard at which we shall ourselves aim in *SCC*, vol. 6.

[1] See Needham & Lu Gwei-Djen (*Hygiene*). [Reprinted, p. 340 above.]

cortisone production), and not along the line of fighting the invading *pneumata* or organisms, venoms or toxins, i.e. not the characteristic 'antiseptic' assault which has naturally dominated in the West since the time of origin of modern bacteriology. This is shown by the significant fact that while Westerners are often prepared to grant value to acupuncture in affections such as sciatica or lumbago (for which modern-Western medicine can do very little anyway), Chinese physicians have never been prepared to limit either acupuncture or the related moxa (mild cautery and heat-treatment) to such fields; on the contrary they have recommended and practised it in many diseases for which we believe we know clearly the invading organisms, e.g. typhoid, cholera, or appendicitis, and they claim at least remission if not radical cure. The effect is thus cortisone-like. It is surely very interesting that both these conceptions (the exhibition of hostile drugs and the strengthening of the body's resistance) have developed in both civilisations, in the medicine of both cultures; and one of the things which any adequate history of medicine in China will have to do will be to elucidate the extent to which these two contrasting ideas dominated in the systems of East and West at different times.

The dichotomy just elucidated, the dichotomy between strengthening the defender of the organic citadel on the one hand, or sallying forth to attack its attackers on the other, is obviously closely related to another dichotomy, another pair of antitheses in pathological thought. The idea of the attacks of external agents, whether parasites or toxins, corresponded well enough with the biological idea of the stimulus, while the strengthening of the patient's resistance corresponded equally well to the biological idea of reactivity. But what if there was no attack from outside at all and therefore no special possibility of reacting to it well or badly? Perhaps illness could be caused by some imbalance of the normal processes going on in the body, some imperfect organisation, in fact what the Greeks understood by an abnormal *krasis*, a failure to achieve the right mixture or combination.[1] The history of these ideas in the West has been brilliantly sketched by Temkin, Jones, Edelstein and others in recent writings, but the extraordinary parallels in Chinese thought, which constantly envisaged failures of balance between the Yang and Yin, and among the Five Elements, have been little commented upon. The topicality of the whole question is extraordinary, for modern endocrinology has abundantly shown the calamities which may ensue upon glandular malfunction. So again, it will be necessary for any adequate study

[1] [Cf. pp. 292, 302 above.]

of the history of Chinese medical thought to do justice to the parallels with the West in the field of ideas of morbid imbalance.

These remarks would not be complete without a reference to the importance of the traditional Chinese pharmacopoeia. I do not think that anyone today is

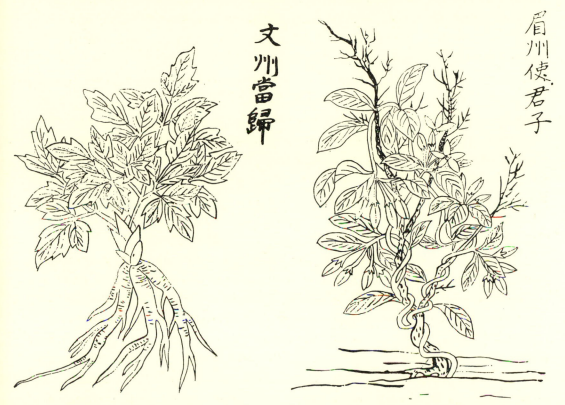

Fig. 97. *Angelica polymorpha* (*tang kuei*), containing an oxytocic active principle; a drawing from the *Shao-Hsing Pên Tshao* (Pharmacopoeia of the Shao-Hsing reign-period), +1159.

Fig. 98. *Quisqualis indica* or Rangoon creeper (*shih chün tzu*), a combretaceous vine with valuable anthelminthic properties, named after the Sung or pre-Sung physician Kuo Shih-Chün; another drawing from the *Shao-Hsing Pên Tshao*, +1159.

inclined to despise pharmacopoeias of a traditional or empirical character developed among non-European peoples.[1] Since the recognition of the use of ephedrine from *Ephedra sinica* in the Chinese pharmacopoeia [Fig. 96], one of its greatest triumphs, there have been many more shocks administered to pharmacologists in the West, as, for example, the famous case of *Rauwolfia* with its

[1] See, for example, the symposium recently edited by Chhen Kho-Khuei, Mukerji & Volicer. Or the monograph of Mosig & Schramm.

numerous powerful and highly peculiar alkaloids. I suppose that the whole modern science of chemo-therapy has been closely connected with, if not directly dependent on, the investigation of naturally occurring drugs of an alkaloidal or otherwise highly complex organic character. The pharmacopoeia in China is in

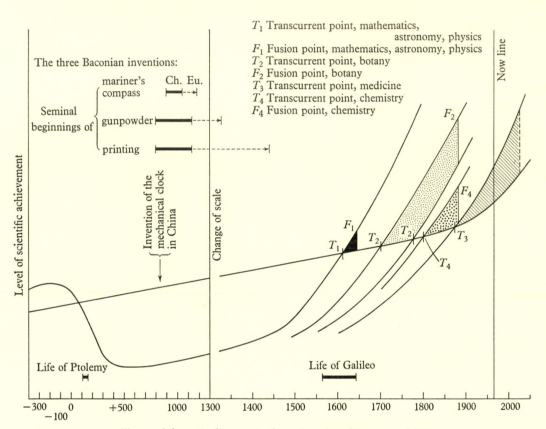

Fig. 99. Schematic diagram to show the roles of Europe and China in the development of oecumenical science.

fact full of things which are of great interest from this point of view [cf. Fig. 97]. When I was in China during the war running a scientific liaison mission between the Chinese and the Western allies, as Director of the British Scientific Mission in China, I had a good deal to do with *Dichroa febrifuga* (in Chinese *chhang shan*). People were looking about urgently for anti-malarials other than quinine, and *chhang shan* was therefore studied a good deal. Various pharmaceutical laboratories in China worked on it, gaining positive results early on in the war; these were doubted in the West, but eventually at the National Institute of Medical Research

414

Dr Thomas Work made a study of it and it has turned out to be a quite powerful anti-malarial.[1] Its interest is indeed somewhat impaired by a variety of side-effects, presumably due to other substances present, but if the active principle could be purified (and I am not quite sure how far this has been done) it could be a valuable medicament.

The naming of plants in the Linnaean system after personal names is often regarded as very modern, but sometimes the names of particular people were given to drug plants in China too. There is one, for example, called *shih chün tzu* named after a physician, Kuo Shih-Chün, who studied and used it in the Wu Tai or Thang period about the +10th century. This is *Quisqualis indica* (or Rangoon creeper), a really valuable anthelminthic, especially in pædiatric use, and still employed today on a wide scale [Fig. 98].

If we may now sum up the results of our meditation so far we can make a very simple Table to assemble certain figures that have been mentioned, and to accompany Fig. 99.

	Transcurrent point	Fusion point	Lag
Mathematics Astronomy Physics	1610	1640	30
Botany	1700 or 1780	1880 1880	180 100
Medicine	1800, 1870 or 1900	not yet	?

From this one might be tempted to deduce quite tentatively a 'law of oecumenogenesis' which would state that the more organic the subject-matter of a science, the higher the integrative level of the phenomena with which it deals, the longer will be the interval elapsing between the transcurrent point and the fusion point, as between Europe and an Asian civilisation. If this were in general principle right one might try to test it by looking into the history of chemistry in East and West, for which one would expect a figure intermediate between those for the physical sciences and for botany.

This subject bristles with difficulties partly because of the contingent historical

---

[1] See Chang Chhang-Shao; Fu Fêng-Yung & Chang Chhang-Shao; Chang, Fu, Huang & Wang; Tonkin & Work; Duggar & Singleton.

trends which intervene. Chemistry as we know it is of course a science like that of the branch of physics which deals with electricity—wholly post-Renaissance, indeed + 18th-century in character. The pre-history of chemistry goes far back into antiquity and the Middle Ages, and it does so in China at least as much as the West.[1] In the West there were first the mystical proto-chemists of Alexandria on the one hand, and the pharmaceutical alchemists of China on the other.[2] There is overwhelmingly strong ground for believing that Arabic alchemy was influenced from China (even the very name is probably Chinese in origin),[3] and that it handed on its alchemical afflatus, coupling the art of making gold with that of finding the elixir of immortality, to the European alchemists of the + 12th to the + 15th centuries whose triumphs included the discovery of alcohol.[4] Vast treasuries of Chinese alchemy from the + 2nd to the + 14th centuries are contained in the *Tao Tsang* or Taoist patrology, and besides all this there are rich texts of other genres which tell of metallurgy and the chemical industries. When Paracelsus in the + 16th century inaugurated iatro-chemistry he was only copying unknowingly in Europe just what had come about in China somewhat earlier, with the difference that in that culture there had never been any prejudice against mineral remedies. So brilliant was the iatro-chemical period in China (+ 11th to + 17th centuries) that it has been possible to show how the adepts of those times were able to prepare mixtures of crystalline steroid sex hormones and use them for therapy in cases for which they are normally prescribed today.[5]

But all this was not theoretical modern chemistry. The foundations of this were laid, as everyone knows, during the late + 18th and early 19th centuries, with the exploration of the nature of gases by Priestley and others (+ 1760 to + 1780), the 'revolution in chemistry' effected by Lavoisier (+ 1789), and then the atomic

[1] Pending *SCC*, vol. 5, a good deal of which is already written, the history of Chinese alchemy and chemistry is still a closed chapter. But I may refer to the pioneer book of Li Chhiao-Phing. Studies of particular fields reach, as usual, a more scholarly level; see e.g. Dubs; Sivin; Ho Ping-Yü & Needham; Tshao Thien-Chhin, Ho Ping-Yü & Needham. The brief chapter in Leicester's handbook is balanced and perceptive.

[2] These terms are 'words of art'. The Hellenistic adepts were aurifictors, imitating gold but not believing that they could make it from other substances. But the essence of alchemy is macrobiotics, the chemical search for means of longevity and material immortality, and the combination of this with belief in aurifaction was fundamentally Chinese.

[3] Etymologies of 'chem-' from Greek or Egyptian roots having long been notoriously unconvincing, its derivation from Chinese *chin* (gold) or *chin i* (gold juice) was suggested independently in 1946 by the writer (Needham, *Reflections*) and Mahdihassan, who since then has elaborated the equation in many papers. This view has now gained wide acceptance (see e.g. Dubs, *Origin*; and Schneider).

[4] Cf. the masterly little books of Sherwood Taylor and Holmyard. The medieval volume of Partington's *magnum opus* is still awaited.

[5] See Lu Gwei-Djen & Needham (*Steroids*). [Cf. p. 312 above.]

theory of Dalton (1810) followed by the far-reaching insights of the founder of organic chemistry, Justus von Liebig (1830 to 1840).[1] This was already the beginning of the Opium Wars and the Thai-Phing Rebellion, but as soon as quiet recurred in China and modern science was able to strike roots again, chemistry in its new forms was introduced. There were no obstacles to fusion, because there had been no competing Chinese theories in the past;[2] the basic facts of chemical change which alchemists, industrial workers, and medical men had long known fitted simply into the new explanations, the superiority of which over the traditional Yin and Yang and Five-Element theories was much more obvious here than in physiology or medicine. Modern chemistry was taught at all the Chinese universities after 1896, and books on it had been published by the Translation Department of the famous Kiangnan Arsenal from the time of its foundation by Ting Jih-Chhang in 1865 onwards. Private institutions such as the Ko Chih Shu Yuan at Shanghai, which opened in 1874, also propagated chemical knowledge.[3] It may therefore be quite fair to set a period of some 80 years, say between 1800 (the approximate transcurrent point) and 1880 (the approximate fusion point), as the time elapsing. This evidently fits in adequately with the general picture, but I should not like to put too much emphasis on it, partly because of the adventitious historical circumstances, and partly because modern chemistry, as a relative late-comer in modern science, had no alternative system to meet with when it reached the Chinese culture-area, as all the other sciences considered had.

In conclusion, then, what we have done here is to examine the time elapsing between the first sprouts of particular natural sciences *in their modern forms* in European culture, and their fusion with the traditional forms as Chinese culture had known them, to form the universal oecumenical body of the natural sciences at the present day. The more 'biological' the science, the more organic its subject-matter, the longer the process seems to take; and in the most difficult field of all, the study of the human and animal body in health and disease, the process is as yet far from accomplished. Needless to say, the standpoint here adopted assumes that in the investigation of natural phenomena all men are potentially equal, that the oecumenism of modern science embodies a universal language that they can

---

[1] Cf. the usual accounts; Thorpe; Lowry; Partington (*Short Hist.*).

[2] This does not mean that there were no theories in Chinese alchemy and chemical industry, but they always remained essentially medieval in type (cf. Ho Ping-Yü & Needham).

[3] This was headed for a time by Wang Thao, the Chinese collaborator of the great sinologist James Legge.

all comprehensibly speak, that the ancient and medieval sciences (though bearing an obvious ethnic stamp) were concerned with the same natural world and could therefore be subsumed into the same oecumenical natural philosophy, and that this has grown, and will continue to grow among men, *pari passu* with the vast growth of organisation and integration in human society, until the coming of the world co-operative commonwealth which will include all peoples as the waters cover the sea.

BIBLIOGRAPHY

CHAPTER 7

The Earliest Snow Crystal Observations

Bentley, W. A. & Humphreys, W. J. (1931). *Snow Crystals*. McGraw-Hill, New York.

Dobrowolski, A. B. (1922). *Historia Naturalna Lodu*. Warsaw.

Dogiel, J. (1879). Ein Mittel, die Gestalten der Schneeflocken künstlich zu erzeugen. *Mélanges Phys. et Chim. de l'Acad. de St. Petersbourg*, **9**, 266.

Doi Toshitsuru (1832). *Sekka Zusetsu* (Illustrations of Snow Blossoms). (1839). *Zoku Sekka Zusetsu* (supplement). Both reproduced in *Nihon Kagaku Koten Zensho* series, ed. Saegusa Hiroto, Tokyo, 1946–, vol. 6. Facsimile reproduction, with commentary by Kobayashi Teisaku, and Eng. résumé, Tsukiji Shokan, Tokyo, 1968.

Fritsch, K. (1853). On snow-flake forms and temperature of precipitation. *Sitzungsber. d.k.k. Akad. d. Wiss. zu Wien* (Math-Naturw. Kl.), **11**, 492.

Glaisher, J. (1855). Snow crystals... *Rep. of Council of Brit. Meteorol. Soc.* 17; abridged in *Quart. J. Mic. Sci.* **3**, 179; **4**, 203.

Guettard, J. E. (1762). On snow-flake forms and temperature of precipitation. *Mém. de l'Acad. de Paris*, p. 402.

Hellmann, G. (1893) (with microphotographs by R. Neuhauss). *Schneekrystalle; Beobachtungen und Studien*. Mückenberger, Berlin.

Ho Ping-Yü & Needham, J. (1959). Ancient Chinese observations of solar haloes and parhelia. *Weather*, **14**, 124.

Mason, B. J. (1961). The growth of snow crystals. *Scientific American*, **204** (no. 1), p. 120.

Mason, B. J. & Maybank, J. (1958). Ice-nucleating properties of some natural mineral dusts. *Quart. J. R. Met. Soc.* **84**, 235.

Nakaya Ukichiro (1954). *Snow Crystals, Natural and Artificial*. Harvard Univ. Press, Cambridge, Mass.

Needham, J. *et al.* (1954–). *Science and Civilisation in China*. Cambridge.

Phei Chien & Chou Thai-Yen (1956). *Chung-Kuo Yao Yung Chih-Wu Chih* (Chinese Botanical Materia Medica), vol. 4, Kho-Hsüeh, Peking. (Publications of the Botanical Institute of Academia Sinica.)

Read, B. E. (1936). *Chinese Medicinal Plants*. Peiping.

Read, B. E. & Pak, C. (1936). *A Compendium of Minerals and Stones used in Chinese Medicine*. Peiping.

Schneer, C. (1960). Kepler's New Year's Gift of a Snowflake. *Isis*, **51**, 531.

Scoresby, W. (1820). *An Account of the Arctic Regions, with a History and Description of the Northern Whale-Fishery*. Edinburgh.

(1825). *Tagebuch einer Reise auf den Wallfischfang, verbunden mit Untersuchungen und Entdeckungen an der Ostküste von Grönland, im Sommer 1822*, translated by F. Kries. Hamburg.

Spencer, J. (1856). *On the Similarity of Form observed in Snow Crystals as compared with Camphor*. London.

Stuart, G. A. (1911). *Chinese Materia Medica.* Shanghai.

Tamura Sennosuke (1958). *Tōyōjin no Kagaku to Gijutsu* (Essays on the History of Science and Technology among East Asian Peoples). Awaji Shobō Shinsha, Tokyo.

Von Laue, E. (1952). *Introduction to the International Tables for X-ray Crystallography.* Kynoch, Birmingham.

Wang Yü-Hu (1957). *Chung-Kuo Nung-Hsüeh Shu Lu* (Bibliography of Chinese Agricultural Books). Chung-Hua, Peking.

Wilcke, J. K. (1761). Rön och Tankar om Snö-Figurers Skiljaktighet. *Kongl. Vetenskaps Acad. Handlingar,* **22**, 1.

   (1769). Nya Rön om Vattnets Frysning til Snölike Is-Figurer. *Kongl. Vetenskaps Acad. Handlingar,* **30**, 90.

Wilhelm, R. (1928). *Frühling und Herbst d. Lü Bu-We* (translation of the *Lü Shih Chhun Chhiu,* with annotations). Diederichs, Jena.

CHAPTER 10

The Pre-Natal History of the Steam-Engine

Afet Inan (1941). *Aperçu Général sur l'Histoire Economique de l'Empire Turc-Ottoman.* Publ. de la Soc. d'Hist. Turque, Ser. VII, no. 6. Maarif Matbassi, Istanbul.

Andrade, E. N. da C. (1957). The early history of the vacuum pump. *END,* **16** (no. 61), 29.

Anon. (1836). The economy of the Chinese illustrated by a notice of the tinkers, with a description of the bellows. *CRRR,* **4**, 37.

Arago, D. F. (1855). Notice historique sur les Machines à Vapeur. *AL,* 1829, repr. 1830 and 1837. Repr. in *Oeuvres,* vol. 5, pp. 1 ff. Gide & Baudry, Paris, and Weigel, Leipzig.

D'Arnal, E. Scipion, Abbé (1783). *Mémoire sur les Moulins à Feu établis à Nîmes.* Nîmes.

Atkinson, F. (1962). The horse as a source of rotary power. *TNS,* **33**, 31.

Balfour, H. (1907). The fire piston. Art. in *Anthropological Essays presented to Edw. Burnett Tylor in honour of his 70th Birthday.*

Beck. T. [BGM] (1900). *Beiträge z. Geschichte d. Maschinenbaues.* Springer, Berlin.

   [HAA] (1909). Herons der älteren Automatentheater. *BGTI,* **1**, 182.

Becker, C. O. & Titley, A. (1930). The valve gear of Newcomen's engine. *TNS,* **10**, 6.

Beckmann, J. (1872). *A History of Inventions, Discoveries and Origins.* 1st German ed., 5 vols. 1786 to 1805. 4th ed., 2 vols. Bohn, London, 1846. Enlarged ed., 2 vols. tr. by W. Johnston. Bell & Daldy, London.

de Bélidor, B. F. (1737–53). *Architecture Hydraulique; ou l'Art de Conduire, d'Elever et de Ménager les Eaux, pour les différens Besoins de la Vie.* 4 vols. Jombert, Paris.

Berthelot, M. (1891). Pour l'Histoire des Arts Mécaniques et de l'Artillerie vers la Fin du Moyen Age (I). *ACP* (6e sér.), **24**, 433.

Blümner, H. (1912). *Technologie und Terminologie der Gewerbe und Künste bei Griechen und Römern.* 4 vols. Teubner, Leipzig and Berlin.

Bossert, H. T. & Storck, W. F. (1912). *Das mittelalterliche Hausbuch...* Seemann, Leipzig.

Bowden, F. P. & Yoffe, A. D. (1952). *The Initiation and Growth of Explosion in Liquids and Solids.* Cambridge. (Cambridge Monographs on Physics.)

Boxer, C. R. (ed.) (1953). *South China in the Sixteenth Century; being the Narratives of Galeote Pereira, Fr. Gaspar da Cruz, O.P., and Fr. Martin de Rada, O.E.S.A. (1505 to 1575).* Hakluyt Society, London. (Hakluyt Society Pubs. 2nd series, no. 106.)

Burstall, A. (1963). *A History of Mechanical Engineering.* Faber & Faber, London.

Cardwell, D. S. L. (1963). *Steam Power in the 18th Century; a Case Study in the Application of Science.* Sheed & Ward, London. (Newman History and Philosophy of Science Series, no. 12.)

(1965). Power technologies and the advance of science, 1700 to 1825. *TCULT*, **6**, 189, 195.

Carter, T. F. (1955). *The Invention of Printing in China and its Spread Westward.* Columbia Univ. Press, New York, 1925, revised ed. 1931. 2nd ed. revised by L. Carrington Goodrich, Ronald, New York.

Chambers, Sir Wm. (1757). *Designs of Chinese Buildings, Furniture, Dresses, Machines and Utensils; to which is annexed, A Description of their Temples, Houses, Gardens, etc.* London.

Cherry, T. M. (1962). Anthony George Maldon Michell, 1870–1959. *BMFRS*, **8**, 91.

Childe, V. Gordon (1954). Rotary motion [down to 1000 B.C.]. Art. In *A History of Technology*, vol. 1, p. 187. Ed. C. Singer, E. J. Holmyard & A. R. Hall. Oxford.

Churchill, A. & Churchill, J. (ed.) (1704). *A collection of Voyages and Travels...* Churchill, London. 2nd ed. 1732–52.

Cline, W. (1937). *Mining and Metallurgy in Negro Africa.* Banta, Menasha, Wisconsin (mimeographed). (General Studies in Anthropology, no. 5, Iron.)

Colladon, M. & Championnière, M. (1835). Note sur les Machines à Vapeur de Savery. *ACP*, **59**, 24.

Coomaraswamy, A. K. (1924). The Treatise of al-Jazarī on Automata [1206]; Leaves from a MS. of the *Kitāb fī Ma'arifat al-Ḥiyal al-Handasīya* in the Museum of Fine Arts, Boston, and elsewhere. Mus. of Fine Arts, Boston. (Communications to the Trustees, no. 6.)

Couvreur, F. S. (tr.) (1914). '*Tch'ouen Ts'iou*' [*Chhun Chhiu*] *et* '*Tso Tchouan*' [*Tso Chuan*]; *Texte Chinois avec Traduction Française*, 3 vols. Mission Press, Hochienfu.

Cranmer-Byng, J. L. (ed.) (1962). *An Embassy to China; being the Journal kept by Lord Macartney during his Embassy to the Emperor Chhien-Lung, 1793 and 1794.* Longmans, London.

Cummins, J. S. (ed. and tr.) (1962). *The Travels and Controversies of Friar Domingo de Navarrete (1618 to 1686)*, 2 vols. Cambridge. (Hakluyt Society, 2nd series, nos. 118, 119.)

Daremberg, C. & Saglio, E. (1875). *Dictionnaire des Antiquités Grecques et Romains.* Hachette, Paris.

Daumas, M. [BPN] (1961). Le Brevet du Pyréolophore des Frères Niepce (1806). *DHT*, **1**, 23.

[LIS] (1953). *Les Instruments Scientifiques aux 17e et 18e Siècles.* Presses Univ. de France, Paris.

Davey, H. (1903) (with appendices by W. G. Norris, Sir Frederick Bramwell, H. W. Pearson, J. H. Crabtree, W. E. Hipkins, Messrs. Thornewill & Warham, W. B. Collis & H. S. Dunn). The Newcomen engine. *PIME*, 655.

Degenhart, B. (1941). *[Antonio] Pisanello.* Schroll, Vienna; Chiantore, Turin.

Dickinson, H. W. [JW] (1936). *James Watt, Craftsman and Engineer.* Cambridge.

[JWE] (1911). John Wilkinson [engineer]. *BGTI*, **3**, 215.

Dickinson, H. W. [SE] (1939). *A Short History of the Steam-Engine.* Cambridge. Photolitho reprint, with unpaged introduction by A. E. Musson and list of corrigenda, Cass, London, 1963.

   [SE 1830] (1958). The steam-engine to 1830. Art. in *A History of Technology*, ed. C. Singer *et al.* vol. 3, p. 168. Oxford.

Dickinson, H. W. & Titley, A. (1934). *Richard Trevithick, the Engineer and the Man.* Cambridge.

Dickmann, H. (1959). *Aus der Geschichte der deutschen Eisen- and Stahlerzeugung.* Stahleisen MBH, Düsseldorf. (Monographien ü. Stahlverwendung, no. 1.)

Dircks, H. (1865). *The Life, Times, and Scientific Labours of the Second Marquis of Worcester, to which is added a Reprint of his 'Century of Inventions' (1663) with a Commentary thereon.* Quaritch, London.

Dobzrensky, J. J. V. (1657 or 1659). *Nova, et amaenior, de Admirando Fontium Genio, Philosophia.* Ferrara.

Drachmann, A. G. [HWM] (1961). Heron's windmill. *CEN*, **7**, 145.

   [KPH] (1948). Ktesibios, Philon and Heron; a Study in Ancient Pneumatics. *AHSNM*, **4**, 1–197.

   [MTGR] (1963). *The Mechanical Technology of Greek and Roman Antiquity; a Study of the Literary Sources.* Munksgaard, Copenhagen.

von Essenwein, A. (1887). *Mittelalterliche Hausbuch; Bilderhandschrift des 15 Jahrh...* Keller, Frankfurt-am-Main.

Esterer, M. (1929). *Chinas natürliche Ordnung und die Maschine.* Cotta, Stuttgart and Berlin. (Wege d. Technik series.)

Ewbank, T. (1842). *A Descriptive and Historical Account of Hydraulic and other Machines for Raising Water, Ancient and Modern...* Scribner, New York. 2nd ed., here used, 1847. (Best ed. the 16th, 1870.)

Farey, J. (1827). *A Treatise on the Steam Engine, Historical, Practical and Descriptive.* Longman, Rees, Orme, Brown & Green, London.

F[eldhaus], F. M. [GBC] (1927). Gebläse in China. *GTIG*, **11**, 310.

Feldhaus, F. M. [MLV] (1954). *Die Maschine im Leben der Völker; ein Überblick von der Urzeit bis zur Renaissance.* Birkhäuser, Basel and Stuttgart.

   [TDV] (1914). *Die Technik der Vorzeit, der Geschichtlichen Zeit, und der Naturvölker* (encyclopaedia). Engelmann, Leipzig and Berlin.

   [ZEP] (1908). Ü. Zweck u. Entstehungszeit d. sog. Püstriche. *MGNM*, 140.

Fitzgerald, Keane (1758). An attempt to improve the manner of working the ventilators [of mines] by the help of the fire-engine [i.e. Newcomen's atmospheric steam-engine]. *PTRS*, **50**, 727.

Forbes, R. J. [MMA] (1956). Metallurgy [in the Mediterranean civilisations and the Middle Ages]. In *A History of Technology*, ed. C. Singer *et al.* vol. 2, p. 41. Oxford.

   [PMA] (1956). Power [in the Mediterranean civilisations and the Middle Ages]. Art. in *A History of Technology*, ed. C. Singer *et al.* vol. 2, p. 589. Oxford.

Forke, A. (tr.) (1911). *Lun-Hêng, Philosophical Essays of Wang Chhung.* Vol. 1, 1907, Kelly & Walsh, Shanghai, Luzac, London, Harrassowitz, Leipzig. Vol. 2 (with the addition of Reimer, Berlin). *MSOS*, Beibände **10** and **14**.

Frémont, C. [EFC] (1903). *Études Expérimentales de Technologie Industrielle, No. 15: Évolution de la Fonderie de Cuivre, d'après les Documents du Temps.* Renouard, Paris.

    [LFM] (1923). *Études Expérimentales de Technologie Industrielle, No. 66: La Forge Maréchale.* Paris.

    [OES] (1917). *Études Expérimentales de Technologie Industrielle, No. 50: Origines et Évolution de la Soufflerie.* Paris.

Galloway, R. L. (1881). *The Steam-Engine and its Inventors; a Historical Sketch.* Macmillan, London.

de Genssane, M. [TFM] (1770). *Traité de la Fonte des Mines par le Feu du Charbon de Terre.* Paris.

    [MEE] (1744). Machine pour Élever l'Eau par le Moyen du Feu, simplifiée par M. de G.... (An automatised Savery steam pumping system using a form of 'cataract' or tipping bucket.) *MIARA*, **7**, 227.

Gerlach, W. (1967). Das Vakuum in Geistesgeschichte, Naturwissenschaft und Technik. *Hochschuhl Dienst*, no. 2.

Gerland, E. & Traumüller, G. (1899). *Geschichte d. physikalischen Experimentierkunst.* Engelmann, Leipzig.

Gille, B. (1956). Machines [in the Mediterranean civilisations and the Middle Ages]. Art. in *A History of Technology*, ed. C. Singer *et al.* vol. 2, p. 629. Oxford.

Goodrich, L. Carrington & Fêng Chia-Shêng (1946). The early development of firearms in China. *Isis*, **36**, 114. With important addendum, *Isis*, **36**, 250.

Gowland, W. [CAP] (1906). Copper and its alloys in prehistoric times. *JRAI*, **36**, 11.

    [EMC] (1899). The early metallurgy of copper, tin and iron in Europe as illustrated by ancient remains, and primitive processes surviving in Japan. *AAA*, **56**, 267.

Granet, M. [DLC] (1926). *Danses et Légendes de la Chine Ancienne.* 2 vols. Alcan, Paris.

    [FACC] (1926). *Fêtes et Chansons Anciennes de la Chine.* Alcan, Paris. 2nd ed. Leroux, Paris, 1929.

    [RDC] (1922). *La Religion des Chinois.* Gauthier-Villars, Paris.

Hart, I. B. (1961). *The World of Leonardo da Vinci, Man of Science, Engineer, and Dreamer of Flight.* McDonald, London.

Hauser, F. (1922). *Ü. d. 'Kitāb al-Ḥiyal', das Werk ü. d. sinnreichen Anordnungen der Banū Mūsā.* Mencke, Erlangen.

de Hautefeuille, Jean (1678). *Pendule Perpetuelle, avec un nouveau Balancier, et la Manière d'élever l'Eau par le moyen de la Poudre à Canon.* Paris.

Hildburgh, W. L. (1951). Aeolipiles as fire-blowers. *AAA*, **94**, 27.

de la Hire, J. N. (1716). Mémoire pour la Construction d'une Pompe qui fournit continuellement de l'Eau dans le Reservoir. *MRASP*, 322.

Ho Ping Yü & Needham, Joseph (1959). The laboratory equipment of the early medieval Chinese alchemists. *AX*, **7**, 57.

Ho Shan (1958). New inventions in irrigation pumps [gas explosion pumps invented by Tai Kuei-Jui, Phêng Ting-I and others]. *PKR*, 14.

Hommel, R. P. (1937). *China at Work; an illustrated Record of the Primitive Industries of China's Masses, whose Life is Toil, and thus an Account of Chinese Civilisation.* Bucks County Historical Society, Doylestown, Pa. 1937; John Day, New York.

Hoover, H. C. & Hoover, L. H. (tr.) (1912). *Georgius Agricola 'De Re Metallica'*, translated from the 1st Latin edition of 1556, with biographical introduction, annotations and appendices upon the development of mining methods, metallurgical processes, geology, mineralogy and mining law from the earliest times to the sixteenth century. 1st ed. Mining Magazine, London. 2nd ed. Dover, New York, 1950.

Hope, W. H. St J. & Fox, G. (1896). Excavations on the site of the Roman city of Silchester (water-pumps). *AAA*, **55**, 215, 232.

Hosie, A. (1890). *Three Years in Western Szechuan; a Narrative of Three Journeys in Szechuan, Kweichow and Yunnan*. Philip, London.

Hough, W. (1926). Fire as an agent in human culture. *BUSNM*, no. 139.

Jenkins, Rhys (ed.) (1930). *R. d'Acres' 'Art of Water-Drawing', published by Henry Brome, at the Gun in Ivie Lane. London, 1659 and 1660*. Newcomen Society (Heffer), Cambridge. (Newcomen Society Extra Pubs. no. 2.)

Johannsen, O. [AWH] (1916). Die erste Anwendung der Wasserkraft im Hüttenwesen. *SE*, **36**, 1226.

    [FAE] (1911). Filarete's Angaben über Eisenhütten; ein Beitrag z. Geschichte des Hochofens und das Eisengusses im 15. Jahrh. *SE*, **31**, 1960 and 2027. (On the *Trattato di Architettura* of Antonio Averlino Filarete, *c.* 1462.)

    [GDE] (1925). *Geschichte des Eisens*. 2nd ed. Verlag Stahleisen MBH, Düsseldorf. 3rd ed. 1953.

Jones, F. D. & Horton, H. L. (1951–2). *Ingenious Mechanisms for Designers and Inventors; Mechanisms and mechanical Movements selected from Automatic Machines and various other forms of Mechanical Apparatus as outstanding examples of Ingenious Design, embodying Ideas or Principles applicable in designing Machines or Devices requiring Automatic Features or Mechanical Control*. 3 vols. Industrial Press, New York. Machinery Pub. Co., Brighton.

Konen, H. (1941). *Physikalischen Plaudereien; Gegenwartsprobleme und ihre technische Bedeutung*. Buchgemeinde, Bonn.

Lach, O. (1965). *Asia in the Making of Europe*. 2 vols. Univ. of Chicago Press, Chicago.

Lardner, Dionysius (1840). *The Steam Engine Explained and Illustrated; with an Account of its Invention and Progressive Improvement, and its Application to Navigation and Railways; including also a Memoir of Watt*. 7th ed. Taylor & Walton, London.

Laufer, B. (1934). The swing in China. *MSFO*, **67**, 212.

Ledebur, A. (1901). Über den japanischen Eisenhüttenbetrieb. *SE*, **21**, 8412.

Leroi-Gourhan, André (1945). *Evolution et Techniques*. Vol. 1, *L'Homme et la Matière*, 1943; vol. 2, *Milieu et Techniques*, 1945. Albin Michel, Paris.

Li Chhung-Chou (1959). Reconstruction of the 'water pusher' or water-powered blowing engine for ironworks, an ancient discovery in applied science. *WWTK* (no. 5), 45. [In Chinese.]

Lindroth, S. H. (1955). *Gruvbrytning och Kopparhantering vid Stora Kopparberget intill 1800-talets Början* [in Swedish]. 2 vols. Almqvist & Wiksell, Uppsala. (Skrifter utgivna av Storakopparbergs Bergslags Aktiebolag.) Résumé, with map, in *Sweden Illustrated*. 1954.

Liu Hsien-Chou (1935). Materials for the history of engineering in China. *CHESJ*, **3**, and

4 (no. 2), p. 27. Reprinted Chhinghua Univ. Press, Peiping. 1935. With Supplement, *CHER*, 1948, **3**, 135. [In Chinese.]

Lockhart, W. (1861). *The Medical Missionary in China; a Narrative of Twenty Years' Experience.* Hurst & Blackett, London.

de Manoury d'Ectot, Marquis (1821). Rapport sur une Nouvelle Machine à Feu presenté à l'Académie et executée aux Abattoirs de Grenelle. *ACP*, **18**, 133.

de la Martinière, Breton (1812, 1813). *China, its Customs, Arts, Manufactures, etc., edited principally from the Originals in the Cabinet of the late Mons. Bertin [1719 to 1792], with Observations Explanatory, Historical and Literary...* (tr. from the French), 3rd ed., Stockdale, London. Repr. 1824.

Medhurst, W. H. (1876). The fire-piston in Yunnan. *CR*, **5**, 202.

Montandon, G. (1934). *L'Ologénèse Culturelle; Traité d'Ethnologie Cyclo-Culturelle et d'Ergologie Systématique.* Payot, Paris.

Morgan, E. (tr.). (1933?). *Tao the Great Luminant; Essays from 'Huai Nan Tzu', with introductory articles, notes and analyses.* Kelly & Walsh, Shanghai, n.d.

Muramatsu, Teijirō (1953). The *tatara* method of iron-smelting in recent times. *JJHS*, (no. 26), p. 30. [In Japanese.]

Nagler, J. (1957). Die erste 'Curieuse Feuer-Maschine' in Österreich; eine Grossleistung von Joseph Emanuel Fischer von Erlach. *AMK*, **2** (nos. 7–8), 26.

Needham, Joseph [DITC] (1958). *The Development of Iron and Steel Technology in China.* Newcomen Society, London. (Second Biennial Dickinson Memorial Lecture, Newcomen Society.) Repr. Heffer, Cambridge, 1964.

　　[LOC] (1964). Science and China's influence on the West. Art. in *The Legacy of China*, ed. R. N. Dawson. Oxford Univ. Press, Oxford.

Needham, Joseph *et al.* (1954). *Science and Civilisation in China.* 7 vols. in 11 parts. Cambridge.

Needham, Joseph, Wang Ling & Price, Derek J. de S. (1960). *Heavenly Clockwork; the great astronomical clocks of medieval China.* Cambridge. (Antiquarian Horological Society Monographs, no. 1.) Prelim. pub. *AHOR*, 1956, **1**, 153.

[Ucelli di Nemi, G.] (ed.) (1956). *Le Gallerie di Leonardo da Vinci nel Museo Nazionale della Scienza e della Tecnica [Milano].* Museo Naz. d. Sci. e d. Tecn., Milan.

Neuburger, A. (1919). *The Technical Arts and Sciences of the Ancients.* Tr. from the Germ. ed. *Die Technik des Altertums*, Voigtländer, Leipzig, by H. L. Brose. Methuen, London, 1930.

Newbould, G. T. (1935). The atmospheric engine at Parkgate. *TNS*, **15**, 225.

Paulinyi, Á. (1962). Príspevok k Technologickému Vývinu Hroneckých Železiarní v Prevj Polovici 19 Storočia (Beitrag zur Gesch. d. technischen Entwicklung der Rohnitzer Eisenwerke in der ersten Hälfte des 19. Jahrhunderts) [in Czech with German summary]. *SDPVT*, **7**, 159.

Poggendorff, J. C. (1879). *Geschichte d. Physik.* Barth, Leipzig.

Pole, W. (1844). *A Treatise on the Cornish Pumping Engine.* London.

Proudfoot, W. J. (1868). *Biographical Memoir of James Dinwiddie, LL.D., Astronomer in the British Embassy to China (1792–4), afterwards Professor of Natural Philosophy in the College of Fort William, Bengal; embracing some account of his Travels in China and Residence in India, compiled from his Notes and Correspondence by his grandson...* Howell, Liverpool.

Raistrick, A. (1953). *Dynasty of Iron Founders; the Darbys and Coalbrookdale.* Longmans Green, London.

Reti, L. (1957). Leonardo da Vinci nella Storia della Macchina a Vapore. *RDI,* 21.

Reuleaux, F. (1876). *Kinematics of Machinery; Outlines of a Theory of Machines* (tr. A. B. Kennedy from *Theoretische Kinematik,* Wieweg, Braunschweig, 1875). London. French tr. by A. Debize, *Cinématique: Principes fondamentaux d'une Théorie générale des Machines.* Savy, Paris, 1877.

Rocher, E. (1879, 1880). *La Province Chinoise du Yunnan.* 2 vols. (incl. special chapter on metallurgy). Leroux, Paris.

Rolt, L. T. C. (1963). *Thomas Newcomen, the Prehistory of the Steam-Engine.* Dawlish and London.

Sarraut, A. & Robequin, C. (1930). *Indochine* (album of photographs). Didot, Paris.

Schafer, E. H. (1950). The camel in China down to the Mongol dynasty. *S,* **2,** 165, 263.

Schlutter, C. A. [Schlüter] (1750–3). *De la Fonte des Mines, des Fonderies, etc.* Tr. M. Hellot from the German. Pissot & Herissant, Paris.

Schmeller, H. (1922). *Beitr. z. Gesch. d. Tech. in d. Antike u. bei den Arabern.* Mencke, Erlangen.

Schroeder, A. (1905). *Études numismatiques en Annam.* Imp. Nat., Paris.

Schuhl, P. M. (1947). *Machinisme et Philosophie.* Presses Univ. de France, Paris.

Sévoz, M. (1876). On the Japanese iron and steel industry, and the *tatara* bellows. *ANM,* **6,** 345.

Shapiro, S. (1964) The origin of the suction pump. *TCULT,* **5,** 566.

Singer, C. *et al.* (ed.)(1954–8). *A History of Technology,* 5 vols. Oxford.

Sisco, A. G. & Smith, C. S. (tr.) (1951). *Lazarus Ercker's Treatise on Ores and Assaying (Prague, 1574).* Translated from the German edition of 1580. Univ. Chicago Press, Chicago.

Smiles, S. (1874). *Lives of the Engineers; Boulton and Watt.* Murray, London.

Smith, A. H. (ed.) (1920). *A Guide to the Exhibition illustrating Greek and Roman Life.* British Museum Trustees, London.

Sollmann, T. (1901). *A Textbook of Pharmacology and some Allied Sciences.* Saunders, Philadelphia and London.

Spencer, J. R. (1963). Filarete's description of a fifteenth-century Italian iron smelter at Ferriere. *TCULT,* **4,** 201. Followed by a discussion including contributions by L. C. Eichner, C. S. Smith, T. A. Wertime, Joseph Needham & J. R. Spencer.

Stuart, R. (1829). *Historical and Descriptive Anecdotes of Steam-Engines and of their Inventors and Improvers.* Wightman, London.

Thurston, R. H. (1939). *A History of the Growth of the Steam-Engine.* 1878. Centennial edition, with a supplementary chapter by W. N. Barnard. Cornell Univ. Press, Ithaca, N. Y.

Tredgold, T. (1827). *The Steam-Engine...* Taylor, London.

Turner, H. D. (1959). Robert Hooke and Boyle's air-pump. *Nature,* **184,** 395.

Ure, A. (1839). *A Dictionary of Arts, Manufactures and Mines* (1st ed.). 2 vols. London; (5th ed.) 3 vols. Longmans Green, London, 1860.

Usher, A. P. (1929). *A History of Mechanical Inventions*. McGraw-Hill, New York. 2nd ed. revised. Harvard Univ. Press, Cambridge, Mass., 1954.

de Waard, C. (1936). *L'Expérience Barométrique; ses Antécédents et ses Applications*. Imp. Nouv., Thouars.

Wailes, R. (1962). James Watt—instrument maker. *CME*, **9**, 136.

Waley, A. (tr.) (1934). *The Way and its Power; a study of the 'Tao Tê Ching' and its Place in Chinese Thought*. Allen & Unwin, London.

Wang Chen-To (1959). On the reconstruction of the metallurgical iron-casting bellows of the Han period. *WWTK* (no. 5), 43. [In Chinese.]

Wang Ling (1947). On the Invention and Use of Gunpowder and Firearms in China. *ISIS*, **37**, 160.

Wang Yü-Chhüan (1957). Iron farm-tools in Ancient China. *CREC*, **6** (no. 2), 10.

Westcott, G. F. (1932). *Pumping Machinery (Pt. I. Historical Notes)*. (Handbooks of the Collections, Science Museum, South Kensington.) HMSO, London.

White, Lynn (1962). *Medieval Technology and Social Change*. Oxford.

Wiedemann, E. & Hauser, F. [UBI] (1915). Über die Uhren im Bereich d. Islamischen Kultur. *NALC*, **100**, no. 5. Addendum, *SPSME*, **47**, 125.

   [VHW] (1918). Über Vorrichtungen zum Heben von Wasser in der Islamischen Welt. *BGTI*, **8**, 121.

Wilkinson, J. G. (1854). *A Popular Account of the Ancient Egyptians*. 2 vols. Murray, London.

Williamson, A. (1870). *Journeys in North China*. London.

Willis, Robert (1841). *Principles of Mechanism*. Parker, London. Deighton, Cambridge. 2nd ed. Longmans Green, London, 1870.

Wilson, George (1849). On the early history of the air-pump in England. *ENPJ*, April.

Wolf, A. [XVII] (1935). (With the co-operation of F. Dannermann & A. Armitage.) *A History of Science, Technology and Philosophy in the 16th and 17th centuries*. Allen & Unwin, London. 2nd ed., revised by D. McKie, London, 1950.

   [XVIII] (1938). *A History of Science, Technology and Philosophy in the 18th Century*. Allen & Unwin, London. 2nd ed., revised by D. McKie, London, 1952.

Woodcroft, B. (tr.) (1851). *The 'Pneumatics' of Heron of Alexandria*. Whittingham, London.

Yang Jen-Khai & Tung Yen-Ming (1962). *Album of Pictures illustrating the Collection of Paintings in the Liaoning Provincial Museum*. [In Chinese.] Peking.

Yang Khuan [FPHB] (1959). Queries on the reconstruction of the (vertical water-wheel) hydraulic blowing-engines in ironworks (of the Sung and Yuan periods). *WWTK* (no. 7), 48. [In Chinese.]

   [OIT] (1956). *The Origins, Inventions, and Development of Iron [and Steel] Technology in Ancient and Medieval China*. Jen-Min, Shanghai. [In Chinese.]

   [TIC] (1955). On the blast furnaces used for making cast iron in ancient China, and the invention of hydraulic blowing engines for them. [In Chinese.] Essay in *Chung-Kuo Kho-Hsüeh Chi-Shu Fa-Ming*, Peking.

427

Key to Abbreviations

AAA	*Archaeologia* (London).
ACP	*Annales de Chimie et de Physique* (Paris).
AHOR	*Antiquarian Horology* (London).
AHSNM	*Acta Historica Scientiarum Naturalium et Medicinalium* (Copenhagen).
AL	*Annuaire du Bureau des Longitudes* (Paris).
AMK	*Alte und Moderne Kunst* (Österr. Zeitschr. f. Kunst, Kunsthandwerk und Wohnkultur).
ANM	*Annales des Mines* (Paris).
AX	*Ambix* (London).
BGTI	*Beiträge z. Geschichte d. Technik und Industrie* (now *Technikgeschichte*).
BMFRS	*Biographical Memoirs of Fellows of the Royal Society.*
BUSNM	*Bull. United States National Museum.*
C	*Chi Kung Tzhu Tien* (Dictionary of Mechanical Engineering) by Thang Hsin-Yü, Sci. & Technol. Press, Shanghai, 1955.
CEN	*Centaurus* (Copenhagen).
CHER	*Chhing-Hua (University) Engineering Reports* (Peking).
CHESJ	*Chhing-Hua (University) Engineering Society Journal* (Peking).
CME	*Chartered Mechanical Engineer.*
CR	*China Review* (Hongkong and Shanghai).
CREC	*China Reconstructs* (Peking).
CRRR	*Chinese Repository* (Hongkong).
D	*Hsin Ting Ying Han Tzhu Tien* (An Abridged English and Chinese Dictionary). Commercial Press, Shanghai, 1911.
DHT	*Documents pour l'Histoire des Techniques* (Cahiers du Centre de Documentation d'Hist. des Tech., Conservatoire Nat. des Arts et Métiers, Paris).
END	*Endeavour* (London).
ENPJ	*Edinburgh New Philosophical Journal.*
GTIG	*Geschichtsblatter f. Technik, Industrie u. Gewerbe* (Vienna).
ISIS	*Isis* (Cambridge, Mass.).
JJHS	*Japanese Journ. History of Science* (Tokyo).
JRAI	*Journ. Royal Anthropological Institute.*
MGNM	*Mitt. a. d. germanisches National Museum* (Nuremberg).
MIARA	*Machines et Inventions approuvés par l'Académie des Sciences* (Paris).
MRASP	*Mémoires de l'Académie Royale des Sciences* (Paris).
MSFO	*Mémoires de la Société Finno-Ougrienne.*
MSOS	*Mitt. d. Seminar f. Orientalischen Sprachen* (Berlin).
NALC	*Nova Acta* (Abhandlungen d. kaiserl. Leop.-Carol. deutsch. Akad. Naturforsch. Halle).
PIME	*Proc. Institution of Mechanical Engineers.*
PTRS	*Philosophical Transactions of the Royal Society.*
RDI	*Rivista d'Ingegneria.*
S	*Sinologica* (Basel).

SDPVT   *Sbornik pro Dějiny Přirodnich Věd a Techniky* (*Acta Historiae Rerum Naturalium necnon Technicarum*, Prague).

SE   *Stahl und Eisen.*

SPMSE   *Sitzungsber. d. physikal.-med. Soc. Erlangen.*

TCULT   *Technology and Culture.*

TNS   *Transactions of the Newcomen Society.*

WWTK   *Wên-Wu Tshan Khao Tzu-Liao* (Peking).

CHAPTER II

The Missing Link in Horological History: a Chinese Contribution

von Basserman-Jordan, E. (1926). *Alte Uhren und ihre Meister.* Leipzig: Diebener.

Beck, T. (1909). Herons (des älteren) Automatentheater. *Beitr. Gesch. Tech. Industr.* **1**, 182.

Bernal, J. D. (1954). *Science in History.* London: Watts. (Beard Lectures at Ruskin College, Oxford.)

von Bertele, H. (1953). Precision time-keeping in the pre-Huygens era. *Horol. J.* **95**, 794.

Berthoud, F. (1802). *Histoire de la Mésure du Temps par les Horloges.* Paris: Impr. de la République.

Bolton, L. (1924). *Time measurement.* London: Bell.

Carlton, W. J. (1911). *Timothe Bright, Doctor of Phisicke; a memoir of the father of modern shorthand.* London: Elliott Stock.

Carter, T. F. (1925). *The Invention of Printing in China and its spread Westwards.* New York: Columbia University Press. (2nd ed. revised by L. Carrington Goodrich. New York: Ronald, 1955.)

de Caus, Isaac (1644). *Nouvelle Invention de lever l'Eau plus Hault que sa Source, avec quelque Machines movantes par le moyen de l'eau, et un discours de la conduite d'ycelle.* London. (English trans. by John Leak. London: Moxon, 1659.)

Desaguliers, J. T. (1734). *A Course of Experimental Philosophy*, 2 vols. London: Innys, Longman, Shewell and Hitch (2nd ed. with Senex, London, 1745).

Drover, C. B. (1954). A medieval monastic water-clock. *Antiquar. Horol.* **1**, 54.

Feldhaus, F. M. (1914). *Die Technik d. Vorzeit, d.geschichtlichen Zeit, und der Naturvölker.* Berlin and Leipzig: Engelmann.

Frémont, C. (1915). *Etudes expérimentales de Technologie industrielle*, no. 47; *Origine de l'Horloge à Poids.* Paris.

Gunther, T. R. (1923–45). *Early Science in Oxford*, 14 vols. Oxford.

Harris, J. (1719). *Astronomical Dialogues.* London.

Hogben, L. (1939). John Wilkins, parliamentarian and pioneer of scientific humanism. Essay in *Dangerous Thoughts*, p. 25. London: Allen and Unwin.

Howgrave-Graham, R. P. (1927). Some clocks and jacks, with notes on the history of horology. *Archaeologia*, **77**, 257.

Liu Hsien-Chou (1956). Chung-Kuo tsai Chi-Shih-Chhi fangmien ti Fa-Ming (Chinese Inventions in Horological Engineering). *Chhing-Hua Engineering Reports*, 1956 (N.S.), **4**, 1; *Thien-Wên Hsüeh Pao*, 1956, **4** (no. 2), 219 (in Chinese).

Lloyd, H. A. (1951). George Graham, horologist and astronomer. *Horol. J.* **93**, 708.

(1954). *Giovanni di Dondi's horological masterpiece of +1364.* pr. pr. (no place, no date).

Lübke, A. (1931). Altchinesische Uhren. *Dtsch. Uhrm-Ztg.* **55**, 197. Chinesische Zeitmesskunde. *Naturw. Kultur* (München), **28**, 45.

Matschoss, C. & Kutzbach, K. (1940). *Geschichte des Zahnrades.* Berlin: VDI-Verlag.

Needham, J. *et al.* (1954–). *Science and Civilisation in China*, 7 vols. Cambridge University Press (in course of publication).

Needham, J., Wang Ling & Price, D. J. de S. (1956). Chinese astronomical clockwork. *Nature, Lond.* **177**, 600; Chinese trans. by Hsi Tsê-Tsung, *Kho-Hsüeh Thung Pao* (no. 6), p. 100.

(1959). *Heavenly Clockwork; the Great Astronomical Clocks of Medieval China.* Cambridge University Press, 1960. (Antiquarian Horological Society Monographs, no. 1.)

Orrery, the Countess of Cork and (1903). *The Orrery Papers*, 2 vols. London: Duckworth.

Planchon, M. (1899). *L'Horloge; son Histoire retrospective, pittoresque et artistique.* Paris: Laurens. (2nd ed. 1912).

Price, D. J. de S. (1955–6). Clockwork before the clock. *Horol. J.* **97**, 810; **98**, 31.

(1956). The prehistory of the clock. *Discovery*, **17**, 153.

Sheridan, P. (1896). Les inscriptions sur ardoise de l'abbaye de Villers. *Ann. Soc. roy. d'Archéol.* (Brussels), **9**, 359, 454; **10**, 203, 404.

Staunton, Sir George T. (1797). *An Authentic Account of an Embassy from the King of Great Britain to the Emperor of China, taken chiefly from the papers of H.E. the Earl of Macartney K.B. etc,...,* 2 vols. London: Bulmer and Nicol. (Repr. 1798). Abridged ed. 1 vol. London: Stockdale, 1797.

Taylor, E. W. & Wilson, J. S. (1945). *At the Sign of the Orrery.* For Messrs Cooke, Troughton and Simms, pr. pr. York, no date.

Thorndike, L. (1941). The invention of the mechanical clock about +1271. *Speculum*, **16**, 242.

Uccelli, A. (1945). *Storia della Tecnica dal Medio Evo ai Nostri Giorni.* Milan: Hoepli.

Wang Chen-To (1958). Chieh Khai Liao Wo-Kuo Thien-Wên-Chung ti Pi-Mi (The Secret of our Medieval Astronomical Clocks—the Escapement). *Wên-Wu Tshan-Khao Tzu-Liao* (no. 9), 5 (in Chinese).

Wiedemann, E. (1913). Ein Instrument das die Bewegung von Sonne und Mond darstellt, nach al-Bīrūnī. *Der Islam*, **4**, 5.

Wiedemann, E. & Hauser, F. (1915). Über die Uhren im Bereich d. Islamischen Kultur. *Nova Acta; Abhandl. d. kaiserl. Leop.-Carol. deutsch. Akad. Naturf. Halle*, **100** (no. 5).

(1918*a*). Die Uhr des Archimedes und zwei andere Vorrichtungen. Halle: Ehrhardt Karras.

(1918*b*). Byzantinische und arabische akustiche Instrumente. *Archiv. f.d. Gesch. d. Naturwiss. u. Technik*, **8**, 140.

Zinner, E. (1954). *Aus der Frühzeit der Räderuhr; von der Gewichtsuhr zur Federzugsuhr.* München: Oldenbourg. (*Dtsch. Museum Abhandl. u. Ber.* **22**, no. 3.)

BIBLIOGRAPHY

CHAPTER 16
Elixir Poisoning in Medieval China

Bull, J. P. (1951). *A Study of the History and Principles of Clinical Therapeutic Trials.* Inaug. Diss. Cambridge.

Dobson, J. F. (1925). Herophilus of Alexandria. *Proc. Roy. Soc. Med.* **18**, 19.

(1927). Erasistratus. *Proc. Roy. Soc. Med.* **20**, 21.

Dubs, H. H. (1947). The beginnings of Alchemy. *Isis*, **38**, 62.

Eliade, M. (1956). *Forgerons et Alchimistes.* Flammarion, Paris.

Feifel, E. (1944). *Pao Phu Tzu, Nei Phien*, ch. 4, translated and annotated. *Monumenta Serica*, **9**, 1.

(1946). *Pao Phu Tzu, Nei Phien*, ch. 11, translated and annotated. *Monumenta Serica*, **11**, 1.

Giles, H. A. (1898). *A Chinese Biographical Dictionary.* 2 vols., Shanghai.

(1924). The *Hsi Yuan Lu* or 'Instructions to Coroners'. *Proc. Roy. Soc. Med.* **17**, 59.

(1923). *Gems of Chinese Literature; Prose.* 2nd edition, Shanghai.

Green, F. H. K. (1954). The Clinical Evaluation of Remedies. *Lancet*, 1085.

Ho Ping-Yü & Needham, J. (1959). The Laboratory Equipment of the early Medieval Chinese Alchemists. *Ambix*, **7**, 57.

(1959). Theories of Categories in Early Medieval Chinese Alchemy. *Journ. Warburg & Courtauld Inst.* **22**, 173.

Hübotter, F. (1927). Zwei berühmte chinesische Ärzte des Altertums, Chouen Yu-J [Shunyü I] und Hoa T'ouo [Hua Tho]. *Mitteilungen d. deutsch. Gesellsch. f. Natur. u. Volkskunde Ostasiens*, **21**, 1.

Jacob, E. F. (1956). John of Roquetaillade. *Bull. John Rylands Library*, **39**, 75.

Kracke, E. A. (1957). *Translation of Sung Civil Service Titles.* Paris.

Liao Wên-Kuei (1939). *The Complete Works of Han Fei Tzu; a Classic of Chinese Legalism.* 2 vols., Probsthain, London.

Lieben, F. (1935). *Geschichte d. physiologische Chemie.* Deuticke, Leipzig & Vienna.

Maspero, H. (1950). Le Taoisme, in *Mélanges Posthumes sur les Religions et l'Histoire de la Chine.* Vol. 1, ed. P. Demiéville. Civilisations du Sud, Paris.

(1937). Procédés de 'Nourrir le Principe Vital' dans la Religion Taoiste Ancienne. *Journ. Asiat.* **229**, 177, 353.

Masson-Oursel, R., de Willman-Grabowska, H. & Stern, P. (1933). *L'Inde Antique et la Civilisation Indienne.* Michel, Paris.

Multhauf, R. P. (1954). John of Rupescissa and the Origin of Medical Chemistry. *Isis*, **45**, 359.

(1956). The Significance of Distillation in Renaissance Medical Chemistry. *Bulletin of the (John Hopkins Institute of the) History of Medicine*, **30**, 329.

(1954). Medical Chemistry and the 'Paracelsians'. *Bulletin of the (John Hopkins Institute of the) History of Medicine*, **28**, 101.

Needham, J. (1954–). *Science and Civilisation in China.* 7 vols. Cambridge.

(1959). *A History of Embryology.* 2nd ed., Abelard-Schuman, New York.

Pagel, W. (1958). *Paracelsus; an Introduction to Philosophical Medicine in the Era of the Renaissance.* Karger, Basel & New York.

Read, B. E. (1931). *Chinese Materia Medica: Animal Drugs.* Peking.

431

Read, B. E. (1936). *Chinese Medicinal Plants from the 'Pên Tshao Kang Mu'*. 3rd ed. Peking. (1941). *Chinese Materia Medica: Insect Drugs*. Peking.

Read, B. E. & Pak, C. (1936). *A Compendium of Minerals and Stones used in Chinese Medicine from the 'Pên Tshao Kang Mu'*. Peking.

Renou, L. & Filliozat, J. (ed.) (1947). *L'Inde Classique; Manuel des Études Indiennes*. Payot, Paris.

Sherlock, T. P. (1948). The Chemical Work of Paracelsus. *Ambix*, **3**, 33.

Stuart, G. A. (1911). *Chinese Materia Medica, Vegetable Kingdom*. Shanghai.

Thorndike, L. (1923–58). *History of Magic and Experimental Science*. 8 vols., Columbia Univ. Press, New York.

Tshao Thien-Chhin, Ho Ping-Yü & Needham, J. (1959). An Early Medieval Chinese Alchemical Text on Aqueous Solutions. *Ambix*, **7**, 122.

Wieger, L. (1911). *Le Canon Taoiste (Patrologie)*. Hsienhsien.

Wu Lu-Chiang & Davis, T. L. (1932). An Ancient Chinese Treatise on Alchemy entitled *Tshan Thung Chhi. Isis*, **18**, 210.

CHAPTER 17

Hygiene and Preventive Medicine in Ancient China

Adam, N. K. & Stevenson, D. G. (1953). Detergent action. *Endeavour*, **12**, 25.

Baynes, C. F. (1950). *The 'I Ching' or 'Book of Changes'* (after R. Wilhelm's tr.). New York, Bollingen-Pantheon, 2 vols.

Biot, E. (1930). *Le 'Tcheou Li' [Chou Li] ou 'Rites des Tcheou'*. Paris, Imp. Nat., 1851; photo-litho. reprod.: Peiping, Wên-tien-ko, 2 vols.

Blake, J. B. (1953). Smallpox inoculation in colonial Boston. *J. Hist. Med.* **8**, 284.

Bodde, D. (1942). Early references to tea-drinking in China. *J. Amer. Orient. Soc.* **62**, 74.

Bretschneider, E. (1882). *Botanicon Sinicum; Notes on Chinese Botany from Native and Western Sources*. London, Trübner, 3 vols.

Chhen Pang-Hsien (1957). *Chung-Kuo I Hsüeh Shih* (History of Chinese Medicine), new ed. Shanghai, Commercial Press.

Chou Tsung-Chhi (1956). The invention of the toothbrush in China. *Chin. J. Stomatol.* **4**, 5. (1956). Toothbrushes made with bristles in the Liao Dynasty. *Chin. J. Stomatol.* **4**, 159.

Chu Chhi-Chhien & Liu Tun-Chen (1936). 'Chê Chiang Lu' (biographies of engineers, architects, technologists and master-craftsmen), pt. 9. *Bull. Soc. Res. Chin. Archit.* **6**, 148.

Cordier, H. (1920). *Histoire générale de la Chine*. 4 vols. Paris, Geuthner.

Couvreur, F. S. (1914). '*Tch'ouen Ts'iou*' [*Chhun Chhiu*] et '*Tso Tchouan*' [*Tso Chuan*]; texte Chinois avec traduction française. 3 vols. Hochienfu, Mission Press.

Dudgeon, J. (1895). 'Kung-fu', or Medical Gymnastics. *J. Peking Orient. Soc.* **3**, 341.

Fan Hsing-Chun (1953). *Chung-Kuo Yü-Fang I-Hsüeh Ssu Hsiang Shih* (History of the conceptions of hygiene and preventive medicine in China). Peking, Jen-Min Wei-Sêng.

Filliozat, J. (1949). *La doctrine classique de la médecine Indienne*. Paris, Imp. Nat.

Forbes, R. J. (1954). Chemical, culinary and cosmetic arts [in early times down to the fall of the Ancient Empires]. In: *A History of Technology*, C. Singer *et al.*, eds., Oxford, vol. 1, p. 238.

Gibbs, F. W. (1957). Invention in chemical industries [+1500 to +1750]. In: *A History of Technology*, C. Singer *et al.*, eds., Oxford, vol. 3, p. 676.

Goodrich, L. C. & Wilbur, C. M. (1942). Additional notes on tea. *J. Amer. Orient. Soc.* **62**, 195.

Haas, P. & Hill, T. G. (1928). *Introduction to the Chemistry of Plant Products*. London, Longmans Green.

Halsband, R. (1953). New light on Lady Mary Wortley Montagu's contribution to inoculation. *J. Hist. Med.* **8**, 390.

Ho Ping-Yü & Needham, J. (1959). Theories of categories in early medieval Chinese alchemy. *J. Warburg & Courtauld Inst.* **22**, 173.

Huang Man (1950). The *Nei Ching*, the Chinese Canon of Medicine. *Chin. med. J.* **68**, 1.

Legge, J. (1926). *The Texts of Confucianism, III; the 'Li Ki' [Li Chi]*. 2 vols. Oxford.

Lu Gwei-Djen & Needham, J. (1951). A contribution to the history of Chinese dietetics. *Isis*, **42**, 13.

Moule, A. C. (1921). The 'Wonder of the Capital', roughly translated. *New China Rev.* **3**, 12, 356.

Needham, J., Wang, L. & Price, D. J. de S. (1960). *Heavenly Clockwork; the Great Astronomical Clocks of Medieval China*. Cambridge. (Antiq. Horol. Soc. Monographs, no. 1.)

Pilarini, J. (1716). Nova et Tuta Variolas excitandi per Transplantationem Methodus... *Philos. Trans. Roy. Soc.* **29**, 393.

Read, B. (1936). *Chinese medicinal plants from the 'Pên Tshao Kang Mu' of +1596;...a botanical, chemical and pharmacological reference list*. Peiping, French Bookstore.

Schafer, E. H. (1956). The development of bathing customs in ancient and medieval China, and the history of the floriate clear palace. *J. Amer. Orient. Soc.* **76**, 57.

Schmauderer, E. (1968). Seife...im klassischen Altertum. *Technik Gesch.* **34**, 300, **35**, 205.

Shih Shêng-Han (1958). *A Preliminary Survey of the Book 'Chhi Min Yao Shu', an Agricultural Encyclopaedia of the +6th Century*. Peking, Science Press.

Shu Shih-Chhêng (1945). Chung-Kuo Ku-Tai I Yao Wei-Sêng Khao. (A study of medical and hygienic practices in ancient China.) *Bull. Chin. Stud.*, Chhêng-tu, **5** (no. 1), p. 85.

Singer, Charles (ed.) (1954-8). *A History of Technology*. 5 vols. Oxford.

Stearns, R. P. & Pasti, G. (1950). Remarks upon the introduction of inoculation for smallpox in England. *Bull. Hist. Med.* **24**, 103.

Stuart, G. A. (1911). *Chinese materia medica, vegetable kingdom; extensively revised from Dr. F. Porter Smith's work*. Shanghai, Presbyterian Mission Press.

Taylor, F. Sherwood (1957). *A History of Industrial Chemistry*. London, Heinemann.

Taylor, F. S. & Singer, Charles (1956). Pre-scientific industrial chemisty [in the Mediterranean civilisations and the Middle Ages]. In: *A History of Technology*, C. Singer *et al.*, eds., Oxford, vol. 2, p. 347.

Thang Yü-Lin (1958). Tsu Kuo Ku-Tai I-Hsüeh tsai Yin Shih Ying Yang Wei-Sêng-Hsüeh fang-mien-ti Kung-Hsien (Contributions of ancient Chinese medicine on diet, nutrition and hygiene). *I-Hsüeh Shih yü Pao-Chien Tsu-Chih*, **1**, 54.

Timoni, E. (1714). An Account, or History, of the Procuring the Small Pox by Incision, or Inoculation; as it has for some time been practised at Constantinople. *Phil. Trans. Roy. Soc.* **29**, 72.

Vogralik, V. G. (1958). Kuan-yü Chung-Kuo Min Tsu I-Hsüeh ti chi-pên Yuan-Tsê (On the basis of Chinese traditional medicine [the *Nei Ching*). Thao Chung-Hsin, tr. *Shanghai Chung I Yao Tsa Chih* (no. 10), p. 442.

Waley, A. (1934). *The Way and its Power; a Study of the 'Tao Tê Ching' and its Place in Chinese Thought*. London, Allen & Unwin.

—— (1938). *The 'Analects' of Confucius, Translated and Annotated*. London, Allen & Unwin.

Wang Hsin-Hua (1958). Wo Kuo Ku-Tai-ti Yü-Fang I-Hsüeh (Preventive medicine in ancient China). *Shanghai Chung I Yao Tsa Chih* (no. 1), p. 6.

Wiedemann, E. (1911). Beiträge z. Gesch. d. Naturwiss. XXVI; Über Charlatane bei den Muslimen nach al-Jaubarī (fl. +1216 to +1221). *Sitzungsber. d. Phys-Med. Soc. Erlangen*, **43**, 206.

—— (1915). Beiträge z. Gesch. d. Naturwiss. XLV; Zahnärztzliches bei den Muslimen. *Sitzungsber. d. Phys-Med. Soc. Erlangen*, **47**, 127.

Wieger, L. (1911). *Taoisme; Bibliographie générale*. Hsienhsien, Mission Press.

Wilhelm, R. (1928). *Frühling und Herbst des Lü Bu-We* [transl. of the *Lü Shih Chhun Chhiu*]. Jena, Diederichs.

Wu Yün-Jui (1947). Chung-O I-Hsüeh Chiao-Liu Shih Lüeh (Historical materials on medical intercourse between China and Russia). *Chin. J. Hist. Med.* **1**, 23.

Yang Yuan-Chi (1953). *Chung-Kuo I Yao Wên Hsien* (References to medicine and pharmacy in Chinese literature). Shanghai, Ta Tê.

CHAPTER 18

China and the Origin of Qualifying Examinations in Medicine

Ball, W. W. R. (1889). *A History of Mathematics at Cambridge*. Cambridge.

Barrow, J. (1804). *Travels in China*. London.

Bazin, M. (1858). Recherches sur l'histoire, l'organisation et les travaux de l'académie Imperiale de Péking. *J. Asiat.* (5e ser.), **8**, 393.

Biot, E. (1851). *Le Tcheou Li*. 3 vols. Paris.

Castiglioni, A. (1947). *A History of Medicine*. New York.

Chhen Pang-Hsien (1957). *Chung-Kuo I-Hsüeh Shih* (History of Chinese Medicine). Shanghai.

Dubs, H. H. (1938–55). *The History of the Former Han Dynasty*. Baltimore.

Elgood, C. (1951). *A Medical History of Persia and the Eastern Caliphate*... Cambridge.

Fujikawa Yu (1911). *Geschichte der Medizin in Japan*. Tokyo.

Galt, H. S. (1951). *A History of Chinese Educational Institutions*. London.

Gernet, J. (1959). *La Vie Quotidienne en Chine à la Veille de l'Invasion Mongole*. Paris.

Greenwood, M. (1942). Medical statistics from Graunt to Farr. *Biometrika*, **32**, 101, 203.

Harper, R. (1904). *The Code of Hammurabi, King of Babylon*. Chicago.

Iskandar, A. Z. (1962). Galen and Rhazes on examining physicians. *Bull. Hist. Med.* **36**, 362.

Lecomte, L. (1698). *Memoirs and Observations...made in a late journey through the Empire of China*... London.

Li Thao (1953). Achievements of Chinese medicine in the Sui and Thang dynasties. *Chin. med. J.* **71**, 301.

—— (1954). Achievements of Chinese medicine in the Northern and Southern Sung dynasties. *Chin. med. J.* **72**, 65, 225.

Lu, G. D. & Needham, J. (1951). A contribution to the history of Chinese dietetics. *Isis*, **42**, 13.

Mettler, C. C. (1947). *History of Medicine*. Philadelphia.

des Rotours, R. (1932). *Le Traité des Examens, traduit de la Nouvelle Histoire des Thang*. Paris. (1947). *Traité des Fonctionnaires et Traité de l'Armée, traduits de la Nouvelle Histoire des Thang*. Leiden.

Têng Ssu-Yü (1943). Chinese influence on the Western examination system. *Harv. J. Asiat. Stud.* **7**, 267.

CHAPTER 19

The Roles of Europe and China in the Evolution of Oecumenical Science

Adanson, Michel (1763). *Familles des Plantes*, 2 vols. Vincent, Paris. 2nd ed. of vol. 1 only, posthumous; with many additions; *Familles Naturelles; Première Partie, comprenant l'Histoire de la Botanique*, ed. A. Adanson & J. Payer, Paris, 1847, 1864.

Anon. (1962). *Chinese Therapeutical Methods of Acupuncture and Moxibustion*. Foreign Languages Press, Peking. (An authoritative statement prepared by the National Academy and Research Institute of Traditional Chinese Medicine in Peking.)

Baratoux, J. (1942). *Précis Élémentaire d'Acuponcture; avec Repérage Anatomique des Points et leurs Applications Thérapeutiques*. Le François, Paris.

Baratoux, J. & Khoubesserian, H. (1945). *Thérapeutique et Acuponcture; Points à employer dans chaque Maladie*. Peyronnet, Paris.

Beau, G. (1965). *La Médecine Chinoise*. Editions du Seuil, Paris (Le Rayon de la Science series, no. 23).

Bretschneider, E. (1882). *Botanicon Sinicum; Notes on Chinese Botany from Native and Western Sources*, 3 vols. Trübner, London (printed in Japan). (Reprinted from *Journ. Roy. Asiatic Soc., North China Branch* (N.S.), 1881, **16**.)

(1871). *On the Study and Value of Chinese Botanical Works; with Notes on the History of Plants and Geographical Botany from Chinese Sources*. Rozario & Marcal, Fuchow, Chinese tr. by Shih Shêng-Han, *Chung-Kuo Chih-Wu-Hsüeh Wên-Hsien Phing-Lun*. Nat. Compilation & Transl. Bureau, Shanghai, 1935. Repr. Com. Press, Shanghai, 1957.

(1880). Early European Researches into the Flora of China. *Journ. Roy. Asiatic Soc., North China Branch* (N.S.), **15**, 1–194. Sep. pub. Amer. Presbyterian Mission Press, Shanghai and Trübner, London, 1881.

(1898). *History of European Botanical Discoveries in China*. 2 vols. Sampson Low & Marston, London. Photolitho reproduction, Koehler, Leipzig, 1935.

Bridgman, R. F. (1955). La Médecine dans la Chine Antique, d'après les Biographies de Pien-ts'io [Pien Chhio] et de Chouen-yu Yi [Shunyü I] (Chapitre 105 des *Mémoires Historiques* de Sseu-ma Ts'ien [Ssuma Chhien]). *Mélanges Chinois et Bouddhiques*, **10**, 1–213.

Brockbank, W. (1954). *Ancient Therapeutic Arts*. Heinemann, London. (Fitzpatrick Lectures, Royal College of Physicians, 1950–1.)

Castiglioni, A. (1947). *A History of Medicine*, tr. and ed. E. B. Krumbhaar, 2nd ed. Knopf, New York.

Chamfrault, A. & Ung Kang-Sam (1954–). (With illustrations by M. Rouhier.) *Traité de Médecine Chinoise; d'après les Textes Chinois Anciens et Modernes*. 5 vols. Coquemard, Angoulême.

Chang Chhang-Shao (1945). The present status of studies on Chinese anti-malarial drugs. *Chinese Medical Jour.* **63** A, 126.

Chang Chhang-Shao, Fu Fêng-Yung, Huang, K. C. & Wang, C. Y. (1948). Pharmacology of *chhang shan* (*Dichroa febrifuga*), a Chinese Anti-malarial Herb. *Nature*, **161**, 400. With comment by T. S. Work.

Chêng Wei (1966). 'Chia Khou Phan Chhê' Thu Chüan. (The Scroll-Painting entitled 'The Horizontal Water-Wheels beside the Sluice-Gate' [by Wei Hsien, *c.* +970].) *Wên Wu Tshan Khao Tzu Liao* (in Chinese) (no. 2), 17.

Chhen Kho-Khuei, Mukerji, B. & Volicer, L. (ed.) (1965). *The Pharmacology of Oriental Plants*. Pergamon Press, London and Czechoslovak Med. Press, Prague. (*Proc. IInd International Pharmacological Meeting*, Prague, 1963, vol. 7.)

Cox, E. H. M. (1945). *Plant-Hunting in China; a History of Botanical Exploration in China and the Tibetan Marches*. Collins, London. Photolitho edition, Scientific Book Guild, London, 1945.

Cranmer-Byng, J. L. (ed.) (1962). *An Embassy to China; being the Journal kept by Lord Macartney during his Embassy to the Emperor Chhien-Lung, +1793 and +1794*. Longmans, London.

Curott, D. R. (1966). Earth deceleration from ancient Solar eclipses. *Astron. Journ.* **71**, 264.

Dubs, H. H. (1947). The beginnings of alchemy. *Isis*, **38**, 62.

——— (1961). The origin of alchemy. *Ambix*, **9**, 23.

Dudgeon, J. (1895). 'Kung-fu', or medical gymnastics. *Journ. Peking Oriental Soc.* **3**, 341.

Duggar, B. M. & Singleton, V. L. (1953). The biochemistry of antibiotics. *Ann. Rev. Biochem.* **22**, 459 (see esp. p. 478).

Edelstein, L. (1952). The relation of ancient philosophy to medicine. *Bull. Hist. Med.* **26**, 299.

d'Elia, Pasquale (1946). Echi delle Scoperte Galileiane in Cina vivente ancora Galileo (1612–1640). *Atti. r. Accad. d. Lincei, Rendiconti Sci. Mor.* (8e ser.), **1**, 125. Republished in enlarged form as 'Galileo in Cina. Relazioni attraverso il Collegio Romano tra Galileo e i gesuiti scienzati missionari in Cina (1610–1640)', *Analecta Gregoriana*, **37** (Series Facultatis Missiologicae A, no. 1), Rome, 1947. Eng. tr. *Galileo in China*, with emendations and additions by R. Suter & M. Sciascia, Harvard Univ. Press, Cambridge, Mass., 1960.

Fang Hsien-Chih, Chou Ying-Chhing, Shang Thien-Yü & Ku Yün-Wu (1963–4). The Integration of Modern and Traditional Chinese Medicine in the Treatment of Fractures. *Chinese Medical Jour.* **82**, 493; **83**, 411, 419, 425.

Fu Fêng-Yung & Chang Chhang-Shao (1948). Chemotherapeutic Studies on *chhang shan* (*Dichroa febrifuga*), III; Potent Anti-malarial Alkaloids from *chhang shan*. *Science and Technology in China*, **1** (no. 3), p. 56.

Gardner, F. F. & Milne, D. K. (1965). The supernova of A.D. 1006. *Astron. Journ.* **70**, 754.

Garrison, Fielding H. (1929). *An Introduction to the History of Medicine; with Medical Chronology, Suggestions for Study, and Bibliographic Data*. 4th edition, Saunders, Philadelphia and London.

Giles, H. A. (1924). The *Hsi Yüan Lu* or 'Instructions to Coroners' (translated from the Chinese). *Proc. Roy. Soc. Med.* **17**, 59.

Goldstein, B. R. (1965). Evidence for a supernova of A.D. 1006. *Astron. Journ.* **70**, 105.

Goldstein, B. R. & Ho Ping-Yü (1965). The 1006 supernova in Far Eastern sources. *Astron. Journ.* **70**, 748.

Ho Ping-Yü & Needham, Joseph (1959). Ancient Chinese observations of solar haloes and parhelia. *Weather*, **14**, 124.

(1959). The laboratory equipment of the early medieval Chinese alchemists. *Ambix*, **7**, 57.

(1959). Theories of categories in early medieval Chinese alchemy (with translation of the *Tshan Thung Chhi Wu Hsiang Lei Pi Yao, c.* +6th to +8th cent.). *Journ. Warburg & Courtauld Institutes*, **22**, 173.

Holmyard, E. J. (1957). *Alchemy*. Penguin, London.

Huard, P. & Huang Kuang-Ming (M. Wong) (1959). *La Médecine Chinoise au Cours des Siècles*. Dacosta, Paris.

Hübotter, F. (1929). *Die chinesische Medizin zu Beginn des XX Jahrhunderts, und ihr historischer Entwicklungsgang* (China-Bibliothek d. Asia Major, no. 1). Schindler, Leipzig.

Hume, E. H. (1940). *The Chinese Way in Medicine*. John Hopkins Univ. Press, Baltimore.

Jones, W. H. S. (1946). *Philosophy and Medicine in Ancient Greece*. Johns Hopkins Univ. Press, Baltimore. (Suppl. Bull. Hist. Med. no. 8.)

Keele, K. D. (1963). *The Evolution of Clinical Methods in Medicine*. Pitman, London. (Fitzpatrick Lectures, Royal College of Physicians, 1960–1.)

Lavergne, M. & Lavergne, C. (1947). *Précis d'Acuponcture Pratique*. Baillière, Paris.

Lavier, J. (1964). *Les Bases Traditionelles de l'Acuponcture Chinoise; les Définitions essentielles de la Bio-énergetique Chinoise dans la Terminologie des Acuponcteurs*. Maloine, Paris.

(1965). *Points of Chinese Acupuncture*. Tr., indexed and adapted by P. M. Chancellor. Health Sci. Press, Rustington, Sussex.

Lawson-Wood, D. & Lawson-Wood, J. (1964). *Acupuncture Handbook*. Health Sci. Press, Rustington, Sussex.

Leicester, H. M. (1965). *The Historical Background of Chemistry*. Wiley, New York.

Li Chhiao-Phing (1948). *The Chemical Arts of Old China*. Journ. Chem. Ed. pub., Easton, Pa.

Lowry, T. M. (1936). *Historical Introduction to Chemistry*. Macmillan, London.

Lu Gwei-Djen & Needham, Joseph (1964). Medieval preparations of urinary steroid hormones. *Medical History*, **8**, 101. Prelim. pub. *Nature*, 1963, **200**, 1047.

Mahdihassan, S. (1946). The Chinese origin of the word chemistry. *Current Science*, **15**, 136.

(1946). Another probable origin of the word chemistry from the Chinese. *Current Science*, **15**, 234.

(1951). The Chinese origin of three cognate words: chemistry, elixir, and genii. *Journ. Univ. Bombay*, **20**, 107.

(1953). The Chinese origin of alchemy. *United Asia*, **5** (no. 4), 241.

(1957). Alchemy and its connection with astrology, pharmacy, magic and metallurgy. *Janus*, **46**, 81.

(1959). Alchemy in its proper setting, with jinn, sufi and suffa as loan-words from the Chinese. *Iqbal*, **7** (no. 3), 1.

(1959). On alchemy, kimiya and iksir. *Pakistan Philos. Journ.* **3**, 67.

(1961). Alchemy in the light of its names in Arabic, Sanskrit and Greek. *Janus*, **49**, 79.

Mahdihassan, S. (1961). Alchemy; its three important terms and their significance. *Medical Journ. of Australia*, 227.

(1961). Der Chino-Arabische Ursprung des Wortes Chemikalie. *Die Pharmazeutische Industrie*, **23**, 515.

(1962). Kimiya and Iksir; Notes on the two fundamental concepts of alchemy. *May & Baker Laboratory Bulletin*, **5** (no. 3), 38.

Mann, F. (1962). *Acupuncture; the Ancient Chinese Art of Healing*. Heinemann, London (with foreword by Aldous Huxley); revised ed., 1962.

(1963). *The Treatment of Disease by Acupuncture*. Heinemann, London.

(1962). *Anatomical Charts of Acupuncture Points, Meridians and Extra Meridians*. Barnet Publications, Barnet, Herts.

Marsden, B. G. (1965). Summary of information on the position of the supposed supernova of A.D. 1006. *Astron. Journ.* **70**, 126.

Minkowski, R. (1965). The suspected supernova of A.D. 1006. *Astron. Journ.* **70**, 755.

de Morant, G. Soulié (1939–). *L'Acupuncture Chinoise*, 4 vols. I. l'Énergie (Points, Méridiens, Circulation), II. Le Maniement de l'Énergie, III. Les Points et leurs Symptômes, IV. Les Maladies et leurs Traitements. Mercure de France, Paris.

(1934). *Précis de la vraie Acuponcture Chinoise; Doctrine, Diagnostique, Thérapeutique*. Mercure de France, Paris.

Morse, W. R. (1934). *Chinese Medicine*. Hoeber, New York. (Clio Medica Series.)

Mosig, A. & Schramm, G. (1955). *Der Arzneipflanzen- und Drogen-Schatz Chinas; und die Bedeutung des 'Pên Tshao Kang Mu' als Standardwerk der chinesischen Materia Medica*. Volk und Gesundheit, Berlin. (Beihefte der *Pharmazie*, no. 4.)

Moss, L. (1964). *Acupuncture and You; a New Approach to Treatment based on the Ancient Method of Healing*. Elek, London.

Nakayama, T. (1934). *Acuponcture et Médecine Chinoise vérifiées au Japon*. Tr. from the Japanese by T. Sakurazawa & G. Soulié de Morant. Éditions Hippocrate (le François), Paris.

Needham, Joseph (1948). The Chinese contribution to science and technology. Art. in *Reflections on our Age (Lectures delivered at the Opening Session of UNESCO at the Sorbonne, Paris, 1946)*, ed. D. Hardman & S. Spender. Wingate, London, p. 211; tr. from the French *Conférences de l'Unesco*. Fontaine, Paris, 1947, p. 203.

(1958). *The Development of Iron and Steel Technology in China*. Newcomen Society, London. (Dickinson Lecture, 1956.) Repr. Heffer, Cambridge, 1964.

(1963). The pre-natal history of the steam engine. *Transactions of the Newcomen Society*, **35**, 3. (Newcomen Centenary Lecture.)

(1964). Science and China's influence on the West. Art. in *The Legacy of China*, ed. R. N. Dawson. Oxford.

Needham, Joseph & Lu Gwei-Djen (1962). Hygiene and preventive medicine in ancient China. *Journ. Hist. Med. and Allied Sciences*, **17**, 429. Abridged in *Health Education Journal*, 1959, **17**, 170.

(1967). The optick artists of Chiangsu. Art. in *Historical Aspects of Microscopy*, ed. S. Bradbury & G. l'E. Turner, Roy. Mic. Soc. London (1966 Oxford Symposium Volume), p. 113. Abstract in *Proc. Roy. Microscop. Soc.* **1** (pt. 2), 59.

(1966). A Korean astronomical screen of the mid-eighteenth century from the royal palace of the Yi dynasty (Chosŏn Kingdom, +1392 to 1910). *Physis*, **8**, 137.

Needham, Joseph, Wang Ling & Price, Derek J. de S. (1960). *Heavenly Clockwork; the Great Astronomical Clocks of Medieval China*. Cambridge. (Antiquarian Horological Society Monographs, no. 1.) Prelim. pub. *Antiquarian Horology*, 1956, **1**, 153.

Needham, Joseph (with the collaboration of Wang Ling, Tshao Thien-Chhin, K. Robinson, Ho Ping-Yü, Lu Gwei-Djen *et al.*) (1954–). *Science and Civilisation in China*. 7 vols. in 11 parts. Cambridge.

Pálos, S. (1963). *Chinesische Heilkunst; Rückbesinnung auf eine grosse Tradition*. Tr. from the Hungarian by W. Kronfuss. Delp, München.

Partington, J. R. (1957). *A Short History of Chemistry*, 3rd ed. Macmillan, London.

(1961–). *A History of Chemistry*. Vol. 1, *Earliest Period to +1500* (in the press); vol. 2, *+1500 to +1700*; vol. 3, *+1700 to 1800*; vol. 4, *1800 to the Present Time*. Macmillan, London.

Poynter, F. N. L. (with the collaboration of J. Barber-Lomax & J. J. Crellin) (1966). *Chinese Medicine; an Exhibition illustrating the Traditional System of Medicine of the Chinese People* (catalogue with introduction). Wellcome Historical Medical Museum and Library, London.

Rall, Jutta (1962). Über die Wärmekrankheiten. *Oriens Extremus*, **9**, 139.

Ruhland, Martin (1612). *Lexicon Alchemiae sive Dictionarium Alchemisticum cum obscuriorum verborum et rerum Hermeticarum tum Theophrast-Paracelsicarum Phrasium, planam explicationem continens*. 2nd ed. Frankfurt, 1661.

Schlegel, G. (1875). *Uranographie Chinoise*... 2 vols. with star-maps in separate folder. Brill, Leyden.

Schneider, W. (1959). Über den Ursprung des Wortes 'Chemie'. *Die Pharmaceutische Industrie*, **21**, 79.

Sivin, N. (1965). *Preliminary Studies in Chinese Alchemy; the 'Tan Ching Yao Chüeh' attributed to Sun Ssu-Mo (+581? to after +672)*. Inaug. Diss., Harvard University. (1968). Harvard Univ. Press, Cambridge, Mass.

Taylor, F. Sherwood (1951). *The Alchemists*. Heinemann, London.

Temkin, O. (1956.) On the relationship between the history and philosophy of medicine. *Bull. Hist. of Med.* **30**, 241.

Thorpe, Sir Edward (1921). *History of Chemistry*. 2 vols. in one. Watts, London.

Tonkin, I. M. & Work, T. S. (1945). A new anti-malarial drug. *Nature*, **156**, 630.

Tshao Thien-Chhin, Ho Ping-Yü & Needham, Joseph (1959). An early medieval Chinese alchemical text on aqueous solutions (the *San-shih-liu Shui Fa*, early +6th century). *Ambix*, **7**, 122.

Wang Chi-Min & Wu Lien-Tê (1932). *History of Chinese Medicine*. Nat. Quarantine Service, Shanghai. 2nd ed., 1936.

Wang Ling (1947). On the invention and use of gunpowder and firearms in China. *Isis*, **37**, 160.

Wu Hui-Phing (1962). *Chinese Acupuncture*. French tr. from the Chinese, with added comments, by J. Lavier. Engl. tr. by P. M. Chancellor. Health Sci. Press. Rustington, Sussex.

(1959). *Formulaire d'Acuponcture; la Science des Aiguilles et des Cautérisations Chinoises*. Tr. from Chinese (Thaipei ed.) and abridged by J. Lavier. Maloine, Paris.

439

CHRONOLOGY OF CHINA

	Hsia kingdom (legendary?)	*c.*−2000/*c.*−1520
	Shang (Yin) kingdom	*c.*−1520/*c.*−1030
Chou dyn. (Feudal Age)	⎰ Early Chou period	*c.*−1030/−722
	⎨ Chhun Chhiu period	−722/−480
	⎱ Warring States period	−480/−221

	Chhin dyn.	−221/−207	
First Unification	Han dyn. ⎰ Earlier or Western Han	−202/+9	
	⎨ Hsin interregnum	+9/+23	
	⎱ Later or Eastern Han	+25/+220	

First Partition	Three Kingdoms Period (San Kuo)	+221/+265	
	Shu (west)		+221/+264
	Wei (north)		+220/+265
	Wu (southeast)		+222/+280

Second Unification	Chin dyn.: Western	+265/+317	
	Eastern	+317/+420	
	Former (or Liu) Sung dyn.	+420/+479	
	[Northern Wei dyn. (Thopa Tartar)		
	later split into Eastern and Western		+386/+556]

Second Partition	Northern and Southern Dynasties		
	(Nan Pei chhao)	+479/+581	
	Chhi (Southern)		+479/+502
	Liang		+502/+557
	Chhen		+557/+589
	Chhi (Northern)		+550/+577
	Chou (Northern)		+557/+581

Third Unification	Sui dyn.	+581/+618
	Thang dyn.	+618/+906

Third Partition	Five Dynasty Period (Wu Tai)	+907/+960	
	Later Liang		+907/+923
	Later Thang		+923/+936
	Later Chin		+936/+946
	Later Han		+947/+950
	Later Chou		+951/+960
	[Liao dyn. (Chhitan Tartar)		+907/+1124]
	[Hsi-Hsia (Tangut Tibetan) State		+986/+1227]

Fourth Unification	Northern Sung dyn.	+960/+1126
Fourth Partition	⎰ Southern Sung dyn.	+1127/+1279
	⎱ Chin (Jurchen Tartar) dyn.	+1115/+1234
Fifth Unification	Yuan (Mongol) dyn.	+1260/+1368
	Ming dyn.	+1368/+1644
	Chhing (Manchu) dyn.	+1644/1911
	Republic	1912/1949
	People's Republic	1949 on

PLATE I

Fig. 3. Kuo Shou-Ching's 'equatorial torquetum' (Simplified Instrument, *chien i*) of +1276, the first equatorial mounting, in the position it now occupies at the Purple Mountain Observatory near Nanking (orig. photo. 1958).

(*facing page* 440)

PLATE II

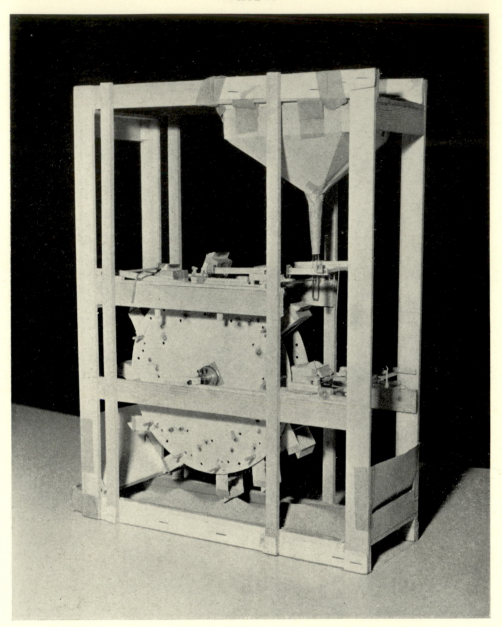

Fig. 5. Working model of the Chinese water-wheel link-work escapement keeping time to ± 20 sec. per hour, built by John Combridge and demonstrated at the History of Science Symposium at Worcester College, Oxford, August 1961 (photo. London Planetarium). Sand reservoir, top right; the two weigh-bridges beside the horizontal scoop at right; the two gate lever locks and the upper balancing lever at top. The chain connecting the latter with the lower weigh-bridge (the coupling tongue) can be clearly seen.

PLATE III

Fig. 7. A derrick for a deep-drilled borehole for brine, depicted on a Szechuanese moulded brick of Han date. The pipeline of bamboo running among the hills includes an inverted siphon, and on the right we see one of the furnaces for the evaporation of the brine. The kneeling figure may be stoking it with wood, but it is equally possible that the lines beside him indicate pipes bringing the natural gas which we know was used already at this early date.

PLATE IV

Fig. 9. The oldest representation of equine collar-harness in the Chhien-fo-tung cave-temple frescoes, painted under the Northern Wei between +477 and +499 (cave no. 257). A green cart with a brown 'sun-bonnet' roof of covered-wagon type, striped red and yellow curtains and a green sun-awning, is being drawn by a large white horse. The arching cross-bar is clear, but the artist failed to draw the cushioning collar behind it, without which it would have been useless. Two thin lines higher up the horse's neck may be a collar drawn in the wrong place but are more probably reins, and the breech strap is indicated by another.

PLATE V

Fig. 10. Collar-harness in a relief at the Yünkang cave-temples, between +466 and +486 (cave no. 9). An equine animal within shafts draws a chariot with a haloed Buddhist saint (photo. Needham & Lu).

PLATE VI

Fig. 11. Shantung freighter off Chefoo under a light breeze (photo. Waters). This is a small prototype of Chêng Ho's +15th-century five- or six-masted Treasure-Ships with a complement of 1,000 men.

PLATE VII

Fig. 12. Deck of a Fuchow freighter at Wei-hai-wei looking aft (photo. Waters). Note the 'lion-head' strut assisting the conveyance of the sail's thrust to the compartmented hull.

PLATE VIII

Fig. 13. A Tolo fishing junk from the New Territories at Hongkong (photo. Waters). Note the typically southern rounded leach of the sail, with its stiff battens, and the multiple sheets which originate from them. The projecting cheeks of the stern enclose the axial rudder.

PLATE IX

Fig. 15. An iron Japanese sword-guard of the Tokugawa period (1603 to 1868). The obverse (right) shows 'cherry-blossoms' (*Prunus pseudo-cerasus* or *yedoensis*, recognisable by their indented petals) in silver against a background dotted with snowflake crystals. The reverse (left) has the new moon among snow crystals (photo. Toledo Museum of Art).

PLATE X

Fig. 16. Sung cast-iron figure of a guardian deity, at the Taoist
temple of Chin-tzhu, Shansi (orig. photo. 1964).

PLATE XI

Fig. 17. Sung cast-iron figure of a monk, at Thai-yuan, Shansi
(orig. photo. 1964).

PLATE XII

Fig. 22. Coupling-rod on the cranks of a ropemaker's frame in Chiangsi (orig. photo. 1964).

PLATE XIII

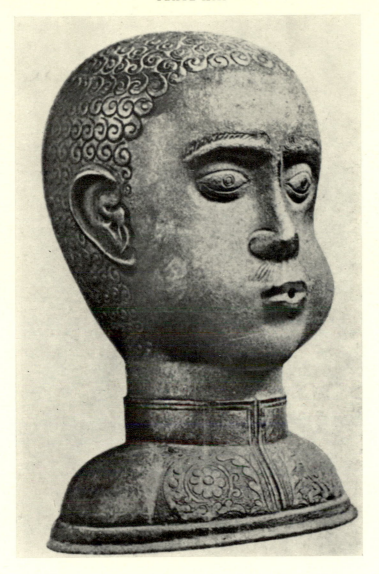

Fig. 26. German or Italian steam fire-blower (*sufflator*),
+15th century (Correr Museum).

PLATE XIV

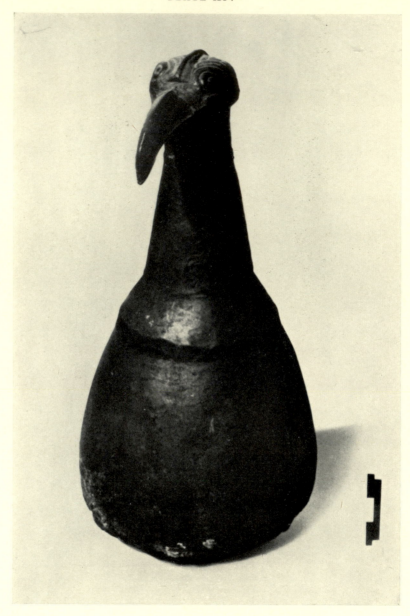

Fig. 27. Tibetan steam fire-blower (*sufflator*), unknown date
(Cambridge Ethnological Museum).

PLATE XV

Fig. 29. Dial water-clock with tipping-bucket mechanism depicted by Isaac de Caus, +1644.

PLATE XVI

Fig. 35. The oldest Chinese representation of the fan or hinged-piston type of forge and furnace bellows, a cave-temple fresco of Hsi-Hsia date (*c.* +960 to *c.* +1220) at Wan-fo-hsia (Yü-lin-khu).

PLATE XVII

Fig. 36. South-east Asian fire-pistons for igniting tinder by air compression (Hough).
Examples from (1) Siam, (2) Lower Siam, (3, 4) the Philippines, (5) Java.

PLATE XVIII

閘 口 盤 車 圖 卷

Fig. 45. The earliest representation of a water-powered pusher, or reciprocator; Wei Hsien's painting
Cha Khou Phan Chhê Thu, of +965 (see p. 186 and Pl. opp.).

PLATE XIX

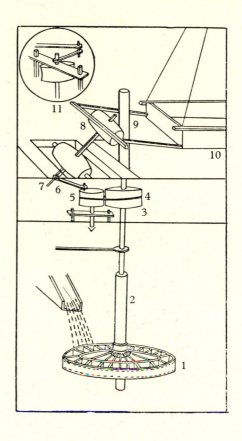

Fig. 46 a. Detail from Wei Hsien's painting to show the reciprocator, used for a sifting or bolting machine.

Fig. 46 b. A reconstruction of its mechanism by Chêng Wei.

PLATE XX

紡車圖　傳未王居正作

Fig. 47. The earliest representation of a spinning-wheel; Wang Chü-Chêng's painting, c. +1035 (from Thien Hsiu). See text.

PLATE XXI

Fig. 48. Quern connecting-rod model, +6th century (Nanking Museum).

Fig. 49. Rubbing of the Thêng-hsien relief of an ironworks, a Han depiction dating from about +50;
on the left, blast-furnace, or, more probably, bellows; centre and right, forging.

PLATE XXII

Fig. 52. Furnace-bellows worked by a vertical water-wheel; a design by Ramelli (+1588). The rocking rollers and bell-cranks set in motion by crank and connecting-rod are remarkably reminiscent of the arrangement in the Chinese water-powered reciprocator.

PLATE XXIII

Fig. 53. Rocking rollers in use at the present time for a sifting or bolting machine at Chia-chia-chuang, Shansi (orig. photo. 1964).

PLATE XXIV

Fig. 55. The earliest European representation of the standard inter-conversion assembly of crank, connecting-rod and piston-rod (with one intermediate lever); a design for water-pumps worked by a vertical water-wheel, due to Antonio Pisanello, c. +1440 (Degenhart).

PLATE XXV

Fig. 59. Bracket verge-and-foliot clock from St Sebald's Church at Nuremberg, c. +1380 (Zinner).

Fig. 58. A 'whimsey engine', i.e. a Newcomen atmospheric steam-engine made rotary by connecting-rod and crank, for colliery winding duty, here used at Coalbrookdale (Davey).

PLATE XXVI

Fig. 64. Model of Su Sung's clock tower in the Science Museum at South Kensington. In the centre below, the driving-wheel, to the left the norias and tanks, to the right the time-annunciator. On the first floor the celestial globe, and on the roof the armillary sphere, both provided with clock-drive.

PLATE XXVII

Fig. 71 *a*. A traditional treadle-operated square-pallet chain-pump for water-raising, in
Chiangsi (orig. photo. 1964).

Fig. 71 *b*. A traditional hand-operated square-pallet chain-pump
in Chiangsi (orig. photo. 1964).

PLATE XXVIII

Fig. 78. The Oxford MS water-clock, perhaps a water-wheel clock (Drover). The manuscript has been dated c. +1285 but some authorities consider it to be much nearer +1250. The prophet Isaiah puts back the sun 10° by the clock for King Hezekiah during his illness (2 Kings xx. 5–11).

PLATE XXIX

Fig. 80. Traditional ship's compass (a late Japanese example from the author's collection, diam. $3\frac{1}{2}$in.). Allowance seems to be made, by means of the meridian line and a fiducial spot, for a declination some 2–4° E. of N.

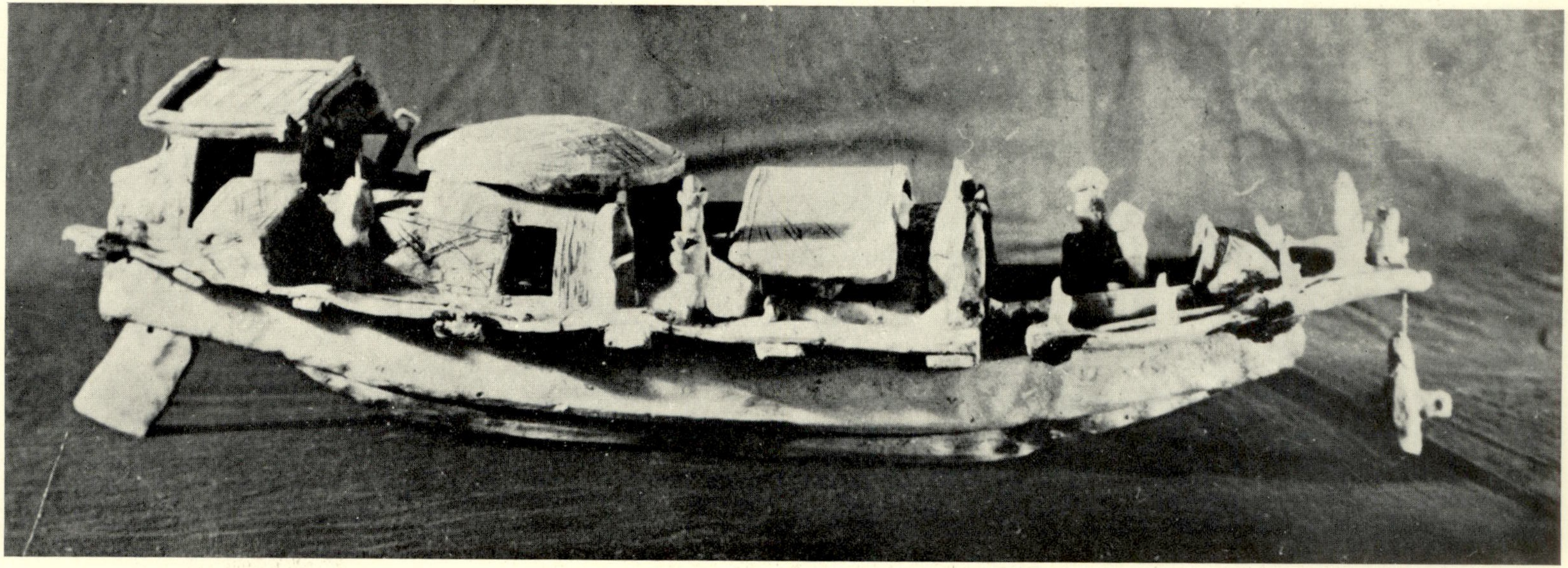

PLATE XXX

Fig. 81a. Canton ship model of the +1st century, abeam (Kuangchow Historical Museum).

PLATE XXXI

Fig. 81 b. Canton ship model, from astern, showing
the rudder (Kuangchow Historical Museum).

PLATE XXXII

Fig. 82. Physician applying moxibustion; a painting by
Li Thang, *Chih Ai Thu*, c. +1150.

PLATE XXXIII

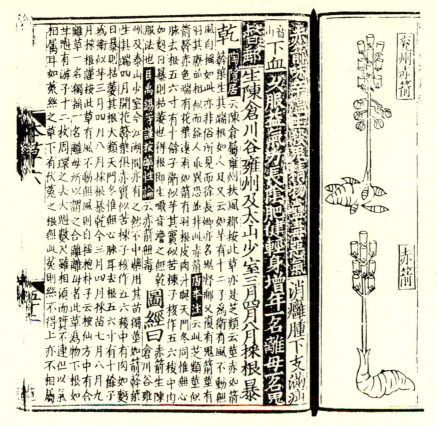

Fig. 83 *a*. A page from the *Chêng Lei Pên Tshao* (Classified and Consolidated Pharma-copoeia) of +1249, showing the entry and illustrations for the orchid *Gastrodia elata*.

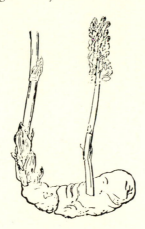

Gastrodia elata Bl.

Fig. 83 *b*. A modern Chinese botanical drawing of the
same plant (Chia Tsu-Chang).

PLATE XXXIV

Fig. 86. Bronze acupuncture figure of Sung date (Peking Historical Museum, orig. photo. 1964).

PLATE XXXV

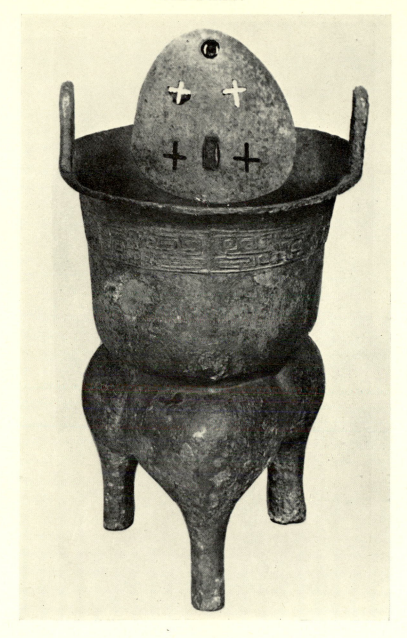

Fig. 87. A bronze *hsien* or steamer from the early
Chou period, *c.* −1100 (Bishop).

PLATE XXXVI

Fig. 88. A bronze *tui*, possibly used as a reaction-vessel, Chou period, *c.* −6th century, from Chia-ko-chuang (National Institute of Archaeology, Peking).

PLATE XXXVII

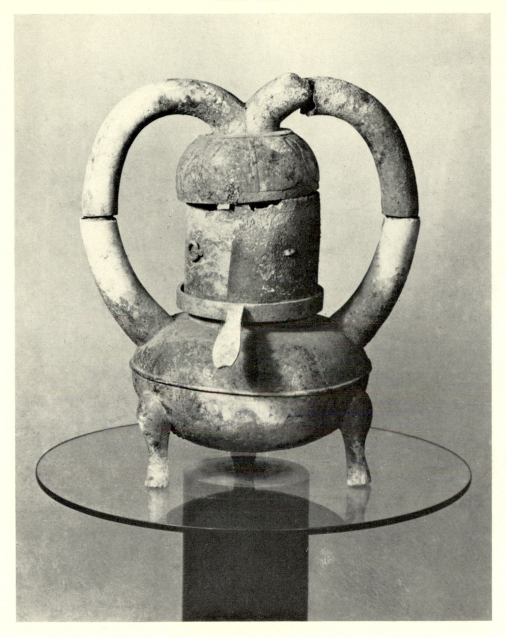

Fig. 89. A later Han bronze 'rainbow *têng*', a piece of alchemical apparatus, probably for sublimation (Nanking University Museum).

PLATE XXXVIII

Fig. 90. Pottery reaction-vessel, Thang period
(Thaiyuan Museum, orig. photo. 1964).

PLATE XXXIX

Fig. 91. A traditional Chinese wine-still at Tung-chêng
in Anhui province (Hommel).

PLATE XL

Fig. 95. Consultation with a pædiatrician; a section from the painting by Chang Tsê-Tuan, *Chhing-Ming Shang Ho Thu*, of +1125. The physician, named Chao Thai-Chhêng (as we know from his sign overhead), is seen examining a child in his shop at the centre, to the right of which people are drawing water from a well, while in front a scholar rides by with his lute carried by a servant behind him. The city depicted is Khaifêng just before its fall to the Chin Tartars.

INDEX

A-Lo-Na-Shun (+648, Indian prince), 20
Abbasid caliphate, 16
Abietic acid, 371
Ablutions. *See* Bathing customs
Academia Sinica, 134
Acids, mineral, 20–1
Acoustics, 114
d'Acres, R. (+17th-century engineer), 139, 143
Acrobats, 58
Acupuncture, 18, 269, 271, 289 ff., 308, 352, 388, 392,
    402, 408–11
  literature on, 409
  technical terms, 289–91
Ad-aqueous and ex-aqueous, classification of water-
    wheels, 26, 115, 126–7
Adanson, Michel (botanist, 1864), 406
Aden, 41, 51
Adulis, 42
Aeolipile, 141
Aerial navigation, 10
Africa, 42, 45, 49, 50, 55, 158
  maps of, 51
Aged, care of the, 358
Agrarian bureaucratism. *See* Feudal bureaucratism
Agricola (Georg Bauer, +1494 to +1555, writer on
    mining and metallurgy), 141, 162, 168
Aḥsan al-Taqāsīm (The Best Divisions for the Know-
    ledge of the Climates), 258
Ai Ti (Chin emperor, r. +361 to +366), 317
Air-pumps. *See* Pumps
Aircraft wings, 260
al-Aiyūbī, Abū'l-Fidā (geographer, +1321), 16
Alagakkonara (king of Ceylon, +1410), 53
Albertus Magnus (natural philosopher, +1206 to
    +1280), 92, 98, 142
Alchemist-Royal, office of, 321
Alchemy and alchemists, 74, 76, 80, 153, 264,
    272, 283–5, 297, 302, 311, 315, 361, 416,
    417
  and elixirs of longevity, 76, 316 ff., 342
  origin of the word, 77, 416
  physiological, 272, 337
  psychological interpretation of, 337
  technical terms, 89, 402–3
Alcohol, 416
Alexander the Great, 40, 144
Alexander of Tralles, 302
Alexandria, 51

Alexandrian mechanicians, 59, 138, 151, 168, 232–3,
    284, 416
Alfonso X (king of Castile, d. +1284), 208
Algebra, 15, 21, 84, 113, 398
  notation, 406
Alkaloids, 407, 414
Allantois, 92
Alloxan, 308
Almagest (Ptolemy), 4
Almoravid dynasty, 66
Altitude, measurements of, 47–8
Ammianus Marcellinus (c. +360), 41
An Lu-Shan (rebel general, d. +757), 309
Anaesthesia, 407
Anamnesis, 269
Anaphoric clocks. *See* Clocks
Anatomy, 17, 78, 407
Anchor escapement. *See* Escapements
Anchors, 69, 256, 257
'Ancients' and 'moderns' controversy, 71
al-Andalusī, Yaḥyā ibn Muḥammad ibn abū' l-Shukr
    al-Maghribī (+13th-century astronomer), 16
Andrew of Perugia (first bishop of Zayton, +1323 to
    +1332), 42
Androgens, 285, 297, 310, 315
Angkor Thom, reliefs of ships at, 256
Angola, 49, 54
Annamese piston-bellows, 159
Anthelminthics, 415
Anti-hogging truss, 65
Anti-Kythera planetarium, 233
Anti-malarials, 414
Antibiotics, 308, 407
Antibodies, 292, 409, 411
Antidotarium (+1080), 393
  The Mysterious, 334
Antidotes, 334, 336, 363
Antimony, 18, 127
Antioch, 228
Antiseptics, 407, 412
Anyang Hsien Chih (Topography of Anyang District),
    188–9
Ao Pho Thu Yung (The Boiling down of the Sea), 118
Apianus, Peter (mathematician, *fl.* +1540), 10
Apotropaics, 388, 392
Appendicitis, 412
Apricot, 100–1
Aquatic pests, 348

29   441   N C A